AR

ANNUAL REVIEW OF PHYTOPATHOLOGY

ANNUAL REVIEW OF PHYTOPATHOLOGY

VOLUME 29, 1991

R. JAMES COOK, *Editor*
Washington State University

GEORGE A. ZENTMYER, *Associate Editor*
The University of California

ELLIS B. COWLING, *Associate Editor*
North Carolina State University

ANNUAL REVIEWS INC. 4139 EL CAMINO WAY P.O. BOX 10139 PALO ALTO, CALIFORNIA 94303-0897

ANNUAL REVIEWS INC.
Palo Alto, California, USA

International Standard Serial Number: 0066–4286
International Standard Book Number: 08243–1329-1
Library of Congress Catalog Card Number: 63-8847

♾ The paper used in this publication meets the minimum requirements of American National Standard for Information Sciences—Permanence of Paper for Printed Library Materials, ANSI Z39.48-1984.

Typesetting by Kachina Typesetting Inc., Tempe, Arizona; John Olson, President
Typesetting Coordinator, Janis Hoffman

PRINTED AND BOUND IN THE UNITED STATES OF AMERICA

Katharine Baker

Kenneth Baker

PREFACE

It is my great pleasure to announce a gift to Annual Reviews Inc. from Katharine and Kenneth Baker designed to provide funds commencing in 1990 to support and advance the profession of plant pathology through the *Annual Review of Phytopathology*.

The gift will be used to make the *Annual Review of Phytopathology* series even more valuable to the scientific and academic community worldwide. As one example, current issues and some back issues (to the extent that annual funds allow) will be made available at no cost to as many libraries as possible that serve plant pathologists in developing countries. In North America, each new PhD recipient whose thesis research is in plant pathology or a plant pathology-related problem will receive a current volume of the *Annual Review of Phytopathology*. The gift may also be used to permit the inclusion of additional authors from outside North America, and to allow use of special illustrations, such as micrographs, four-color inserts, half-tones, and other graphics.

(*continued*)

Kenneth Baker served the *Annual Review of Phytopathology* from its beginning, first as Associate Editor from 1963 until 1971, and then as Editor from 1971 to 1977, a total of 15 years. Ken's high professional standards in writing and editing helped to set the course for the journal, which is now so widely used by researchers, teachers, and students in plant pathology throughout the world. From my close personal acquaintance with the Bakers, I can add that Katharine Baker, a botanist by training, also served the *Annual Review of Phytopathology* for those 15 years, through assistance to Ken while exercising her own commitment to high standards in writing and to professionalism within the plant sciences. The Bakers' gift is both a tangible and symbolic expression of their lasting commitment to the advancement of plant pathology through the enhancement of the quality and the further dissemination of the widely read *Annual Review of Phytopathology*.

R. James Cook

Annual Review of Phytopathology
Volume 29 (1991)

CONTENTS

(*continued*)

SOME RELATED ARTICLES IN OTHER *ANNUAL REVIEWS*

From the *Annual Review of Entomology,* Volume 36, 1991

Induction of Defenses in Trees, *Erkki Haukioja*
Trap Cropping in Pest Management, *Heikki M. T. Hokkanen*
Transmission of Retroviruses by Arthropods, *L. D. Foil and C. J. Issel*
Environmental Impacts of Classical Biological Control, *Francis G. Howarth*

From the *Annual Review of Microbiology,* Volume 45, 1991

Polymerase Chain Reaction: Applications in Environmental Microbiology
R. J. Staffan and R. M. Atlas
Putative Virulence Factors of *Candida albicans, Jim E. Cutler*
Plant Genetic Control of Nodulation, *Gustavo Caetano-Anolls and Peter M. Gresshoff*

From the *Annual Review of Plant Physiology and Plant Molecular Biology,* Volume 42, 1991

Control of Nodulin Genes in Root-nodule Development and Metabolism, *Federico Sánchez, Jaime E. Padillia, Héctor Pérez and Miguel Lara*
Gene Transfer to Plants: Assessment of Published Approaches and Results, *I. Potrykus*

ANNUAL REVIEWS INC. is a nonprofit scientific publisher established to promote the advancement of the sciences. Beginning in 1932 with the *Annual Review of Biochemistry*, the Company has pursued as its principal function the publication of high quality, reasonably priced *Annual Review* volumes. The volumes are organized by Editors and Editorial Committees who invite qualified authors to contribute critical articles reviewing significant developments within each major discipline. The Editor-in-Chief invites those interested in serving as future Editorial Committee members to communicate directly with him. Annual Reviews Inc. is administered by a Board of Directors, whose members serve without compensation.

ANNUAL REVIEWS OF
Anthropology
Astronomy and Astrophysics
Biochemistry
Biophysics and Biophysical Chemistry
Cell Biology
Computer Science
Earth and Planetary Sciences
Ecology and Systematics
Energy and the Environment
Entomology
Fluid Mechanics
Genetics
Immunology
Materials Science
Medicine
Microbiology
Neuroscience
Nuclear and Particle Science
Nutrition
Pharmacology and Toxicology
Physical Chemistry
Physiology
Phytopathology
Plant Physiology and Plant Molecular Biology
Psychology
Public Health
Sociology

SPECIAL PUBLICATIONS

Excitement and Fascination of Science, Vols. 1, 2, and 3

Intelligence and Affectivity, by Jean Piaget

For the convenience of readers, a detachable order form/envelope is bound into the back of this volume.

George W. Bruehl

Annu. Rev. Phytopathol. 1991. 29:1–12

PLANT PATHOLOGY, A CHANGING PROFESSION IN A CHANGING WORLD

G. W. Bruehl

Department of Plant Pathology, Washington State University, Pullman, Washington 99164

KEY WORDS: basic and applied research, teaching, extension, financing research

---

INTRODUCTION

My intent is to present perceptions of changes, some subtle and some not subtle, in the profession (not the science) of plant pathology in the United States. Perceptions are personal and those of another person would differ. Some of my impressions are somber. Overall, my impression is that the profession is strong and progressing; I know of no other that could have more completely satisfied me. Few other careers combine study, classroom, laboratory, greenhouse, and field activities. In preparation for this article, I read most of the prefatory chapters of the Annual Review of Phytopathology and was surprised by how many authors, starting with Walker (16) in 1963, have expressed similar impressions. Are these valid judgments or the nostalgic reflections of aged authors?

Plant pathology encompasses more than is included in dictionary definitions, and it should remain so. Pathologists in human and veterinary medicine study tissues, fluids, etc, working primarily to determine the nature and effects of disease, with emphasis upon structural and functional changes. Medical doctors diagnose, treat, cure, or prevent disease. Plant pathologists do all of the above and more, including pertinent studies that do not include the host, i.e. studies on the survival of pathogens in soil, vectors, symptomless carriers, plant quarantines, etc. For plant pathology to be most effective,

0066-4286/91/0901-0001$02.00

it must go beyond the concepts of the dictionary, but I suggest no change in the definition given in the dictionary.

Muskett (11) defined plant pathology narrowly as the study of the plant in sickness or ill-health and then proceeded to use it in the broad manner to which we are accustomed. Sequeira (13) considered the interaction of host and parasite (pathogen) as the main emphasis of plant pathology—the same narrow concept. Grogan (7) combined the dictionary meanings of pathology and medicine, referring to the former as the science and to the latter as the art of plant pathology. The breadth of plant pathology requires persons with varied talents, training, and interests.

BASIC AND APPLIED RESEARCH

Our purpose, whether as scientists or artists, is to control diseases of economic plants, producing tangible benefits that justify expenditure of public funds (8, 14). Basic research increases understanding, leading to more effective and safer controls, but the value of plant pathology to the public comes from successful application of disease controls in forests, fields, gardens, greenhouses, orchards, or landscapes. Applied research depends upon basic research for new tools and understanding. Basic research depends upon applied research to justify its support. Neither flourishes alone.

Scientists were commonly viewed as dilettantes when I was young. A favorite cartoon depicted an entomologist with a net frantically chasing a butterfly through the jungle. This form of humor is now rare. After the sugarcane industry of Puerto Rico had been saved from extinction by cultivars resistant to mosaic, Carlos E. Chardón (5) in 1927 wrote the following:

> No where [sic] else, has a calamity brought about such a rational change in the cultivation and improvement of a crop; indeed, it has been, as Rosenfeld said, a blessing in disguise. It has been for us an education, for now, every *colono* (farmer) in the island, has awakened to the fact that after all, only science applied to crop production can in the end solve his problems. He knows mosaic and fears it; he knows about cane varieties and knows that the ones he has now have originated through the application of botanical science; he uses more fertilizer, he now considers the government expert as his true and unselfish friend, not the 'crazy bum running after bugs'.

We not only teach, we indoctrinate, and most of us neglect control of plant diseases in our courses. Teachers strive to impart understanding, and in doing so stress basic research. In my classes, we discussed fusarium wilt in detail, except for control, which was passed quickly by. If we as teachers spend less than 10 percent of lecture time on control of plant diseases, how can we expect our students to have a balanced perspective of applied and basic research? It is only natural, being scientists, to focus on the excitement of new knowledge, but bringing a disease under control is also exciting, and we should give greater recognition to workers who actually achieved disease control.

SPECIALIZATION

Increased knowledge, resources, and advancing technology have made specialization essential, but in doing so our sense of oneness as a profession is diminished. In 1963, Walker (16) listed the names of specialties some pathologists applied to themselves, and he was concerned that some would lose plant pathology. McCallan (10) found that in 1963 more workers gave virology than virus diseases as their main interest. The American Phytopathological Society (APS) is still cohesive, but I regret the almost total loss of our nematologists. We must be aware that the publication of more and increasingly specialized journals is a divisive force within plant pathology.

In the earliest days of American agricultural experiment stations, generalists were essential. A state experiment station often had a single Botanist, responsible for botany, mycology, and plant pathology. Full-time plant pathologists were hired as faculties of agricultural colleges expanded. They read all the pertinent literature, but about 1920 specialization began, with pathologists devoted to disease of field crops, vegetables, fruits, or forest trees. At least as late as 1950, Professors J. G. Dickson in field crops, J. C. Walker in vegetable crops, and many of their contemporaries, were competent in broad areas. They matured as scientists during the expansion of the literature, and they kept up with advances within the profession. I was a graduate student at the University of Wisconsin during the 1940s, and there were no courses based on etiology, such as virology, nematology, or bacteriology in relation to plant diseases.

At some point after 1950, it became increasingly difficult to be a competent generalist, a "complete" pathologist in the sense of Sequeira (13), but R. G. Grogan, D. J. Hagedorn, and J. T. Slykhuis, all now retired, were complete pathologists. Hopefully, they will not be the last. I did not specialize; H. H. McKinney, about 1958, advised me to specialize in either virus or fungal diseases, but the fun of varied activities outweighed, in my mind, the loss of competence.

Hierarchies of prestige, or pecking orders, exist within any science. Glenn Pound, in an address sometime during the 1960s before the national meeting of APS, said that the future of plant pathology was in the study of the diseased plant, meaning physiology. I was no physiologist, and was by this definition therefore already obsolete. Disturbed, I approached my department chairman, George W. Fischer. His response was to ask if I thought I could do anything useful? I thought so, and felt better. That night, the vision of all future pathologists in white laboratory coats kept me awake. On the way to breakfast, I was still concerned about the future of the profession, and Ralph Caldwell laughed. During the late 1920s, Caldwell was the only graduate student at Wisconsin whose thesis did not involve soil temperature tanks, and he, in his words, was at the bottom of the "pecking order". We all believe, for one reason or another, that what we do is at least as valuable as what others do. Saint Paul, in 1 Corinthians, chapter 12, eloquently declared that various parts of the human body are worthy of respect, and so are different activities within plant pathology. In an address before the annual APS meeting in 1977, I expressed dismay because some pathologists referred to themselves as being *just* plant pathologists, implying lower status than virologist, physiologist, bacteriologist, etc. We do not need hierarchies of prestige or implied competence.

Some of our most gifted young scientists should nurture broad interests in contrast to excessive specialization. The complex research required to study soilborne viruses, for instance, involving exotic vectors, the virus, and host, require multiple skills. Pathologists who cooperate with plant breeders will be most useful if they are able to work with many different kinds of pathogens. All who enter extension should be as broadly trained as possible. General practitioners should be honored members of the profession, even as their numbers diminish.

TEACHING

Until recently, many pathology courses were based on the study of *whole* diseases, not on parts or aspects of diseases. Each disease is a distinct and complete entity, and differences among them are important. Courses in plant pathology should not obscure these differences. Verticillium wilt differs from Fusarium wilt, and the rusts of wheat (stripe, leaf, stem) differ in their epidemiology. Studies of whole diseases (including history, symptomatology, distribution, host ranges, etiology, environment, control) reveal the contributions of cytology, histology, physiology, genetics, etc, partially insuring that even after they specialize, pathologists will appreciate the contributions of others and recognize better where their personal contributions fit into the total research effort on plant diseases and their control. Luttrell (9) noted that the only comprehensive course in *plant pathology* that a graduate student might take is the introductory course. The necessity of specialized courses is obvious, but every graduate curriculum should include at least one substantive course based on whole, individual diseases.

Emphasis upon whole diseases, including history, forces study of early research and imparts perspective and appreciation of outstanding pioneers. Such a course should instill the habit of in-depth scholarship, not just perusal of the most recent advances or review articles. Some professors dismiss studies over one or two years old as not worth reading. It is my hope that students reject this attitude. Although pathology literature is now so ponderous that thorough study is neither practical nor possible, enough depth should be attained to minimize the chances of reporting discoveries as new that are not new (see ref. 16, but particularly 2).

Garrett (6) and Bailey (1) both feared too much specialization, and Bailey stated that too many courses were chosen to support the thesis effort. Bawden (3) feared that separating bacteriology, virology, mycoplasmology, etc, would lead to concentration on minutiae of the pathogen and to less and less pathology. According to Luttrell (9), "All departments through selfish self-interest fragment and multiply their courses and force a full schedule on their own students to justify offering the courses." My own view is that the special courses are justified because we offer degrees in *plant pathology,* not in molecular biology, physiology, etc.

Formal teacher evaluation by students came into being during my career. At Washington State University, for example, the teacher passed out the evaluation forms, left the room, and the students filled them out. A student collected them, put them in an envelope, sealed the envelope and delivered them to the department office. The teacher did not see the evaluations until the final grades were recorded. The students were assured by this technique that the teacher could not retaliate for an unfavorable rating. Administrators wanted tangible evidence of teacher performance and this was the chosen instrument but the implied distrust in this procedure wounded my pride.It was the process, not the evaluation, that seemed to degrade the profession.

None of my plant pathology teachers was scintillating, but all were good. Even though some improvement can occur, most of us (regardless of effort) are not gifted teachers.

Computers with graphics have provided a great boon to the teaching of plant pathology. It is possible to make excellent charts and graphs in a few minutes that can illustrate more cogently than anything drawn on the blackboard. However, the graphics should be simple. I once attended a lecture and admired the multicolor graphics so much that I missed the message of the lecture.

EMPLOYMENT

Students should seize every opportunity to improve their verbal skills, because presenting a seminar is now a part of the interview process in applying for employment. An excellent seminar, with well-prepared, properly used visual aids, furthers the candidate, but an unimpressive performance essentially disqualifies. During the 1940s and 1950s presenting a seminar was not part of the interview. Aspiring plant pathologists today must have this added skill for success, not only to land a job, but as an on-going and increasingly more important medium of communicating in science.

A new scientist in the university system is on probation until obtaining tenure. Tenure in research is based on publications in reviewed journals. All too commonly, the number of publications is more important than their significance, depending upon the ability of administrators to discern quality. The need to establish a record of productivity favors short-term experiments, such as with bacteria or seedlings, so that experiments can be quickly repeated. This course of action leads to prompt production of valid data, with a minimum of field work that depends upon weather and that usually produces only one set of data per year. Will selecting short-term experiments during the first years of employment favor continuation of this practice? Fewer but more comprehensive papers would reduce the work of reviewing literature and indexing, but emphasis upon number of papers favors fragmented publication. Comprehensive works were abundant 40 or 50 years ago, but Baker (2) points out that our predecessors were under less pressure to publish then. Short-term grants, with the necessity of documenting progress, also lead to an increased number of papers.

Some individuals, especially those in the modern biotechnology, believe their efforts should be published with great rapidity. I personally know of no publication in plant pathology that could not have been published one, two, or even three months later without loss to the economy. Speed becomes important when several have the same specific goal, resulting in a struggle for priority. There is a "thundering-herd" element in science, particularly in "big" science, but most plant pathologists should have enough originality to avoid working as part of a thundering herd, lessening anxiety over speedy publication. However, no scientist should procrastinate in publishing results.

EXTENSION

It is pleasant and informative to talk to farmers and we as scientists learn much from them, but most extension should be done by extension specialists. Some scientists are exceptional speakers, and their enthusiasm for their subject often leaves the impression that the problem they are studying is more important than it really is. An extension specialist, without special interest in a particular disease or control practice, often presents the information in a more balanced way. Furthermore extension specialists must be able to deal tactfully with zealots within their client audience without engendering animosity.

Some "cutting-edge" scientists view extension specialists with disdain, revealing their own ignorance. It is difficult to be a competent generalist.

Farm leaders are well-informed, capable of detecting shallowness. Poor extension presents the college of agriculture to its most important clientele in an unfavorable way, ultimately leading to reduced public support.

The founders of US land grant institutions stressed research, teaching, and extension. They wanted knowledge extended to producers.

H. Fenwick, an extension specialist in Idaho and personal friend, toured Idaho to inform one and all about the alfalfa stem nematode. He had a program that differed from being purely reactive, i.e. waiting for specimens and queries to come to the office. Designing and executing extension programs fosters initiative and leadership in the industry, and surely reduces queries sent to the office.

Tomiyama (15) did much extension in Hokkaido during World War II, when food supplies were inadequate in Japan. After making repetitive diagnoses and recommendations, he began to lose his love of plant pathology. Tomiyama later turned successfully to applied and basic research. He was a research scientist by inclination. Successful, dedicated, long-term productivity in extension requires special personality traits, whatever they may be, as does research.

FINANCING RESEARCH

In a study of the past 100 years of plant pathology at Washington State University (WSU), I found that for the past 20–30 years, the cost of research per full-time scientist increased faster than inflation, that state support decreased in relation to inflation, and that support from outside agencies ("soft" money) increased faster than inflation. The main sources of our outside grants are agricultural commodity commissions, the National Science Foundation (NSF), competitive USDA grants, and private industry (chemical companies). On a national scale, the NSF and the National Institutes of Health (NIH) have been important sources of research funds since about 1950 (2). Changes in sources of funds have influenced plant pathology as a profession more than any other factor, and the changes are still in progress.

Part of the recent increased cost per pathologist is due to increased technology. When McCallan (10) started about 1928, at Boyce Thompson Institute, he had glassware, a second-hand microscope, and a balance; but in 1968, he commented, "One cannot hire a young PhD fresh out of graduate school without expecting to buy some $10,000 worth of equipment immediately." Today at Washington State University the figure is at least $100,000. Plant pathologists today, as in the past, want to work with the best and latest tools available to them, which now means the tools to understand diseases at the molecular level, but every pathologist should not require ever-expanding laboratory space to contain ever-increasing amounts of equipment. Many worthwhile investigations, not requiring expensive equipment, can be done with "initiative, imagination, observation, and industry" (3). Bawden (3) commented that the use of modern equipment guarantees neither originality nor meaningful results.

Young scientists sometimes comment that the easy things have all been done. Those with this attitude are usually laboratory scientists. Roderick Sprague, who studied epidemiology of *Pseudocercosporella herpotrichoides* in Oregon and Washington in the late 1920s and 1930s, traveled in an unheated Model-T car, spent much time in chilly, rainy, snowy fields, and caught severe pneumonia. His observations of epidemiology were not easily made, but all the "easy" things have not been done. We really know little about the longevity of *Typhula* sclerotia in nature, as an example. Many important "simple" discoveries that would require little laboratory equipment remain to be made.

Prior to 1940–1950 the number of assistantships and fellowships for financial support of graduate students within the United States was small, and most pathologists were trained at a few major universities. When outside funds, especially from NSF, began to be substantial, some were used to improve teaching and research facilities, but most went toward increased support for graduate students; in a few years more US universities offered the PhD in plant pathology. Eventually the number of PhDs exceeded opportunities for full-time employment as professional plant pathologists. Few departments abandon doctoral programs voluntarily, and no formal mechanism balances PhD production with demand. A surplus of plant pathologists exists in the United States, and it greatly increases competition for good jobs. On the positive side, this situation of intense competition theoretically has the potential to enhance the profession through higher quality PhDs and higher quality work.

One response to the abundance of PhDs has been the use of postdoctoral fellows. Employment of PhDs in temporary, soft-money positions uses funds that could have been used for graduate assistantships, reducing the production of surplus graduates. These appointments enable new PhDs to obtain research experience as full-time scientists, usually in productive research programs. Luttrell (9) referred to postdoctoral workers as members of a cast of migratory workers exploited by entrepreneurial senior scientists. This characterization seems harsh, in that employment within your chosen field, even in temporary positions, is useful. In addition, postdoctorates bring advanced training to the employing laboratories, benefitting research. Another discernible trend has been for plant pathologists to look beyond the traditional academic and/or extension arenas for employment, serving as self-employed consultants. In summary, however, I view the surplus of plant pathologists with sadness and see no solution.

NSF and NIH support basic research, and accordingly they increase the proportion of basic to applied research. Allocation of money is a means of directing research because, by law, the funds must be used according to the approved research plan.

Until recently most products of public research were public property, provided without fee to the public. Today many public scientists and their host institutions patent selected research results. Will we lose the spirit of working for the public welfare only to be satisfied by what is gained in fees? We must strengthen the ethic of public service.

The grant system, along with stronger relationships with private industry through patents and consulting, can eventually change the type of person entering plant pathology, for better or worse. The contradictions in my thoughts are apparent: I pleaded for greater emphasis upon applied research, and now express fears of an overly close association with some private sources of support.

I have seen *opportunism,* the policy or habit of adapting one's actions, thought, and utterances to funding opportunities to further one's immediate interests, without regard for basic principles or eventual consequences (the dictionary is eloquent), rewarded by the current federal grant system. Research goals are set by the granting agencies, and the adaptable, successful scientist will ride whatever horse (fund) is available, an inevitable outcome of the grants method of funding research.

Dependence upon grants also diminishes the ability of directors of experiment stations to direct research. Directors influence the selection of positions to fill, abandon, or create, but once hired, the scientist essentially controls his or her own program. Freedom of inquiry within a general area is ideal, provided that the scientist has common sense, and provided that funds are available for what needs to be done. Over 50% of plant pathology funds at W.S.U. are from grants, less than 50% from Washington state-appropriated funds. About 90% of the state funds go for salaries, secretarial help, technical aides, office equipment, etc, leaving about 10% or less for discretionary research theoretically controllable by the experiment station director. The 50% from grants is at least partially controlled by the granting agency. Washington State University is fortunate in that direct support from agricultural commodity groups (tree fruit, wheat, barley, potato, and other grower commodity groups) is unusually great.

Fortunately, federal grants increased when state support diminished, but I would prefer a system in which a greater percentage of federal funds is given to the state experiment stations, placing more funds under the control of the directors of research in the 50 states. Experiment station directors should be responsive to real needs, add stability to funding, reduce the stress on scientists engendered by continually competing for funds, and increase the diversity of the research. Under all situations the directors should have funds adequate to insure completion of worthwhile studies when grants are lost.

In the "old days" few professors had research empires. Today, many do. We still need a few quiet, reserved "small shop" operators like C. Drechsler (mycology), H. H. Storey (virology), H. H. Flor (flax rust), and R. Sprague (fungi on grass)—persons who worked much of their careers in academic isolation, but who used dedication, talent, and a minimum of support to add greatly to our knowledge. We need true scholars with time and ability to read, think, reflect, and bring order out of the chaos of ever-expanding research results. Maybe this ordering could be done by senior scientists with tenure.

The grant system distorts salaries within universities, with the highest nonadministrative salaries going to scientists successful in obtaining substantial grants, on the assumption that receipt of these grants is evidence of superior research. Teachers observe these discrepancies and interpret them as evidence of the true priorities of the university, with teaching a lesser activity than research. Excellence in research as measured by success in obtaining grants, however, is often accompanied by excellence in teaching.

The NSF grant system requires great skill on the part of scientists interested in applied research. They must tailor their objectives (2) to make them palatable to NSF review panels, or they must obtain additional grants from agencies that support applied research. I was lucky. The Washington State Wheat Commission, an agency supported financially by wheat farmers, supplied dependable funds for applied research, but they encouraged the use of these funds for those basic studies that supported applied objectives. Grant proposals and reports to the Washington Wheat Commission were short and easy to make. These funds plus regular experiment station funds were adequate for a modest research program, and I never developed the skills required for big-time success. My first job, in 1948, was with the United States Department of Agriculture, Agricultural Research Service, and for at least ten years I never applied for a grant of any kind. Times have also changed for scientists employed by this agency.

Most enterprises measure product in relation to cost. Workers differ greatly in this regard, yet I have observed no effort to reward those who produce much with little, and I have seen some who spend much and produce little. This is unfortunate, yet I offer no solution. Research in pure science is particularly hard to evaluate, but as costs per scientist increase, some approach to accountability should be found.

In applied research a tendency to be economical may in reality be expensive. If a solution to a serious problem takes one year longer with an "economical" research effort, and a disease strikes with severity in the year prior to research success, the cost of delay could far exceed any possible saving.

Conventional Breeding

The excitement over genetic engineering and the concurrent rapid deployment and redirection of research funds to this activity could reduce progress in improving cultivated plants. Genetic engineering may add specific genes for disease resistance, or other characters with simple inheritance, but the most important crop characteristics, such as hardiness, general adaptation, and yield will be improved by conventional breeding. H. Marshall Ward in 1901 (in Stakman, 14) stated that we ". . . must regard the plant as a living machine, the purpose of which is to produce a product." The general genome determines the horsepower of the machine. Specific genes seldom increase horsepower. They are comparable to protective paint, important in prolonging the useful life of the machine. Starving conventional breeding will reduce improvements dependent upon complex inheritance. Moreover, even foreign genes, once expressed in a crop plant, will need to be moved from variety to variety by conventional breeding along with the many other desirable traits that are controlled genetically.

It is tragic that screening for disease resistance is held in such low regard within academia. Determining the presence or absence or degree of resistance within a species is basic research, providing information of value to generations to come. In complex species, such as hexaploid wheat, it is almost impossible, and surely impractical, to thoroughly examine all wheats. R. E. Allan, USDA-ARS geneticist at Pullman, Washington, in contemplating mutation research, commented that our exploitation of existing genes in wheat is far from complete. S. H. Ou (12) stressed the use of resistance to important tropical diseases of rice that can be found within rice. What modern university scientist would devote his life to such mundane, pedestrian research as screening a species for disease resistance; who would fund the work for an indeterminate number of years; and what would be the chances of personal advancement with a poor publication record?

Before leaving exploitation of genetics in plant pathology I wish to mention a comment by L. M. Black (4) that genetics is "easily studied in plant pathology." It caused me no pain to subject large populations of plants to disease, or to cut or mutilate them as I saw fit. In contrast, I was once on a program with an animal scientist who studied emphysema in dairy cows. It was sad to see the suffering cow, but the study was essential to determine the cause of the disease. We are blessed by the low cost of wheat or tomato plants in contrast to the cost of most experimental animals. In addition, we can be unabashedly elitist: developing improved varieties thrives on inequality.

FINANCING SCIENTIFIC SOCIETIES

The membership of APS (American Phytopathological Society) expects leadership and a degree of pioneering, but to provide this potential APS must be well-financed. During most of the time I served on the APS council (1967–1977) the biggest problems of the society were financial. In an address to APS at the annual meeting held in East Lansing, Michigan, in 1977, I told the membership that we owed strong finances to our professional staff at headquarters, and that, if we expected bold leadership, we needed to send Christopher Columbus to sea with full water barrels and many beans. Publication of the *Corn Disease Compendium,* under the leadership and vision of M. C. Shurtleff, extension pathologist of the University of Illinois, financially strengthened APS, and led to its role as a low cost, quality publisher of plant pathology. I applaud those with courage and vision who moved APS in this direction. The Canadian Phytopathological Society and the British Society for Plant Pathology are now also being strengthened in this way.

Literature Cited

1. Bailey, D. L. 1966. Whither pathology. *Annu. Rev. Phytopathol.* 4:1–8
2. Baker, K. F. 1982. Meditation on fifty years as an apolitical plant pathologist. *Annu. Rev. Phytopathol.* 20:1–25
3. Bawden, F. C. 1970. Musings of an erstwhile plant pathologist. *Annu. Rev. Phytopathol.* 8:1–12
4. Black, L. M. 1981. Recollections and reflections. *Annu. Rev. Phytopathol.* 19:1–19
5. Chardón, C. E. 1927. The varietal revolution in Puerto Rico. *J. Dep. Agric. P. R.* 11:9–24
6. Garrett, S. D. 1972. On learning to be a plant pathologist. *Annu. Rev. Phytopathol.* 10:1–7
7. Grogan, R. G. 1987. The relation of art and science of plant pathology for disease control. *Annu. Rev. Phytopathol.* 25:1–8
8. Horsfall, J. G. 1975. Fungi and fungicides, the story of a nonconformist. *Annu. Rev. Phytopathol.* 13:1–13
9. Luttrell, E. S. 1989. The package approach to growing peanuts. *Annu. Rev. Phytopathol.* 27:1–10
10. McCallan, S. E. A. 1969. A perspective on plant pathology. *Annu. Rev. Phytopathol.* 7:1–12
11. Muskett, A. E. 1967. Plant pathology and the plant pathologist. *Annu. Rev. Phytopathol.* 5:1–16
12. Ou, S. H. 1984. Exploring tropical rice diseases: A reminiscence. *Annu. Rev. Phytopathol.* 22:1–10
13. Sequeira, L. 1988. On becoming a plant pathologist: The changing scene. *Annu. Rev. Phytopathol.* 26:1–13
14. Stakman, E. C. 1964. Opportunity and obligation in plant pathology. *Annu. Rev. Phytopathol.* 2:1–12
15. Tomiyama, K. 1983. Research on the hypersensitive response. *Annu. Rev. Phytopathol.* 21:1–12
16. Walker, J. C. 1963. The future of plant pathology. *Annu. Rev. Phytopathol.* 1:1–4

NATHAN COBB

Annu. Rev. Phytopathol. 1991. 29:15–26

NATHAN AUGUSTUS COBB: The Father of Nematology in the United States*

R. N. Huettel and A. M. Golden

Nematology Laboratory, USDA ARS, Beltsville Agricultural Research Center, Beltsville, Maryland 20705

KEY WORDS: nematodes, nematode taxonomy

INTRODUCTION

Nathan Augustus Cobb, referred to as the "Father of Nematology in the United States" (56), was a Renaissance man and a man of humble beginnings. His early life experiences instilled in him the creativity, fortitude, and self-reliance that were needed to raise himself alone from age 14, to complete college with little formal pre-college education, receive his Ph.D. in Germany, and go on to found a new discipline of science called "Nematology". His scientific and technical accomplishments were impressive, as indicated by the over 250 first-authored manuscripts he produced during his career. He identified over 1000 species of nematodes, including animal parasitic, plant parasitic, free-living, fresh water, and marine forms. He made many innovative technical contributions to Nematology, including: (*a*) fixation and preservation methods; (*b*) the Cobb metal mounting slide; (*c*) adaptation of photographic equipment, light filters, and improvements in the camera lucida for microscopic use with nematodes; and (*d*) development of the first flotation device for removing nematodes from soil. A self-taught and gifted artist, his illustrations were of the highest quality. He made great contributions in many other areas of science, such as botany and plant pathology, and to the cotton industry as well.

EARLY LIFE

Nathan Augustus Cobb was born on June 30, 1859, in Spencer, Massachusetts. He was an only child. His father, a jack-of-all-trades, moved the family from place to place as various jobs became available. This disruptive lifestyle, instead of being detrimental, gave Cobb the opportunity to learn many skills that would be carried over into his future life. His early responsibilities ranged from working the night shift at a shingles mill when he was eight years old to running a 150-acre farm alone for weeks in the absence of his father. His formal education was spotty, with only an occasional winter term available that did not interfere with his work responsibilities.

When Cobb was fourteen, his father died and his mother could no longer support him. He took a job as groundskeeper and stableboy for Mr. Charles Prouty in Spencer, Massachusetts. He was fascinated by an advertisement for a microscope in a magazine sold by N. Waldstein, New York, and managed to save the $25.00 needed for its purchase, one third of his total yearly income. Entranced by what he could see with the microscope, he shared his observations with Mr. Prouty's children. Mr. Prouty, impressed with this young man's self-taught knowledge, encouraged Cobb to take the local teacher's examination, aware that he had little formal schooling. Cobb easily passed the examination and within a year was headmaster, in charge of a small school with 65 students of ages 11–18 years. During this time a friendship developed with Prouty's son, Fred, and Joseph H. Greenwood, who introduced Cobb to drawing and painting and who eventually gained fame as a New England landscape artist. This exposure to sketching and drawing would give more than pleasure to Cobb: it was to serve at times as a means of livelihood and was invaluable in his scientific illustrations.

When Cobb started his early schoolteaching career, he moved in with a family named Proctor, whose daughter, Alice, would eventually become his wife. She, too, was a teacher and very interested in botany. They spent much time together studying plants and sketching their findings. They married four years after they met and over the years had seven children.

HIGHER EDUCATION

In 1878, at the age of 19, Cobb entered the Worcester Polytechnic Institute in Massachusetts. He graduated three years later at the top of his class, majoring in chemistry since no biology was taught. His undergraduate thesis was entitled "Mathematical Crystallography" and combined analytical geometry and crystallography. The mathematical knowledge he acquired in optical applications from his undergraduate training later would aid him in developing novel microscope skills and microscope accessories that so enhanced his career.

After graduation in 1881, he taught at Williston Seminary in Easthampton, Massachusetts. His teaching duties soon included zoology, physiology, botany, and geology. He developed several comprehensive drawing courses in architectural design, mechanical drawing, and topography. In spite of his prowess in the physical sciences, his interests turned to the natural sciences, particularly to systematic botany. He published his first paper entitled "A list of plants found growing wild within thirty miles of Amherst" (3).

After six years at Williston Seminary, Cobb looked towards postgraduate education. Failing to win a scholarship at Johns Hopkins University because of his age (28 years), he pursued graduate studies in Germany. In 1887, he set off with his wife and three young children to the University of Jena. Here he was introduced to nematodes. It was during this time that Cobb's talents in mathematics, optics, and art came together with his new knowledge of biology. These combined abilities allowed him to move forward in studies of microscopic organisms. He wrote a thesis on nematodes associated with whales, entitled "Beiträge zur Anatomie und Ontogenie der Nematoden" (4). Here, also, the University published his first new nematode species in "Neue Parasitische Nematoden" (5). There is no way to know how fluent he was in the German language, but he did complete the Ph.D. requirements in just ten months!

EARLY PROFESSIONAL CAREER

Interest in nematodes and marine fauna led him to apply for a position through the British Association for the Advancement of Science at the Zoological Research Station in Naples, Italy. Surprisingly, however, it was not his knowledge of marine fauna but his artistic talents that led to his recommendation for the position. Sir John Murray, a famous oceanographer, was so impressed with a series of Cobb's water colors that he helped him secure the Naples appointment. At Naples, Cobb named his first marine nematode genus, *Tricoma,* but did not publish on it until 1894 (17).

After obtaining tenure in Naples, Cobb seized the opportunity to go to Australia even though no science position was available. Arriving in Sydney in 1889, he was unable to find suitable employment, so turned to his artistic skills and secured a job advertising oils and soaps. Putting his chemistry skills to use, he analyzed the oil and soap's contents to assure the "purity" of the company's product and included it in the advertisement. By the following year however, he had secured a part-time job as a Consulting Pathologist in the newly formed Department of Agriculture in New South Wales. Without formal training in agriculture, he relied heavily on his childhood farm experiences and quickly developed a knowledge not only of plant and animal parasitic nematodes, but also of fungal and bacterial diseases of plants. After six months, he had a permanent position as Vegetable Pathologist, the first full-time Plant Pathologist position in Australia.

Duties were extensive in this new position and his responsibilities included studies of nematodes, flukes, and tapeworms of ranch and farm animals. He conducted studies on various grains, with a major concentration on wheat and its diseases. Breeding experiments and research for handling and storage of the grains were undertaken. Acting as a de novo extension agent, he responded to letters and questions from ranchers and farmers and conducted research in their behalf. He managed the Governmental Experimental Farm at Wagga Wagga, which was 300 miles away from his office in Sydney.

Most field experiments in Wagga Wagga were on wheat. His studies on many aspects of this grain resulted in over 25 publications on the subject, including "Improving wheats by selection", 1894 (16); "Notes on the form and size of the grain in different varieties of wheat", 1895 (18); and "Notes on the threshing of wheat", 1896 (20). These studies of wheat culminated in 1905 when he produced the "Universal nomenclature of wheat" (26).

His studies of fungal pathogens, especially the rusts on wheat, revived botantical skills learned in Germany. He published not only many papers on fungi but also on all aspects of plant pathology. Some of his noted works were: "Contributions to an economic knowledge of the Australian rusts (Uredineae)", 1892 (14), "The hot-air treatments of bunt or stinking smut", 1896 (21), and "The cause of an important apple disease", 1897 (22). He also contributed new research to the bacterial disease of sugar cane called "gumming disease", sometimes referred to as Cobb's disease (19).

About the time of his appointment to the position of plant pathologist, Cobb became interested in parasitic nematodes and published his first paper on plant parasitic nematodes, "*Tylenchus* and root-gall", 1890 (10). He continued to publish on nematodes of every type. Some major contribution from 1890–1891 were on: plant parasitic nematodes, "Strawberry bunch, A new disease caused by nematodes" (13); human parasitic nematodes; "*Oxyuris* larvae hatched in the human stomach under normal conditions" (8); animal parasitic nematodes, "Parasites in the stomach of a cow" (12); marine nematodes, "*Anticoma,* a genus of free-living marine nematodes" (7); and free-living soil nematodes, "*Onxy* and *Dipeltis,* new nematode genera, with a note on *Dorylaimus*" (11).

During 1890–1891, he conducted a major study on nematodes that he isolated from diseased bananas and soil sent to him from Fiji. About 30 species of nematodes were identified, the most significant of which were *Tylenchus granulosus* and *T. similis* (15). Some future confusion arose over the identity of these species but they were finally synonymized, with *T. granulosus* being the nomen oblitum, *Tylenchus similis* being valid and finally becoming *Radopholus similis,* the type species (61). This confusion came about because the male and female of this species are sexually dimorphic, so each sex was identified as a separate species. However, this species is still known as the most important plant parasitic nematode on bananas worldwide (60). Another important paper on this nematode and its host associations, "*Tylenchus similis,* the cause of a root disease of sugar cane and banana," (41) was published after Cobb's return to the United States.

While in Australia, Cobb began the development of instrumentation, microscope techniques, and photographic adaptations that were to aid in future studies of nematodes. One such invention was an instrument called a "Differentiator" (6). The purpose of this device was to use a gradient approach to fixation and preservation of delicate organisms. This allowed for gentle dehydration in alcohol, staining in borax-carmine, and a final embedding in balsam. The oldest known slide of Cobb's, dated June, 1890, of *Mononchus longicaudatus* was embedded by this method (57). This technique was used throughout his career, only the stains used were changed.

He also developed a tool for the handling of human parasitic eggs and larvae of *Oxyuris* spp. For handling these small organisms, he developed a suction capsule that held eggs of the human parasitic nematodes and that could be swallowed. The capsule containing the nematode eggs would suck in stomach juices but not release the eggs into the stomach. Once passed in the feces, it could be recovered. Egg hatch proved that the human body could be a reservoir of these pests and that the acid stomach juices would not destroy them. Cobb states in his paper, "Two new instruments for biologists", that he tried the capsule many times with great success (9).

In 1905, Cobb left Australia for Hawaii where he helped to establish and direct the Division of Pathology and Physiology of the Hawaiian Sugar Planters Association Experiment Station in Honolulu. Here he studied and published on the diseases of sugar cane, especially fungal diseases, "Fungus maladies of sugar cane", 1906, 1909 (28, 29). In his first fungal paper, he outlined the approaches to field studies, made valuable suggestions on how to collect field data, to set up plots and to use nematode sampling techniques. He stated that:

> It would be folly to spend large sums of money on the production of new varieties, then fail, through poorly conducted trails or bad reasoning, to reap the good results. . . . (28).

The taxonomic studies of nematodes continued while in Hawaii and many new species of nematodes associated with sugar cane were described. In the 1906 paper, he identified a "protozoan" parasite of *Dorylaimus,* which was the first known identification of a parasite of nematodes (28).

His optical and mathematical skills led him to the development of instrumentation that provided new approaches in microscopic techniques. As early as 1897 (23), he proposed novel microscopic techniques and would continue to publish his many improvements on the microscope itself, new ways to filter light, and major improvements on the camera lucida system. In 1905, while in Hawaii, he proposed substituting of a 45° prism for the mirror that was commonly used at that time for camera lucida. This approach greatly reduced damage to eyesight. Also, the prism had the advantages of increasing magnification of the object as well as helping to reduce the effect of a second image or reflection (25). Cobb also designed microscope rooms (27), anti-vibration tables (42), and his most famous revolving microscope table pictured in Figure 1. This table allowed him to have several specimens available for comparison at the same time.

UNITED STATES DEPARTMENT OF AGRICULTURE

In 1907, at the request of Erwin F. Smith, Cobb accepted the job of Crop Technologist for the United States Department of Agriculture in Washington, D.C. This title was soon changed to Agricultural Technologist. Though he was hired to work on nematodes, a great deal of this time (until 1915) was spent on cotton and quality of its fibers. His knowledge of so many scientific and technical subjects allowed him to work not only on the process of spinning and milling cotton, but also the storage and marketing of this commodity. He contributed several inventions to the cotton industry that were of great benefit worldwide. One of these studies was entitled "An accurate

Figure 1 Cobb working at his revolving microscope table, USDA, Division of Plant Industry, Washington, DC

method of measuring cotton staple", 1912 (30). Two very important government pamphlets were authored by Cobb, "Memorandum of information concerning official cotton grades" (32) and "United States official cotton grades" (35) that helped to set standards for US cotton grades. He was presented a medal by the National Cotton Manufacturers Association for his contributions to the cotton fiber industry.

In 1910, he was called upon as the USDA nematologist to aid in the inspection of cherry trees sent to Washington, D.C. in 1909 as a gift from the Japanese government. First Lady, Helen H. Taft had proposed a beautification project along the Potomac River in the area that would become known as the Tidal Basin. As a gesture of friendship, the city of Tokyo offered to send 2000 young cherry trees to its sister city. When the trees arrived they were severely infested with several species of insects as well as root-knot nematodes. The Department of Agriculture recommended the destruction of these trees, which presented an embarrassing political situation (see Ref. 58 for a complete history).

In a letter to the Secretary of Agriculture on January 19, 1910, C. L. Marlatt, Acting Chief of the Bureau of Entomology, stated:

> *Root gall worm:* Doctor Cobb's inspection indicates that about 72% of the different lots of trees are infested with root gall worm. He reports that in all probability the vast majority of the trees are infested with gall worm, and that the soil attached to the roots contains large numbers of various species of nematodes, among which are a number of very injurious species. He (Cobb) concludes: "I have no hesitation in saying that in a country where a proper inspection of disease material was legally in force with the object of protecting agriculture, the importation of these trees would not be permitted. Root gall is a very serious disease which attacks scores of species of cultivated and wild plants. . . ." (59).

This statement by Cobb was to have a major impact on the quarantine regulations of the United States. At the time of the first shipment of cherry trees, there were no national quarantine laws. In 1912, when the first Plant Quarantine Act was passed Cobb served as one of its authors (62).

Many of his publications on nematodes during his first few years at the USDA were from studies conducted in Naples and Australia but never published. He continued his work on *Tricoma,* "Further notes on *Tricoma*", 1912 (31) and on "*Draconema,* A remarkable genus of marine free-living nematodes", 1913 (33). In 1914, he published a paper on marine nematodes that were collected on an expedition from Australia to the Antarctic. In this paper, "Free-living nematodes of the Shackleton Expedition" (38), he first proposed that Nematology be named such and recognized as a separate science:

> *Nematology*—A contraction of Nematodology. The founding of this branch of science, on a par with Entomology for example, is fully justified by the fact that the Nematodes

> constitute such a distinct and highly characteristic group of organisms, containing an enormous number of species readily susceptible of division into definite orders, some of which are of great economic importance.

The following year in another publication, "Nematodes and their relationships" (39), he again described Nematology. This particular publication is most often cited as his first definition. However, this lengthy essay covers the history, habitat, relationships in the soil, as well as morphology and physiology of all types of nematodes. He redescribes Nematology at the end of this publication as follows:

NEMATOLOGY

> The foregoing fragmentary sketch may indicate to the student, as well as to the general reader, the vast number of nematodes that exist, the enormous variety of their forms, and the intricate and important relationships they bear to mankind and the rest of creation. They offer an exceptional field of study, and probably constitute almost the last great organic group worthy of a separate branch of biological science comparable with entomology—Nematology.

Throughout the rest of his career Cobb would contribute scientific names and descriptions of over 1000 nematode species. He studied morphology as well as physiology, host relationships, and habitats. In 1913, he described 26 new species of nematodes in "New nematode genera found inhabiting fresh water and non-brackish soils" (34). In this publication he identified the first ectoparasitic plant nematodes, *Tylenchorhynchus cylindricus, Trichodorus obtusus,* and *Xiphinema americanum*. It would be years before the importance of this group of nematodes would be appreciated. Another of his major studies entitled simply, "The mononchs", 1917 (45), contained over 75 illustrations plus a glossary of abbreviations for the labeled drawings and a descriptive key to the genus. His publications ranged from the description of only one new nematode species to, in one case, "One hundred new nemas", 1920 (50). He covered a large number of plant parasitic nematodes and their associated hosts. For example, he investigated a variety of plant hosts, such as "Nemic diseases of narcissus", 1926 (53), "A new parasitic nema found infesting cotton and potatoes", 1917 (43), and "Losses in slash pine seedlings due to nemas", 1930 (54).

He conducted many studies throughout his career on the root-gall nematode. His first description of this nematode was in 1890 (10) while in Australia, not knowing it had already been described as *Heterodera radicicola* (Greef) Muller. He continued his studies on this nematode, "Root-gall", 1901 (24), "The control of root-knot", 1915 (40), and even used it as an experimental tool, "Effect of X-rays on *Heterodera radicicola*", 1920 (48).

He was the first to recognize that *H. radicicola* was different from the other *Heterodera* spp. and named the new genus, *Caconema* (51). He based this new genus on the presence of two lateral cheeks on the head and two testes in males. This definition was noted in a publication by Chitwood who renamed the genus in 1949 (2). Chitwood recognized that the little-known name first proposed by Goeldi in 1887 as *Meloidogyne* was the nomen a priori and therefore synonomous with *Caconema radicicola,* Cobb 1924.

Cobb's first studies on development and morphology of nematodes were published in his dissertation in 1888 (4) and continued throughout his career. Cobb's proficiency in light microscopy gave him a distinct advantage in conducting his morphological studies. Organs and organelles difficult to observe in nematodes were differentiated and he began to assign form to function. In 1913, descriptions of the amphids in the head regions on nematodes were offered. He defined them as "paired cephalic structures of specialized (unknown) function" (36). In 1917, he proposed that the amphids appear to be more like "ducts" based on their histology than nerve organs, as previously proposed by other workers (44). Cobb pointed out that it was difficult to separate lateral papillae from amphids and care should be taken in descriptive studies. He contributed detailed illustrations of the reproductive structures of nematodes, including many papers on spermatogenesis. In the 1925 publication "Nemic spermatogenesis", intricate studies were conducted on meiosis in the sperm of *Spirina parasitifera* (52), presented as drawings of chromosome divisions in sperm, in the female reproductive system, and in fertilization of the ova.

Cobb often took the opportunity to redescribe many genera of nematodes and add new species to a genus that helped to strengthen its taxonomic position. His classification system, started in 1919, proposed a high-level hierarchy, "The orders and classes of Nemas" (47). He divided nematodes into two subphylla; Alaimia, those without a distinct pharynx, and Laimia, those with a distinct pharynx and with a onchia (spear). He continued with class, subclass, order, and genus. In 1935, three years after his death, his daughter, Margaret, and Corrine Cooper published his most comprehensive work, "The key to the genera of free-living nematodes" (55). This publication represented the accumulation of forty years of data that he had almost completed at the time of his death. This epitome uses the taxonomic scheme he developed in 1919 and it contains over 1200 entries, close to the number of described nematodes known at that time. Even though the nomenclature has drastically changed, the extensive literature citations for all named species is invaluable.

Cobb introduced many methods into nematology that are still used today. Any student of nematology would be able to describe the "Cobb slide" he developed in 1917 (44). Its metal frame and the two-coverslip system easily allow for observation of both sides of the specimen and easy remounting of old specimens. Another still commonly used technique is the use of glycerine jelly for *en face* mounts, which he developed in 1920 (49). He first suggested using the Syracuse Dish for counting nematodes in 1918, "Estimating the nema populations of soil" (46). In this publication, he first introduced a soil-sampling tube, a device for the separation of nematodes from soil, and directions for the first cross-hair eyepiece for counting nematodes. This remarkable device was made with spider webs (46).

Death came unexpectedly on June 4, 1932, during a routine health check-up at Johns Hopkins Hospital, Baltimore, Maryland. It marked the end to an insightful mind that had spawned the new discipline, Nematology. The impetus for his life's devotion to the study of these organisms is best reflected in his often quoted description of their world, from "Nematodes and Their Relationships" (39):

> In short, if all the matter in the universe except the nematodes were swept away, our world would still be dimly recognizable, and if, as disembodied spirits, we could then investigate it, we should find its mountains, hills, vales, rivers, lakes, and oceans represented by a film of nematodes. The location of towns would be decipherable, since for every massing of human beings there would be a corresponding massing of certain nematodes. Trees would still stand in ghostly rows representing our streets and highways. The location of the various plants and animals would still be decipherable, and, had we sufficient knowledge, in many cases even their species could be determined by an examination of their erstwhile nematode parasites.

OTHER SIGNIFICANT INFORMATION

About 1918–1920 (exact date unknown), the Bureau of Plant Industry recognized Cobb's contributions to nematology and started a new laboratory called the Division of Nematology with Cobb as its director. This division continues today and is now called the Nematology Laboratory. It is one the oldest units in the Agricultural Research Service, United States Department of Agriculture. The laboratory was relocated at the Beltsville Agricultural Research Center, Beltsville, Maryland, in 1940. The Nematology Laboratory still houses Cobb's slide collection, an extensive collection of his literature, personal letters and drawings, many of his microscopes, prisms, and light filters as well as other items he developed for the study of nematodes.

It can be said that Cobb prepared the second generation of American nematologists. Gotthold Steiner, J. R. Christie, G. Thorne, B. G. Chitwood, A. L. Taylor, and Edna Buhrer all worked with Cobb. Gotthold Steiner took over the Nematology Laboratory after Cobb's death.

Cobb was a charter member of the Helminthological Society of Washington, which was started in 1910 and at the time called the "Worm Club". This society is now in its 81st year. Cobb served as president of the Helminthological Society of Washington, American Microscopic Society, American Society of Parasitology, and Washington Academy of Science.

ACKNOWLEDGMENT

We are indebted to his daughter, Frieda Cobb Blanchard, who, in 1957, published "Nathan A. Cobb, botantist and zoologist, a pioneer scientist in Australia" (1). This booklet gives great human insight into a man not only recognized by society as a master in science but obviously well loved and respected by his children.

Literature Cited

1. Blanchard, F. C. 1957. Nathan A. Cobb, botanist and zoologist, a pioneer scientist in Australia. *Asa Gray Bull.* 3:205–72
2. Chitwood, B. G. 1949. "Root-knot Nematodes"—Part I. A revision of the genus *Meloidogyne* Goeldi 1887. *Proc. Helminthol. Soc. Wash.* 16:90–104
3. Cobb, N. A. 1887. *A List of Plants Found Growing Wild Within Thirty Miles of Amherst.* Northampton, MA: S. E. Bridgeman & Co. 151 pp.
4. Cobb, N. A. 1888. Beitrage zur Anatomie und Ontogenie der Nematoden. Dissertation. 36 pp.
5. Cobb, N. A. 1889. Neue parasitische Nematoden. (In Kukenthal, *Beitrage zur Fauna Spitsbergens). Arch. Naturgesch.* 55. 1:149–59
6. Cobb, N. A. 1889. The differentiator. *Am. Nat.* 23:745–47
7. Cobb, N. A. 1890. *Anticoma:* a genus of free-living marine nematodes. *Proc. Linn. Soc. NSW* 5:765–74
8. Cobb, N. A. 1890. *Oxyuris* larvae hatched in the human stomach under normal conditions. *Proc. Linn. Soc. NSW* 5:168–85
9. Cobb, N. A. 1890. Two new instruments for biologists. *Proc. Linn. Soc. NSW* 5:157–67
10. Cobb, N. A. 1890. *Tylenchus* and root-gall. *Agric. Gaz. NSW* 1:155–84
11. Cobb, N. A. 1891. *Onyx* and *Dipeltis:* new nematode genera, with a note on *Dorylaimus. Proc. Linn. Soc. NSW* 6: 143–58
12. Cobb, N. A. 1891. Parasites in the stomach of a cow. *Agric. Gaz. NSW* 2:614–15
13. Cobb, N. A. 1891. Strawberry bunch, a new disease caused by nematodes. *Agric. Gaz. NSW* 2:390–400
14. Cobb, N. A. 1892. Contributions to an economic knowledge of Australian rusts *(Uredineae). Agric. Gaz. NSW* 3:181–212
15. Cobb, N. A. 1893. Nematodes, mostly Australian and Fijian. *Dept. Agric. NSW, Misc. Publ.* No. 13:3–59
16. Cobb, N. A. 1894. Improving wheats by selection. *Agric. Gaz. NSW* 5:239–57
17. Cobb, N. A. 1894. *Tricoma* and other new nematode genera. *Proc. Linn. Soc. NSW* 8:389–421
18. Cobb, N. A. 1895. Notes on the form and size of the grain in different varieties of wheat. *Agric. Gaz. NSW* 6:744–51
19. Cobb, N. A. 1895. The cause of gumming disease in sugar cane. *Agric. Gaz. NSW* 6:683–89
20. Cobb, N. A. 1896. Notes on the threshing of wheat. *Agric. Gaz. NSW* 7:82–83
21. Cobb, N. A. 1896. The hot-air treatments of bunt or stinking smut. *Agric. Gaz. NSW* 7:204–8
22. Cobb, N. A. 1897. The cause of an important apple disease. *Agric. Gaz. NSW* 8:126–27
23. Cobb, N. A. 1897. A method of using the miscroscope. *Agric. Gaz. NSW* 8:1–5
24. Cobb, N. A. 1901. Root-gall. *Agric. Gaz. NSW* 12:1041–42
25. Cobb, N. A. 1905. Miscroscope, camera-lucida and solar projector. *J. R. Microsc. Soc.*, pp. 496–508
26. Cobb, N. A. 1905. Universal nomenclature of wheat. *Dept. Agric. NSW, Misc. Publ.* No. 539
27. Cobb, N. A. 1906. Construction and fittings of a microscope room. (Extracts from Rept. Exp. Stn. Commun. Hawaiian Sugar Plant. Assoc. 1904/05: 39–95) *J. R. Microsc. Soc.*, pp. 496–508
28. Cobb, N. A. 1906. Fungus maladies of the sugar cane with notes on associated

insects and nematodes. *Hawaiian Sugar Plant. Assoc., Div. Pathol. Physiol.*, Bull. 5. 254 pp.
29. Cobb, N. A. 1909. Fungus maladies of the sugar cane. *Hawaiian Sugar Plant. Assoc., Div. Pathol. Physiol.*, Bull. 6. 110 pp.
30. Cobb, N. A. 1912. An accurate method of measuring cotton staples. *Natl. Assoc. Cotton Manuf.*, Boston. 26 pp.
31. Cobb, N. A. 1912. Further notes on *Tricoma. J. Wash. Acad. Sci.* 4:480–84
32. Cobb, N. A. 1912. Memorandum of information concerning official cotton grades. *USDA, Bur. Plant Ind.*, Doc. 720
33. Cobb, N. A. 1913. *Draconema:* a remarkable genus of marine free-living nematodes. *J. Wash. Acad. Sci.* 3:145–49
34. Cobb, N. A. 1913. New nematodes genera found inhabiting fresh water and non-brackish soils. *J. Wash. Acad. Sci.* 3:432–44
35. Cobb, N. A. 1913. United States official cotton grades. *USDA, Bur. Plant Ind.*, Circ. 109
36. Cobb, N. A. 1913. New terms for the lateral organs and ventral gland. *Science NS* 37:498
37. Cobb, N. A. 1914. Antarctic marine free-living nematodes of the Shackleton Expedition. *Contrib. Sci. Nematol.* 1:1–31
38. Cobb, N. A. 1914. Note on nematodes of the Shackleton Expedition. *J. Parasitol.* 1:52–53
39. Cobb, N. A. 1915. Nematodes and their relationships. *Year Book Dept. Agric. 1914*, pp. 457–90. Washington, DC: Dept. Agric.
40. Cobb, N. A. 1915. The control of root-knot. *USDA Bull.* 648
41. Cobb, N. A. 1915. *Tylenchus similis,* the cause of a root disease of sugar cane and banana. *J. Agric. Res.* 4:561–68
42. Cobb, N. A. 1916. Masonary bases for the installation of miscroscopes and their accessories, including the camera lucida and the microscope camera. *Trans. Am. Microsc. Soc.* 35:7–22
43. Cobb, N. A. 1917. A new parasitic nema found infesting cotton and potatoes. *J. Agric. Res.* 11:27–33
44. Cobb, N. A. 1917. Notes on nemas. *Contrib. Sci. Nematol.* 5:117–28
45. Cobb, N. A. 1917. The mononchs (*Mononchus* Bastian 1866), a genus of free-living predatory nematodes. *Soil Sci.* 3:431–86
46. Cobb, N. A. 1918. Estimating the nema population of the soil, with special reference to the sugar-beet and root-gall nemas, *Heterodera schachtii* Schmidt and *Heteodera radicicola* (Greef) Muller, and with a description of *Tylencholaimus aequalis* n. sp. *USDA, Agr. Tech. Circ.* 1. 47 pp.
47. Cobb, N. A. 1919. The orders and classes of nemas. *Contrib. Sci. Nematol.* 8:213–16
48. Cobb, N. A. 1920. Effects of X-rays on *Heterodera radicicola. J. Parasitol.* 7: 101
49. Cobb, N. A. 1920. Micro-technique, suggestions for methods and apparatus. *Trans. Am. Microsc. Soc.* 39:231–42
50. Cobb, N. A. 1920. One hundred new nemas. *Contrib. Sci. Nematol.* 9:215–343
51. Cobb, N. A. 1924. The amphids of *Caconema* (nom. nov.) and other nemas. *J. Parasitol.* 11:118–21
52. Cobb, N. A. 1925. Nemic spermatogenesis. *J. Hered.* 16:357–59
53. Cobb, N. A. 1926. Nemic diseases of narcissus. *USDA Off. Rec.* 5:3
54. Cobb, N. A. 1930. Losses in slash pine seedlings due to nemas. *J. Parasitol.* 17:52
55. Cobb, N. A. 1935. A key to the genera of free-living nemas, ed. M. V. Cobb, C. Cooper. *Proc. Helminthol. Soc. Wash.* 2:1–40
56. Esser, R. P., Tarjan, A. C., Perry, V. G. 1989. Jesse Roy Christie: The gentleman nematologist. *Annu. Rev. Phytopathol.* 27:41–45
57. Golden, A. M., Huettel, R. N. 1990. The USDA nematode collection and its oldest slide. *Nematol. Newsl.* 36:20–21
58. Jefferson, R. M., Fusonie, A. E. 1977. The Japanese flowering cherry trees in Washington, D.C. *ARS, USDA, Washington, DC* 66 pp.
59. National Archives, RG 42. 1910. Charles Lester Marlatt, Acting Chief of the Bureau of Entomology, US Dept. Agric. to the Hon. James Wilson, Sec. Agric., Jan. 19
60. Taylor, A. L. 1969. The Fiji banana-root nematode, *Radopholus similis. Proc. Helminthol Soc. Wash.* 36:157–63
61. Thorne, G. 1949. On the classification of the Tylenchida, new order (Nematoda:Phasmidia). *Proc. Helminthol Soc. Wash.* 16:37–73
62. Wiser, V. 1977. Protecting American agriculture inspection and quarantine of imported plants and animals. *Agric. Econ. Rep. No. 266. Econ. Res. Ser. USDA*. 58 pp.

ALBERT E. DIMOND

Annu. Rev. Phytopathol. 1991. 29:29–33

ALBERT EUGENE DIMOND, 1914 TO 1972: One of the Bright Lights of Plant Pathology

James G. Horsfall

The Connecticut Agricultural Experiment Station, New Haven, Connecticut 06504

KEY WORDS: biography, disease physiology, fusarial wilt, Dutch elm disease, vascular disease

---

On February 4, 1972, a brilliant light in plant pathology flickered out. Albert Eugene Dimond died. My good friend was gone. I still feel a deep personal loss. We worked and debated plant pathology together for 31 years. I was best man at his wedding in 1956.

I was flattered when Editor R. James Cook asked me to do a personal portrait of my distinguished colleague for the Annual Review of Phytopathology. Dr. Cook asked me to delve into my memory and recall interesting events that occurred during my aquaintance with Dimond.

In January 1940, Al came as a callow PhD to The Connecticut Agricultural Experiment Station from the laboratory of Professor B. M. Duggar at the University of Wisconsin. He was the eleventh professional plant pathologist to be appointed to the staff of the Station. He was next in line after Thaxter, Sturgis, Clinton, Stoddard, McCormick, Zundel, Dunlap, Horsfall, Heuberger, and Sharvelle.

I would like to think that when Duggar sent Al to the Station, he was repaying a debt incurred in 1902 when he declined an offer by Director E. H. Jenkins to come to the Station as the third plant pathologist. Clinton came

*My story could not have been written without the enthusiastic assistance of Kenneth F. Baker, John Dimond, and Louise M. Dimond. To each of you my thanks.

E0066-4286/91/0901-0029$02.00

instead. In his old age Duggar used to come to New Haven to visit his daughter, and our friend, Mrs. Nancy Adams. I like to tease him about his turning us down. He would just grin.

When Al came to the Experiment Station, we had no permanent post available. In modern jargon we had no "tenure track position" available. We were able to give him a job as a fellow of the Crop Protection Institute to continue the fellowship just vacated by Eric Sharvelle and supported by the U.S. Rubber Co., now Uniroyal.

During Sharvelle's tenure, the fellowship had developed "Spergon", the first organic fungicide to sell in tonnage lots. It was used for protecting seeds of peas and lima beans from decay in the soil.

Although the fellowship had begun to sicken by the time Dimond took over, it took another year to die.

By 1940 the Röhm and Haas Company had expanded its cuprous oxide fellowship in the Institute to include the development of organic fungicides. Dimond was moved to that project. From that project came the ethylene bisdithiocarbamates now known as mancozeb, zineb, and "their sisters, and their cousins, and their uncles, and their aunts". That led on into his senior authorship of the first paper in that series (1).

I have to say that Dimond was not altogether happy during his first years at the Experiment Station. He had been trained in basic plant physiology by Duggar in Wisconsin. Here he had to meld some practical with the theoretical.

Eventually, his unhappiness boiled over and in August 1942, Dimond left the Station to accept a post as plant physiologist at the University of Nebraska. Perhaps, I made a mistake, but I deliberately allowed him to go. At the time I said to him in effect, Al, why don't you go to Nebraska? From that distance you can look back on the Station with a better perspective. If you would like to return, let me know. He eventually did let me know and he did come back.

During the war in the South Pacific, the armed forces encountered considerable difficulty with fungal deterioration of their electrical and photographic equipment in the hot, steaming jungles. S. A. Waksman, later to win a Nobel Prize, was in charge of a segment of the research at the University of Pennsylvania. I became a consultant. Waksman needed a person knowledgeable in fungicides. At my suggestion, he hired Al from Nebraska.

In 1946 we invited Al back to the Station for a "tenure track position" in the department of Plant Pathology and Botany. He rose fast, becoming chief of the department in 1949. Twenty years later in 1969, he was selected as one of the first three members of the Station staff to be made S. W. Johnson Distinguished Scientists. In the fall of 1971, he was named Vice-Director, but he died before he could exercise his duties.

Al was born in Spokane, Washington on May, 11, 1914, the son of Maud Nichols Dimond and John Dimond. This was two years after his older brother, John, was born. When Al was a small boy, his mother became a stenographer and brought up her two bright sons.

As Al and I were both second sons, I can understand Al's efforts to establish his individuality. He called attention to himself by leaving his hair uncombed. This, of course, distressed his mother. One morning she took a comb and followed him to school where she found him with a passel of boys waiting for the doors to open. She marched up to him, comb in hand, and without a word proceeded to comb his hair. After she left, the boys asked him who she was. Al replied, "I don't know; never saw her before in my life."

Although I always enjoyed Al's quips, for the life of me, I can't recall many of them. His brother gave me another great story as we shall see. At one point Al served on a biological review panel of the National Science Foundation. About this time his mother came down with cancer and son John moved her to a nursing home in Urbana Illinois where he lived. When Al visited his sick mother there, John decided that Al might like to see a little biology being done there by one of John's friends in the University. It had to do with what areas in the cat brain perform what functions. The professor, being busy writing, was a little annoyed at being interrupted. Without hardly looking up, he dispatched a graduate student to show Al and John around his laboratory. John wrote that when they returned to the office, the professor was just short of furious at the second interruption. Al led him on with a few stupid tourist questions, which were answered curtly. Then Al cleared his throat and said, "I want to thank you for the opportunity to see this research. I am a member of the NSF panel that awarded this contract and I have found it interesting." John wrote that he had never before seen a man click his heels while sitting down. He immediately launched into an explanation of his work.

In 1932 Al began his undergraduate work at Washington State University in Pullman, a hundred miles south of Spokane. There he met Kenneth F. Baker who was finishing his PhD there. From there Baker went to Wisconsin to work under B. M. Duggar on a National Research Council fellowship. On Baker's recommendation Duggar took Dimond and gave him a fellowship offered by the Wisconsin Alumni Research Foundation, the first undergraduate to receive one.

Al earned his BS in 1936, his MA in 1937, and his PhD in 1939 with a thesis on the effects of ultraviolet radiation on fungus spores.

While at Wisconsin he married Naomi Sorkin, a fellow student under Duggar. In due course, Susan was born.

During his 30 years of research at the Station, Dimond worked intently, effectively, and quietly, (except for his clattering typewriter) to advance the frontiers of science. He was an avid supporter of what we call here the

Johnson principle as laid down in 1853 by the first-director-to-be, S. W. Johnson, who wrote that "theory and practice must march together". They surely marched together for Al Dimond.

Dimond's science was the science of the diseased plant, the physiology and pathology of plants dying from thirst brought on by pathogens in the vascular system. His favorite diseases were the Dutch elm disease and fusarial wilt of tomato.

He worked hard on the basics of the wilt diseases of plants, water flow in the xylem, the role of conjugated phenols in the blackening reaction in the vessels, nature of the epinastic symptom, toxins, extracellular enzymes, the effects of ionizing radiation, and others.

He worked equally hard in practical plant pathology - developing methods for testing fungicides, aiding in the discovery of the ethylenebisdithiocarbamate fungicides, and pioneering in the chemotherapy of plant diseases, for example. Thinking as a chemist, he showed a greenhouse grower how to prevent damage to his roses from a mercury-containing paint that had been applied to the sash bars.

All this work brought him a stream of visiting scientists, 20 in all, that came to work with him. They came from Argentina, Brazil, Egypt, England, Germany, Greece, Holland, India, Japan, the Phillipines, and Russia. During those years the foreign workers in his laboratory totalled more than all the foreign workers in the rest of the Experiment Station put together. He was invited to symposia in Argentina, England, Italy, India, and Japan to say nothing of his lectures in American universities. For several years he was a lecturer on the graduate faculty of the Yale School of Forestry.

As you would expect, this world recognition brought him many kudos listed here by year: 1950 to 1952 Treasurer and Business Manager of the American Phytopathological Society, 1955 to 1961 Regulatory Biology Panel of N.S.F., 1964 President of the American Phytopathological Society, 1966 Fellow of the American Phytopathological Society, 1970 Bronze Medal of the XI International Botanical Congress. In addition he was a fellow of the AAAS and a councilor of the International Society of Plant Pathology.

Practical people honored him, too. He was made an honorary member of the American Carnation Society, member of the Research Committee of the National Shade Tree Conference, and Vice President of the Connecticut Tree Protective Association. After he died, the latter group planted a memorial tree by the library of the Experiment Station.

It seems to me that every time I walked into his office, his typewriter was chattering away. He would be working on one of his 207 publications. Their range covered such topics as *Dutch Elm disease* (2), the toxicity to greenhouse roses of paints containing mercury fungicides (3), through the reduction in water flow in vessels (4) to the role of cellulolytic enzymes in pathogenesis by *Fusarium oxysporum lycopersici* (5). In addition he was co-editor of the 3-volume work, *Plant Pathology,* an Advanced Treatise (6).

Dimond is survived by his second wife, Louise W. Dimond, by his brother, John, by a daughter, Susan Dimond Block, and by two grandchildren.

As one of his friends said; "All of us recall his friendliness, his sense of humor, and his respect for every person. Of any decent man it may be said, we shall not see his like again. Of Albert Dimond we can also truly say, how fortunate we are to have known this kindly man."

I cannot say it better.

Literature Cited

1. Dimond, A. E., Heuberger, J. W., Horsfall, J. G. 1943. A water-soluble protectant fungicide with tenacity. *Phytopathology* 33:1095–97
2. Dimond, A. E., Plumb, G. H., Stoddard, E. M., Horsfall, J. G. 1949. An evaluation of chemotherapy and vector control by insecticides for combating Dutch elm disease. *Conn. Agric. Exp. Stn. Bull.* 531. 69 pp.
3. Dimond, A. E., Stoddard, E. M. 1955. Toxicity to greenhouse roses from paints containing mercury fungicides. *Conn. Agric. Exp. Stn. Bull.* 595. 19 pp.
4. Waggoner, P. E., Dimond, A. E. 1954. Reduction in water flow by mycelium in vessels. *Am. J. Bot.* 41:637–40
5. Husain, A., Dimond, A. E. 1960. Role of cellulolytic enzymes in pathogenesis by *Fusarium oxysporum* f. *lycopersici*. *Phytopathology* 50:329–31
6. Horsfall, J. G., Dimond, A. E., eds. 1959, 1960. *Plant Pathology, An Advanced Treatise*. New York: Academic. 3 Vols.

Annu. Rev. Phytopathol. 1991. 29:35–63

PLANT DISEASES AND THE USE OF WILD GERMPLASM

Jillian M. Lenné and David Wood

International Germplasm Associates, 26 Watersedge, Milnthorpe, Cumbria LA7 7HN, UK

KEY WORDS: genetic resources, crop relatives, breeding for disease resistance, international agricultural research

---

INTRODUCTION

The Increasing Uniformity of Crops

In a pioneer and wide-ranging review of the use and conservation of variability in crop plants, Simmonds (139) noted a general tendency in breeding programs toward the rapid elimination of variability, coupled with the notion that strictly uniform crop populations were the universal ideal. Unfortunately, but predictably, as uniform varieties were grown over wider areas, their vulnerability to disease epidemics increased. The Southern corn blight outbreak in the US maize crop in 1970 drew attention to this danger and occasioned a report (113) on the phenomenon of *genetic vulnerability*. Browning (13) argued that the epidemic was "the greatest biomass loss of any biological catastrophe" and that it was "a man-made epidemic caused by excessive homogeneity of the USA's tremendous maize hectarage." A survey by Duvick in 1984 (44) showed that genetic vulnerability was still regarded by plant breeders in the USA as potentially dangerous.

The Need to Broaden the Genetic Base

Arguing for a better use of plant variability, Simmonds (139) noted that the "materials of adaptability—that is, the variability immediately available to the

0066-4286/91/0901-0035$02.00

plant breeder—may be called the genetic base of the crop and it is upon this that plant breeding depends". Since then, the need to broaden the genetic base of crops has been widely recognized (26, 44). As the variation within and between the varieties of traditional agriculture is progressively replaced by more uniform and higher yielding varieties, evolution of the crop in the farmer's field is no longer possible. The responsibility of breeders in managing variation to combat genetic vulnerability is now paramount.

The Potential of Wild Germplasm

Three main germplasm resources are available to plant breeders: commercial varieties; landraces (traditional varieties); and a range of wild ancestral species and other wild relatives. The increasing awareness of genetic vulnerability of major crops has encouraged collection and conservation of a multitude of germplasm samples as a resource for future breeding. Most samples in larger collections are landraces. Details of these collections and of their management are given in Plucknett et al (125). Apparent needs of breeders for additional characters and increased accessibility to traits from wild species have led to wide recommendations for the increased collection, conservation, and use of wild germplasm (10, 54, 68).

By reviewing the characteristics of wild species (and associated pathogens), their evaluation and management as a genetic resource, and their use as sources of disease resistance in crop improvement, we indicate some of the advantages and problems of using wild germplasm in breeding programs. Emphasis is placed on major food crops. We have not attempted to gaze into the crystal ball of genetic engineering, although advancing technology is expected to increase both the efficiency and range of use of wild species (45, 55). Some requirements for the successful use of wild species as sources of disease resistance are listed.

WILD GERMPLASM

The simplest way of defining wild germplasm is that it is not grown as a crop. The often long process of domestication differentiates crops in many ways from their wild ancestors: nonancestral wild relatives will be yet more different from crops. Although the details of the domestication of each crop are different, there are some remarkable parallels. Crops differ from their wild ancestors by a syndrome of characters (59, 129). Those of relevance to the use of wild relatives as sources of disease resistance include: (*a*) geographical and environmental expansion of range and changed ecological preference; (*b*) loss of dispersal mechanisms; (*c*) conversion of perennials into annuals; (*d*) changes in breeding system from cross- to self-fertilization; (*e*) loss of defensive adaptations; (*f*) increased susceptibility to diseases and pests; and (*g*) multiplication by vegetative propagation. The closeness of relation to a crop has hitherto determined the value of wild germplasm in breeding. Relative closeness indicates possible resistances to comparable diseases, and further, that these resistances may be available to breeders using conventional crossing.

The interface between crops and their wild relatives and often weedy ancestors is an evolutionary dynamic one, with introgression, polyploidization, and selection by farmers (71). Similarly, pasture species are often polymorphic species of immense variability—"taxonomic nightmares" (36, 106). Taxonomists have tended to neglect crops and crop relatives. While classification based on morphology will provide an initial estimate of which species to collect and where they are to be found, the most important indicator of the usefulness of wild germplasm as sources of disease resistance will continue to be experimental evidence of resistance to disease and of crossability.

The "gene pool" concept of Harlan & de Wet (65), defining relationship to a crop in terms of primary, secondary, and tertiary gene pools, is an example of a classification based on experimental evidence of the degree to which genes can be exchanged between populations. The need for the classification of crops and relatives in a way relevant to the needs of breeders (39) has led to agricultural taxonomists using classifications based on a survey of genomes. Kimber & Feldman (87) note that the genomic analysis of wheat and its relatives provides both an insight into the evolutionary past of the genus and a sound basis for the pragmatic use of both wild and cultivated forms. The utilitarian value of genome-based classifications is high. Cauderon (25) gives examples of successful genome surveys of *Triticum, Brassica, Gossypium, Nicotiana,* and *Coffea*. Hexaploid *Triticum aestivum* provides an excellent example of genome analysis, with a full chromosomal complement of three different diploid species (*Triticum monococcum, Aegilops searsii,* and *Ae. squarrosa*).

The choice of wild species for domestication was not a random process. This is evident from the taxonomic groupings of crops. There are obvious concentrations of crops within the Gramineae and Leguminosae, and within these families, in certain tribes including Triticeae in the grass family, with *Triticum, Hordeum,* and *Secale;* the Phaseoleae in the legume family with *Vigna, Phaseolus, Glycine,* and *Dolichos*; and the tribe Viciëae, with *Vicia, Cicer, Lens, Lathyrus,* and *Pisum*. Other concentrations of crops are to be found for example in the Solanaceae (tobacco, potato, tomato, chili pepper, eggplant), and Cucurbitaceae (*Citrullus, Cucumis, Cucurbita, Sechium*). Genera within families may also show a concentration of cultivated species: *Allium, Citrus, Coffea, Dioscorea, Diospyros, Gossypium, Passiflora* and several other genera in different families contain many domesticated species.

Such groupings of crops may share diseases (3), some examples being the mosaic viruses of the Solanaceae, and the rusts and mildews of the Triticeae. The systematic clustering of crops reduces the number of groups of wild relatives to be investigated as useful sources of disease resistance. We can follow the demonstrably correct perceptions of early farmers on what was of potential value and target work more intensively to taxonomically related groups of crop relatives.

The ideas on centers of origin and crop diversity, developed by Vavilov, and subjected to extensive comment and modification (62, 69, 166) are only partly relevant to the distribution of wild relatives and their use as a source of characters for crop improvement, as the distribution and adaptive strategies of wild species predate crops by millions of years. Crops often originated in one continent as members of large genera with a distribution in two or more continents (for example, *Amaranthus, Elaeis, Fragaria, Glycine, Gossypium, Ipomoea, Lupinus, Nicotiana, Panicum, Piper, Solanum, Sorghum, Vigna, Vitis*). Some genera have species domesticated in more than one continent (for example, *Dioscorea, Oryza*). Although at least one ancestral wild relative must have been present in the area of origin of each crop, areas of crop origin may not correspond to areas of diversity of wild relatives and pathogens. Several examples of this are given in FAO (46) including: *Vitis,* the grape, with crop diversity in Central and West Asia and the Mediterranean, and wild species diversity in North America and East Asia; and soybean, with crop diversity in East Asia, but wild diversity in Australia.

DISEASES OF WILD GERMPLASM

Only a limited number of wild germplasm-pathogen systems have been extensively studied. The best documented studies are of wild relatives of barley, oats, and wheat and their associated rusts and mildews in Israel (11, 40, 42, 101, 137, 159); wild *Glycine* spp. and soybean rust (*Phakopsora pachyrhizi*) populations (17, 22); and wild populations of *Linum marginale* and flax rust (*Melamspora lini*) (19, 21) in Australia. Populations of tropical pasture legumes *Stylosanthes* spp. and anthracnose (*Colletotrichum gloeosporioides*) have also been studied in South America (93, 107). Wild germplasm-pathogen systems have generally been ignored by plant pathologists (12, 67). Extensive surveys of the occurrence and distribution of wild relatives of many crops (29, 37, 82, 123, 145, 156, 162) have neglected associations with diseases. Most studies related to diseases of wild germplasm of use in crop improvement have been done ex situ.

Considerable variability for resistance has been recognized in coevolved systems of wild germplasm and their pathogens. Genetic mechanisms include race-specific or vertical resistance, race-nonspecific or polygenic resistance, and tolerance whereas avoidance and escape commonly support population survival (11, 17). Although different types of resistance are present in wild germplasm populations, studies of wild species and their pathogens have almost invariably been focused on race-specific reactions. The occurrence of race-specific resistance in seedlings derived from wild germplasm populations has been demonstrated in barley, oat, and wheat relatives to rusts and mildews (40, 42, 108, 109, 119, 137, 159); in wild *Glycine* species populations to soybean rust (17); and in populations of *Linum marginale* to flax rust (19, 21). A few studies of race-nonspecific resistance have documented slow mildewing in *Hordeum spontaneum* in Israel (47) and slow rusting in *Avena sterilis* in Israel (136). More than sixty years ago, Ridley (134) observed in wild vegetation that "Where plants of one kind are separated by those of other kinds, the pest, even if present, cannot spread, and will itself die out, or at least become negligible." Ridley also argued that plant dispersal mechanisms were important to allow plants to escape from diseases. In coevolved systems, defense depends on genetic, spatial, and demographic variability (17, 53).

Successful disease resistance in wild germplasm populations is related to both individual and population survival. The former has an identifiable genetic base while the latter results primarily from the spatial and temporal dynamics of wild populations and their pathogens (8, 20). In comparing crops and wild populations, Harper (66) notes "it would be surprising if the behaviour that favoured one individual was also the behaviour that maximized the performance of the population as a whole". Key wild germplasm defense mechanisms such as escape and avoidance are of limited value in present day agroecosystems. Successful disease resistance in crops must support high productivity through dense, uniform plant populations and synchronization of growth. Disease resistance relevant to crop improvement might be more efficiently sought in crop landraces, rather than in wild germplasm.

Knowledge of diseases and pathogens of wild germplasm allows construction of maps of geographic patterns of disease resistance (5) that can facilitate more effective collection of resistant germplasm. A thorough knowledge of the epidemiology of the disease would also aid in identifying areas with environmental characteristics conducive to high disease pressure and associated concentrations of resistance. Accumulated knowledge of the distribution of diseases and their wild relatives has allowed mapping of resistance to *Puccinia coronata* in *Avena sterilis* (40, 42, 159) and to *P. recondita* in *Hordeum spontaneum* (103) in different regions of Israel and resistance to several potato diseases among *Solanum* species in Central and South America (69). Similar maps have not yet been prepared for wild germplasm of value in the improvement of other crops. Areas with a rich diversity of pathogens and a corresponding diversity of resistant plants ("pathogen parks" (128)) would be excellent sites for in situ studies of diseases and pathogens of wild crop relatives. Surveys of wild relatives of cereals have identified subcenters where resistance is more common due to specific environmental conditions (15, 17, 102). Resistance to three races of *Erysiphe graminis hordei* in *H. spontaneum* and to *E. graminis tritici* in *Triticum dicoccoides* was common in humid areas where the pathogens were also most common (108, 109). Resistance to wheat leaf rust in *Aegilops, Agropyron* and *Triticum* spp. was defined by ecogeographical parameters such as latitude, altitude, annual rainfall, and soil type (102, 117–119). Resistance to a wide range of tomato diseases was found in *Lycopersicon esculentum* var. *cerasiforme* under very humid conditions (132, 133). Often concentration of resistance to certain pathogens is extremely localized. High levels of resistance to potato late blight (*Phytophthora infestans*) are restricted to *Solanum* spp. in the Toluca valley of Mexico (121). Levels of resistance to *E. graminis hordei* in *H. spontaneum* varied between populations separated by only 6 km in Israel (109); levels of anthracnose (*Colletotrichum gloeosporioides*) resistance in *Stylosanthes capitata* varied between populations less than 50 m apart in Brazil (93); while resistance to flax rust varied among populations of *Linum marginale* separated by < 100 m to 2.7 km (19). Efficient collecting of resistance for use in crop improvement is dependent on an understanding of spatial variability for disease resistance in wild germplasm populations.

Coexistence of wild relatives of crops and their indigenous pathogens in the center of origin of the crop is only one of many systems where field studies are needed to facilitate effective use of resistance. Following domestication, the widespread movement of crops and pathogens can produce a complicated pattern of escapes, encounters and re-encounters (13a). There is not necessarily a parallel between the center of diversity of the crop and the center of diversity of the pathogen. The best sources of resistance may be found in centers of diversity of geographically distant wild germplasm. High levels of resistance to both grape powdery mildew, *Uncinula necator,* and downy mildew, *Plasmopara viticola,* have been found in *Vitis* spp. originating in North America and not among *V. vinifera* relatives in its center of diversity in central and west Asia and the Mediterranean basin (91, 100). High levels of resistance to soybean rust have been found in *Glycine* spp. in Australia and not in the center of diversity of soybean in east Asia (9, 16). Searches for disease resistance should therefore consider centers of diversity of geographically distant wild germplasm that coevolved with the specific pathogens.

Crop introduction is a major feature of modern agriculture (88, 163). As crops often originated in one continent as members of large genera with much wider distribution, introduction to other regions would allow escape from coevolved diseases in the center of origin but could place the crop in contact with geographically distant but taxonomically related wild species and their pathogens. Some examples include the movement of banana to tropical

America where *Heliconia* spp. are wild hosts of moko, caused by *Pseudomonas solanacearum* (152); introduction of the South-east Asian legume *Desmodium ovalifolium* as a pasture species to Colombia, where native *Desmodium* spp. host stem gall nematode, *Pterotylenchus cecidogenus* (97); and introduction of cocoa to West Africa where *Cola chlamydantha,* a common West African forest tree, hosts Cocoa Swollen Shoot Virus (126, 151). It is sometimes assumed that only by long association of wild species and pathogens can useful resistance evolve. However, crops in secondary centers of diversity have proven value as sources of disease resistance. The only source of resistance to Barley Yellow Dwarf Virus was found in Ethiopian barley landraces and not in the region of origin of barley in South West Asia (41, 130); valuable resistance to Bean Common Mosaic Virus, conferred by the *bc-3* gene, came from USDA PI 181954, collected in Homs, Syria; and a much higher frequency of resistance to the root-knot nematode, *Meloidogyne incognita,* is found in sweet potato landraces in Papua New Guinea than in landraces in South America (138). Resistance to pathogens newly encountered by the crop following introduction may also be found in landraces and wild germplasm in the crop center of origin. The major source of resistance to African Cassava Mosaic Virus is *Manihot glaziovii,* a wild relative of cassava from Brazil (58); putative sources of resistance to *Elsinoe phaseoli,* the causal agent of African bean scab, have been found in South American bean landraces (D. J. Allen, personal communication); while good sources of resistance to African Maize Streak Virus occur in Mexican landraces (86).

Crop-weed complexes offer an area of special interest to in situ study and the efficient use of wild germplasm. Weeds can exchange genes with the crop through occasional bursts of hybridization (68, 69). Associations of crops with weeds and wild germplasm will allow persistence of diseases. Pathotypes could move from wild germplasm to the crop or vice versa; crop epidemics could threaten the survival of wild and weedy species; and new virulent races could develop in contiguous crop-weed-wild associations. Studies in India have traced the emergence of virulent rust strains and epidemics on wheat in India and Pakistan to wheat and wild germplasm in the Himalayan foothills (83). Natural introgressive hybridization between wild, weedy, and cultivated species is especially evident in maize, wheat, barley, rice, oats, and sorghum (61) and has been studied in traditional potato production in the Andes (81). Important resistance to Grassy Stunt Virus of rice was obtained from *Oryza nivara,* a member of the crop-weed complex of *O. sativa*. It is notable that breeders have followed nature in utilizing relatives of some of these cereals as sources of useful characters. The identification and a greatly enhanced in situ study of crop-weed-wild species complexes should provide useful information for their manipulation to allow exchange of genetic material, including disease resistance.

Understanding of disease processes in wild germplasm populations requires long-term studies of diseases, pathogens, and their effects on the genetic structure of plant populations (1). The effort required may be partly responsible for the lack of such studies before wild germplasm is collected. Knowledge of relationships between wild germplasm and associated pathogens in indigenous habitats should, however, be sought before any major collection of disease resistance for use by crop scientists.

COLLECTION, MAINTENANCE, AND EVALUATION OF WILD GERMPLASM FOR DISEASE RESISTANCE

Collection and Maintenance

Most experience of the scientific management of germplasm has been gained with landraces. At all stages in the collection and maintenance of wild germplasm there are problems and possibilities not encountered with landrace germplasm (68). There may be many species of wild relatives of a crop, and, compared to landraces, greater population differentiation in wild germplasm on both geographic and microgeographic scales; greater genetic variability especially in flowering characteristics; greater variability in population diversity, both inter- and intraspecifically; greater range of breeding systems; and a complex age structure because many wild species are perennials (105, 106). Breese (6) adds a high variation in seed production, dormancy, and shattering. These features indicate that collecting and management strategies for wild species must differ from those used for landraces. In addition to these practical problems in the management of wild germplasm, there may be a difference in objectives. Bunting (14, 106) identified two broadly different objectives in genetic resources work: (*a*) to conserve diversity in plants and (*b*) to collect material for specific practical uses in agricultural development. Bunting argued that a sampling strategy designed to conserve a reasonable sample of the genes present (105) will be inappropriate to find material with specific adaptations, such as disease resistance, from which new cultivars can be developed.

Although the management of wild germplasm is in many ways more problematical than landrace management, in one area wild germplasm shows an advantage. Taxonomic botany is one of the oldest branches of science and a vast quantity of information is available on the relationships, natural crossing, distribution, phenology, ecological preferences, and ethnobotany of wild species. Harlan (60) considered that taxonomy "provides us with a vast catalog of materials through which we might search in our armchairs for likely sources of germplasm. An enormous amount of basic information may be obtained by poring over the great regional floras of the world." As an example of the predictive value of taxonomic studies, 11 intergeneric hybrids have been recognized in the Triticeae, a tribe containing wild relatives of wheat and barley that have been extensively used in resistance breeding. Use can be made of herbarium and field surveys, termed "ecogeographical surveying" (76). Unfortunately, pathologists are not involved in these ecogeographical surveys, although their input would be valuable in targeting characters of interest of wild germplasm. Existing germplasm collections should be reviewed. Many samples of potential for crop breeding, for example, *Arachis, Phaseolus, Pennisetum,* and many wheat relatives, are maintained in pasture germplasm collections.

In contrast to the threat of genetic erosion of landraces, many wild species of value to agriculture are aggressive annual colonizers of disturbed habitats (See ref. 38, for cereals). Chapman (33) pointed out for wild relatives of crops that few are in great danger and many are quite abundant. The past emphasis on conserving threatened diversity was and still is appropriate to landraces: more targeted strategies are needed both to capture useful characters and to reduce the burden of maintenance with wild species. As wild germplasm has mainly been used to date as a source of disease resistance, this must provide the main justification for collection. This objective determines the strategy of collecting, subsequent maintenance, and evaluation.

As far as possible, collecting of wild germplasm for disease resistance should concentrate in areas most favorable for disease development (63). If the objectives include collecting resistance to several pathogens, collecting in well-defined areas may be difficult. Host density and distribution should also be considered. At the collecting site, the biology of the host species (life span, ecology, and breeding system) and the biology and ecology of the pathogen(s) will determine the pattern of distribution of disease resistance between and within host populations (20). The spatial scale at which coevolution (including for disease resistance) occurs may differ for different host-pathogen associations (19). As much of this type of information as possible should be available for precollecting planning, which should involve pathologists in consultation with breeders. Predictions on the presence or absence of hosts and pathogens will need to be verified in the field, necessitating a pathologist with the collecting mission. Disease evaluation data for previously collected germplasm should be used to target collecting sites. When collection locations of germplasm of *Desmodium ovalifolium* with resistance to wart (*Synchytrium desmodii*) were superimposed on maps of distribution of this legume in South-east Asia, clusters of resistant accessions were evident and sites were targeted for future collection (35). A wide combination of experience at the planning stage and in the collecting and evaluation team is a major factor in the success of the breeding programs of the International Agricultural Research Centers.

Sampling intensity during collection will be determined by population density and breeding systems. Autogamous species occupying ecologically diverse ranges, for example, tropical pasture legume species, should be sampled more intensively than open-pollinated, highly outcrossing species such as many tropical pasture grasses occupying ecologically uniform ranges, for example, *Andropogon gayanus*. One major constraint to collecting wild species is heterogeneity of seed maturing time. A single collection trip may yield only a small proportion of the available variation (75). The need for multiple visits for comprehensive collecting will also aid documentation of diseases over time and, subsequently, collection of a higher proportion of useful, resistant germplasm.

Collected samples of wild species can be maintained as seed, in tissue culture, and in field collections. The most common form of storage of germplasm is as seed under cold, dry conditions. The technology of this storage is well known for crops, but the diversity of wild germplasm indicates a need for research. Seedborne disease readily persists under cold storage: indeed, useful information on the world distribution of *Ascochyta lentis* on lentil was gained by study of a large germplasm collection (84). The spread of Soybean Mosaic Virus (Centrosema strain) between samples during multiplication proved to be a significant factor in the regeneration of *Centrosema macrocarpum* (35). Previously healthy samples could be infected during regeneration, negating costly attempts to free samples of virus. Major germplasm collections of crops, such as those at International Agricultural Research Centers of the CGIAR, now have associated seed health testing laboratories to help manage the collections.

Samples need to be multiplied to provide sufficient seed for distribution, evaluation, and storage. The extensive review by Breese (6) emphasizes the problems of regeneration of wild germplasm, particularly for the outcrossers (24). Breese also emphasizes the need for "much more collation of relevant information by crop species/specialists" before any attempt can be made to define specific standards for seed regeneration. It is virtually impossible to preserve the genetic content of highly polymorphic germplasm to correspond exactly to the original. Other problems during multiplication include shattering (64, 73), lack of seed production (146, 157), and contamination of samples (77). Loss of samples can be particularly severe during field maintenance. Kambuou (85) reported the loss of 75% of a banana field collection, including many wild species. Seventy-nine percent of a collection of six species of *Dioscorea*, and 72% of a germplasm collection of *Saccharum* had also been lost. Even with considerable care, much of the peanut germplasm collected in the 1950s and 1960s has been lost (144). Harlan (64) concluded that maintenance problems caused rampant genetic erosion within collections of wild species.

Problems of ex situ maintenance can be avoided by in situ conservation of germplasm. This strategy is being attempted for several groups of wild relatives. Wilkes (161) reports the conservation of *Zea* in several areas in Mexico, Noy-Meir et al (122) wild wheat in Israel, and Valls (78, 154, 155) wild *Arachis* species. Conserved populations can be monitored regularly and seed harvested as needed. Monitoring for disease incidence is apparently not included in these studies. Prescott-Allen & Prescott-Allen (128) broadly review advantages and disadvantages of in situ conservation. Some polarity can be discerned between the views of genetic resource scientists, who emphasize the value of ex situ storage in genebanks whenever possible, and conservationists, attempting to justify the conservation of ecosystems. From a strictly utilitarian position, in situ conservation is very appropriate for wild species of potential as a source of disease resistance. The often intractable maintenance problems ex situ can be avoided, populations can be studied, sampled, and evaluated for a fuller understanding of their potential, and continuing evolution of hosts and pathogens is possible.

Ex Situ Evaluation of Wild Germplasm for Disease Resistance

Evaluation is a prerequisite for use of germplasm (27). With few exceptions, all evaluations for disease resistance in wild germplasm have been done ex situ using previously collected samples. Although in situ evaluation of wild germplasm is considered the most effective and efficient method of evaluating disease resistance for use in crop improvement, recommendations can be made to improve the efficiency of ex situ evaluation. Large collections should first be evaluated in field trials, under high pathogen pressure and against as varied a range of pathogen races as possible before controlled evaluations. Screening in the centers of diversity of major pathogens, disease hotspots, and in the site where the germplasm will ultimately be used will enhance this objective (98). Although International Agricultural Research Centers commonly evaluate collections of crop germplasm in hotspots for major diseases, few do so for wild germplasm. CIAT has, however, evaluated wild tropical pasture legume germplasm in disease hotspots since 1978 (96).

Sources of resistance identified in the field should be evaluated under controlled conditions with specific pathogenic races, to characterize resistance and assess its potential for crop improvement. Controlled studies of resistance in wild germplasm have almost always focused on race-specific resistance (17, 40, 42, 137, 159). Disease evaluation scales adapted from crops, however, may not be sufficiently discriminative. Infection responses to hemibiotrophic or necrotrophic pathogens, such as the reaction of *Stylosanthes* species to *Colletotrichum gloeosporioides*, will not be discriminated as easily as responses to biotrophic pathogens such as rusts. Appropriate differential sets may have to be developed with representatives of wild species if crop cultivars do not contain sufficient genetic diversity. A differential set of perennial *Glycine* species was developed to distinguish races of soybean rust because insufficient variation was found among soybean cultivars (23). Appropriate differential sets of wild *Stylosanthes* species were developed by trial and error by researchers in South America (92) to distinguish races of *C. gloeosporioides*. Analyses of race-specific resistance in wild germplasm may show a continuum of infection responses (17). Miles & Lenné (107) found a continuum of infection responses, reflecting the combined expression of race-specific and race-nonspecific resistance mechanisms in a wild population of *S. guianensis* to *C. gloeosporioides* from the same wild population in Colombia. A complexity of 18 different types of reactions of one population of *Avena* species to *Puccinia coronata* was found in Israel (42). The number and choice of isolates used for studies of race-specific resistance will determine the diversity of resistance identified (17). Incomplete, simplistic pictures of the resistance structure of wild germplasm populations may be obtained if the isolates used do not reflect the spectrum of existing variability (15). This will reduce the effective use of disease resistance from wild germplasm.

Provided methodological problems are addressed, race-specific resistance can be evaluated in wild germplasm under controlled conditions. Race-nonspecific resistance, however, is difficult to evaluate in wild germplasm populations (11, 17). Evaluation is restricted to documentation of simply determined parameters such as latent period, size of sporulating lesions, number of spores produced, or rate of spore production (17). Generally little is known about the frequency of occurrence or the genetic control of race-nonspecific resistance in populations of wild species (17). A proportion of the genetic defense mechanisms present in wild germplasm of potential use in crop improvement is therefore not being used at present.

It is widely believed that wild germplasm in crop centers of origin and diversity is a concentrated source of disease resistance genes because of the long association of hosts and pathogens (99, 100, 167). Many studies have shown, however, that susceptible plants are more common than resistant plants in wild populations (11, 13, 15, 17, 40, 42, 93, 137, 159). Random sampling of wild germplasm, even from areas of host-pathogen coevolution, will collect a considerable proportion of susceptible plants. In situ evaluation followed by targeted collection of resistant germplasm is a more efficient strategy for use of wild germplasm than ex situ evaluation of large collections.

In Situ Evaluation of Wild Germplasm for Disease Resistance

Although available literature reveals a limited experimental basis, the concept of in situ collection of disease-resistant wild germplasm has been criticized because of lack of information about pathogen races present; lack of control of disease pressure; and because of its time-consuming nature, including the need to revisit sites to survey and tag plants (40, 41, 50). However, an analysis of Dinoor's results (40) clearly shows that the most efficient *single* method of identifying adult *Avena sterilis* plants resistant to *Puccinia coronata* was by in situ evaluation. Alternative methods failed to detect at least 40% of the locations where resistant plants were found by the in situ method (40).

Provided that objectives are well-defined, evaluation in situ could provide basic information on disease resistance to facilitate collection of manageable samples of wild germplasm. Precollection pathogen and host surveys are needed to explore the range and distribution of pathogenic races and host resistance present before collecting. A large-scale general survey of a region would indicate which localities should be more thoroughly investigated (40). The feasibility of such studies is supported by results from the many ecological and genetical studies of wild germplasm populations completed in Israel and Australia (17, 40, 101, 137, 159). Unfortunately, their objective was not collecting resistant germplasm for use in crop improvement. Monitoring disease incidence and severity in the target area for several seasons should ensure that collections are made under high disease pressure. Brown (7) recommended adoption of measures that speed up divergence in situ, such as encouraging disease development. To increase disease selection pressure, inoculum of common local pathogenic races could be multiplied ex situ and introduced into wild populations the season before sampling. This procedure could help to reduce collection of susceptible escapes. Periodic visits to collecting sites to tag and survey apparently resistant plants would allow collection of samples more relevant as sources of disease resistance. The high costs in time and funds involved in evaluating, processing, and maintaining large collections ex situ would be reduced by in situ evaluation.

Seasonal conditions, such as dry seasons, will prevent in situ surveys of diseases in the field year round. Herbarium specimens, which often show disease symptoms, can be used to obtain valuable information on disease distribution especially for wild species for which no other pathological information exists. During 1984, surveys of diseases of tropical pasture grasses were made in Africa by the first author. Because it was impossible to visit every country at the most appropriate time to observe diseases in the field, herbarium specimens were surveyed in countries visited during the dry season. A picture was thereby built up on the incidence, host range, and distribution of diseases of *Brachiaria* in Ethiopia and Zimbabwe. Several locations were identified for future surveys and collection. Although herbarium surveys provide limited information on disease severity, their rapidity, low cost, and year-round potential can aid in planning more intensive field surveys.

International Germplasm Movement and Quarantine

Germplasm movement is vital for successful crop improvement programs. However, movement of germplasm poses serious hazards for crop production worldwide through dissemination of pathogens by seed. The most adverse effect of seedborne pathogens is the introduction of new pathogens and/or new races of pathogens to previously disease-free sites. Seedborne diseases also reduce seed quality in storage and cause pre-emergent losses and seedling and plant mortality. Germplasm collected from centers of host diversity, including wild germplasm, could be of high risk (100, 115).

Viruses are considered the greatest risk to collection, multiplication, and conservation of wild germplasm. They may not be recognized at collection due to symptomless colonization and may spread rapidly in new areas and through collections during initial multiplication and evaluation if conditions are conducive and/or vectors are numerous. Wild relatives of Citrus (52), potato (82, 151), and sugar beet (165) have been implicated as hosts of important viruses of these crops. The need to isolate and purify new viruses before development of detection methods creates further difficulties for quarantine. Although viruses are considered the greatest risk in international germplasm movement, most seedborne infections are fungal. Internally borne fungi can escape surface treatment and may remain viable in seed for long periods (115). Although international germplasm exchange through tissue culture is one method of reducing pathological risks, techniques need to be developed for many wild species.

Because much wild germplasm of interest to breeders is poorly understood pathologically, it is almost impossible to assess the risks involved in its collection and international movement. However, information is often available in published country and region host-disease indices and in unpublished records of institutes such as the International Mycological Institute, Kew, England, which offer worldwide diagnostic services. Such information proved invaluable in the preparation of the *World List of Fungal Diseases of Tropical Pasture Species,* all of which are wild species (94). Considerable information is readily available on seedborne fungi and bacteria of crops (131) that could be used as a data base of potential problems for wild species. There is a great need to collate and increase accessibility to available information before collection of wild germplasm. Lack of published information on diseases and seedborne pathogens of wild germplasm also emphasizes the need for the participation of plant pathologists in defining objectives and during germplasm collecting trips. As part of a multidisciplinary team, plant pathologists could document any diseases observed on collected plants, even if the collected part is healthy, and later ensure application of appropriate phytosanitary standards, especially for germplasm destined for international movement (114). Wild relatives of some crops are being intensively collected by IBPGR (77) but usually no plant pathologist is associated with collecting expeditions.

USE OF WILD GERMPLASM FOR DISEASE RESISTANCE

Use of wild species for crop improvement has been very successful for a few crops but disappointing for numerous others (144). The most common reason for using wild germplasm is as a source of disease resistance. Although many successful transfers of single gene resistance have been achieved, much of the literature reports identification of resistance and production of interspecific hybrids but rarely the actual release of a new cultivar and its use by farmers (63, 68, 143).

Comprehensive reviews of the use of disease resistance from wild germplasm for crop improvement have been numerous (28, 51, 55, 63, 64, 68, 74, 90, 127, 128, 132, 143, 144, 146, 160). The greatest impacts of disease resistance from wild germplasm in food crops have been in wheat, potato, and tomato. This is mainly due to the ease of use of their wild relatives in breeding programs; the presence of many polyploid species among the wild germplasm; and the extensive background research effort and knowledge specific to these crops and their wild germplasm (143). The proven value of wild species as sources of disease resistance for some crops has fostered a growing belief that many wild relatives of crops contain valuable disease resistance and collection of wild germplasm has increased dramatically since 1976 (77). Wild germplasm has also been credited with resistance potentially more valuable than that in crops (158). Moreover, the more reportable nature of research on wild germplasm relative to traditional research on cultivars and landraces (32) has contributed to exaggerated opinions of the value of wild germplasm. The following examples portray a realistic appraisal of the value of disease resistance from wild germplasm in selected food crops.

Resistance to the most important rice diseases has been found in landraces from Asia and Africa within rice's natural distribution (28, 29). Both *Oryza sativa* and *O. glaberrima* are sufficiently diverse to continue to dominate screening programs for disease resistance. Outstanding sources of disease resistance have been identified in landraces from Bangladesh, India, Indonesia, the Philippines, and Vietnam (31). It has rarely been necessary to resort to wild germplasm. The only striking example of use of disease resistance from wild germplasm is resistance to Grassy Stunt Virus (GSV) (30). After screening 17,000 accessions of *O. sativa* and over 100 wild taxa in the *O. sativa* species complex, a few plants of a single population of *O. nivara*—believed by some taxonomists to be a weedy *O. sativa* (43)—were found to be highly resistant to GSV (30). Other sources of resistance were later found in four wild species that were available when a new highly virulent strain of the virus appeared in the 1980s (30).

Since the first wide crosses were made in the Triticeae over 100 years ago, disease resistance from *Triticum, Aegilops* and *Agropyron* species has made a considerable contribution to wheat improvement (112). The potential of wide hybridization in the Triticeae is probably greater than in most other groups of crop relatives because of the ease of hybridization, extensive taxonomic, genetic, and cytogenetic understanding of wheat and its relatives, and because wheat is an hexaploid and thus well-buffered against introgression of alien material (112, 168). In CIMMYT, where wide crossing with wheat has been in progress for more than 20 years, traditional breeding techniques and wide crossing research are complementary (111). Many genes for disease resistance have been transferred from wild germplasm to wheat (55). At least 15 genes for resistance to stem rust, *Puccinia graminis tritici,* are available from wild *Triticum* species and other relatives (135). Some of the most effective and durable single gene resistances used worldwide against stem rust are *Sr 26* from perennial *Agropyron elongatum* and *Sr 36* from wild tetraploid *Triticum timopheevii* (135). Similarly impressive feats can be attributed to landrace germplasm. PI 178383 wheat from Turkey had near immunity to 4 races of *P. striiformis* and good resistance to 35 races of *Tilletia caries* and *T. foetida* and to 10 races of *T. contraversa* (63).

Wheat is a frequently cited example of the common and successful use of wild germplasm, especially for disease resistance. Easily inherited and identifiable single gene resistance has accounted for most of this (32). Such single gene resistance to wheat diseases has frequently been overcome (89). Wide-ranging epidemics of stem and leaf rust have repeatedly decimated wheat fields in the USA during the past 75 years (44).

Steady increases in maize yields, especially in the USA, from the use of North American landraces have precluded the need to seek out wild species. Exotic germplasm (both landrace and wild) accounts for less than 1% of the US maize germplasm base (54). Hybrids of maize x annual teosinte, *Zea mexicana,* and maize x tetraploid *Tripsacum* species have been produced (143) and some disease resistance transferred from the wild relatives (28). Tolerance and immunity to seven tropical viruses have been identified in perennial teosinte, *Zea diploperennis* (74), and may eventually be incorporated into maize, but they are of no immediate consequence to improving maize production (54). Even attempts to use tropical landrace germplasm have been limited, ineffective, and static to date (54), although it is a rich source of disease resistance (86). It is presently believed that the pool of elite inbred lines in the USA has a great diversity of genes for disease resistance of priority use in maize improvement (56).

Wild species have played (28, 70) and continue to play a key role in the development of high-performance potato cultivars with resistance to fungal, bacterial, viral, and nematodal diseases (79). The first interspecific hybrids, between *Solanum demissum* and *S. tuberosum* for late blight resistance, were made more than 100 years ago (91). As for wheat, the long history of interspecific hybridization and the complex of highly polyploid species give access to great potential genetic variability (72). Studies on the parentage of many cultivars from Europe and the USA have shown that 80% and 33%, respectively, have wild species or primitive cultivars in their pedigrees. The most significant example of the use of wild germplasm for disease resistance is to late blight. Single gene resistance was identified in *S. demissum,* transferred to potato, and proved to be an expensive and protracted failure (116). Successful polygenic resistance was later found (116, 121). Wild species have been successfully used as sources of both race-specific and race-nonspecific resistance to fungal pathogens, viruses, and nematodes in potatoes (79) but some race-specific resistance has been overcome.

No more than 12 of the total of about 175 *Solanum* wild species have actually been used in potato cultivars (72). The high percentage of cultivars with wild germplasm is due to the frequent use of derivatives from *S. demissum* and *S. tuberosum* ssp. *andigena,* which occur in the parentage of 60–70% of cultivars with wild genes. Many collections of wild *Solanum* species have therefore not yet been of value as sources of disease resistance for potato improvement.

Tomato has been described as the most outstanding example of the use of wild germplasm for disease resistance (132, 133). A large pool of wild tomato germplasm is available to breeders and *Lycopersicon esculentum* can be crossed with eight species to form fertile hybrids (132). Prebreeding, by a series of backcrosses to the recurrent, cultivated parent alternated with selfed generations in which desired combinations of parental characteristics are selected, is very important in using wild germplasm in tomato improvement. Combined resistance to several major pathogens has been achieved. Satisfactory resistance to at least 30 different diseases has been found in wild forms and identified genes for 14 diseases have been introgressed into productive widely grown cultivars (132, 133). No comparable figures for the contribution from landraces were found. Determination of the actual impact and value of the use of wild germplasm for disease resistance in tomato is hampered by difficulties of tracing back pedigrees of varieties (132), a problem not unique to tomato.

Since the 1960s, attempts to use wild germplasm of peanut have increased greatly because of a perceived lack of good sources of resistance to major pathogens such as early and late leaf spots (*Cercospora arachidicola* and *Phaeoisariopsis personata*), rust (*Puccinia arachidis*), and viruses in available landrace germplasm and the existence of high levels of resistance and immunity to these diseases in wild germplasm (110, 144, 146, 147). High levels of disease resistance to major fungal pathogens, viruses, and nematodes have been identified in wild *Arachis* species in the sections *Arachis, Caulorhizae, Extranervosae,* and *Rhizomatosae* (78, 146–149). Although some of this resistance has been known for more than 20 years (146), difficulties of gene transfer from wild species to *A. hypogaea* have precluded use of much of this germplasm. *Arachis hypogaea* is reproductively isolated from all species except those in the section *Arachis* (57). In this section, although *A. monticola* has been successfully used, the success rate of interspecific hybrids with diploid species of the section has been low. Resistance to early and late leaf spots and rust was eventually transferred from diploid *A. cardenasii* but it has been necessary to self many times to recover hybrids with 40 chromosomes (146). Although the hybrids have been used as breeding lines (146), stable cultivars with high resistance to these diseases have not yet been released.

In recent years, researchers have shown that cultivated peanut germplasm is of more value than was previously suspected (142, 164). Levels of resistance to all important diseases, of use in breeding programs, have been found (48, 164). Valuable concentrations of resistance to late leaf spot and rust have been identified in landraces from the Tarapoto region of Peru (150), the potential of which was documented 30 years ago (2). Such a concentration of high levels of resistance to major peanut diseases in one of the centers of diversity of peanut would be a valuable focus for in situ studies of host-pathogen associations for targeting further collection of resistant germplasm.

"Until more information is collected on circumventing sterility barriers and on the durability of resistance in *Arachis,* it would be wise for most programs to concentrate on utilizing genetic resources available within *A. hypogaea*" (164). If "efforts to utilize distantly related species should only be conducted when desirable traits are not found in the cultivated species or species closely related to *A. hypogaea*" (146), it is difficult to justify the time and resources necessary to collect, evaluate, and maintain wild *Arachis* germplasm that cannot be used for peanut improvement in the foreseeable future, especially when sources of resistance to major diseases have been identified in usable landrace germplasm.

Perennial diploid *Glycine* species are valuable sources of disease resistance of potential use in soybean improvement (16, 104). Rust is a major constraint to soybean production in much of Asia and resistance breeding is severely hampered by scarcity of known sources of resistance in the crop (9). Both quantitative and qualitative resistance has been identified in six *Glycine* species, including *G. canescens* and *G. clandestina* (22). Resistance in *G. canescens* is dominant and controlled by major genes (18). Such resistance would be of great value to soybean if it could be transferred. Although some diploid *Glycine* species have been hybridized with soybean (120), transfer of disease resistance has not yet been accomplished. Landraces have offered much more than wild species, to date (28).

In contrast to crops, there are no landraces of pasture species. Most cultivars of pasture legumes and grasses, particularly tropical species, are unbred selections of wild species. That susceptible plants predominate in wild populations and that valuable single character traits, such as disease resistance, are very diluted, is strongly supported by results of efforts to develop tropical pasture legume cultivars in South America, the center of diversity of many legume genera with pasture potential (96). *Stylosanthes guianensis* has been a source of tropical pasture cultivars since the early 1960s. Cultivars, selected in Australia and reintroduced into South America, were highly susceptible to anthracnose (107). A collection of more than 1500 accessions was screened under high anthracnose pressure by CIAT in Colombia at a cost of $120,000 to $200,000 (153). Only two accessions were identified with high resistance (34). Similarly, *S. capitata* was widely collected in central Brazil, its center of diversity. Almost 90% of a large collection screened near Brasilia was highly susceptible to the most common race of *Colletotrichum gloeosporioides* in the region (98). An extensive collection of 250 accessions of *Centrosema brasilianum*, from throughout the native range of the species, showed no usable resistance to Rhizoctonia foliar blight (95). Experience with tropical pasture plants clearly shows that considerable effort, time, and funds will be needed to collect and evaluate wild germplasm to identify resistant genotypes for use in crop improvement.

A general emphasis on wild relatives (70), and on reconsidering germplasm collecting priorities to favor wild relatives (49), needs justification. Before investment of the effort needed for use of resistance from wild germplasm for crop improvement, the most effective and relevant sources of resistance should be identified. This identification will require a knowledge of the availability and accessibility of sources of resistance from crop germplasm; the extent of taxonomic, cytogenetic, and pathological knowledge of the wild relative; the availability and accessibility of sources of resistance from wild germplasm; the comparative value of sources of resistance from crop and wild germplasm; and a realistic estimate of the amount of effort involved in using the resistance.

INCREASING THE EFFICIENCY OF WILD GERMPLASM USE

The move by breeders to wild germplasm as a source of disease resistance is at least partly a response to inadequacies in the evaluation of landraces for disease resistance (124)—often a result of the enormous size of collections. Past collecting of landraces emphasized diversity (6, 105), and not particular characters of interest. For efficient collecting of disease resistance, a more practicable approach is needed. Breeders should define their needs to make sure that study and collecting of wild species efficiently targets resistance:

sampling of the kind designed to capture common alleles (105) is not indicated. Targeting is possible:

1. Between crops: effort should concentrate on crops whose wild relatives show a high value as sources of disease resistance.
2. Within large existing germplasm collections of wild relatives (wheat, 12,000; potatoes, 1,400 samples), which should be efficiently screened to capture resistance and to allow future emphasis on targeted wild species and areas of importance.
3. Between and within field populations identified from preliminary surveys as meriting further study.

Evaluation of field populations before collecting, and of samples after distribution, should enable continuing feedback to indicate regions of particular value as sources of resistance. Evaluation results should be pooled internationally: this should be insisted on before germplasm is distributed for evaluation.

To increase the efficiency of wild germplasm use basic knowledge of wild germplasm-disease systems is needed. For example, wild relatives of cereals and soybean and their pathogens have been the target of detailed ecological, genetic, and pathological studies for at least 20 years. Considerable information has been generated on the resistance structure of host populations and the virulence structure of pathogen populations. This is essential preparatory research to future use of this germplasm. It has had, however, minimal impact on present methods of collecting and evaluating wild relatives of other crops. A comprehensive understanding of disease distribution and pathogen variability is essential to the efficient and effective collection and evaluation of wild germplasm for crop improvement.

The movement of crops and pathogens from their centers of diversity to other regions has important implications to the search for wild germplasm as potential sources of disease resistance. For many important crops, we are largely ignorant of the regions of diversity or even distribution of important pathogens within and outside centers of crop diversity. Even when crop and pathogen centers of diversity coincide, useful resistance has often been found in highly localized subcenters (17, 101) and, notably, in landraces in secondary centers of diversity (130). Resistance is also commonly found in wild and weedy relatives outside the center of diversity of the crop (16, 121). The best sources of disease resistance are not necessarily found in wild relatives in the center of diversity of the crop. Comparative evaluation of resistance is needed from wild relatives in the center of diversity of the crop; from geographically distant wild relatives; from landraces in secondary centers of diversity; and from elite lines from modern plant breeding efforts. Results will depend on a complex of interrelated factors involving pathogen biology, host biology, environmental factors, and the history of movement of the crop and the pathogen. Recognition of valuable concentrations of resistance in wild relatives is not possible for most crops.

The efficiency of methods of ex situ evaluation is dependent on present recommendations for collecting wild germplasm (6, 105) that stress the need for collecting variation per se, particularly between-sample variation. Using this strategy, a considerable proportion of susceptible plants will be collected, even from areas of host-pathogen coevolution, as susceptible plants are more common than resistant plants in wild populations. Considerable input is therefore needed during ex situ screening to identify sources of disease resistance in large, often difficult to manage collections with low frequencies of resistance. Efficiency of ex situ evaluation of wild germplasm for disease resistance in the field, even with large collections, can be increased by the right choice of field location for initial screening. Screening in pathogen centers of diversity, disease hotspots, and multilocationally will greatly facilitate identification of widely resistant germplasm for subsequent controlled screening. Elimination of inferior samples before movement of germplasm will also reduce the associated risks of pathogen dissemination, especially serious for pathologically unknown wild species collected in pathogen centers of diversity. Risks of spreading pathogens by international movement of wild germplasm can be minimized by seeking as much precollection information as possible; by the involvement of plant pathologists during collecting; by post-collecting, pre-entry, and post-entry seed health testing; by appropriate seed-treatment; by frequent inspection of plants resulting from field collections; and by third country quarantine, especially for viruses and internally seedborne fungi and bacteria.

In situ evaluation before collection can be an important strategy for increasing the efficiency of use of wild germplasm for disease resistance. Essential initial input includes precollection herbarium studies together with host and pathogen field surveys to explore the range and distribution of pathogenic races and host resistance present; monitoring disease incidence and severity in the target area to ensure that collections are made under high disease pressure; and periodic visits to collecting sites to survey and tag resistant plants. Information from in situ evaluations will enable collection of small, manageable samples of wild germplasm with higher frequency of resistance for later controlled, ex situ evaluations. Controlled screening of wild germplasm for disease resistance has almost exclusively targeted race-specific resistance. Screening methodologies used for crops have been directly applied to wild species without appreciation of the problems created by the unknown genetic backgrounds of these plants. Improved methodology for controlled evaluation of disease resistance will increase the efficiency of use of wild germplasm. Information from in situ evaluation should be used to develop differential sets appropriate to analyzing the range of variability for resistance present in wild germplasm, while the choice and number of isolates used should reflect knowledge of pathogen variability obtained from crop and wild population studies.

The key process to increase the efficiency of use of wild germplasm is to study populations in the field to identify specific characters. Only after such study can an intelligent sampling from the wild increase the effectiveness of supplying germplasm relevant to the use of breeders. The importance of preserving wild germplasm populations within which the interaction of adaptive genetic variability and natural selection continues to give rise to new genotypes has been stressed for more than 25 years (4, 12, 80, 128). The impossibility of maintaining representative collections of the diversity of wild germplasm ex situ; the difficulties of collecting, processing, maintaining, and rejuvenating wild germplasm ex situ; and the advantages of maintaining wild germplasm in an environment where evolution for disease resistance and other characters can continue (128) clearly support the value of in situ evaluation. The in situ populations that form the basis of such studies would be the source of all necessary seed samples for further evaluation. In situ and controlled evaluations would provide feedback to redefine and better target known areas of pathogen diversity for further collection. Collecting only a carefully defined sample would greatly reduce the expense of germplasm collecting and subsequent management and increase the efficiency of identifying of resistance.

The most effective and efficient approach to collecting and evaluating wild germplasm for disease resistance is proposed as:

1. Definition of objectives;
2. Pre-collecting host and pathogen surveys to identify geographic areas for study;
3. In situ evaluation of disease resistance and monitoring of populations;
4. Initial collecting of promising samples;
5. Ex situ field evaluation in disease hotspots, preferably in centers of pathogen diversity;
6. Selection and targeted collection of manageable samples of wild germplasm with higher frequencies of disease resistance; and
7. Controlled evaluation of disease resistance using improved methodologies to allow selection and further collection of the most relevant sources of resistance for use in crop improvement.

THE VALUE OF WILD GERMPLASM FOR DISEASE RESISTANCE

The presence of a wide spectrum of genetic resistance mechanisms in wild germplasm of potential value in crop improvement has been clearly demonstrated (17, 137, 146). Lack of information on the frequency of occurrence and genetic control of race-nonspecific resistance in wild germplasm (17), difficulties of extracting it from wild sources, and reluctance of breeders to use such resistance (89) have resulted in the use of wild germplasm almost exclusively as sources of race-specific resistance. The claim that resistance genes, which the pathogen cannot match, are more likely to be found in foreign germplasm (158) has not been substantiated (13, 89, 140, 141), nor is there any proof that race-specific resistance from the wild will be any more durable than resistance from crop germplasm (140, 141). Although race-specific resistance has been the basic strategy for disease control in most agricultural crops, many pathogens have overcome such resistance, including that from wild species. In a survey of crops in the USA receiving genes from wild species, almost all single gene resistance had failed but six identified cases of polygenic resistance were durable (127). Wild species have been used in desperate situations to salvage a crop and prevent its failure commercially (64). Germplasm resources of wild species have not provided a miracle cure in combating plant diseases (143).

Race-specific resistance from wild germplasm will continue to be used in the future and will demand a continuous high-input breeding effort (140). While the identification of usable resistance in wild germplasm will broaden the genetic base available to breeders, it will not solve the problem of genetic vulnerability in crops unless deployed intelligently. Complementary approaches are possible and are needed, based on increasing the diversity for resistance. Browning (13) has argued that "Diversity is the only defense against the unknown, as against a future disease threat". Strategies could include:

1. Spatial deployment of race-specific resistance from wild germplasm in varietal mixtures;
2. Development of contrasting genetic families of varieties to those used at present; and
3. Increased effort on characterization and manipulation of race-nonspecific resistance from wild germplasm.

Literature Cited

1. Alexander, H. M. 1990. Dynamics of plant-pathogen populations in natural plant communities. See Ref. 21, pp. 31–45
2. Allen, D. J. 1983. *The Pathology of Tropical Food Legumes: Disease Resistance in Crop Improvement*. Chichester: Wiley. 413 pp.
3. Anikster, Y., Wahl, I. 1979. Coevolution of the rust fungi on Gramineae and Liliaceae and their hosts. *Annu. Rev. Phytopathol*. 17:367–403
4. Bennett, E. 1965. Plant introduction and genetic conservation: genecological aspects of an urgent world problem. *Scott. Plant Breed. Stn. Rec*. 7–113
5. Braverman, S. W., Leppik, E. E. 1972. Origins of cultivated plants and the search for disease resistance. *NY Food Life Sci*. 5:15–18
6. Breese, E. L. 1989. *Regeneration and Multiplication of Germplasm Resources in Seed Banks: the Scientific Background*. Rome: IBPGR. 69 pp.

7. Brown, A. H. D. 1990. Population divergence in wild relatives. *Proc. Int. Wksh. Dynamic In Situ Conservation of Wild Relatives of Major Cultivated Plants, Israel, 1990* (Abstr.)
8. Brown, A. H. D., Frankel, O. H., Marshall, D. R., Williams, J. T., eds. 1989. *The Use of Plant Genetic Resources*. Cambridge: Cambridge Univ. Press. 382 pp.
9. Brown, A. H. D., Grant, J. E., Burdon, J. J., Grace, J. P., Pullen, R. 1985. Collection and utilization of wild perennial Glycine. *Proc. World Soybean Research Conf., 3rd*, pp. 345–52. Boulder: Westview Press Inc.
10. Brown, A. H. D., Marshall, D. R. 1986. Wild species as genetic resources for plant breeding. In *Proc. Plant Breed. Symp., DSIR, Spec. Publ. No. 5*, pp. 9–15. Palmerston North: NZ Agron. Soc.
11. Browning, J. A. 1974. Relevance of knowledge about natural ecosystems to development of pest management programs for agro-ecosystems. *Proc. Am. Phytopathol. Soc.* 1:191–99
12. Browning, J. A. 1980. Genetic protective mechanisms of plant pathogen populations: their co-evolution and use in breeding for resistance. See Ref. 67, pp. 52–75
13. Browning, J. A. 1988. Current thinking on the use of diversity to buffer small grains against highly epidemic and variable foliar pathogens: problems and future prospects. See Ref. 141, pp. 76–90
13a. Buddenhagen, I. W. 1977. Resistance and vulnerability of tropical crops in relation to their evolution and breeding. *Ann. NY Acad. Sci.* 287:309–26
14. Bunting, A. H. 1983. Review and prospect. See Ref. 106, pp. 313–21
15. Burdon, J. J. 1985. Pathogens and the genetic structure of plant populations. In *Studies on Plant Demography: A Festschrift for John Harper*, ed. J. White, pp. 313–25. London: Academic
16. Burdon, J. J. 1986. The potential of Australian *Glycine* as sources of resistance to soybean leaf rust (*Phakopsora pachyrhizi*). *Proc. Int. SABRAO Congr., 5th., Bangkok, Thailand, 1985*. pp. 823–32
17. Burdon, J. J. 1987. *Diseases and Plant Population Biology*. Cambridge: Cambridge Univ. Press. 208 pp.
18. Burdon, J. J. 1987. Phenotypic and genetic patterns of resistance to the pathogen *Phakopsora pachyrhizi* in populations of *Glycine canescens*. *Oecologia (Berl.)* 73:257–67
19. Burdon, J. J., Brown, A. D. H., Jarosz, A. 1990. The spatial scale of genetic interactions in host-pathogen co-evolved systems. See Ref. 21, pp. 233–47
20. Burdon, J. J., Jarosz, A. 1989. Wild relatives as sources of disease resistance. See Ref. 8, pp. 280–96
21. Burdon, J. J., Leather, S. R., eds. 1990. *Pests, Pathogens and Plant Communities*. Oxford: Blackwell. 344 pp.
22. Burdon, J. J., Marshall, D. R. 1981. Inter- and intra-specific diversity in the disease response of *Glycine* species to the leaf-rust fungus *Phakopsora pachyrhizi*. *J. Ecol.* 69:381–90
23. Burdon, J. J., Speer, S. S. 1984. A set of differential *Glycine* hosts for the identification of *Phakopsora pachyrhizi*. *Euphytica* 33:891–96
24. Burton, G. W., Ellis Davies, W. 1984. Handling germplasm of cross-pollinated forage crops. See Ref. 73, pp. 180–90
25. Cauderon, Y. 1986. Cytogenetics in breeding programmes dealing with polyploids, interspecific hybridization and introgression. In *Genetic Manipulation in Plant Breeding, EUCARPIA Symp. 1985*, pp. 83–104. Berlin: Gruyter
26. Chang, T. T. 1985. Principles of genetic conservation. *Iowa Stn. J. Res.* 59:325–48
27. Chang, T. T. 1985. Evaluation and documentation of crop germplasm. *Iowa Stn. J. Res.* 59:379–97
28. Chang, T. T. 1985. Germplasm enhacement and utilization. *Iowa Stn. J. Res.* 59:399–424
29. Chang, T. T. 1985. Crop history and genetic conservation: rice—a case study. *Iowa Stn. J. Res.* 59:425–56
30. Chang, T. T. 1989. The case for large collections. See Ref. 8, pp. 123–35
31. Chang, T. T., Ou, S. H., Pathak, M. D., Ling, K. C., Kauffman, H. E. 1975. The search for disease and insect resistance in wild germplasm. See Ref. 50, pp. 183–200
32. Chapman, C. G. D. 1986. The role of genetic resources in wheat breeding. *Plant Genet. Res. Newsl.* 65:2–5
33. Chapman, C. G. D. 1989. Collection strategies for the wild relatives of field crops. See Ref. 8, pp. 263–79
34. CIAT. 1984. *Annual Report, Tropical Pastures Program*, Cent. Int. Agric. Trop., CIAT, Cali, Colombia
35. CIAT. 1988. *Annual Report, Tropical Pastures Program*, Cent. Int. Agric. Trop., CIAT, Cali, Colombia
36. Clayton, W. D. 1983. Tropical grasses. See Ref. 106, pp. 39–46
37. Cubero, J. I. 1984. Taxonomy, distribution and evolution of lentil and its

wild relatives. See Ref. 162, pp. 187–203
38. de Wet, J. M. J. 1981. Grasses and the culture history of man. *Ann. Mo. Bot. Gdn.* 68:87–104
39. de Wet, J. M. J., Harlan, J. R., Brink, D. E. 1986. Reality of infraspecific taxonomic units in domesticated cereals. In *Infraspecific Classification of Wild and Cultivated Plants,* ed. Styles, B. T., pp. 211–22. Oxford: Clarendon
40. Dinoor, A. 1970. Sources of oat crown rust resistance in hexaploid and tetraploid wild oats in Israel. *Can. J. Bot.* 48:153–61
41. Dinoor, A.1975. Evaluation of sources of disease resistance. See Ref. 50, pp. 201–10
42. Dinoor, A. 1977. Oat crown rust resistance in Israel. *Ann. NY Acad. Sci.* 287:357–66
43. Duistermaat, H. 1987. A revision of *Oryza* (Gramineae) in Malaysia and Australia. *Blumea* 32:157–93
44. Duvick, D. N. 1984. Genetic diversity in major farm crops on the farm and in the reserve. *Econ. Bot.* 38:161–78
45. Ellis, J. G., Lawrence, G. J., Peacock, W. J., Pryor, A. J. 1988. Approaches to cloning plant genes conferring resistance to fungal pathogens. *Annu. Rev. Phytopathol.* 26:245–63
46. FAO. 1984. *In Situ Conservation of Wild Plant Genetic Resources: a Status Review and Action Plan.* FAO For. Res. Div. Rome: FAO/FORGEN/MISC/84/3. 83 pp.
47. Fischbeck, G., Schwarzbach, E., Sobel, Z., Wahl, I. 1976. Types of protection against powdery mildew in Germany and Israel selected from *Hordeum spontaneum.* In *Proc. Int. Barley Genet. Symp., 3rd, Garching, 1975,* pp. 412–17. Munich: Verlag Karl Thiemig
48. Foster, J. D., Stalker, H. T., Wynne, J. C., Beute, M. K. 1981. Resistance of *Arachis hypogaea* L. and wild relatives to *Cercospora arachidicola* Hori. *Oleagineux* 36:139–43
49. Frankel, O. H. 1989. Perspectives on genetic resources. In *CIMMYT 1988 Annual Report,* pp. 10–17. Mexico: CIMMYT
50. Frankel, O. H., Hawkes, J. G., eds. 1975. *Crop Genetic Resources for Today and Tomorrow.* Cambridge: Cambridge Univ. Press. 492 pp.
51. Frey, K. J. 1983. Genes from wild relatives for improving plants. *Proc. Int. SABRAO Congr., 4th, Kuala Lumpur, Malaysia, 1981.* pp. 1–20
52. Garnsey, S. M. 1988. Detection of citrus diseases and prevention of their spread during international exchange of citrus germplasm. In *Conservation and Movement of Vegetatively Propagated Germplasm: In Vitro Culture and Disease Aspects,* pp. 39–41. Rome: IBPGR. 60 pp.
53. Gillett, J. B. 1962. Pest pressure, an underestimated factor in evolution. *Syst. Assoc. Publ.* 4:37–46
54. Goodman, M. M. 1985. Exotic maize germplasm: status, prospects, and remedies. *Iowa St. J. Res.* 59:497–528
55. Goodman, R. M., Hauptli, H., Crossway, A., Knauf, V. C. 1987. Gene transfer in crop improvement. *Science* 236:48–54
56. Gracen, V. E. 1986. Sources of temperate maize germplasm and potential usefulness in tropical and subtropical environments. *Adv. Agron.* 39:127–72
57. Gregory, M. P., Gregory, W. C. 1979. Exotic germplasm of *Arachis* L. interspecific hybrids. *J. Hered.* 70:185–93
58. Hahn, S. K., Terry, E. R., Leuschner, K. 1980. Breeding cassava for resistance to cassava mosaic virus disease. *Euphytica* 29:673–83
59. Hammer, K. 1984. Das Domestikationssyndrom. *Kulturpflanze* 32:11–34 (In German)
60. Harlan, J. R. 1959. Plant exploration and the search for superior germplasm for grasslands. In *Grasslands,* ed. H. B. Sprague, pp. 1–11. Washington, DC: AAAS, Publ. 53
61. Harlan, J. R. 1961. Geographic origin of plants useful to agriculture. In *Germplasm Resources,* ed. R. E. Hodgson, pp. 3–19. Washington DC: AAAS, Publ. 66
62. Harlan, J. R. 1975. *Crops and Man.* Madison: Am. Soc. Agron. 295 pp.
63. Harlan, J. R. 1976. Diseases as a factor in plant evolution. *Annu. Rev. Phytopathol.* 13:31–51
64. Harlan, J. R. 1984. Evaluation of wild relatives of crop plants. See Ref. 73, pp. 212–22
65. Harlan, J. R., de Wet, J. M. J. 1971. Towards a rational classification of cultivated plants. *Taxon* 20:509–17
66. Harper, J. L. 1977. *Population Biology of Plants.* London: Academic. 892 pp.
67. Harris, M. K. 1980. *Biology and Breeding for Resistance to Arthropods and Pathogens of Agricultural Plants.* Texas A & M Agr. Exp. Stn. Misc. Publ. 1451
68. Hawkes, J. G. 1977. The importance of wild germplasm in plant breeding. *Euphytica* 26:615–21
69. Hawkes, J. G. 1983. *The Diversity of Crop Plants.* Cambridge, MA: Harvard Univ. Press. 184 pp.

70. Hawkes, J. G. 1985. *Plant Genetic Resources: the Impact of the International Agricultural Research Centers*. CGIAR Study Paper No. 3. Washington DC: World Bank. 115 pp.
71. Hawkes, J. G. 1986. Problems of taxonomy and nomenclature in cultivated plants. *Acta Horticulturae* 182:41–52
72. Hermsen, J. G. Th. 1989. Current use of potato collections. See Ref. 8, pp. 68–87
73. Holden, J. H. W., Williams, J. T., eds. 1984. *Crop Genetic Resources: Conservation and Evaluation*. London: Allen & Unwin. 296 pp.
74. Hoyt, E. 1988. *Conserving the Wild Relatives of Crops*. Rome: IBPGR, IUCN, WWF. 45 pp.
75. IBPGR.1983. *Practical Constraints Affecting the Collection and Exchange of Wild Species and Primitive Cultivars*. Rome: IBPGR. 11 pp.
76. IBPGR. 1985. *Ecogeographical Surveying and In Situ Conservation of Crop Relatives*. Rome: IBPGR. 27 pp.
77. IBPGR. 1990. *Annual Report 1989*. Rome: IBPGR. 64 pp.
78. ICRISAT, ed. 1985. *Cytogenetics of Arachis. Proc. Int. Wksh., ICRISAT, India, 31 Oct.–2 Nov., 1983*. Pantancheru, India: ICRISAT. 191 pp.
79. Iwanaga, M., Schmiediche, P. 1989. Using wild species to improve potato cultivars. *CIP Circular* 17: 1–7
80. Jain, S. K. 1975. Population structure and the effects of breeding system. See Ref. 50, pp. 379–96
81. Johns, T., Keen, S. L. 1986. Ongoing evolution of the potato on the Altiplano of Western Bolivia. *Econ. Bot*. 40:409–24
82. Jones, R. A. C. 1981. The ecology of viruses infecting wild and cultivated potatoes in the Andean region of South America. See Ref. 151, pp. 89–107
83. Joshi, L. M. 1986. Perpetuation and dissemination of wheat rusts in India. In *Problems and Progress of Wheat Pathology in South Asia*, ed. L. M. Joshi, D. V. Singh, K. D. Srivastava, pp. 41–68. New Delhi: Malhotra Publ. House
84. Kaiser, W. J., Hanna, R. M. 1986. Incidence of seedborne *Ascochyta lentis* in lentil germplasm. *Phytopathology* 76: 355–60
85. Kambuou, R. N. 1988. The current status of plant genetic resources activities in Papua New Guinea. *Proc. Int. Wksh. Crop Genet. Res. East Asia, Tsukuba, Jpn, Nov. 1987*, pp. 59–65. Rome: IBPGR
86. Kim, S. K., Brewbaker, J. L., Hallauer, A. R. 1989. Insect and disease resistance from tropical maize for use in temperate zone hybrids. *Proc. Ann. Corn Sorghum Res. Conf., 43rd., Chicago, 1988*, pp. 194–226. ASTA, Washington DC: Dolores Wilkinson
87. Kimber, G., Feldman, M. 1987. *Wild Wheat: An Introduction*. Coll. Agric., Univ. Miss., Columbia, Spec. Rep. 353. 143 pp.
88. Kloppenberg, J. R., Kleinman, D. L. 1988. *Seeds and Sovereignty: the Use and Control of Plant Genetic Resources*. Durham: Duke Univ. Press. 368 pp.
89. Knott, D. R. 1988. Using polygenic resistance to breed for stem rust resistance in wheat. See Ref. 141, pp. 39–47
90. Knott, D. R., Dvorak, J. 1976 Alien germplasm as a source of resistance to disease. *Annu. Rev. Phytopathol*. 14: 211–35
91. Large, E. C. 1940. *The Advance of the Fungi*. London: Johnathan Cape. 488 pp.
92. Lenné, J. M. 1986. Recent advances in the understanding of anthracnose of *Stylosanthes. Proc. Int. Grassl. Congr., 15th, Kyoto, Jpn, 1985*. pp. 773–75
93. Lenné, J. M. 1988. Variation in reaction to anthracnose within native *Stylosanthes capitata* populations in Minas Gerias, Brazil. *Phytopathology* 78:131–34
94. Lenné, J. M. 1990. A world list of fungal diseases of tropical pasture species. *Phytopathol. Pap*. 31. 161 pp.
95. Lenné, J. M., Olaya, G., Miles, J. W. 1989. Importance of Rhizoctonia foliar blight of the promising tropical pasture legume genus *Centrosema. Proc. Int. Grassl. Congr., 16th, Nice, France*, pp. 697–98
96. Lenné, J. M., Pizarro, E. A., Toledo, J. M. 1986. Importance of diseases as constraints to pasture legumes in the tropical American lowlands. *Proc. Int. Grassl. Congr., 15th., Kyoto, Jpn., 1985*, pp. 810–12
97. Lenné, J. M., Stanton, J. M. 1990. Diseases of *Desmodium* species—a review. *Trop. Grassl*. 24:1–14
98. Lenné, J. M., Thomas, D., de Andrade, R. P., Vargas, A. 1984. Anthracnose of *Stylosanthes capitata:* implications for disease evaluations of indigenous tropical pasture legumes. *Phytopathology* 74:1070–73
99. Leppik, E. E. 1965. A pathologist's viewpoint on plant exploration and introduction. *Plant Introd. Genet. Res. Investig. Pap*. 1:1–5
100. Leppik, E. E. 1970. Gene centers of plants as sources of disease resistance. *Annu. Rev. Phytopathol*. 8:323–44

101. Manisterski, J., Anikster, Y., Brodny, U, Wahl, I. 1990. The fertile crescent as a source of genes for disease resistance. *Proc. Int. Wksh. Dynamic In Situ Conservation of Wild Relatives of Major Cultivated Plants, Israel, 1990.* (Abstr.)
102. Manisterski, J. Segal, A., Levy, A., Feldman, M. 1988. Evaluation of Israeli *Aegilops* and *Agropyron* species for resistance to wheat leaf rust. *Plant Dis.* 72:941–44
103. Manisterski, J., Segal, A., Wahl, I. 1978. Rust resistance in natural ecosystems of *Avena sterilis. Proc. Int. Congr. Plant Pathol., 3rd, 1978,* pp. 301. (Abstr.)
104. Marshall, D. R., Broúe, P. 1981. The wild relatives of crop plants indigenous to Australia and their uses in plant breeding. *J. Aust. Inst. Agric. Sci.* 47: 149–54
105. Marshall, D. R., Brown, A. H. D. 1983. Theory of forage plant collection. See Ref. 106, pp. 135–48
106. McIvor, J. G., Bray, R. A., eds. 1983. *Genetic Resources of Forage Plants.* Melbourne: CSIRO. 337 pp.
107. Miles, J. W., Lenné, J. M. 1984 Genetic variation within a natural *Stylosanthes guianensis, Colletotrichum gloeosporioides* host-pathogen population. *Aust. J. Agric. Res.* 35:211–18
108. Moseman, J. G., Nevo, E., El Morshidy, M. A., Zohary, D. 1984. Resistance of *Triticum dicoccoides* to infection with *Erysiphe graminis tritici. Euphytica* 33:41–47
109. Moseman, J. G., Nevo, E., Zohary, D. 1983. Resistance of *Hordeum spontaneum* collected in Israel to infection with *Erysiphe graminis hordei. Crop Sci.* 23:1115–19
110. Moss, J. P. 1980. Wild species in the improvement of peanuts. In *Advances in Legume Science,* ed. R. J. Summerfield, A. H. Bunting, pp. 525–35. Kew: Royal Bot. Gdn.
111. Mujeeb-Kazi, A., Jewell, D. D. 1985. CIMMYT's wide cross program for wheat and maize improvement. *Biotechnology in International Agricultural Research, Proc. Inter-Center Semin. IARCs and Biotech., Apr. 1984,* pp. 219–26. Philippines: IRRI
112. Mujeeb-Kazi, A., Kimber, G. 1985. The production, cytology and practicality of wide hybrids in the Triticeae. *Cereal Res. Commun.* 13:111–24
113. Natl. Acad. Sci. 1972. *Genetic Vulnerability of Major Crops.* Washington, DC: USA Natl. Acad. Sci.
114. Neergaard, P. 1984. Seed health in relation to the exchange of germplasm. In *Seed Management Techniques for Genebanks, Proc. Wksh., Royal Bot. Gdn Kew,* July 1982, pp. 1–21. Rome: IBPGR. 294 pp.
115. Neergaard, P. 1986. Screening for plant health. *Annu. Rev. Phytopathol.* 24:1–16
116. Nelson, R. R. 1978. Genetics of horizontal resistance to plant diseases. *Annu. Rev. Phytopathol.* 16:359–78
117. Nevo, E., Beiles, A., Gutterman, Y., Storch, N., Kaplan, D. 1984. Genetic resources of wild cereals in Israel and vicinity. I. Phenotypic variation within and between populations of wild wheat, *Triticum dicoccoides. Euphytica* 33: 717–35
118. Nevo, E., Beiles, A., Krogman, T. 1988. Natural selection of allozyme polymorphism: a microgeographic climatic differentiation in wild emmer wheat *(Triticum dicoccoides). Theor. Appl. Genet.* 75:529–38
119. Nevo, E., Moseman, J. G., Beiles, A., Zohary, D. 1985. Patterns of resistance of Israeli wild emmer wheat to pathogens. I. Predictive method by ecology and alloenzyme genotypes for powdery mildew and leaf rust. *Genetica* 67:209–22
120. Newell, C. A., Delannay, X., Edge, M. E. 1987. Interspecific hybrids between soybean and diploid perennials. *J. Hered.* 78:301–6
121. Niederhauser, J. S., Cobb, W. C. 1959. The late blight of potatoes. *Sci. Am.* 200:100–12
122. Noy-Meir, I., Anikster, Y., Waldman, M., Ashri, A. 1989. Population dynamics research for in situ conservation: wild wheat in Israel. *Plant Genet. Res. Newsl.* 75/76:9–11
123. Ochoa, C. 1975. Potato collecting expeditions in Chile, Bolivia, and Peru, and the genetic erosion of indigenous cultivars. See Ref. 50, pp. 167–73
124. Peeters, S. P., Galwey, N. W. 1988. Germplasm collections and breeding needs in Europe. *Econ. Bot.* 42:503–21
125. Plucknett, D. L., Smith, N. J. H., Williams, J. T., Anishetty, N. M. 1987. *Gene Banks and the World's Food.* Princeton: Princeton Univ. Press
126. Posnette, A. F. 1981. The role of wild hosts in cocoa swollen shoot virus. See Ref. 151, pp. 71–78
127. Prescott-Allen, R., Prescott-Allen, C. 1986. *The Vanishing Resource.* Newhaven: Yale Univ. Press
128. Prescott-Allen, R., Prescott-Allen, C. 1988. *Genes from the Wild.* London: Earthscan Publ. Ltd., 2nd ed. 111 pp.
129. Purseglove, J. W. 1981. *Tropical*

Crops—Dicotyledons. Harlow: Longman. 719 pp.
130. Qualset, C. O.1975. Sampling germplasm in a center of diversity: an example of disease resistance in Ethiopian barley. See Ref. 50, pp. 81–96
131. Richardson, M. J. 1979. An annotated list of seed borne diseases. *Phytopathol. Pap*. 23. 3rd ed. 320 pp.
132. Rick, C. M. 1973. Potential genetic resources in tomato species: clues from observations in native habitats. In *Genes, Enzymes and Populations*, ed. A. M. Srb, pp. 255–69. NY: Plenum
133. Rick, C. M. 1986. Germplasm resources in the wild tomato species. *Sci. Hortic*. 200:45–55
134. Ridley, H. N. 1930. *The Dispersal of Plants Throughout the World*. Ashford: Reeve. 744 pp.
135. Roelfs, A. P. 1988. Resistance to leaf and stem rusts in wheat. See Ref. 141, pp. 10–22
136. Segal, A., Dorr, K. H., Fischbeck, G., Zohary, D., Wahl, I. 1987. Genotypic composition and mildew resistance in a natural population of wild barley, *Hordeum spontaneum*. *Plant Breed*. 99: 118–27
137. Segal, A., Manisterski, J., Fischbeck, G., Wahl, I. 1980. How plant populations defend themselves in natural ecosystems. In *Plant Disease*, ed. J. G. Horsfall, E. B. Cowling, 5:75–102. NY: Academic. 534 pp.
138. Shiga, T., Takemata, T. 1981. Distribution of sweet potato clones with resistance to root-knot nematode in the Pacific Islands. *Proc. Res. Plann. Conf. Root-knot Nemat., Jakarta, Indonesia, 1981*, pp. 64–68
139. Simmonds, N. W. 1962. Variability in crop plants, its use and conservation. *Biol. Rev*. 37:442–65
140. Simmonds, N. W. 1988. Synthesis: the strategy of rust resistance breeding. See Ref. 141, pp. 119–36
141. Simmonds, N. W., Rajaram, S., eds. 1988. *Breeding Stategies for Resistance to the Rusts of Wheat*. Mexico, DF: CIMMYT. 151 pp.
142. Smith, D. H. 1979. Disease resistance in peanut. See Ref. 67, pp. 431–47
143. Stalker, H. T. 1980. Utilization of wild species for crop improvement. *Adv. Agron*. 33:111–47
144. Stalker, H. T. 1989. Utilizing wild species for crop improvement. See Ref. 145, pp. 139–54
145. Stalker, H. T., Chapman, C., eds. 1989. *Scientific Management of Germplasm: Characterization, Evaluation and Enhancement*. Rome: IBPGR. 194 pp.
146. Stalker, H. T., Moss, J. P. 1987. Speciation, cytogenetics, and utilization of *Arachis* species. *Adv. Agron*. 41:1–39
147. Subrahmanyam, P., Ghanekar, A. M., Nolt, B. L., Reddy, D. V. R., McDonald, D. 1985. Resistance to groundnut diseases in wild *Arachis* species. See Ref. 78, pp. 49–55
148. Subrahmanyam, P., Moss, J. P., McDonald, D., Subda Rao, P. V., Rao, V. R. 1985. Resistance to leaf spot caused by *Cercosporidium personatum* in wild *Arachis* species. *Plant Dis*. 69:951–54
149. Subrahmanyam, P., Moss, J. P., Rao, V. R. 1983. Resistance to peanut rust in wild *Arachis* species. *Plant Dis*. 67: 209–15
150. Subrahmanyam, P., Ramanantha Rao, V., McDonald, D., Moss, J. P., Gibbons, R. W. 1989. Origins of resistances to rust and late leaf spot in peanut (*Arachis hypogaea*, Fabaceae). *Econ. Bot*. 43:444–55
151. Thresh, J. M., ed. 1981. *Pests, Pathogens and Vegetation*. London: Pitman. 517 pp.
152. Thurston, H. D. 1984. *Tropical Plant Diseases*. St. Paul, Minn.: Am. Phytopathol. Soc. 208 pp.
153. Toledo, J. M., Lenné, J. M., Schultze-Kraft, R. 1989. Effective utilization of tropical pasture germplasm. In *FAO Plant Prod. Prot. Pap. 94, Utilization of Genetic Resources: Suitable Approaches, Agronomical Evaluation and Use*, pp. 27–57. Rome: FAO
154. Valls, J. F. M. 1985. Groundnut germplasm management in Brazil. See Ref. 78, pp. 43–45
155. Valls, J. F. M., Ramanatha Rao, V., Simpson, C. E., Krapovickas, A. 1985. Current status of collection and conservation of South American groundnut germplasm with emphasis on wild species of *Arachis*. See Ref. 78, pp. 15–35
156. van der Maesen, L. J. G. 1984. Taxonomy, distribution and evolution of the chickpea and its wild relatives. See Ref. 162, pp. 95–104
157. van der Maesen, L. J. G., Pundir, R. P. S. 1984. Availability and use of wild *Cicer* germplasm. *Plant Genet. Res. Newsl*. 57:19–24
158. Vanderplank, J. E. 1982. *Host-Pathogen Interactions in Plant Disease*. NY: Academic
159. Wahl, I. 1970. Prevalence and geographic distribution of resistance to crown rust in *Avena sterilis*. *Phytopathology* 60:746–49
160. Watson, I. A. 1970. The utilization of wild species in the breeding of cultivated

crops resistant to plant pathogens. In *Genetic Resources in Plants—Their Exploration and Conservation,* ed. O. H. Frankel, E. Bennett, pp. 441–60. Oxford: Blackwell

161. Wilkes, G. 1988. Teosinte and the other wild relatives of maize. In *Recent Advances in the Conservation and Utilization of Genetic Resources, Proc. Global Maize Germpl. Wksh.*, pp. 70–80. CIMMYT, Mexico DF
162. Witcombe, J. R., Erskine, W., eds. 1984. *Genetic Resources and Their Exploitation—Chickpeas, Faba Beans and Lentils*. The Hague: Martinus Nijhoff. 256 pp.
163. Wood, D. 1988. Introduced crops in developing countries: a sustainable agriculture? *Food Policy* 13:167–77
164. Wynne, J. C., Halward, T. M. 1989. Germplasm enhancement in peanut. See Ref. 145, pp. 155–74
165. Zeven, A. C. 1979. Collecting genetic resources in highly industrialized Europe, especially the Netherlands. *Broadening the Genetic Base of Crops. Proc. Conf., Wageningen, 1978,* pp. 49–58. Wageningen: Pudoc
166. Zeven, A. C., de Wet, J. M. J. 1982. *Dictionary of Cultivated Plants and their Centres of Diversity*. Wageningen: Pudoc
167. Zhukovsky, P. M. 1959. Interrelation between host and fungus parasite in their habitat and beyond it. *Vestn. Sel'skokhoz. Nauki Moscow* 4:25–34 (In Russian)
168. Zohary, D., Harlan, J. R., Vardo, A. 1969. The wild diploid progenitors of wheat and their breeding value. *Euphytica* 18:58–65

Annu. Rev. Phytopathol. 1991. 29:65–87

BIOLOGY AND EPIDEMIOLOGY OF BACTERIAL WILT CAUSED BY *PSEUDOMONAS SOLANACEARUM*

A. C. Hayward

Department of Microbiology, The University of Queensland, 4072, Australia

KEY WORDS: Phylogeny and classification, new hosts and geographical distribution, environmental interactions, latent infections, disease management strategies

---

INTRODUCTION

Bacterial wilt caused by *Pseudomonas solanacearum* is a disease widely distributed in tropical, subtropical, and some warm temperate regions of the world, and a major constraint on production of many crop plants. This wide distribution is reflected in a literature that is scattered and of variable status. Kelman (70) evaluated the early literature on bacterial wilt most comprehensively and his review is a first point of reference for the experienced worker as well as the uninitiated. There have been two later reviews of a general nature (12, 136) and reports from several planning conferences and workshops. The first of these held in North Carolina, USA (128) was followed by several others that were either regionally based (108) or devoted to particular crops such as potato (60, 61) or peanut (93). These meetings were held in response to the recognition by plant pathologists and plant breeders of the slow progress in reducing the serious depredations caused by bacterial wilt, and of the need to establish priorities for future research and a more coordinated effort among interested workers.

The purpose of this chapter is to review fundamental aspects of the pathogen and its taxonomic relationships as revealed by modern techniques of molecular biology as well as host range and geographical distribution, en-

0066-4286/91/0901-0065$02.00

vironmental interactions and epidemiology, and the strategies that have been developed to control the disease. Some aspects that will not be covered include the genetics of pathogenesis that has attracted increasing attention in North America, Europe, Australia and China (10, 10a, 21, 23, 57, 58a, 59, 81, 156), and the biochemical and physiological aspects of host-parasite relationships (12, 70).

PROPERTIES, RELATIONSHIPS AND GEOGRAPHICAL DISTRIBUTION

Phylogeny

Modern techniques of molecular biology enable the construction of trees or dendrograms depicting evolutionary relationships at different levels or depths. On the basis of sequence analysis of the 16S fragment of ribosomal RNA it has been determined that, at the highest level, *Pseudomonas solanacearum* is a member of the beta-subdivision of the class Proteobacteria (13, 135). At a lower level similarities between species and between genera have been determined using RNA/DNA hybridization. This showed that the genus *Pseudomonas* is phylogenetically heterogeneous consisting of five more or less distantly related homology groups. This separation of species of *Pseudomonas* into five groups has been confirmed in several complementary investigations (24, 25, 106, 155). *P. solanacearum* is contained within rRNA homology group II, together with other plant pathogens *P. andropogonis, P. caryophylli, P. cepacia, P. gladioli, P. glumae*, and *P. rubrisubalbicans,* the human and animal pathogens *P. mallei* and *P. pseudomallei*, and *P. pickettii*.

Within rRNA homology groups closer relationships between species can be determined by the technique of DNA/DNA hybridization. In group II few of the species show a significant degree of DNA homology in reciprocal tests (65, 105). The only species that show a relationship with *P. solanacearum* are *P. pickettii* (115) and *P. syzygii* (120). The former species has apparently never been recognised from plants but is an uncommon pathogen of relatively low virulence found in hospital-acquired infections originating in contaminated water supplies. Although differing in pathogenesis and ecology, there are close similarities in phenotypic properties, subdivisions of the two species have been based on similar criteria (48, 50, 110), and they are similar in being able to multiply in distilled water solutions or solutions of low salt concentration (152, 154). *P. syzygii* is a plant pathogen causing Sumatra disease of clove (*Syzygium aromaticum*), which is related to *P. solanacearum* on serological grounds as well as DNA/DNA homology (120), although clearly distinct in cultural and physiological properties. *P. syzygii* does not grow, or grows only very poorly, on many of the media on which *P. solanacearum* is routinely grown. The pseudomonad that causes blood disease, a bacterial disease of banana and other Musaceae in Indonesia (29), reacts with DNA probes prepared for *P. solanacearum* (20; L. Sequeira, unpublished data) but the full significance of this finding awaits further study.

Secondary Metabolites

Membership of a particular rRNA homology group may be predictive of shared phenotypic properties. For example, rRNA group II pseudomonads have proved to be a prolific source of biologically active secondary metabolites (94, 153), most commonly substances with antifungal activity. Twelve isolates of *P. solanacearum* representing three biovars, and the wild type as well as avirulent mutants, were shown to produce substances with antifungal activity in cell-free culture fluid, apparently due to inhibition of fungal cell wall synthesis (146). In other studies several isolates were shown to produce the cytokinin trans-zeatin at concentrations of up to one mg per liter (6). Examination of the secondary metabolites of *P. solanacearum* has been a relatively neglected topic that is worthy of further study, particularly in view of the unexpected findings in other species representing rRNA homology group II (94).

Subspecific Classification

Strains of *P. solanacearum* differ in host range, geographical distribution, pathogenicity, epidemiological relationships, and physiological properties (12, 70, 105). It is therefore important to have a classification of strains that is sufficient in information content to have predictive value in the context of epidemiology and control of bacterial wilt. For almost the past three decades a binary system has been in use reflecting two different approaches to differentiation, one placing emphasis on host affinity and establishment of races, the other making use of selected biochemical properties as the basis for separation into biovars (12, 48, 54). Races and biovars are informal groupings at the infrasubspecific level that are not governed by the Code of Nomenclature of Bacteria (77). Five races have been described, designated according to the host or hosts primarily affected (8, 11, 12, 54), and five biovars (48, 54) differentiated according to ability to utilize and/or oxidize several hexose alcohols and disaccharides.

There may be sufficient evidence to justify progression to a formal system in which *P. solanacearum* is divided up into two or more subspecies governed by the Code. Present evidence may be summarized as follows: Biovars 1 and 2 are less nutritionally versatile than biovars 3 and 4 (49, 105) and this difference is reflected in separation into distinct phenons by numerical taxonomy (30, 133); the electrophoretic pattern of the membrane proteins differs somewhat between biovars (27); and biovars 1 and 2 are distinct from biovars 3, 4, and 5 on the basis of DNA probes and RFLP analysis (20). There are also marked differences in the geographical distribution of biovars suggestive of separate evolutionary origin. In general, biovar 1 is predominant in the Americas and biovar 3 in Asia. In the USA only biovar 1 has been reported from Florida to North Carolina (45, 83, 90, 151). By contrast, biovar 1 is absent from most parts of Asia. Until its introduction, and presumed eradication, on *Heliconia* (5) biovar 1 had never been found in Australia. Biovars 2, 3, and 4 occur in Australia (111), China (together with biovar 5) (52, 54), India (132), Indonesia (82), Papua New Guinea (142), and Sri Lanka (125). Only in the Philippines (149) have all of biovars 1–4 been found and here as elsewhere in Asia biovar 3 is the predominant biovar in lowland regions. The wide distribution of biovar 2, from its presumed origin in South America as a pathogen of potato, probably reflects the ease with which it is carried as latent infections in seed tubers (11, 20). Very probably, the nutritionally most versatile biovar 3 of the lowland tropics of Asia is the same in character as the tobacco isolate from Sumatra that Erwin F. Smith recognized as distinct from isolates from the USA and that he termed "var. *asiaticum*" (70). If future work supports these generalizations it may be anticipated that *P. solanacearum* will be divided into two or more subspecies.

The relationship between host specialization and phenotype is most clearly evident between race 3, the potato race, and biovar 2. In general race 3 and biovar 2 are equivalent. Important generalizations concerning the epidemiology and control of biovar 2 (race 3) of *P. solanacearum* set it apart from other races and biovars. In many countries of southern Europe and the Mediterranean area (48), and in Argentina, Chile, and Uruguay (83), biovar 2 is the sole biovar. At the altitudinal as well as the latitudinal limit of distribution of *P. solanacearum*, biovar 2, in association with its primary host potato, is the only biovar found. Biovar 2 has a limited host range and is found mainly on potato and sometimes on tomato and a few weed hosts; many strains are adapted to growth and pathogenesis at low temperatures (16, 32, 140) and several studies suggest a more limited capacity for survival in fallow soil than for biovar 3 (3, 95, 116). Strains of biovar 2 from potato in Australia formed a uniform phenotype whereas those from South America showed some heterogeneity and were divisible into several similar but distinct phenotypes (50). This is consistent with the concept that pests and pathogens are most variable near the center of origin of their primary hosts, and also with the proposal of Martin et al (83) that the genetic diversity of *P. solanacearum* is less at greater distances from the equator.

HOST RANGE

The very extensive host range of *P. solanacearum* includes several hundred species representing 44 families of plants, and many newly recognized hosts

(54, 93, 108). *P. solanacearum* biovar 3 has been recently described on several woody perennial hosts, including cashew (*Anacardium occidentale*) in Indonesia (131) and custard apple (*Annona* spp.) (85) and the Alexandra palm (*Archontophoenix alexandrae*) (4) in Queensland, Australia.

There are several apparent anomalies in the distribution of bacterial wilt on certain hosts. For example, the disease on sweet potato (*Ipomoea batatas*) has been reported only in some regions of China, and only since the 1950s (52, 54); bacterial wilt of cassava is confined to Indonesia (100), although the host is widely cultivated in countries where bacterial wilt is endemic; and *Eucalyptus* is affected only in Brazil (26) and China but not in Australia. Strawberry is a host in Japan (36) and Taiwan but not in the southeastern USA, where this host is widely grown. Bacterial wilt is a severe disease on peanut only in China and Indonesia whereas it is a relatively unimportant or nonexistent problem in many other countries, including India (93). The reason for these differences is not obvious. Specific strains pathogenic for certain hosts may have evolved only in certain parts of the world and are not found elsewhere or, these hosts may only be susceptible where a number of environmental factors conducive to disease expression coincide, such as temperature regime, rainfall, soil type, inoculum potential, and other soil biological factors such as nematode populations. Until isolates from different hosts can be directly compared under standard conditions that use infectivity titration as a measure of resistance of the host and virulence of the pathogen (31), this discrepancy will not be understood.

MODES OF DISPERSAL AND SOURCES OF INOCULUM

Some aspects of the epidemiology of bacterial wilt such as root-to-root transmission (71), movement of soil and dissemination by farm implements, as in the clipping of tomato plants (87, 88), and insect transmission of Moko disease of banana, are well understood (12, 70); others require reinvestigation using the more sensitive methods of pathogen detection that are now available.

Movement of Planting Material

Bacterial wilt can be carried over long range on vegetative propagating material. Although this fact has been known in some aspects for many years (12), the importance to plant quarantine of traffic in plant material potentially capable of carrying *P. solanacearum* has probably been greatly underestimated. Fortunately there are now a few well-documented cases (Table 1) of long-range dispersal that should serve to increase awareness of the potential for spread in this manner. The unregulated use of infected ginger rhizomes as planting material within China, Indonesia, and Malaysia (79) has

Table 1 Some examples of long-range movement of *Pseudomonas solanacearum* on plant material

Plant material	Source of material	Recipient	References
Tomato transplants	Georgia, USA	Kent County, Southwestern Ontario, Canada	78
Heliconia rhizomes[a]	Hawaii, USA	Queensland, Australia	5
Banana Corms	Central America	Philippines	11, 119
Potato tubers for processing	Mediterranean region	Sweden	102, 103

[a]Confirmed as *P. solanacearum* race 2 (L. Sequeira, M. Gillings, unpublished data) and as biovar 1 (5).

amplified and extended a severe disease problem, and international spread on planting material has probably also occurred in the tropics and subtropics.

Without doubt, bacterial wilt of potato has been spread locally and internationally on latently infected potato tubers (11, 17, 18, 101, 139). Latent infection has emerged in the past two decades as a most insidious aspect of disease biology of great concern to plant breeders. Cruza 148 (CIP 720118) is an example of a potato selection that is particularly prone to act as a vehicle (33). Obviously, this type of tolerance is unsafe as it helps to spread the disease with apparently healthy material. In other circumstances the cultivar may be susceptible but the disease remain unexpressed because the temperatures are too low, as at high elevation in the tropics, and symptoms are only shown when harvested tubers are incubated or planted at temperatures favorable for disease expression (37, 101). Work in India has shown that *P. solanacearum* is present on the surface, in lenticels and within the vascular tissue of latently infected potato tubers, and that there is differential survival at these sites with storage temperature (139, see also 18). At all storage temperatures surface populations declined most rapidly, and decrease in pathogen population both in lenticels and vascular tissues was less, at 10–15°C, than at either higher or lower temperatures of storage.

The possibility of spread on infested or infected true seed needs more study with the use of sensitive techniques such as DNA probes or monoclonal antibodies. Some indirect evidence suggests that such long-range dispersal may occur, although most opinion has been to the contrary (70). An isolate of *P. solanacearum* biovar 1 was obtained from wilted tomato plants grown from imported seed in a glasshouse in Finland (E. R. French, unpublished data). Aldrick (7) considered that bacterial wilt of Townsville Stylo (*Stylosanthes humilis*) in the Northern Territory of Australia may have been transmitted in seed, particularly in uncleaned samples containing plant trash. The expansion of the distribution of bacterial wilt of peanut from a few provinces in southern China in the middle 1960s to 17 provinces or regions between latitudes 19° N to 39° N suggests seed transmission for which, however, there is no experimental evidence from work conducted in China (53).

Epiphytic Survival

Some evidence suggests that an epiphytic phase in the life cycle may contribute to survival and provide another source of inoculum for the renewal of soil populations. Although *P. solanacearum* has rarely been reported as the cause of a leaf spot infection (51, 70), transmission of antibiotic-resistant mutants on contaminated seed of capsicum and tomato to the emerging cotyledons has been shown (97) under conditions of high relative humidity. Epiphytic colonization of the true leaves of capsicum occurred at 92 and 73% relative humidity (r.h.) and lesion development was observed on the leaves at the highest r.h. More direct evidence of an epiphytic phase and of aerial transmission through rain splash dispersal has come from work on tobacco in Japan. The occurrence of high populations of the pathogen in surface soil near infected plants after rainfall, and during rainfall of large numbers in surface furrow water near slightly or severely infected plants suggested an aboveground source of inoculum (104). *P. solanacearum* was shown to exude from diseased tobacco plants and to be distributed widely to adjacent healthy plants under conditions of high wind and rain. Secondary dispersal from the aboveground parts of plants, particularly at a late stage of cultivation, was more important than other modes of transmission to underground parts. When the pathogen was spray-inoculated to the upper surface of tobacco leaves, survival was for more than 15 days at 95% r.h., whereas the pathogen died within a short time at 50% r.h., or under sunlight (46, 47).

Weed Hosts and Sheltered Survival Sites

The factors determining the persistence of *P. solanacearum* in infested soil require more attention. Sheltered sites occur where the bacterium may survive between successive plantings of a susceptible crop. These include plant debris and latently infected potato tubers (37), the deeper soil layers (40, 73), and in the rhizosphere of roots of weed hosts (42, 63, 96, 113). The range and variety of weed hosts is very extensive (70) but their significance also varies greatly in different environments and cropping systems. Some are symptomless carriers (41, 157). In India, *Ageratum conyzoides* and *Ranunculus sceleratus* showed no symptoms but the pathogen was readily isolated from surface sterilized roots (139). In a cool temperate region of Australia, the perennial weed *Solanum cinereum* was identified as a symptomless carrier of *P. solanacearum* biovar 2 in potato cropping land, which could have significance in long-term survival of the pathogen (39). In Sweden, another perennial weed *Solanum dulcamara* serves as a symptomless carrier of *P. solanacearum* biovar 2 (102, 103).

Other weeds reported to act as symptomless hosts include *Solanum carolinense* in the USA (28), *Phyllanthus niruri* (Euphorbiaceae) (117), and *Lagasca mollis* (Asteraceae) (74) in India. In the Philippines, *Portulaca oleracea* (Portulacaceae) was reported to be a symptomless host and in Queensland (Australia) a susceptible host to *P. solanacearum* biovar 3 (96, 113).

In the lowland tropics, annual weed hosts are likely to be available year round and thereby increase the population of bacteria in the soil.These weeds may also serve as alternate hosts whenever nonhost plants are cultivated, thus making crop rotation ineffective in reducing the bacterial population in the soil (1). In this context, unrecognized symptomless hosts may also be important. Some weed hosts serve as sources of infection as virgin land in the lowland tropics is cleared for agricultural use (12). For example, the weed host *Croton hirtus* is thought to have particular significance in relation to bacterial wilt on the susceptible cassava cultivar Kuning in Lampung Province, South Sumatra, Indonesia (82). Both the weed host and cassava (*Manihot esculenta*) are members of the Euphorbiaceae, and this weed is thought to have been the primary source of infection for cassava bacterial wilt (K. Nakagawa, unpublished data). *Croton hirtus* has been reported to be a significant weed host in other parts of the lowland tropics including Malaysia (1) and Sri Lanka (126).

Few studies have looked critically at the rhizosphere of weeds as a protected survival site (42, 113). In the Philippines, the population of *P. solanacearum* gradually declined in both rhizosphere and non-rhizosphere soils of various crop and weed hosts, except in the rhizosphere soil of *Portulaca oleracea*, where it gradually but slowly increased. Suppression of the population was greater with rice and corn. In bioassays with a wilt-susceptible tomato cultivar, the highest level of wilting occurred in soils previously planted to *P. oleracea* and the lowest in soils previously planted to corn and rice, thus confirming the field observation that rice and corn are reliable rotation crops with susceptible species (113).

SOME ENVIRONMENTAL INTERACTIONS

Temperature and Light Intensity

Temperature is the most important factor affecting the host-pathogen interaction as well as survival in soils. In general, increase in ambient temperature to between 30–35° C increases the incidence and rate of onset of bacterial wilt on hosts such as tomato, for many but not all strains of the pathogen. Plants that are resistant at moderate temperature may become susceptible at a higher temperature. Resistance is temperature-sensitive and strain-specific (76, 84, 92, 114, 148). Whether this is primarily a function of virulence factors only fully expressed at high temperatures (148), or primarily an effect of lack of expression of resistance genes in the host at high temperature, is not clear.

Krausz & Thurston (76) showed that temperatures of 32° C in controlled environment chambers significantly increased severity of bacterial wilt in two tomato lines resistant to bacterial wilt. The degree of resistance of cultivar Venus to isolate K-60 from the United States was not significantly affected by temperature, but this line expressed no resistance to isolate LB-6 from the Philippines at either 26 or 32° C. Similar effects have been observed in tomato cultivars grown at different soil temperatures (92, 114).

Similar temperature sensitivity has been shown in resistant potato cultivars and there is a strong interaction with strain of the pathogen (17, 35, 129, 140). Resistant and susceptible potato cultivars were inoculated by stem puncture and by soil infestation, the latter without wounding, and maintained in warm and cool regimes (28/16 and 20/8° C day/night, respectively). Stem puncture and warm treatments each increased susceptibility. Latent infection developed at cool temperatures in cultivars inoculated by stem puncture, but not soil infestation, and at warm temperatures with soil infestation. Those plants inoculated by stem puncture were all rated susceptible. Resistance expressed following infestation of the soil without wounding appeared to be more representative of field resistance (35). Tung et al (148) tested eight true potato seed progenies and four clones for resistance to three potato isolates of *P. solanacearum* under two temperature regimes. Highly significant interactions of all types: progeny x isolate, isolate x temperature, progeny x temperature, and progeny x isolate x temperature, were observed. Changes of pathogenicity of isolates with temperature were the major source of the total variation. The host-pathogen interaction effect was small as compared to the main (progeny, isolate) effects. An important finding was that the two strains classified as biovar 3 and biovar 4 increased in virulence with ambient temperature increase whereas a biovar 2 (race 3) isolate gave significantly more wilt at lower temperatures. Similarly, when Russet Burbank potato plants were stem-inoculated with several strains of *P. solanacearum* and then held at 16, 20, 24, and 28° C for 15 days only the biovar 2 (race 3) isolates killed all of the inoculated plants (16) at the three lower temperature regimes. These results support the concept that specific strains of *P. solanacearum* are adapted to pathogenesis on potato at lower ambient temperatures such as obtained in the Andes at high elevation.

Soil temperatures at different locations are also a factor in survival. In the cool hill country of Sri Lanka, for example, soil temperatures are such that no great stress is imposed and there is prolonged survival of *P. solanacearum*, whereas in the lowland plains near to sea level soil temperatures rise above

43° C and remain above that temperature for several hours daily. It has been shown experimentally that soils exposed to 43° C continuously for periods of four days and over were free of the pathogen (126). High soil temperatures probably account for the relative unimportance of bacterial wilt in the Jaffna Peninsula of Sri Lanka. In tobacco fields of Japan, the relatively short-term survival of *P. solanacearum* in upper soil layers contrasted with relatively long-term survival at depth (46). Similarly, in two naturally infested potato fields *P. solanacearum* was detected at all soil layers between 15 and 75 cm with the use of indicator plants (38, 40). Long-term survival in deeper soil layers is likely to be a function of lower soil temperatures and decreased microbial activity owing to a paucity of indigenous soil microorganisms (40). Survival of up to 673 days in naturally infested soils stored in plastic bags at 4° C has been reported (43).

Few studies have investigated the effect of light intensity and photoperiod on host resistance to bacterial wilt (70). Reduced light intensity did not reduce resistance to isolate LB-6 in tomato line 1169 at 26.6° C but significantly decreased resistance at 29.4° C, and reduced photoperiod significantly decreased resistance of line 1169 to isolate LB-6, independent of temperature (76). In potato resistance was expressed more frequently at high light intensity (2000 ft-c.) than at low intensities (1300 ft-c.) at 24 and 28° C (129).

Soil Moisture

High soil moisture accumulations resulting from either a high water table or heavy rainfall usually favor development of bacterial wilt (2, 70). Survival of the pathogen is greatest in wet but well-drained soils, whereas survival is affected adversely by soil desiccation and by flooding (12, 70). This conclusion has been supported by survival studies under conditions of controlled soil moisture (99). Soil moisture significantly affected reproduction and survival of *P. solanacearum* in unsterile soils, regardless of soil type; colony counts increased within 7–10 days of introduction into the soil at the highest soil moistures (from flooded to -1 bar), but did not increase in drier soils (from -5 to -15 bars). The most favorable soil moisture for survival in two suppressive and two conducive North Carolina soils was -0.5 to -1 bars. Moffett et al (95) measured survival of *P. solanacearum* biovars 2 and 3 in clay, clay loam, and sandy loam soil at pressure potentials of -0.003, -0.05 and -0.15 kPa, and biovar 2 was shown to decline more rapidly than biovar 3. The rate of population decline of each biovar at -0.003 and -0.05 kPa was greater in clay loam than in sandy loam and at all pressure potentials it was greater in clay loam and sandy loam than in clay. Furthermore, the rate of population decline of both biovars tended to decrease in the drier soil treatments. Although numbers of *P. solanacearum* declined most in the driest soil treatments, as the soil adjusted to the required pressure potential at the beginning of the experiment, at the end, the driest soil contained generally higher number of viable pathogens than did the wetter treatments. Thus, provided the pressure potential remains constant, some strains of *P. solanacearum* in certain soils are not sensitive to dry soil conditions. Under natural conditions soil microorganisms are subjected to the stress of cyclical wetting and drying. Dry conditions are a factor contributing to reduction in numbers of *P. solanacearum* in fallow land (127).

Soil Type

Some soils are conducive to bacterial wilt and others suppressive. Some soils in otherwise wilt-endemic areas have never supported the disease; and in some areas where the disease has been introduced there are contrasting reports on persistence in the soil (70, 86, 98, 99).

More systematic investigation is required of the soils in which bacterial wilt does not occur, or in which the disease does not persist once introduced. A detailed description of these soils would assist in the analysis of the complex of physical, chemical, and biological factors involved in disease suppression. In general, the information that is available is scattered, fragmentary, and sometimes contradictory.

Observations made by Vander Zaag in Southeast Asia (150) suggest that flooded soils, sugarcane lands with a pronounced dry season (ustic moisture regime), and river flood plains, which are flooded annually and experience extreme heat when dry, are free of bacterial wilt. Bacterial wilt introduced into a sugarcane field consisting of a clay loam having a ustic moisture regime, i.e. a mean annual soil temperature of 22° C or higher, and isohyperthermic temperatures with soil temperatures approaching 40° C when dry and exposed, had disappeared after one year. In Surinam bacterial wilt of tomato and related hosts was absent on the sea shell ridges of the coastal plain in contrast to sandy ridge and clay soils where disease outbreaks occurred frequently. The resistance to the disease on sea shell-containing soil was attributed to a complex of processes in the host, of which calcium nutrition was supposed to play a key role. Bacterial growth, maceration of tissues, and symptom development in the early stages of bacterial wilt of tomato was primarily in plant tops of low calcium content, rather than in vessels of older tissue high in calcium content (112). Persistence of *P. solanacearum* in four North Carolina soils differed and it was suggested that soil factors (possibly of biological origin) prevented persistence of the pathogen from season to season (98). Soil moisture and antagonistic microorganisms were the most important factors. Soil type, which influenced soil moisture, determined the size of antagonistic populations, which in turn affected the survival of *P. solanacearum* in soil (99).

Reports on the relation between soil type and incidence of bacterial wilt of peanut have contrasted strongly. Abdullah et al (2) found that although infection does occur in sandy soils, it is highest in heavy clay soil. Similarly, in Indonesia the disease has been most severe on peanuts growing in heavy clay soils (70). By contrast, in China bacterial wilt of peanut is prevalent in sandy soil, particularly in gritty soil, but not in heavy clay or loam (53).

In Taiwan, suppressiveness to bacterial wilt was measured by inoculating soils with a known quantity of a streptomycin-resistant strain of *P. solanacearum* and measuring recovery rate by soil dilution plating after three days exposure in soil (56). Among the 10 soil samples that tested suppressive to *P. solanacearum*, four were strongly suppressive and one moderately suppressive to tomato wilt. The remaining five soils were conducive to the disease. Among the 10 pathogen-conducive soils tested, one was moderately suppressive and the rest were conducive to tomato wilt caused by the pathogen. Unknown factors in the five pathogen-suppressive soils, which were conducive to tomato wilt, may have rendered the host susceptible to the pathogen and nullified the gain through pathogen suppression. The single soil that was conducive to the pathogen but suppressive to the disease might reflect an increase in host resistance brought about by a certain component of soil such as calcium (56).

Nematode Populations

The synergistic interaction between root-knot nematode (*Meloidogyne* spp.) and *P. solanacearum* on a variety of hosts is widely recognized (64, 70, 84, 118, 137). Root infection by nematodes as expressed by root galling index generally correlates with bacterial wilt symptoms as expressed by percent plants wilted. This is interpreted as primarily an effect of increased wounding of the root system providing points of ingress of the pathogen. However, the nematode may also act as a modifier of plant tissue in such a way that the tissue becomes more suitable for bacterial colonization. Breakdown of the resistant eggplant cultivar Pusa Purple Cluster in India was attributed, at least in part, to such an effect (118). Considering the global importance of bacterial wilt of potato and its codistribution with root-knot nematodes, breeding programs should include the incorporation of resistance to both organisms (64).

CONTROL STRATEGIES

Various strategies have been developed for the control of bacterial wilt, but many are limited in general application, being often crop- or site-specific, or severely limited in application by socioeconomic conditions. Since potato farmers in the developing world are often among the poorest in resources, the disease may have to be contained instead of, ideally, using tolerant cultivars, clean seed, and adequate crop rotations. Similarly, the strategy for control of bacterial wilt in cool temperate regions where *P. solanacearum* biovar 2 alone occurs, will be very different from control of bacterial wilt in the lowland tropics where strains of wide host range are endemic. No universal solutions exist, but only principles that can be applied and adapted in particular situations.

Host-Plant Resistance

The important strategy of selection or development of resistant cultivars for the control of bacterial wilt has had some success in the case of tobacco and peanut (1, 70, 122, 123, 141). In potato no immunity has been identified, and tolerance has been shown to be unsafe in cultivars that harbor latent infections (33, 123). *Solanum phureja* (121) has been used in potato breeding as the major source of resistance to bacterial wilt but this tactic has not been sufficient in all environments. Resistance genes from other diploid potato species have been identified and used in breeding (122, 123) and even in *S. tuberosum* where substantial resistance had not previously been found (147). Tung et al (147) have made the important finding that genes for adaptation are involved in conferring resistance to bacterial wilt. The effect of a particularly resistant parent clone on the resistance capability of its progeny depended on the resistance, adaptation, or the combination of both characteristics of its mate. A heat-tolerant parent gave a higher frequency of resistant offspring in combination with an ascertained source of resistance. Combining ability of the parents was an apparent feature of resistance to bacterial wilt. Widening of the genetic base for both resistance and adaptation to the environment, particularly heat tolerance, is very important in breeding for resistance to bacterial wilt in potato.

Genetic engineering techniques are being used for the introduction of lysozyme, cecropins, and other potent antibacterial proteins derived from insects into potato, as a way of augmenting resistance to bacterial wilt and other bacterial diseases. Although the short-term probability of success may be low, the limitations to conventional breeding approaches and the magnitude of the disease problem (124) justify the attempt.

Although many cultivars in tomato have been developed with useful levels of resistance for certain environments (55, 70, 91, 109, 140) it has proved difficult to obtain cultivars with stable resistance under conditions of high temperature and humidity in the lowland tropics. In tobacco, somaclonal variation has been used to develop somaclones, of which some show higher resistance to bacterial wilt than their parents (22).

Recent workshops have emphasized the importance of improved screening methods to evaluate resistant germplasm. In tomato and many other hosts, young plants are more susceptible than older plants, but natural infection in the field is usually observed at the flowering stage and onward. Resistance is therefore not necessarily comparable between seedlings and adult plants. For this reason Mew & Ho (91) consider that segregating populations are best screened under field conditions at the flowering stage rather than by artificial infection. By contrast, in Japan bacterial wilt of strawberries is found on younger seedlings in nurseries, and rarely on mature plants in the field. The resistance of mature plants has been attributed to the synthesis in leaves and translocation of a strongly antibacterial substance β-D-glucogallin that is a normal component of healthy tissues (69).

Although limited in application, losses from bacterial wilt on tomato in the lowland tropics can be greatly reduced by grafting on to a resistant eggplant rootstock (80) or *Solanum torvum* (107, 141).

Cropping Systems

In some developing countries where the use of clean seed and long crop rotations are not practical solutions to the problem of bacterial wilt of potato caused by *P. solanacearum* biovar 2 (race 3), intercropping has been used as a means of reducing soil populations of the pathogen and root-to-root transmission (19, 62, 75).

In Burundi, the lower incidence of disease in potato grown with bean indicated that a bean intercrop, which develops and matures quickly and has a dense root system, was preferable to a slower-maturing crop with a more diffuse root system, such as corn (9). Similar experiments with intercropping of potatoes in the Philippines showed that corn and cowpea were the most effective of several intercrops in reducing disease incidence in potato (75). In Fiji, growing of tomatoes as an intercrop with sugar cane resulted in a lower incidence of bacterial wilt and this method may have potential for potato under a similar cropping system (62). The role of crop rotations in the reduction of losses due to bacterial wilt has been well covered in recent reviews (32, 34, 53, 70, 130).

Soil Amendment

In Taiwan, a soil amendment called S-H mixture has been developed that has a broad spectrum of activity in suppression of a range of soil-borne diseases, including bacterial wilt (138). Under greenhouse conditions, S-H mixture suppressed bacterial wilt of tomato when added to soil at 0.5–1.0% (w/w) 7 days prior to transplanting (14). S-H mixture became ineffective when urea or mineral ash high in silicon dioxide and calcium oxide was taken out. Amendment of soil with urea in combination with mineral ash, but not with each of the other organic or inorganic components of S-H mixture, resulted in disease control efficacy equivalent to that of S-H mixture. The reason why the effect of urea is greatly enhanced by mineral ash is not known. In Surinam, following the observation that bacterial wilt never occurs on the seashell ridges of the coastal plain, sea-shell grit (42% CaO) was mixed into the tilled layer at a rate of approximately 1 m^3 per 15 m^3 of sandy or clay bacterial wilt-infested soil with favorable results (112).

Further study is needed of the mechanism underlying these promising reports of disease control following soil amendment.

Disease Avoidance

Since high soil temperatures and soil moisture are conducive to bacterial wilt development (70), losses can be minimized by manipulating the date of planting to accommodate seasons that are less favorable for disease development. For example, in Florida none of five tomato accessions planted after November 30 was killed by bacterial wilt, and lines with intermediate tolerance to the disease were transplanted safely into infested soil earlier in the fall in south Florida than lines with little or no tolerance (134). In Nagasaki, Japan, a highly significant correlation was found between losses from bacterial wilt on potato and the temperature in the last ten days of October (67). A three-week delay in the planting time resulted in reduction of disease primarily because the mean temperature was lower. Similarly, in parts of the north-western and north-eastern hills of India bacterial wilt does not appear in the autumn and losses on potato can be avoided by appropriately adjusting planting and harvesting dates. However, this approach is desirable only for the ware crop, since although the disease is not expressed in the autumn, the produce cannot be used as seed because it carries latent infection of the bacterium and gives rise to disease epidemics when planted under conditions favorable for disease development (130). In the lowland tropics potato production has been most successful where soils have been selected in which bacterial wilt either does not occur or is not maintained if introduced because of a pronounced dry season (150).

Integrated Control

An integrated approach to bacterial wilt control has been advocated, particularly for potato, by combining resistance with proper agronomy. French (34) has stressed the different weighting to be given to control measures under upland conditions where the pathogen is *P. solanacearum* biovar 2 (race 3) and under lowland conditions where other strains of wider host range and greater longevity in soil may be involved. In Nagasaki, Japan, integrated control of bacterial wilt on potatoes in severely infested fields was achieved in the fall crop by combining soil fumigation with chloropicrin, use of a relatively resistant cultivar, and late planting to avoid higher temperatures in the planting season (66, 68). Replacement of the susceptible cv. Dejima by the relatively resistant cv. Meihou enabled a reduction in the rate of application of chloropicrin.

Biological Control

The agents used to control bacterial wilt biologically include antagonistic rhizobacteria (19) and avirulent mutants of *P. solanacearum* (72, 89, 145). The mechanisms involved may include induced resistance or depend upon active colonization of the rhizosphere with antagonistic soil bacteria or bacteriocin- and bacteriophage-producing strains of *P. solanacearum* (15), or protection by competitive exclusion (89). Present indications are that protection dependent on root surface colonization holds less promise than that based on the use of avirulent mutants of *P. solanacearum*. Some of the latter have been generated by Tn5 mutagenesis, followed by selection of those that were still able to invade unwounded root systems and penetrate to the leaf three level without producing symptoms of wilt (144). Some of these mutants were effective in preventing subsequent colonization by wild-type strains, but the degree of protection depended on maintaining the ratio of virulent to avirulent cells in the inoculum at equal to or lower than 0.1 and the timing of inoculation (145). It should be stressed that none of these approaches to biological control of bacterial wilt has reached a point of commercial application and much more work is needed.

Avirulent strains of *P. solanacearum* have also been used in the biological control of *Fusarium* wilt of tomato. Mutants were selected that were resistant to and capable of detoxifying fusaric acid, a nonspecific wilt-inducing toxin produced by *Fusarium* species. Intact tomato plants were protected from wilt when they were pretreated with the mutant before inoculation with the pathogenic fungus (143).

There has been little study of the potential of vesicular-arbuscular mycorrhizae (VAM) for protection of plants from bacterial wilt. In the Philippines, VAM increased growth and yield of tomatoes and reduced infection by *P. solanacearum*. This may be due to competition or the mechanical barrier in the form of VAM vesicles and hyphae that inhibit the bacterial pathogen from deeper penetration into host tissues (44). In view of the potentially broad application of this approach, further investigation is warranted.

CONCLUSION AND PROSPECT

The relatively slow progress in understanding and controlling bacterial wilt is partly a reflection of the geographical distribution of the disease, which has important consequences in terms of the amount of research effort directed at the disease in research centers of Europe and North America. With regard to diseases of potato, for example, the research on late blight and potato viruses in these centers is considerably greater in volume and is assured of greater continuity, than is the case with bacterial wilt. This problem is recognized and recent workshops have laid the groundwork for future international collaboration in which the International Centers would play a leading part (60, 61, 93, 108, 128). There is a need to broaden the genetic base of potentially wilt-resistant germplasm and to improve the mechanisms for its exchange; for improved standardized screening techniques; for improved methods of pathogen detection and differentiation; and for greater understanding of ecological relationships, soil survival, weed hosts, and cropping systems.

If predictions on global warming due to the greenhouse effect are fulfilled, the distribution of this disease may be expected to expand to higher latitudes. The rate of this expansion will depend on the magnitude and rate of temperature increase, which for North America has been estimated at 150–200 km/1° C rise in temperature. A consequence of global warming is that significantly different climatic regions will be created from those to which major crop and pasture plants are currently adapted. The expected impact of these changes gives urgency to greater research effort on bacterial wilt.

ACKNOWLEDGMENTS

I appreciate the help of workers in many countries who have provided their published and unpublished data. In particular, I wish to thank A. Kelman and D. S. Teakle for their helpful comments and review of the manuscript.

Literature Cited

1. Abdullah, H. 1982. Resistance of winged bean (*Psophocarpus tetragonolobus*) to *Pseudomonas solanacearum*. *Malays. Appl. Biol.* 11:35–39
2. Abdullah, H., Maene, L. M. J., Naib, H. 1983. The effects of soil types and moisture levels on bacterial wilt disease of groundnut (*Arachis hypogaea*). *Pertanika* 6:26–31
3. Akiew, E. B. 1986. Influence of soil moisture and temperature on the persistence of *Pseudomonas solanacearum*. See Ref. 108, pp. 77–79
4. Akiew, E. B., Hams, F. 1990. *Archontophoenix alexandrae,* a new host of *Pseudomonas solanacearum* in Australia. *Plant Dis.* 74:615
5. Akiew, E. B., Hyde, K., Diatloff, L., Peterson, R. 1990. Bacterial wilt of *Heliconia* plants from Oahu, Hawaii. *ACIAR Bacterial Wilt Newsl.* 6:5
6. Akiyoshi, D. E., Regier, D. A., Gordon, M. P. 1987. Cytokinin production by *Agrobacterium* and *Pseudomonas* spp. *J. Bacteriol.* 169:4242–48
7. Aldrick, S. J. 1971. Bacterial wilt (*Pseudomonas solanacearum*) of *Stylosanthes humilis* in the Northern Territory.*Trop. Grassl.* 5:23–26
8. Aragaki, M., Quinon, V. L. 1965. Bacterial wilt of ornamental gingers (*Hedychium* spp.) caused by *Pseudomonas solanacearum*. *Plant Dis. Rep.* 49: 378–79
9. Autrique, A., Potts, M. J. 1987. The influence of mixed cropping on the control of potato bacterial wilt. *Ann. Appl. Biol.* 111:125–33
10. Boucher, C. A., Barberis, P. A., Arlat, M. 1988. Acridine orange selects for deletion of *hrp* genes in all races of *Pseudomonas solanacearum*. *Mol. Plant-Microbe Interact.* 1:282–88

10a. Brumbley, S. M., Denny, T. P. 1990. Cloning of wild-type *Pseudomonas solanacearum phcA*, a gene that when mutated alters expression of multiple traits that contribute to virulence. *J. Bacteriol.* 172:5677–85

11. Buddenhagen, I. W. 1986. Bacterial wilt revisited. See Ref. 108, pp. 126–43
12. Buddenhagen, I. W., Kelman, A. 1964. Biological and physiological aspects of bacterial wilt caused by *Pseudomonas solanacearum*. *Annu. Rev. Phytopathol.* 2:203–30
13. Busse, J., Auling, G. 1988. Polyamine pattern as a chemotaxonomic marker within the Proteobacteria. *System. Appl. Microbiol.* 11:1–8
14. Chang, M. L., Hsu, S. T. 1988. Suppression of bacterial wilt of tomato by soil amendments. *Plant Prot. Bull. (Taipei)* 30:349–59
15. Chen, W. Y., Echandi, E. 1984. Effects of avirulent bacteriocin-producing strains of *Pseudomonas solanacearum* on the control of bacterial wilt of tobacco. *Plant Pathol.* 33:245–53
16. Ciampi, L., Sequeira, L. 1980. Influence of temperature on virulence of race 3 strains of *Pseudomonas solanacearum*. *Am. Potato J.* 57:307–17
17. Ciampi, L., Sequeira, L. 1980. Multiplication of *Pseudomonas solanacearum* in resistant potato plants and the establishment of latent infections. *Am. Potato J.* 57:319–29

18. Ciampi, L., Sequeira, L., French, E. R. 1980. Latent infection of potato tubers by *Pseudomonas solanacearum.*: *Am. Potato J.* 57:377–86
19. Ciampi-Panno, L., Fernandez, C., Bustamante, P., Andrade, N., Ojeda, S., et al. 1989. Biological control of bacterial wilt of potatoes caused by *Pseudomonas solanacearum. Am. Potato J.* 66:315–32
20. Cook, D., Barlow, E., Sequeira, L. 1989. Genetic diversity of *Pseudomonas solanacearum*: detection of restriction fragment length polymorphisms with DNA probes that specify virulence and the hypersensitive response. *Mol. Plant-Microbe Interact.* 2:113–21
21. Coplin, D. L., Cook, D. 1990. Molecular genetics of extracellular polysaccharide biosynthesis in vascular pathogenic bacteria. *Mol. Plant-Microbe Interact.* 3:271–79
22. Daub, M. E., Jenns, A. E. 1989. Field and greenhouse analysis of variation for disease resistance in tobacco somaclones. *Phytopathology* 79:600–5
23. Denny, T. P., Carney, B. F., Schell, M. A. 1990. Inactivation of multiple virulence genes reduces the ability of *Pseudomonas solanacearum* to cause wilt symptoms. *Mol. Plant-Microbe Interact.* 3:293–300
24. De Vos, P., De Ley, J. 1983. Intra- and intergeneric similarities of *Pseudomonas* and *Xanthomonas* ribosomal ribonucleic acid cistrons. *Int. J. Syst. Bacteriol.* 33:487–509
25. De Vos, P., Goor, M., Gillis, M., De Ley, J. 1985. Ribosomal ribonucleic acid cistron similarities of phytopathogenic *Pseudomonas* species. *Int. J. Syst. Bacteriol.* 35:169–84
26. Dianese, J. C., Dristig, M. C. G., Crüz, A. P. 1990. Susceptibility to wilt associated with *Pseudomonas solanacearum* among six species of Eucalyptus growing in equatorial Brazil. *Australas. Plant Pathol.* 19:71–76
27. Dristig, M. C. G., Dianese, J. C. 1990. Characterization of *Pseudomonas solanacearum* biovars based on membrane protein patterns. *Phytopathology* 80: 641–46
28. Dukes, P. D., Morton, D. J., Jenkins, S. F. Jr. 1965. Infection of indigenous hosts by *Pseudomonas solanacearum* in south Georgia. *Phytopathology* 55:1055 (Abstr.)
29. Eden-Green, S. J., Sastraatmadja, H. 1990. Blood disease present in Java. *FAO Plant Prot. Bull.* 38:49–50
30. Engelbrecht, M. C., Hattingh, M. J. 1989. Numerical analysis of phenotypic features of *Pseudomonas solanacearum* isolated from tobacco and other hosts in South Africa. *Plant Dis.* 73:893–98
31. Ercolani, G. L. 1984. Infectivity titration with bacterial plant pathogens. *Annu. Rev. Phytopathol.* 22:35–52
32. French, E. R. 1986. Interaction between strains of *Pseudomonas solanacearum*, its hosts and the environment. See Ref. 108, pp. 99–104
33. French, E. R. 1988. Field evaluation of clones bred for resistance to *Pseudomonas solanacearum*. See Ref. 61, pp. 109–12
34. French, E. R. 1988. Strategies for bacterial wilt control. See Ref. 61, pp. 133–41
35. French, E. R., De Lindo, L. 1982. Resistance to *Pseudomonas solanacearum* in potato: specificity and temperature sensitivity. *Phytopathology* 72:1408–12
36. Goto, M., Shiramatsu, T., Nozaki, K., Kawaguchi, K. 1978. Studies on bacterial wilt of strawberry caused by *Pseudomonas solanacearum* (Smith) Smith. I. Strains of the pathogen and disease tolerance of strawberry plants. *Ann. Phytopathol. Soc. Jpn.* 44:270–76
37. Graham, J., Jones, D. A., Lloyd, A. B. 1979. Survival of *Pseudomonas solanacearum* race 3 in debris and latently infected potato tubers. *Phytopathology* 69:1100–03
38. Graham, J., Lloyd, A. B. 1978. An improved indicator plant method for the detection of *Pseudomonas solanacearum* race 3 in soil. *Plant Dis. Rep.* 62:35–37
39. Graham, J., Lloyd, A. B. 1978. *Solanum cinereum* R. Br., a wild host of *Pseudomonas solanacearum* biotype II. *J. Aust. Inst. Agric. Sci.* 44:124–26
40. Graham, J., Lloyd, A. B. 1979. Survival of potato strain (race 3) of *Pseudomonas solanacearum* in the deeper soil layers. *Aust. J. Agric. Res.* 30:489–96
41. Granada, G. A. 1988. Latent infections induced by *Pseudomonas solanacearum* in potato and symptomless plants. See Ref. 61, pp. 93–107
42. Granada, G. A. Sequeira, L. 1983. Survival of *Pseudomonas solanacearum* in soil, rhizosphere and plant roots. *Can. J. Microbiol.* 29:433–40
43. Granada, G. A., Sequeira, L. 1983. Survival of *Pseudomonas solanacearum* at low temperatures. *Fitopatologia* 18:22–24
44. Halos, P. M., Zorilla, R. A. 1979. Vesicular-arbuscular mycorrhizae increase growth and yield of tomatoes and

reduce infection by *Pseudomonas solanacearum*. *Philipp. Agric.* 62:309–15

45. Haque, M. A., Echandi, E. 1984. Characteristics of strains of *Pseudomonas solanacearum* from tobacco in North Carolina. *Phytopathology* 74:858 (Abstr.)
46. Hara, H., Ono, K. 1983. Ecological studies on the bacterial wilt of tobacco, caused by *Pseudomonas solanacearum* E. F. Smith. II. Survival and movement of *Pseudomonas solanacearum* in the soil of fields. *Bull. Okayama Tob. Exp. Stn.*. 42:139–47
47. Hara, H., Ono, K. 1985. Ecological studies in the bacterial wilt of tobacco, caused by *Pseudomonas solanacearum* E. F. Smith. VI. Dissemination in infected field and survival on tobacco leaf of the pathogen exuded from the upper part of infected tobacco plants. *Bull. Okayama Tob. Exp. Stn.*. 44:87–92
48. Hayward, A. C. 1964. Characteristics of *Pseudomonas solanacearum*. *J. Appl. Bacteriol.* 27:265–77
49. Hayward, A. C. 1988. Classification and identification of *Pseudomonas solanacearum*. See Ref. 61, pp. 113–21
50. Hayward, A. C., El-Nashaar, H. M., Nydegger, U., De Lindo, L. 1990. Variation in nitrate metabolism in biovars of *Pseudomonas solanacearum*. *J. Appl. Bacteriol.* 69:269–80
51. Hayward, A. C., Moffett, M. L. 1978. Leaf spot on capsicum and tomato caused by *Pseudomonas solanacearum*. *Plant Dis. Rep.* 62:75–78
52. He, L. Y. 1986. Bacterial wilt in the People's Republic of China. See Ref. 108, pp. 40–48
53. He, L. Y. 1990. Control of bacterial wilt of groundnut in China with emphasis on cultural and biological properties. See Ref. 93, pp. 22–25
54. He, L. Y., Sequeira, L., Kelman, A. 1983. Characteristics of strains of *Pseudomonas solanacearum* from China. *Plant Dis.* 67:1357–61
55. Henderson, W. R., Jenkins, S. F. 1971. New tomatoes resistant to bacterial wilt. *Res. Farming (NC Agric. Res. Serv.)* 29:10
56. Ho, W. C., Chern, L. L., Ko, W. H. 1988. *Pseudomonas solanacearum*-suppressive soils in Taiwan. *Soil Biol. Biochem.* 20:489–92
57. Holloway, B. W., Bowen, A. R. St. G., Kerr, A. 1986. Genetics of *Pseudomonas solanacearum* and prospects for biological control. See Ref. 108, pp. 93–98
58. Hsu, S. T., Chang, M. L. 1989. Effect of soil amendments on survival of *Pseudomonas solanacearum*. *Plant Prot. Bull. (Taipei)* 31:21–33

58a. Huang, Y., Sequeira, L. 1990. Identification of a locus that regulates multiple functions in *Pseudomonas solanacearum*. *J. Bacteriol.* 172:4728–31

59. Huang, Y., Xu, P., Sequeira, L. 1990. A second cluster of genes that specify pathogenicity and host response in *Pseudomonas solanacearum*. *Mol. Plant-Microbe Interact.* 3:48–53
60. International Potato Center (CIP). 1980. Report of the planning conference on the development in the control of bacterial diseases of potato, 1979.
61. International Potato Center (CIP). 1988. Bacterial Diseases of the Potato. Report of the planning conference on bacterial diseases of the potato, 1987.
62. Iqbal, M., Kumar, J. 1986. Bacterial wilt in Fiji. See Ref. 108, pp. 25-27
63. Jackson, M. T., Gonzalez, L. C. 1981. Persistence of *Pseudomonas solanacearum* (race 1) in a naturally infested soil in Costa Rica. *Phytopathology* 71:690–93
64. Jatala, P., Martin, C., Mendoza, H. A. 1988. Role of nematodes in disease expression by *Pseudomonas solanacearum* and strategies for screening and breeding for combined resistance. See Ref. 61, pp. 35–37
65. Johnson, J. L., Palleroni, N. J. 1989. Deoxyribonucleic acid similarities among Pseudomonas species. *Int. J. Syst. Bacteriol.* 39:230–35
66. Katayama, K., Kimura, S. 1987. Ecology and protection of bacterial wilt of potato 2. Some control methods and their integration. *Bull. Nagasaki Agric. For. Exp. Stn.*. 15:29–57
67. Katayama, K., Kimura, S., Hama, K. 1983. Influence of temperature on development of bacterial wilt on potato in fall cropping. *Proc. Assoc. Plant Prot. Kyushu* 29:15–18
68. Katayama, K., Kimura, S., Sawahata, H. 1986. Integrated control of bacterial wilt of potato by the relation of appropriate soil fumigant, cultivar and planting date. *Proc. Assoc. Plant Prot. Kyushu* 32:20–23
69. Kawaguchi, K., Ohta, K., Goto, M. 1981. Studies on bacterial wilt of strawberry plants caused by *Pseudomonas solanacearum* 2. β-D-Glucogallin, the antibacterial substance detected in the tissues of strawberry plants. *Ann. Phytopathol. Soc. Jpn.* 47:520–27
70. Kelman, A. 1953. The bacterial wilt caused by *Pseudomonas solanacearum*. *NC Agric. Exp. Sta. Tech. Bull.* 99. 194 pp.

71. Kelman, A., Sequeira, L. 1965. Root-to-root spread of *Pseudomonas solanacearum*. *Phytopathology* 55:304–9
72. Kempe, J., Sequeira, L. 1983. Biological control of bacterial wilt of potatoes: attempts to induce resistance by treating tubers with bacteria. *Plant Dis.* 67:499–503
73. Kishun, R., Sohi, H. S. 1982. Vertical distribution of *Pseudomonas solanacearum* in cultivated and fallow lands. *Zentralbl. Mikrobiol.* 137:643–45
74. Kishun, R., Sohi, H. S., Rao, M. V. B. 1980. Two new collateral hosts for *Pseudomonas solanacearum*. *Curr. Sci.* 49:639
75. Kloos, J. P., Tulog, B., Tumapon, A. S. 1987. Effects of intercropping potato on bacterial wilt. *Philipp. Agric.* 70:83–90
76. Krausz, J. P., Thurston, H. D. 1975. Breakdown of resistance to *Pseudomonas solanacearum* in tomato. *Phytopathology* 65:1272–74
77. Lapage, S. P., Sneath, P. H. A., Lessel, E. F., Skerman, V. B. D., Seeliger, H., et al., eds. 1975. *International Code of Nomenclature of Bacteria*. Washington, DC: Am. Soc. Microbiol. 180 pp.
78. Layne, R. E. C., McKeen, C. D. 1967. Southern bacterial wilt of field tomatoes in south western Ontario. *Can. Plant Dis. Surv.* 47:94–98
79. Lum, K. Y. 1973. Cross inoculation studies of *Pseudomonas solanacearum* from ginger. *MARDI Res. Bull.* 1:15–21
80. Lum, K. Y., Wong, H. K. 1976. Control of bacterial wilt of tomatoes in the lowlands through grafting. *MARDI Res. Bull.* 4:28–33
81. Ma, Q-S. 1990. Molecular biology and research on *Pseudomonas solanacearum*. See Ref. 93, pp. 32–38
82. Machmud, M. 1986. Bacterial wilt in Indonesia. See Ref. 108, pp. 30–34
83. Martin, C., French, E. R., Nydegger, U. 1982. Strains of *Pseudomonas solanacearum* affecting *Solanaceae* in the Americas. *Plant Dis.* 66:458–60
84. Martin, C., Nydegger, U. 1982. Susceptibility of *Cyphomandra betacea* to *Pseudomonas solanacearum*. *Plant Dis.* 66:1025–27
85. Mayers, P. E., Hutton, D. G. 1987. Bacterial wilt a new disease of custard apple: symptoms and etiology. *Ann. Appl. Biol.* 111:135–41
86. McCarter, S. M. 1976. Persistence of *Pseudomonas solanacearum* in artificially infested soils. *Phytopathology* 66:998–1000
87. McCarter, S. M., Jaworski, C. A. 1968. Greenhouse studies on the spread of *Pseudomonas solanacearum* in tomato plants by clipping. *Plant Dis. Rep.* 52:330–34
88. McCarter, S. M., Jaworski, C. A. 1969. Field studies on spread of *Pseudomonas solanacearum* and tobacco mosaic virus in tomato plants by clipping. *Plant Dis. Rep.* 53:942–45
89. McLaughlin, R. J., Sequeira, L. 1988. Evaluation of an avirulent strain of *Pseudomonas solanacearum* for biological control of bacterial wilt of potato. *Am. Potato J.* 65:255–68
90. McLaughlin, R. J., Sequeira, L. 1989. Phenotypic diversity in strains of *Pseudomonas solanacearum* isolated from a single potato field in northeastern Florida. *Plant Dis.* 73:960–64
91. Mew, T. W., Ho, W. C. 1976. Varietal resistance to bacterial wilt in tomato. *Plant Dis. Rep.* 60:264–68
92. Mew, T. W., Ho, W. C. 1977. Effect of soil temperature on resistance of tomato cultivars to bacterial wilt. *Phytopathology* 67:909–11
93. Middleton, K. J., Hayward, A. C. 1990. *Bacterial Wilt of Groundnuts*. Proc. ACIAR/ICRISAT collaborative res. plan. meet., Genting Highlands, Malaysia, 18–19 March. ACIAR Proc. No. 31
94. Mitchell, R. E., Frey, E. J. 1988. Rhizobitoxine and hydroxythreonine production by *Pseudomonas andropogonis* strains, and the implications to plant disease. *Physiol. Plant Pathol.* 32:335–41
95. Moffett, M. L., Giles, J. E., Wood, B. A. 1983. Survival of *Pseudomonas solanacearum* biovars 2 and 3 in soil: effect of moisture and soil type. *Soil Biol. Biochem.* 15:587–91
96. Moffett, M. L., Hayward, A. C. 1980. The role of weed species in the survival of *Pseudomonas solanacearum* in tomato cropping land. *Australas. Plant Pathol.* 9:6–8
97. Moffett, M. L., Wood, B. A., Hayward, A. C. 1981. Seed and soil: sources of inoculum for the colonization of the foliage of solanaceous hosts by *Pseudomonas solanacearum*. *Ann. Appl. Biol.* 98:403–11
98. Nesmith, W. C., Jenkins, S. F. Jr. 1983. Survival of *Pseudomonas solanacearum* in selected North Carolina soils. *Phytopathology* 73:1300–4
99. Nesmith, W. C., Jenkins, S. F. Jr. 1985. Influence of antagonists and controlled matric potential on the survival of *Pseudomonas solanacearum* in four North Carolina soils. *Phytopathology* 75:1182–87

100. Nishiyama, K., Achmad, N. H., Wirtono, S., Yamaguchi, T. 1980. Causal agents of cassava bacterial wilt in Indonesia. *Contr. Centr. Res. Inst. Agric. Bogor* 59. 19 pp.
101. Nyangeri, J. B., Gathuru, E. M., Mukunya, D. M. 1984. Effect of latent infection on the spread of bacterial wilt of potatoes in Kenya. *Trop. Pest Manage.* 30:163–65
102. Olsson, K. 1976. Experience of brown rot caused by *Pseudomonas solanacearum* (Smith) Smith in Sweden. *EPPO Bull.* 6:199–207
103. Olsson, K. 1976. Overwintering of *Pseudomonas solanacearum* in Sweden. See Ref. 128, pp. 105–9
104. Ono, K. 1983. Ecological studies on the bacterial wilt of tobacco caused by *Pseudomonas solanacearum* E. F. Smith III. Distribution and spread of the pathogen in infected tobacco field under rainfall. *Bull. Okayama Tob. Exp. Stn.* 42:149–53
105. Palleroni, N. J., Doudoroff, M. 1971. Phenotypic characterization and deoxyribonucleic acid homologies of *Pseudomonas solanacearum. J. Bacteriol.* 107:690–96
106. Palleroni, N. J., Kunisawa, R., Contopoulou, R., Doudoroff, M. 1973. Nucleic acid homologies in the genus Pseudomonas. *Int. J. Syst. Bacteriol.* 23:333–39
107. Peregrine, W. T. H., Ahmad, K. B. 1982. Grafting— a simple technique for overcoming bacterial wilt in tomato. *Trop. Pest Manage.* 28:71–76
108. Persley, G. J. 1986. Bacterial wilt disease in Asia and the South Pacific. *Proc. Int. Workshop, PCARRD, Los Banos, Philippines, 8–10 Oct. 1985.* ACIAR Proc. 13. 145 pp.
109. Peterson, R. A., Inch, A. J., Herrington, M. E., Saranah, J. 1983. Scorpio— a tomato resistant to bacterial wilt biovar 3. *Australas. Plant Pathol.* 12:8–10
110. Pickett, M. J., Greenwood, J. R. 1980. A study of the Va-1 group of pseudomonads and its relationship to *Pseudomonas pickettii. J. Gen. Microbiol.* 120: 439–46
111. Pitkethley, R. N. 1981. Host range and biotypes of *Pseudomonas solanacearum* in the Northern Territory. *Australas. Plant Pathol.* 10:46–47
112. Power, R. H. 1983. Relationship between the soil environment and tomato resistance to bacterial wilt (*Pseudomonas solanacearum*) 4. Control methods. *Surinaamse Landbouw.* 31:39–47
113. Quimio, A. J., Chan, H. H. 1979. Survival of *Pseudomonas solanacearum* E. F. Smith in the rhizosphere of some weed and economic plant species. *Philipp. Phytopathol.* 15:108–21
114. Quinon, V. L., Aragaki, M., Ishii, M. 1964. Pathogenicity and serological relationships of three strains of *Pseudomonas solanacearum* in Hawaii. *Phytopathology* 54:1096–9
115. Ralston, E., Palleroni, N. J., Doudoroff, M. 1973. *Pseudomonas pickettii*, a new species of clinical origin related to *Pseudomonas solanacearum. Int. J. Syst. Bacteriol.* 23:15–19
116. Ramos, A. H. 1976. Comparison of survival of two *Pseudomonas solanacearum* strains in soil columns under constant perfusion and in field devoid of host cover. See Ref. 128, pp. 123–31
117. Rao, M. V. B., Sohi, H. S. 1976. Additional hosts for *Pseudomonas solanacearum* Smith. *Curr. Sci.* 45:75–76
118. Reddy, P. P., Singh, D. B., Kishun, R. 1979. Effect of root-knot nematode on the susceptibility of Pusa Purple Cluster Brinjal to bacterial wilt. *Curr. Sci.* 48:915–16
119. Rillo, A. R. 1979. Bacterial wilt of banana in the Philippines. *FAO Plant Prot. Bull.* 27:105–8
120. Roberts, S. J., Eden-Green, S. J., Jones, P., Ambler, D. J. 1990. *Pseudomonas syzygii* sp. nov., the cause of Sumatra disease of Cloves. *Syst. Appl. Microbiol.* 13:34–43
121. Rowe, P. R., Sequeira, L., Gonzalez, L. C. 1972. Additional genes for resistance to *Pseudomonas solanacearum* in *Solanum phureja. Phytopathology* 62: 1093–94
122. Schmiediche, P. 1986. Breeding potatoes for resistance to bacterial wilt caused by *Pseudomonas solanacearum.* See Ref. 108, pp. 105–11
123. Schmiediche, P. 1988. Breeding for resistance to *Pseudomonas solanacearum.* See Ref. 61, pp. 19–27
124. Schmiediche, P., Jaynes, J. M., Dodds, J. H. 1988. Genetic engineering for bacterial disease resistance in potatoes. See Ref. 61, pp. 123–32
125. Seneviratne, S. N. de S. 1969. On the occurrence of *Pseudomonas solanacearum* in the hill country of Ceylon. *J. Hortic. Sci.* 44:393–402
126. Seneviratne, S. N. de S. 1988. Soil survival of *Pseudomonas solanacearum.* See Ref. 61, pp. 85–91
127. Sequeira, L. 1962. Control of bacterial wilt of banana by crop rotation and fallowing. *Trop. Agric. (Trinidad)* 39: 211–17

128. Sequeira, L., Kelman, A., eds. 1976. *Proc. Int. Plann. Conf. Workshop Ecol. Control Bact. Wilt, 1st, Raleigh*. NC State Univ. 166 pp.
129. Sequeira, L., Rowe, P. R. 1969. Selection and utilization of *Solanum phureja* clones with high resistance to different strains of *Pseudomonas solanacearum*.: *Am. Potato J.* 46:451–62
130. Shekhawat, G. S., Gadewar, A. V., Bahal, V. K., Verma, R. K. 1988. Cultural practices for managing bacterial wilt of potatoes. See Ref. 61, pp. 65–84
131. Shiomi, T., Mulya, K., Oniki, M. 1989. Bacterial wilt of cashew (*Anacardium occidentale*) caused by *Pseudomonas solanacearum* in Indonesia. *Ind. Crops Res. J.* 2:29–35
132. Sinha, S. K. 1986. Bacterial wilt in India. See Ref. 108, pp. 28–29
133. Sneath, P. H. A., Stevens, M., Sackin, M. J. 1981. Numerical taxonomy of *Pseudomonas* based on published records of substrate utilization. *Antonie van Leeuwenhoek J. Microbiol. Serol.* 47: 423–48
134. Sonoda, R. M. 1978. Effect of differences in tolerance of tomato to *Pseudomonas solanacearum* and time of planting on incidence of bacterial wilt. *Plant Dis. Rep.* 62:1059–62
135. Stackebrandt, E., Murray, R. G. E., Trüper, H. G. 1988. Proteobacteria classis nov., a name for the phylogenetic taxon including the "purple bacteria and their relatives". *Int. J. Syst. Bacteriol.* 38:321–25
136. Stapp, C. 1965. Die bakterielle Schleimfaule und ihr erreger *Pseudomonas solanacearum* eine Zusamenstellung neuerer Literatur. *Zentralbl. Bakteriol. Parasitenkd. Infektionskr. Hyg. Abt.* 2 119: 166–90
137. Suatmadji, R. W. 1986. Complex diseases involving nematodes and *Pseudomonas solanacearum* in potatoes in the tropics and subtropics. See Ref. 108, pp. 120–25
138. Sun, S-K., Huang, J-W. 1985. Formulated soil amendment for controlling *Fusarium* wilt and other soil borne diseases. *Plant Dis.* 69:917–20
139. Sunaina, V., Kishore, V., Shekhawat, G. S. 1989. Latent survival of *Pseudomonas solanacearum* in potato tubers and weeds. *Pflanzenkr. Pflanzenschutz* 96:361–64
140. Swanepoel, A. E. 1990. The effect of temperature on the development of wilting and on progeny tuber infection of potatoes inoculated with South African strains of biovar 2 and 3 of *Pseudomonas solanacearum*. *Potato Res.* 33: 287–90
141. Thurston, H. D. 1976. Resistance to bacterial wilt (*Pseudomonas solanacearum*). See Ref. 128, pp. 58–62
142. Tomlinson, D. L., Gunther, M. T. 1986. Bacterial wilt in Papua New Guinea. See Ref. 108, pp. 35–39
143. Toyoda, H., Hashimoto, H., Utsumi, R., Kobayashi, H., Ouchi, S. 1988. Detoxification of fusaric acid by a fusaric acid-resistant mutant of *Pseudomonas solanacearum* and its application to biological control of Fusarium wilt of tomato. *Phytopathology* 78:1307–11
144. Trigalet, A., Demery, D. 1986. Invasiveness in tomato plants of Tn5-induced avirulent mutants of *Pseudomonas solanacearum*. *Physiol. Mol. Plant Pathol.* 28:423–30
145. Trigalet, A., Trigalet-Demery, D. 1990. Use of avirulent mutants of *Pseudomonas solanacearum* for the biological control of bacterial wilt of tomato plants. *Physiol. Mol. Plant Pathol.* 36:27–38
146. Tsuyumu, S., Tsuchida, S., Nakano, T., Takikawa, Y. 1989. Antifungal activity in cell-free culture fluid of *Pseudomonas solanacearum*. *Ann. Phytopathol. Soc. Jpn.* 55:9–15
147. Tung, P. X., Rasco, E. T. Jr., Vander Zaag, P., Schmiediche, P. 1990. Resistance of *Pseudomonas solanacearum* in the potato. I. Effects of sources of resistance and adaptation. *Euphytica* 45:203–10
148. Tung, P. X., Rasco, E. T. Jr., Vander Zaag, P., Schmiediche, P. 1990. Resistance to *Pseudomonas solanacearum* in the potato. II. Aspects of host-pathogen-environment interaction. *Euphytica* 45:211–15
149. Valdez, R. B. 1986. Bacterial wilt in the Philippines. See Ref. 108, pp. 49–56
150. Vander Zaag, P. 1986. Potato production under *Pseudomonas solanacearum* conditions: sources and management of planting material. See Ref. 108, pp. 84–88
151. Vellupillai, M., Stall, R. E. 1985. Variation among strains of *Pseudomonas solanacearum* from Florida. *Proc. Fl. State Hortic. Soc.* 97:209–13
152. Verschraegen, G., Claeys, G., Meeus, G., Delanghe, M. 1985. *Pseudomonas pickettii* as a cause of pseudobacteremia. *J. Clin. Microbiol.* 21:278–79
153. Wakimoto, S., Hirayae, K., Tsuchiya, K., Kushima, Y., Furuya, N., et al. 1986. Production of antibiotics by plant pathogenic pseudomonads. *Ann. Phytopathol. Soc. Jpn.* 52:835–42

154. Wakimoto, S., Utatsu, I., Matsuo, N., Hayashi, N. 1982. Multiplication of *Pseudomonas solanacearum* in pure water. *Ann. Phytopathol. Soc. Jpn.* 48:620–7
155. Woese, C. R., Blanz, P., Hahn, C. M. 1984. What isn't a pseudomonad: the importance of nomenclature in bacterial classification. *Syst. Appl. Microbiol.* 5:179–95
156. Xu, P., Iwata, M., Leong, S., Sequeira, L. 1990. Highly virulent strains of *Pseudomonas solanacearum* that are defective in extracellular polysaccharide production. *J. Bacteriol.* 172:3946–51
157. Zehr, E. T. 1969. Studies of the distribution and economic importance of *Pseudomonas solanacearum* E. F. Smith in certain crops in the Philippines. *Philipp. Agric.* 53:218–23

Annu. Rev. Phytopathol. 1991. 29:89–107

INTEGRATION OF MOLECULAR DATA WITH SYSTEMATICS OF PLANT PARASITIC NEMATODES

Bradley C. Hyman

Department of Biology, University of California, Riverside, California 92521

Thomas O. Powers

Department of Plant Pathology, University of Nebraska, Lincoln, Nebraska 68583-0722

KEY WORDS: nematode phylogenetics, taxonomy, chemotaxonomy, root-knot nematodes, cyst nematodes, biochemical systematics

INTRODUCTION

The early history of nematology is dominated by zoologists with strong taxonomic interests who, in turn, trained an entire generation of plant nematologists. Despite their considerable activity, today it is almost a cliché to say the state of nematode taxonomy is in flux. This perspective of nematode taxonomy has become a fixture in invertebrate zoology textbooks and is often echoed by nematode systemists themselves.

Several reasons contribute to this condition, including the relatively small size of the organisms, culturing difficulties, and the availability of a limited number of reliable morphological characters that can be measured by practical laboratory approaches. Taxonomic goals have often focused upon host-parasite relationships rather than evolutionary affinities. The strong influence of the "authoritarian" school of taxonomy has also hindered the development of nematode systematics. In part, these problems have been overcome to a

0066-4286/91/0901-0089$02.00

significant degree with the accessibility of biochemical techniques that involve protein, serological, and nucleic acid analysis. The power of these new methods resides in their requirement for a vanishingly small amount of biological material (often derived from a single individual), but results in a large number of characters in the form of polypeptides, protein antigens, or DNA fragments and their constituent nucleotides. However, techniques alone cannot guarantee reliable taxonomic inferences. The recent emergence in nematology of an operational philosophy in the form of phylogenetic systematics has led to the construction of explicitly stated hypotheses of taxonomic relationships. This combination of new techniques applied in the framework of traditional phylogenetic approaches (5a, 23a) promises to enhance the utility and consistency of nematode systematics.

The question then arises as to how and where in phytonematology these systematic approaches are to be implemented. This review attempts to consider possible points of departure for the integration of classical and biochemical taxonomy of the plant nematodes. That these two approaches can be successfully merged has now been established in numerous systematic studies of other organisms (53). Consider the case of the African black jackal (81), where morphology suggests that several nearly identical species live in sympatry and that genetic separation may be recent, whereas DNA analysis tells us that speciation has been occurring over lengthy evolutionary time periods. Another striking example is the Lake Victorian cichlids (46), where nucleotide analysis suggests very recent speciation despite the occurrence of numerous morphological variants that, taken out of context, would indicate protracted divergence times.

This review considers the conflicts among phylogenetic inferences derived for phytopathogenic nematodes, usually based on information accumulated from traditional approaches. Although biochemical systematics is in its infancy relative to available taxonomy derived from morphological and cytological methods, we have attempted to review this literature and place it in the context of well-established classification schemes. We have focused upon representatives of the root-knot *(Meloidogyne)* and cyst *(Heterodera, Globodera)* nematodes, two groups that have been characterized at both the morphological and molecular levels, thereby presenting the most complete picture. Their degree of importance to worldwide agriculture further supports our choice.

As this article considers nematode taxonomy, literature that employs methodologies for detection and identification (diagnosis) of nematode infection has been largely ignored. Detailed reviews describing classification schemes for these nematode groups (5, 20, 42, 43, 68, 72, 85, 86) and biochemical taxonomy (5a, 7, 28, 36, 37, 55, 62) are available.

RELATIONSHIPS WITHIN THE GENUS *MELOIDOGYNE*

In 1949, Chitwood separated the genus *Meloidogyne* from *Heterodera* (11). Since then, *Meloidogyne* has been among the most intensively studied group of plant-parasitic nematodes. This is not surprising, in light of their cosmopolitan distribution and destructive effect on virtually all crop plants (63). Chitwood's revision recognized five species of *Meloidogyne,* characterized primarily on the basis of morphological variations in perineal patterns and dimensions of the male and female stylet. Over the intervening forty years, nematologists have added approximately 70 species and expanded the range of morphological observations typically used to characterize species. However, in spite of this intense taxonomic scrutiny, the focus of most systematic efforts in *Meloidogyne* is species identification and the definition of species boundaries.

Surprisingly few studies have addressed species relationships. The biology of *Meloidogyne* may, in part, explain this lack of attention. The major economic pest species, *M. arenaria, M. hapla, M. incognita,* and *M. javanica,* are parthenogenetic and most likely polyploid. These obligatory parthenogenetic species comprise 90% of the world's root-knot populations. (The amphimictic species are generally parasites of noneconomic, perennial plants.) As parthenogenetic organisms, they tend to pose problems to systematics in the application of a well-delineated species definition. Some nematologists have recommended abandoning all attempts to resolve relationships (52) or to relegate all parthenogenetic forms to subspecific groupings (71). Yet most nematologists believe that species in *Meloidogyne* are recognizable, discrete taxa with unique evolutionary histories. Morphological, biochemical, and host relationships are in general agreement with perceived species boundaries. It was the cytological studies conducted by Triantaphyllou that first added an evolutionary perspective to species relationships (72, 75–78). Based on chromosome numbers and mode of reproduction, evolutionary scenarios were constructed. Accumulated information suggests that facultative meiotic parthenogenetic forms gave rise to obligatory mitotically parthenogenetic nematodes, possibly through hybridization between species, resulting in polyploidy. Therefore, current species with mixtures of reproductive forms *(M. hapla)* are viewed as ancestral to obligatory parthenogens *(M. incognita, M. javanica, M. arenaria).*

Biochemical approaches to establishing phylogenetic relationships in nematodes include the separation of proteins by one-dimensional polyacrylamide gel electrophoresis (1D-PAGE), protein fractionation using isoelectric focusing (IEF), electrophoretic resolution of proteins in two dimensions using IEF followed by PAGE (2D-gel electrophoresis), immunological (serological) strategies, and molecular (DNA) technology. Studies addressing the "chemotaxonomy" of phytonematodes have been conducted over the course of the past three decades, although molecular approaches have only recently been employed.

Biochemial experimentation can be divided into two quite distinct disciplines. The first can be classified by the terms "identification" or "diagnostics," in which genera, species, and perhaps subspecific populations can be discriminated from one another. Studies involving identification contribute to most of this literature. "Biochemical taxonomy," as we define it here, considers the phytogenetic relationships between nematode groups and is the topic emphasized in this review.

The first studies examining protein variation in *Meloidogyne* included a systematic analysis of the four main species in the genus (16). Population homogenates of whole nematodes were compared on the basis of the total number of matching protein bands for all enzymes examined and clustered according to overall similarity. The resulting species relationships are pictured in Figure 1. *M. javanica* and *M. arenaria* are the most similar, and both share more protein bands with *M. incognita* than with *M. hapla*. Based on the

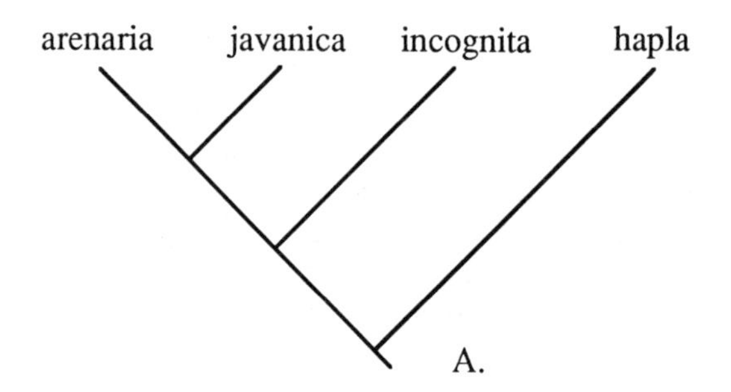

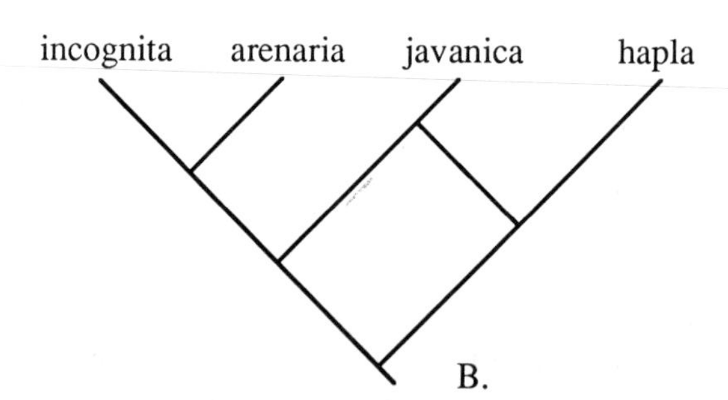

Figure 1 Relationships of the mitotically parthenogenetic species of *Meloidogyne*. (A) Relationships inferred from cytological and isoenzyme data (references 70a, 77). (B) Relationships inferred from mitochondrial DNA (T. O. Powers & L. J. Sandall, unpublished).

large number of nonmatching protein bands, *M. hapla* was considered more primitive relative to the other species. Although this interpretation of species relationships assumes equal evolutionary rates and independence of characters, it is generally accepted by nematologists.

Dalmasso & Berge extended isoenzyme analyses to single female nematodes (13, 14). These efforts eliminated difficulties inherent in comparing patterns derived from population homogenates, which tend to pool allelic variants, thereby reducing measurable diversity. However, systematic comparisons were still complicated by multiple protein bands for particular enzymes. These banding patterns are presumably the result of related enzyme forms produced by several different alleles at the same locus on homologous chromosomes, or the products of duplicated genes. When the major species were evaluated on the basis of the total number of different bands observed for five enzyme systems, *M. javanica* and *M. arenaria* were judged to be closely related and species relationships were in agreement with earlier studies (13, 14). In the most extensive comparison to date, 27 enzymes were evaluated in 30 populations of *Meloidogyne* (18). As in previous analyses, each protein band was evaluated as an independent character, and evolutionary relationships so generated were consistent with previous schemes.

Relationships derived from isoenzyme data must be interpreted with caution. To date, all of the analyses include assumptions and apply algorithms that may not be appropriate for the data. The general approach to analysis of isoenzyme data in nematology has been to treat each protein band as an independent character, use a coefficient of similarity to construct a matrix of pairwise comparisons, and then employ a phenetic clustering technique to arrange groupings according to overall similarity. Assumed independence of protein bands as characters may be compromised by heterozygosity and polyploidy. The reduction of two isozyme profiles to a single distance metric may be influenced by the selection of different similarity coefficients (54). The use of distance metrics in systematics itself is controversial (19). Many taxonomists prefer to use data that code for the presence or absence of particular alleles (47, 48). Phenetic clustering methods rely on the assumption of equal rates of protein evolution, a long-debated concern in systematics (83). Finally, phenetic methods of analysis will be misleading if *Meloidogyne* species have arisen from interspecies hybridization.

Analysis of relationships by comparing mitochondrial DNA restriction patterns has advantages and disadvantages (1, 35). Powers & Sandall (56) examined fourteen populations of *Meloidogyne*, which included the four major species. Purified mitochondrial DNA was digested with ten different restriction enzymes and similarity coefficients were calculated according to the formula of Nei (51). The distance metrics were then clustered by the commonly used algorithm, UPGMA (65). However, these analyses can be criticized using the same set of concerns regarding distance data as argued above. Although not used in the mtDNA study, a method of higher resolution would compare restriction sites on a presence or absence basis and treat the data in a phylogenetic fashion. MtDNA does have a distinct advantage in the analysis of polyploid organisms because no matter how many nuclear genome copies may be present per cell, the mitochondrial genome always acts as a haplotype. A second advantage is the potential in resolving events of interspecifc hybridization by virtue of its maternal mode of inheritance. Offspring of such a hybridization will contain one half of the nuclear genome of each parent, but the entire complement of mtDNA will be derived from the mother. Preliminary nucleotide sequencing data suggests that *M. javanica* may have resulted from a hybridization between a female *M. hapla* and an uncharacterized male progenitor (T. O. Powers, unpublished data; figure 1). The hybrid origin of *M. javanica* has been previously suggested (14), based on the relatively high degree of fixed heterozygosity in this species.

A comparison of phenograms derived from mtDNA restriction fragment polymorphisms (56) and isoenzymes display marked differences. To a certain extent, the dissimilarity can be explained by inadequate species identification (T. O. Powers, unpublished data). Subsequent host race tests and isozyme analysis have demonstrated that the three populations of *M. hapla* included in the mtDNA study were actually other species. Importantly, two of these (isolates K and L in ref. 56) were later determined to be *M. arenaria*. This finding results in a reinterpretation of the data in which a greater estimated nucleotide sequence divergence is observed between two genomes of *M. arenaria* than is seen in any interspecific comparison. Still, major differences remain in the two phenograms. The mitochondrial data do not support a closer relationship between *M. arenaria* and *M. javanica* than *M. incognita,* nor do they suggest that *M. arenaria* is a discrete genetic unit. Clearly, a thorough assessment of relationships among these parthenogenetic species will require evaluation of a combination of data sets together with a sound method of data analysis. It is also highly likely that a resolution of species relationships in *Meloidogyne* will depend on a taxonomic method that can detect reticulate evolution.

Despite many careful investigations designed to establish species boundaries among root-knot nematodes, stringent systematic analysis has not been applied to generate testable phylogenetic relationships within the genus *Meloidogyne*. Rather, hypothesized relationships within the Heteroderinae are probably the most explicit and testable among all the phytonematode taxa. This is primarily due to the efforts of a few taxonomists who have rigorously applied the principles of phylogenetic systematics to the group. Consequently, the Heteroderinae stand as the most propitious collection of nematodes to validate the integration of morphological and molecular data.

RELATIONSHIPS WITHIN THE HETERODERINAE

Generic Relationships

Representatives of the subfamily Heteroderinae *sensu* Luc, et al; (43) are among the most thoroughly studied collection of nematodes. Taxonomic relationships are particularly relevant here because the pathogenic cyst nematodes are encompassed by this group, representing the most economically important phytonematodes of temperate agriculture (69). The Heteroderinae subfamily is defined by as many as 110 recognized species that comprise some 21 identifiable genera, including *Heterodera* and *Globodera* with 57 and 12 species, respectively. Especially important are those pests that pose a worldwide threat to agriculture, including the closely related species *H. glycines* (soybean cyst nematode; SCN) and *H. schachtii* (sugar beet cyst nematode; SBCN), as well as *G. rostochiensis* and *G. pallida* (potato cyst nematodes; PCN).

Members of the Heteroderinae have adapted to a wide assortment of hosts and habitats; this accommodation is reflected by morphological diversity. Given the relative uniformity of the life histories and cytogenetics (basic haploid chromosome number N = 9; ref. 73) exhibited by the cyst nematodes, the most important approach to establishing phylogenetic relationships within Heteroderinae has been based on anatomical characters.

Before the use of cladistic methodologies (34), members of the Heteroderinae were grouped according to a combination of morphological similarity and character weighting (5a, 23a). A study by Stone (67) typifies this approach. Juvenile lip morphology, judged to be a reliable character, was used to group Heteroderinae *sensu lato* into four basic clusters. The single trait examined by Stone was not analyzed as to its ancestral or derived state; lip patterns were often obtained from a single species, raising the possibility of miscoding due to inadequate sampling. It is not surprising, then, that Stone's groupings, based exclusively on lip structures, differ to some degree from phylogenetic schemes available at that time (85, 87). As such, taxonomic relationships among the Heteroderinae have been judged inconsistent and unstable; this inconsistency underscores the requirement for a systematic approach based on a nonarbitrary scientific operational philosophy.

Employment of the cladistic method by Ferris (20) provided just such a scientifically rigorous foundation for future systematic studies involving relationships within Heteroderinae. The success of her insight is evidenced by several major studies that subsequently evaluated additional taxa and characters. These investigations have been described in detail (6, 22, 42, 86); here we review some of their salient features

Ferris generated two similar cladograms relating generic relationships among members of Heteroderinae (20, 22). Importantly, the relationships so derived required only a single evolution of the cyst, thereby grouping selected genera into cyst and noncyst nematodes. However, within this same tree, parallel evolution must be invoked to explain the ability of two separate nematode groups to induce a multinucleate syncytium of plant host cells during infection.

Wouts (86) interpreted his data in a similar context and derived a phylogenetic tree depicting relationships among the Heteroderinae by coding 22 different derived characters. Major groupings have been developed based on the presence or absence of a "D-layer" (64), one of several strata that comprise the body wall cuticle of females within the Heteroderinae. The relationships so generated require the presentation of monophyletic groups containing mixtures of cyst and noncyst-forming nematodes. The cyst, as a terminal character, must have independently evolved several times to generate this scheme.

Baldwin & Schouest (6) scored what they considered to be the 19 most reliable morphological characters and compared several algorithms for generating phylogenetic trees. Importantly, their analysis strongly supports monophyly of all the cyst-forming genera (tribe Heteroderini), contrary to repeated evolution of the cyst as hypothesized by Wouts (86). Moreover, syncytium induction is confined to closely related sister groups derived from a single, common ancestor. Aside from these considerations, the schemes presented by both Ferris (20) and Baldwin & Schouest (6) are convincingly similar and support the employment of parsimony in generating phylogenetic relationships. Krall & Krall (42) attempted a classification based on coevolution with plant hosts that, to a limited extent, agreed with those based on morphology.

Tree comparisons are useful in that areas of controversy are exposed and provide the foundation for future experimentation. Many of the differences observed among schemes describing relationships among the Heteroderinae can result from an absence of reliable characters, miscoded characters, or naive interpretations regarding the form, function, and ontology of these morphological markers. The body wall D-layer provides a useful example of these problems. Groupings within the Heteroderinae have been guided by the presence (e.g. *Globodera*) or absence (e.g. *Heterodera*) of the D-layer (86). The parsimony arguments presented by Baldwin & Schouest (6) suggested that *Globodera* and *Heterodera* were monophyletic, with the Heteroderids undergoing a secondary loss (or reversal) of the D-layer. More recently, it has been demonstrated that a thin D-layer is present between the midbody and cone of females of *H. schachtii* about four weeks after the final molt (12). This result supports the monophyly of *Heterodera* and *Globodera* (6) and alleviates the requirement for parallel evolution of the cyst (86). Emerging detailed information on additional characters and refined assignments as to pleisiomorphy or apomorphy will undoubtedly serve to clarify additional controversies of Heteroderinae phylogeny at the genus level.

Species Relationships

The large number of species and broad morphologic diversity that characterize the Heteroderinae suggest that this nematode group would be a useful model for investigation of phylogenetic relationships among species. Members of the *Heterodera* have traditionally been assigned to one of several species groups that include the *schachtii, goettingiana,* and *avenae*. However, there is disagreement as to the membership of these groups, and it is not known if these groups are monophyletic. Taxonomy within genera of the subfamily Heteroderinae remains confounded, as species have been assembled in a variety of ways resulting in several subgeneric classification schemes (20).

Relationships among species groups of *Heterodera* were depicted in a tree derived from cladistic analysis of five different character states (20). Results from this treatment were directly compared with four independently derived schemes developed by consideration of a single or very few morphological traits, including vulval cone (45), cone top (50), vaginal (30), and lip structures (67, 68). Although each classification scheme is different, the tree derived by Ferris in this landmark study may be the most parsimonious, as character states for each of the multiple morphological markers were defined. Unfortunately, little additional information has been compiled in the past decade, and the taxonomy of subgeneric Heteroderinae remains controversial. The comparative study by Ferris (20) illustrates the variety of ways in which species have been grouped, resulting in competing classification schemes.

Taxonomic relationships among *Globodera* species are equally elusive. Due to apparent widespread incidence of both inter- and intraspecific hybridization (49, 70) a continuum of morphological variants occurs, and the number of classical characters that may be unambiguously assigned to a specific species is limited. Stone (70) attempted to address relationships among six defined *Globodera* species by assessing overall similarity using principle co-ordinate (cluster) analysis of characters describing second-stage juveniles, cyst morphology, and host range. It was suggested that some discrimination of *G. pallida, G. rostochiensis,* and *G. tabacum* can be recognized depending on the traits analyzed, but assignment as to which characters are primitive or derived and assessment of possible parallel and convergent evolution was not considered. Biologically significant conclusions predictive of relationships among PCN are difficult to extract from these types of studies.

Sibling Species, Races, and Pathotypes

Introduction into a new environment of a small nematode sampling originally derived from a larger community may result in local, reproductively isolated populations, perhaps carrying a subset of the original gene pool. Sibling species, biological races, or pathotypes are subspecific designations that may be a direct consequence of genetic drift or exposure to unique selective pressures within these new locations. Subspecific populations are separated by narrow genetic distances but may often exhibit behavioral or morphological differences as adaptions to these varying environments.

Phytonematologists have considered *H. glycines* Ichinohe (SCN) and *H. schachtii* Schmidt (SBCN) to be sibling species within the same taxonomic subgroup (schachtii) (5). Although some morphological traits have been considered diagnostic for the two species (5), substantial overlap of these characters have made discrimination difficult, further exemplifying their close taxonomic relationship.

Phylogenetic relationships within the Heteroderinae have also been informally defined at the intraspecific level. Designations of "races" or "pathotypes" have been based on the ability of individual nematode populations to propagate on a standardized collection of plant host cultivars, thereby overcoming certain host genes that would otherwise confer resistance to a particular nematode population. For *H. glycines,* as many as 16 races have been characterized (29, 61). Three pathotypes for *G. pallida* (Pa1-Pa3) and five for *G. rostochiensis* (Ro1-Ro5) have been described (41). Genetic analysis of races of *H. glycines* (74) indicates that race designation may be a consequence of variable resistance-gene frequencies (or their alleles) between otherwise related populations. Ferris (21) suggests that subspecific designations such as these may in fact be a taxonomic convenience. If race separation is a result of authentic evolutionary change, at present no reliable method exists to assess whether races are of a single origin (monophyletic) or have arisen on several occasions. Given the degree of morphological variation among isolates and the considerable overlap of numerical characters, biochemical approaches to nematode identification may have a considerable impact in establishing relationships at the subspecies level.

Biochemical Taxonomy of Heterodera

Early work on the *Heterodera* species complex (extensively reviewed in ref. 28) focused upon the electrophoretic behavior of total soluble proteins extracted from population homogenates, migration properties of allozymes with assayable function on "activity gels" and by serological techniques that identified common protein antigens by crossreactivity of antibodies generated from nematode protein extracts. However, each study analyzed various combinations of species or different populations of individual species, and consistent chemotaxonomic relationships among the Heteroderids were not generated.

Using improved electrophoretic conditions, Pozdol & Noel (57) achieved enhanced resolution of total Heteroderid proteins under both native and denaturing conditions. Protein patterns derived from *H. schachtii, H. trifolii, H. lespedezae,* and *H. glycines* exhibited an overall similarity but were easily distinguished; races of *H. glycines* were not separable. Polypeptide spectra from *H. trifolii* and *H. lespedezae* exhibited a striking resemblance and suggested an apparent subspecific relationship consistent with the close phylogenetic alignment inferred using serological techniques (60). In a broader context, the small number of differences in overall protein banding patterns exhibited by these four species reflects a close taxonomic association among representatives of the *Heterodera*.

Improved resolution of Heteroderid proteins has been achieved by employing 2D-gel electrophoresis. Bakker & Bouwman-Smits (3) used this technique to separate total proteins from *H. glycines* and *H. schachtii*. Despite their close phylogenetic relationship, the two species could be discriminated by differential electrophoretic behavior of 59% of their polypeptides. This large genetic divergence is in stark contrast to the conserved morphology exhibited by these same species, and suggests that the pace of genetic and morphological evolution are asynchronous in these nematodes. This observation has been recently confirmed by molecular approaches, where significant nucleotide divergence between the mitochondrial genomes of *H. glycines* and *H. schachtii* has been observed (58).

At the subspecific level, early serological work demonstrated that *H. glycines* races 3 and 4 elaborated different polypeptide antigens and could be distinguished from each other (32). This suggested that different isolates may be separable based on their constituent proteins. When total proteins were extracted from the "strict" and "Gotland" strains of the cereal cyst nematode *H. avenae,* a striking 75% of the proteins exhibited differences after fractionation by 2D-gel electrophoresis (23). It was suggested that these two strains may be different species. Similar results were obtained by using the same technique to analyze total proteins prepared from several independent isolates of *H. glycines* (race 3) (24). Although the overall constellation of protein "spots" was similar, enough differences were observed so that genetic similarities could be estimated. Similarity measurements ranged from 74–88% among the three different populations of race 3, values lower than that scored among both intraspecific and congener comparisons in rodents. Recent isozyme studies have also revealed extensive intraspecific variability among populations of *H. glycines* (59). Attempts have also been made to discriminate races of *H. glycines* by scoring differences in restriction enzyme-treated nuclear DNA banding patterns (39), although convincing differences at the DNA level have not yet been demonstrated.

The observation of recognizable genetic dissimilarity among isolates of the same race of *H. glycines* indicates that extensive genetic variation may occur among nematode populations, but is masked by convenient classification schemes represented by subspecific races (21). This reinforces the premise that the race concept in phytonematology may require some review. It has recently been suggested that all intraspecific variants be redefined as "pathotypes" that denote propagation behavior on differential cultivars without inference of genetic variation (17). Populations themselves may vary with respect to allele frequencies and number of virulent individuals inhabiting a population (74), as well as other characters unrelated to growth performance. Therefore, heterogeneity among and between isolates must be carefully considered in interpreting relationships based upon classical or "chemotaxonomic" approaches.

Biochemical Taxonomy of Globodera

The potato cyst nematode (PCN) species *G. rostochiensis* and *G. pallida* exhibit exceptionally similar morphologies, and were historically considered pathotypes of a single species, *Heterodera rostochiensis* (38, 66). Importantly, 1D-electrophoretic analysis of PCN proteins from several isolates contributed to elevating these pathotypes to species status (31, 79, 80) before achieving independent generic ranking as *Globodera*.

Early serological approaches coupled with cluster analysis of shared antigens were used to partially distinguish several populations of PCN (82). However, use of 2D-gel electrophoresis has permitted estimates of genetic distance between *G. rostochiensis* and *G. pallida* (3). Differences in protein patterns between these sibling species was estimated to be 70%, a value higher than that observed for analysis of polypeptides from *H. glycines* and *H. schactii*, and provides yet another example of the discordance between genetic and morphological evolution. Similar conclusions have been drawn from allozyme analysis of anatomically related *Caenorhabditis* species (10).

Biochemical systematics of PCN has largely focused on relationships among the three pathotypes of *G. pallida* (Pa1-Pa3) and five subspecific variants of *G. rostochiensis* (Ro1-Ro5). Although intraspecific variation in polypeptide profiles was not observed using conventional 1D-gel electrophoresis (4), the enhanced resolution available with 2D gels allowed discrimination of Ro1 and Ro5 (2). By employing IEF to fractionate several allozymes, Marks & Fleming (25, 44) generated a dendogram that discriminated isolates of *G. rostachiensis* from isolates of *G. pallida*. Fox & Atkinson (26) also employed IEF to resolve both total proteins and some 40 different allozymes. Genetic similarity measurements and cluster analysis revealed a distinct separation of Pa from Ro pathotypes. Moreover, Ro4 and Ro5 were clustered in a group distinguishable from Ro1, Ro2, and Ro3. Species-specific antigens were recognized within IEF-fractionated total protein preparations from fourteen *Globodera* field populations and eight reference samples (27). Cluster analysis of shared antigens produced a dendogram in which *G. rostochiensis* and *G. pallida* were clearly separated, but grouping into pathotypes was not definitive.

As was discussed for the Heteroderids, data describing phylogenetic relationships among subspecific pathotypes must be interpreted within the context of the ongoing dialogue surrounding the race concept. There is only one example where a valid attempt has been made to obtain biochemical information from nematode race populations characterized with respect to frequencies of virulent genotypes (4). Establishment of biologically correct relationships among pathotypes will require routine integration of behavioral and morphological characters with biochemical markers. Each approach alone appears too "coarse" to establish definitive classification schemes at the subspecific level.

Molecular analysis has provided an opportunity to survey the abundance of information sequestered in the entire nematode genome, in contrast to the small window of possible genetic variation as reflected in the polypeptide endproducts. To date, DNA technology is being evaluated for detection and identification (diagnostics) of PCN with only limited application to establishing phylogenetic relationships. However, restriction enzyme analysis of PCN nuclear genomes has revealed extensive differences between the repeated sequence component of chromosomal DNAs derived from *G. rostochiensis* and *G. pallida* (15). The polymorphic banding patterns visualized in these studies directly reflect the genetic separation of these sibling species first identified by protein gel electrophoresis, and supports the asynchronous pace of genetic and morphological evolution in these nematodes. A significant genetic distance of .87 was estimated to separate *G. pallida* from *G. rostochiensis*, but the accuracy of this value, relative to that obtained with protein data, awaits confirmation by resolution of additional restriction fragment length polymorphisms. Based on the substantial differences between the nuclear genomes of thse two related PCN species, DNA-based molecular diagnostic assays have been developed (8, 9).

USE OF MOLECULAR DATA IN NEMATODE PHYLOGENY

Despite the substantial accumulation of information obtained with biochemical experimentation, little attempt has been made to integrate protein or

DNA characters into nematode taxonomic schemes initially constructed by morphologists. This is due, in part, to the diagnostic focus of molecular investigations rather than the availability of suitable characters. Given this present situation, it might be useful to consider what types of molecular studies could be employed to test established classification systems.

DNA molecules derived from nuclear and mitochondrial genomes may be informative to varying degrees, depending upon the taxonomic questions to be addressed. Consider for example the significant degree of nucleotide sequence divergence exhibited by mtDNAs of *H. glycines* and *H. schactii,* as reflected in the differences of their restriction fragment banding patterns (58). Extensive sequence variation among these closely related species indicates that the probability of incurring homoplasious loss of restriction sites would be high, and that simple restriction analysis such as scoring of shared restriction fragments would be ineffective in establishing relationships at the species or generic levels. In addition, known differences in the sizes of nematode mitochondrial genomes would require analytical approaches that would still permit coding of fragments as homologous, even if their size differences were due to a simple loss (or gain) of a restriction site or to small insertions or deletions that generated restriction fragment size polymorphism. Given the extensive and rapid divergence among related mtDNA molecules, relationships at the intraspecific level may be best suited to mtDNA analysis.

If mtDNA is to be successfully employed for inferring relationships among genera and species, comparative restriction maps of the mitochondrial genomes under study could be constructed and the data treated by cladistics on the basis of the presence or absence of restriction enzyme target sites. However, this approach is quite laborious and does not entirely exclude the possibility of miscoding due to site loss from nonhomologous mutation events. One straightforward alternate approach would be to study a well-conserved region of the mitochondrial genome, such as coding sequences for protein subunits of the electron transport and oxidative phosphorylation complex. Such loci define a collection of orthologous genes present in mtDNAs from widely separated taxa. These gene regions could be readily targeted by universal nucleotide "primers" complementary to highly conserved mtDNA sequences, readily isolated by enzymatic amplification using the polymerase chain reaction (33), and studied in detail by comparative nucleotide sequence analysis, as has been accomplished for several mammalian species groups (40).

Analysis of representative sequence elements comprising the nuclear genome may also provide insight into phylogenetic relationships at the generic and species levels. For example, well-characterized gene families encoding ribosomal RNA (termed rDNA) has been useful in classification at all taxonomic levels (84). Highly conserved rRNA coding sequences should be useful to infer relationships among genera; rapidly evolving noncoding spacers may provide information as to species and subspecific groupings. The utility of each of these cellular DNA components in helping to assess relationships among nematode groups can then best be addressed by integrating these data with established taxonomic schemes.

CONCLUDING REMARKS

With the continued refinement of protein and serological analysis and the increasing accessibility of molecular techniques, we anticipate a significant escalation and interest in the biochemical systematics of plant pathogenic nematodes. No longer will the size and availability of nematodes or number of numerical characters be a constraint to making phylogenetic inferences, as contemporary techniques for allozyme, immunological, and molecular analysis can be conducted on individual nematodes. Studies addressing individual and population variability, essential components of delimiting species and subspecific boundaries, are now greatly facilitated by the exquisite sensitivity of these techniques. In addition, molecular biology has provided a ready source of reagents in the forms of gene sequence data bases and DNA probes from widely separated animal taxa that can be directly tested on nematode systems and used as a foundation for developing relationships via nucleotide sequence analysis.

Studies on the African black jackals and cichlid populations in Lake Victoria document the need to integrate morphological and biochemical data to establish credible phylogenetic relationships. Currently, most biochemical and molecular information obtained by phytonematologists has focused upon detection and identification. With a shift in emphasis towards taxonomic inferences, this information must be integrated with conflicting classification systems deduced by morphological and cytological analysis. Unfortunately, available scientific expertise is distributed among researchers studying a wide variety of nematode groups. To fully exploit the potential of this integrative approach, it may be beneficial to focus on selected nematode systems and coordinate the information therein.

ACKNOWLEDGMENTS

We thank Drs. J. G. Baldwin and V. R. Ferris for their encouragement and advice during preparation of this manuscript. This work was supported by USDA grants 85-CRCR-1-1629 (to B.C.H.) and 90-37263-5257 (to T.O.P.).

Literature Cited

1. Avise, J. C. 1986. Mitochondrial DNA and the evolutionary genetics of higher animals. *Phil. Trans. R. Soc. London Ser. B* 312:325–34
2. Bakker, J., Bouwman-Smits, L. 1988. Genetic variation in polypeptide maps of two *Globodera rostochiensis* pathotypes. *Phytopathology* 78:894–900
3. Bakker, J., Bouwman-Smits, L. 1988. Contrasting rates of protein and morphological evolution in cyst nematode species. *Phytopathology* 78:900–4
4. Bakker, J., Schots, A., Bouwman-Smits, L., Gommers, F. J. 1988. Species-specific and thermostable proteins from second-stage larvae of *Globodera rostochiensis* and *G. pallida*. *Phytopathology* 78:300–5
5. Baldwin, J. G., Mundo-Ocampo, M. 1991. Heteroderinae, cyst and non-cyst forming nematodes. In *Manual of Agricultural Nematology*, ed. W. R. Nickle. New York: Dekker. In press

5a. Baldwin, J. G., Powers, T. O. 1987. Use of fine structure and nucleic acid analysis in systematics. In *Vistas on Nematology: A Commemoration of the Twenty-fifth Anniversary of the Society of Nematologists*, ed. J. A. Veech, D. W. Dickson, pp. 336–45. Maryland: Soc. Nematol., Inc.

6. Baldwin, J. G., Schouest, L. P. Jr. 1990. Comparative detailed morphology of the Heteroderinae Filip'ev and Schuurmans Stekhoven, 1941, *sensu* Luc *et al.* (1988): phylogenetic systematics and revised classification. *Syst. Parasitol.* 15:81–106
7. Burrows, P. R. 1990. The use of DNA to identify plant parasitic nematodes. *Nematol. Abstr.* 59:1–8
8. Burrows, P. R. 1990. The rapid and sensitive detection of the plant parasitic nematode *Globodera pallida* using a non-radioactive biotinylated DNA probe. *Rev. Nematol.* 13:185–90
9. Burrows, P. R., Perry, R. N. 1988. Two cloned DNA fragments which differentiate *Globodera pallida* from *G. rostochiensis*. *Rev. Nematol.* 11:441–45
10. Butler, M. H., Wall, S. M., Luehrsen, K. R., Fox, G. E., Hecht, R. M. 1981. Molecular relationships between closely related strains and species of nematodes. *J. Mol. Evol.* 18:18–23
11. Chitwood, B. G. 1949. "Root-knot nematodes"—Part I. A revision of the genus *Meloidogyne* Goeldi; 1887. *Proc. Helminthol. Soc. Wash.* 16:90–104
12. Cordero-Clark, D. A., Baldwin, J. G. 1990. Effect of age on bodywall cuticle morphology of *Heterodera schachtii* Schmidt females. *J. Nematol.* 22:356–61
13. Dalmasso, A., Bergé, J. B. 1978. Molecular polymorphism and phylogenetic relationship in some *Meloidogyne* spp.: Application to the taxonomy of *Meloidogyne*. *J. Nematol.* 10:323–32
14. Dalmasso, A., Bergé, J. B. 1983. Enzyme polymorphism and the concept of parthenogenic species, exemplified by *Meloidogyne*. See Ref. 70a, pp. 187–96
15. DeJong, A. J., Bakker, J., Roos, M., Gommers, F. J. 1989. Repetitive DNA and hybridization patterns demonstrate extensive variability between the sibling species *Globodera rostochiensis* and *G. pallida*. *Parasitology* 99:133–38
16. Dickson, D. W., Huisingh, D., Sasser, J. N. 1971. Dehydrogenases, acid and alkaline phosphatases, and esterases for the chemotaxonomy of selected *Meloidogyne, Ditylenchus, Heterodera,* and *Aphelenchus* spp. *J. Nematol.* 3:1–16
17. Dropkin, V. H. 1988. The concept of race in phytonematology. *Annu. Rev. Phytopathol.* 26:145–61
18. Esbenshade, P. R., Triantaphyllou, A. C. 1987. Enzymatic relationships and evolution in the genus *Meloidogyne* (Nematoda: Tylenchida). *J. Nematol.* 19:8–18
19. Farris, J. S. 1981. Distance data in phylogenetic analysis. See Ref. 28a, pp. 3–23
20. Ferris, V. R. 1979. Cladistic approaches in the study of soil and plant parasitic nematodes. *Am. Zool.* 19:1195–215
21. Ferris, V. R. 1983. Phylogeny, historical biogeography and the species concept in soil nematodes. See Ref. 70a, pp. 143–61
22. Ferris, V. R. 1985. Evolution and biogeography of cyst-forming nematodes. *Bull. OEPP/EPPO Bull.* 15:123–29
23. Ferris, V. R., Faghihi, J., Ireholm, A., Ferris, J. M. 1989. Two-dimensional protein patterns of cereal cyst nematodes. *Phytopathology* 79:927–33

23a. Ferris, V. R., Ferris, J. M. 1987. Phylogenetic concepts and methods. See Ref. 5a, pp. 346–53

24. Ferris, V. R., Ferris, J. M., Murdock, L. L. 1985. Two-dimensional protein patterns in *Heterodera glycines*. *J. Nematol.* 17(4):422–27

25. Fleming, C. C., Marks, R. J. 1983. The identification of the potato cyst nematodes *Globodera rostochiensis* and *G. pallida* by isoelectric focusing of proteins on polyacrylamide gels. *Ann. Appl. Biol.* 103:277–81
26. Fox, P. C., Atkinson, H. J. 1984. Isoelectric focusing of general protein and specific enzymes from pathotypes of *Globodera rostochiensis* and *G. pallida*. *Parasitology* 88:131–39
27. Fox, P. C., Atkinson, H. J. 1985. Enzyme variation in pathotypes of the potato cyst nematodes *Globodera rostochiensis* and *G. pallida*. *Parasitology* 91:499–506
28. Fox, P. C., Atkinson, H. J. 1986. Recent developments in the biochemical taxonomy of plant-parasitic nematodes. *Agric. Zool. Rev.* 1:301–31
28a. Funk, V. A., Brooks, D. R., eds. 1981. *Advantages in Cladistics. Proc. 1st Meet. Willi Henning Soc.* New York: New York Bot. Gard.
29. Golden, A. M., Epps, J. M., Riggs, R. D., Duclos, L. A., Fox, J. A., Bernard, R. L. 1970. Terminology and identity of infraspecific forms of the soybean cyst nematode *(Heterodera glycines)*. *Plant Dis. Rep.* 54:544–46
30. Green, C. D. 1975. The vulval cone and associated structures of some cyst nematodes (genus *Heterodera*). *Nematologica* 21:134–44
31. Greet, D. N. 1972. Electrophoresis and morphometrics of the round cyst nematodes. *Ann. Appl. Biol.* 71:283–86
32. Griffith, T. W., Koehler, N. J., Coker, S. G., Riggs, R. D. 1982. Race comparisons of *Heterodera glycines* using crossed immunoelectrophoresis. *J. Nematol.* 14:598–99
33. Harris, T. S., Sandall, L. J., Powers, T. O. 1990. Identification of single *Meloidogyne* juveniles by polymerase chain reaction amplification of mitochondrial DNA. *J. Nematol.* 22: 518–24
34. Henning, W. 1966. *Phylogenetic Systematics*. Transl. D. Davis, R. Zangrel. Urbana: Univ. Ill. Press
35. Hillis, D. M., Moritz, G., eds. 1990. *Molecular Systematics*. Sunderland, Mass.: Sinauer. 588 pp.
36. Hussey, R. S. 1979. Biochemical systematics of nematodes—a review. *Helminth. Abstr., Ser. B. Plant Nematol.* 48:141–48
37. Hyman, B. C. 1990. Molecular diagnosis of *Meloidgyne* species. *J. Nematol.* 22:24–30
38. Jones, F. G. W., Carpenter, J. M., Parrott, D. M., Stone, A. R., Trudgill, D. L. 1970. Potato cyst nematode: One species or two? *Nature* 227:83–84
39. Kalinski, A., Huettel, R. N. 1988. DNA restriction fragment length polymorphism in races of the soybean cyst nematode, *Heterodera glycines*. *J. Nematol.* 20:532–38
40. Kocher, T. D., Thomas, W. K., Meyer, A., Edwards, S. V., Pääbo, S., et al. 1989. Dynamics of mitochondrial DNA evolution in animals: Amplification and sequencing with conserved primers. *Proc. Natl. Acad. Sci. USA* 86:6196–200
41. Kort, J., Ross, H., Rumpenhorst, H. J., Stone, A. R. 1977. An international scheme for identifying and classifying pathotypes of potato cyst-nematodes *Globodera rostochiensis* and *G. pallida*. *Nematologica* 23:333–39
42. Krall, E. L., Krall, H. A. 1978. The revision of plant nematodes of the family Heteroderidae (Nematoda: Tylenchida) by a comparative ecological method of studying their phylogeny. In *Printsipy i methody izucheniya vzaimootnoshenii mezhdu parazitichenskimi nematodami i rasteniyami. Phytohelminthol. Proc. Akad. Nauk. Estonoski SSR. Tartu*, pp. 39–56
43. Luc, M., Maggenti, A. R., Fortuner, R. 1988. A reappraisal of Tylenchim (Nemata). The family Heteroderidae Filip'ev and Schuurmans Stekhoven, 1941. *Rev. Nematol.* 11:159–76
44. Marks, R. J., Fleming, C. C. 1982. The use of isoelectric focusing as a tool in the identification and management of potato cyst nematode populations. *Bull. OEPP/EPPO Bull.* 15:289–97
45. Matthews, H. J. P. 1971. Morphology of the nettle cyst nematode, *Heterodera urticae* Cooper, 1955. *Nematologica* 16:503–10
46. Meyer, A., Kocher, T. D., Basasibwaki, P., Wilson, A. C. 1990. Monophyletic origin of Lake Victoria cichlid fishes suggested by mitochondrial DNA sequences. *Nature* 347:550–53
47. Mickevich, M. F., Johnson, M. S. 1976. Congruence between morphological and allozyme data in evolutionary inference and character evolution. *Syst. Zool.* 25:260–70
48. Mickevich, M. F., Mitter, C. 1981. Treating polymorphic characters in systematics: A phylogenetic treatment of electrophoretic data. See Ref. 28a, pp. 45–58
49. Miller, L. I. 1983. Diversity of selected taxa of *Globodera* and *Heterodera* and their interspecific and intergeneric hybrids. See Ref. 70a, pp. 207–20

50. Mulvey, R. H. 1972. Identification of Heterodera cysts by terminal and cone top structures. *Can. J. Zool.* 50:1277–92
51. Nei, M., Li, W-H. 1979. Mathematical model of studying genetic variation in terms of restriction endonucleases. *Proc. Natl. Acad. Sci. USA* 76:5269–73
52. Netscher, C. 1983. Problems in the classification of *Meloidogyne* reproducing by mitotic parthenogenesis. See Ref. 70a, pp. 197–206
53. Patterson, C. 1988. Introduction. In *Molecules and Morphology in Evolution: Conflict or Compromise,* ed. C. Patterson, pp. 1–22. Cambridge: Cambridge Univ. Press
54. Pielou, E. C. 1984. *The Interpretation of Ecological Data. A Primer on Classification and Ordination.* New York: Wiley. 263 pp.
55. Platzer, E. G. 1981. Potential use of protein patterns and DNA nucleotide sequences in nematode taxonomy. In *Plant Parasitic Nematodes,* ed. B. M. Zuckerman, R. A. Rhode, 3:3–21. London/New York: Academic
56. Powers, T. O., Sandall, L. J. 1988. Estimation of genetic divergence in *Meloidogyne* mitochondrial DNA. *J. Nematol.* 20:505–11
57. Pozdol, R. F., Noel, G. R. 1984. Comparative electrophoretic analysis of soluble proteins from *Heterodera glycines* races 1–4 and three other *Heterodera* species. *J. Nematol.* 16:332–40
58. Radice, A. D., Powers, T. O., Sandall, L. J., Riggs, R. D. 1988. Comparisons of mitochondrial DNA from the sibling species *Heterodera glycines* and *H. schachtii. J. Nematol.* 20:443–50
59. Radice, A. D., Riggs, R. D., Huang, F. H. 1988. Detection of intraspecific diversity of *Heterodera glycines* using isozyme phenotypes. *J. Nematol.* 20:29–39
60. Riggs, R. D., Rakes, L., Hamblen, M. L. 1982. Morphometric and serologic comparisons of a number of populations of cyst nematodes. *J. Nematol.* 14:188–98
61. Riggs, R. D., Schmidt, D. P. 1988. Complete characterization of the race scheme for *Heterodera glycines. J. Nematol.* 20:392–95
62. Rollinson, D., Walker, T. K., Simpson, A. J. G. 1986. The application of recombinant DNA technology to problems of helminth identification. *Parasitology* 91:S53–S71
63. Sassar, J. N. 1980. Root-knot nematodes: A global menace to crop production. *Plant Dis.* 64:36–41
64. Shepherd, A. M., Clark, S. A., Dart, P. J. 1972. Cuticle structure in the genus *Heterodera. Nematologica* 18:1–17
65. Sneath, P. H. A., Sokol, R. R. 1973. *Numerical Taxonomy.* San Francisco: Freeman. 573 pp.
66. Stone, A. R. 1973. *Heterodera pallida* n. sp. (Nematoda: Heteroderidae), a second species of potato cyst nematode. *Nematologica* 18:591–606
67. Stone, A. R. 1975. Head morphology of second-stage juveniles of some Heteroderidae (Nematoda: Tylenchoidea). *Nematologica* 21:81–88
68. Stone, A. R. 1977. Recent developments and some problems in the taxonomy of cyst-nematodes, with a classification of the Heteroderoidea. *Nematologica* 23:273–88
69. Stone, A. R. 1977. Cyst nematodes—most successful parasites. *New Sci.* 10:355–56
70. Stone, A. R. 1983. Three approaches to the status of a species complex, with a revision of some species of *Globodera* (Nematoda: Heteroderidae). See Ref. 70a, pp. 221–33
70a. Stone, A. R., Platt, H. M., Khalil, L. F., eds. 1983. *Concepts in Nematode Systematics.* London/New York: Academic
71. Sturhan, D. 1983. The use of the subspecies and the superspecies categories in nematode taxonomy. See Ref. 70a, pp. 41–53
72. Triantaphyllou, A. C. 1966. Polyploidy and reproduction in the root-knot nematode *Meloidogyne hapla. J. Morphol.* 118:403–14
73. Triantaphyllou, A. C. 1975. Oogenesis and the chromosomes of twelve bisexual species of Heterodera (Nematoda: Heteroderidae). *J. Nematol.* 7:34–40
74. Triantaphyllou, A. C. 1975. Genetic structure of races of *Heterodera glycines* and inheritance of ability to reproduce on resistant soybeans. *J. Nematol.* 7:356–64
75. Triantaphyllou, A. C. 1979. Cytogenetics of root-knot nematodes. In *Root-knot Nematodes (Meloidogyne species): Systematics, Biology and Control,* ed. F. Lamberti, C. E. Taylor, pp. 85–109. London/New York: Academic
76. Triantaphyllou, A. C. 1983. Cytogenetic aspects of nematode evolution. See Ref. 70a, pp. 55–71
77. Triantaphyllou, A. C. 1985. Cytogenetics, cytotaxonomy and phylogeny of rootknot nematodes. In *An Advanced Treatise on Meloidogyne, Biology and Control,* ed. J. N. Sasser, C. C. Carter, 1:113–26. Raleigh: NC State Univ. Graphics

78. Triantaphyllou, A. C. 1987. Cytogenetic status of *Meloidogyne (Hypsoperine) spartinae* in relation to other *Meloidogyne* species. *J. Nematol.* 19:1–7
79. Trudgill, D. L., Carpenter, J. M. 1971. Disc electrophoresis of proteins of *Heterodera* species and pathotypes of *H. rostochiensis*. *Ann. Appl. Biol.* 69:35–41
80. Trudgill, D. L., Parrott, D. M. 1972. Disc electrophoresis and larval dimensions of British, Dutch and other populations of *H. rostochiensis*, as evidence of the existence of two species, each with pathotypes. *Nematologica* 18:141–48
81. Wayne, R. K., Van Valkenburgh, B., Kat, P. W., Fuller, T. K., Johnson, W. E., O'Brien, S. J. 1989. Genetic and morphological divergence among sympatric canids. *J. Hered.* 80:447–54
82. Wharton, R. J., Storey, R. M. J., Fox, P. C. 1983. The potential of some immunological and biochemical approaches to the taxonomy of potato cyst-nematodes. See Ref. 70a, pp. 235–48
83. Wiley, E. O. 1981. *Phylogenetics. The Theory and Practice of Phylogenetic Systematics*. New York: Wiley. 439 pp.
84. Woese, C. R. 1988. Macroevolution in the microscopic world. See Ref. 53, pp. 177–202
85. Wouts, W. M. 1973. A revision of the family Heteroderidae (Nematoda: Tylenchida). *Nematologica* 19:279–84
86. Wouts, W. M. 1985. Phylogenetic classification of the family Heteroderidae (Nematoda: Tylenchida). *Syst. Parasitol.* 7:295–328
87. Wouts, W. M., Sher, S. A. 1971. The genera of the subfamily Heteroderinae (Nematoda: Tylenchoidea) with a description of two new genera. *J. Nematol.* 3:129–44

Annu. Rev. Phytopathol. 1991. 29:109–36

CITRUS GREENING DISEASE

J. V. da Graça

Department of Microbiology and Plant Pathology, University of Natal, Pietermaritzburg, South Africa

KEY WORDS: prokaryotic plant pathogens, phytobacteriology, psylla vectors of plant disease, biological control

---

INTRODUCTION

Citrus greening disease is a major cause of crop and tree loss in many parts of Asia and Africa. Before it was identified as one disease, it became known by various names: yellow shoot (huanglungbin) in China; likubin (decline) in Taiwan; dieback in India; leaf mottle in the Philippines; vein phloem degeneration in Indonesia; and yellow branch, blotchy-mottle, or greening in South Africa. As it became clear that all these were similar diseases the name "greening" was widely adopted.

Many reviews of citrus greening have appeared, but with few exceptions (9,186), are either brief, restricted to one country, or by now out of date. This review aims to present an overview of greening worldwide. There is no shortage of literature; Ôtake (186) lists 556 papers in his 1990 bibliography, and the present author found a further 86, although many do not advance our understanding of the disease.

HISTORY AND GEOGRAPHICAL DISTRIBUTION

Although citrus dieback was documented in India in the eighteenth century (192), this disease may not have been greening-induced decline. Indian dieback was first accurately described in 1929 and attributed to poor drainage (190). Yellow shoot disease was well known in south China in the 1890s

0066-4286/91/0901-0109$02.00

(278), and likubin was identified in Taiwan 60 years ago as a nematode-associated problem (186). The first mention of it in English may have been in 1919 when Reinking (196) described a yellowing and a leaf mottle of citrus in southern China; by 1935 it had become a serious problem there (132). In 1921 Lee (129) described mottle leaf disease in the Philippines, which he likened to a disease in California, probably stubborn, and attributed the former to zinc deficiency. It was not a serious problem there until 1957 (143). In Indonesia the disease was noticed in the 1940s (16).

When the disease was first described in South Africa in 1937 (254), it was presumed to be a mineral toxicity. It was known as yellow branch in the western Transvaal province, where it had been observed in 1928/1929 (185), and greening in the eastern Transvaal.

Greening has now also been confirmed in the following countries: Bangladesh (53), Burundi (15), Cameroun (15), Central African Republic (125), Comoros (220), Ethiopia (247), Hong Kong (48), Japan (163), Kenya (230), Madagascar (125), Malawi (15), Malaysia (120), Mauritius (175), Nepal (119), Pakistan (118), Réunion (175), Rwanda (15), Saudi Arabia (25, 28), Somalia (J. M. Bové, personal communication), Swaziland (52), Tanzania (241), Thailand (223), Yemen (25, 28) and Zimbabwe (229). A greening-like disease occurs in Australia (29, 30), but the known insect vectors of greening are absent (30). The disease probably exists in other countries that border on the above, although surveys failed to confirm its presence in Gabon, where evidence of the insect vector was found, and Zambia and Namibia, where no signs were encountered (15).

No evidence of greening has been found in Brazil (10), despite the occurrence of one of the vectors. No reports of greening have been confirmed from Mediterranean areas of Europe, North Africa or the Middle East, or California, all areas where the symptomatically similar citrus stubborn disease occurs. The two diseases have so far not been found to occur together in one area.

ECONOMIC IMPORTANCE

Losses due to greening are not easy to assess. Sometimes only sectors of a tree are affected and losses are small, but in other cases the entire tree is infected and crop loss is total (155). No detailed loss studies have been published, but the severity of the disease is substantiated in the literature. In India Fraser et al (76) commented on catastrophic losses. In the Philippines greening affected an estimated 7 million trees in 1962 (146), was largely responsible for reducing the area planted to citrus by over 60% between 1961 and 1970 (3), and in 1971, killed over 1 million trees in one province (194). In Thailand up to 95% of the trees in the northern and eastern provinces are severely affected (23).

In Taiwan the disease has caused widespread citrus destruction (236), and in Indonesia not less than 3 million trees were destroyed between 1960 and 1970 (244), with groves in most regions of Java and Sumatra being abandoned by 1983 (201). Practically all sweet orange and mandarin trees in south-western Saudi Arabia have declined and disappeared, leaving only limes (25).

In South Africa the incidence was severe from 1932 to 1936, 1939 to 1946, and again after 1958 (185). Crop losses of 30–100% were recorded in some areas (212). Incidence in the late 1960s and early 1970s declined slightly (226), but its overall effect is still marked, with citrus production eliminated in three major production areas (170).

SYMPTOMOLOGY

Symptoms can occur throughout the tree, especially if the infection occurs at or soon after propagation (151). If infection occurs later the symptoms and the causal organism are often partially confined. Earlier reports suggesting that the organism's movement is restricted to new growth with little downward movement are challenged by recent findings that it can move 30–50 cm downwards in 12 months (S. P. van Vuuren, personal communication). Infected trees or branches suffer heavy leaf drop followed by out-of-season flushing and blossoming, with dieback occurring in severe cases (141).

In general leaf symptoms are of two types: Primary symptoms are characterized by yellowing of normal-sized leaves along the veins and sometimes by the development of a blotchy-mottle (209). With secondary symptoms the leaves are small, upright, and show a variety of chlorotic patterns resembling those induced by zinc and iron deficiencies. Analysis of symptomatic leaves shows a higher potassium content and lower calcium, magnesium, and zinc concentrations (7, 121).

Infected fruit are small, lopsided, and have a bitter taste (155), probably because of higher acidity and lower sugars (114). Many fall prematurely, while those that remain on the tree do not color properly, remaining green on the shaded side (155), hence the name of the disease. Any seeds in severely affected fruit are often abortive.

These descriptions are of greening in Africa. Other investigators described very similar symptoms in the Philippines (146), India (201), and China (278), but their papers contain descriptions of more extensive yellowing, dieback, and decline than are found in Africa. In China the disease was reported to kill young trees in 1–2 years (133).

Root systems of both forms are poorly developed with relatively few fibrous roots, possibly because of root starvation (9, 199). New root growth is suppressed and the roots often start decaying from the rootlets (278).

The causes of the symptoms are unknown. Amino acid concentrations are generally lower (70, 94, 240, 266) except for L-proline, which is higher in the leaves and flower buds (70). There are also increases in amylase activity and glucose and fructose production (149), increased respiration early in infection (265), and the appearance of gentisoyl-β-D-glucose (74, 211), which seems to inhibit peroxidase activity (257). These changes do not, however, appear to be involved in symptom production.

Cytopathic studies have revealed pockets of necrotic phloem, excessive phloem formation, abnormal cambial activity and accumulation of starch in plasmids (209, 242, 246). Cytoplasmic membranes formed from invaginations of the plasmalemma, aberration of the chloroplast thylakoids and mitochondrial collapse have also been observed (272). These changes indicate significant metabolic disturbances, possibly via toxins and/or hormones.

Often citrus trees are infected by more than one pathogen, with consequent influence on symptom expression. In the Philippines (141), India (21), and Taiwan (107), plants infected with both greening and seedling yellows (tristeza) develop more severe leaf symptoms, stunting, and increased plant collapse. Such dual infection is common as tristeza is widespread in Africa and Asia, and was previously thought to cause yellow shoot in China (65). Furthermore, it has been noted that *Fusarium* spp. (136, 232), *Colletotrichum gloeosporioides,* and *Diplodia natalensis* (136) augment the symptoms of greening.

TYPES AND STRAINS

Although Asian greening symptoms are more severe than African, they can also be clearly distinguished on the basis of temperature tolerance. With African greening, severe symptoms were obtained under cool glasshouse conditions (22°C for 8 h nights, 24°C for 16 h days), whereas no symptoms appeared at 27–30°C (27). The symptoms of the Asian form are as pronounced at both temperatures. The higher temperature for an extended period appears to completely inactivate the African form (124).

This difference relates to their areas of natural occurrence. In South Africa, leaf symptoms are more pronounced in the cool areas than in the low-lying hot areas, and are more pronounced in winter (215). The most strongly developed symptoms are found at an altitude of 900 m, whereas very poor symptoms occur at or below 360 m (221). In Kenya greening is found only above 700 m, and in the Arabian peninsula the African form occurs in the higher areas of Yemen, and the Asian form in the lower altitudes of neighboring Saudi Arabia (28). Réunion and Mauritius possess both forms, similarly occurring at different altitudes (19).

Within the African type some strain differences are discernible. Isolates from the central Transvaal in South Africa have a higher graft-transmission rate than those from the eastern Transvaal (219). An isolate from one area induced a blotchy-mottle whereas two others caused zinclike deficiencies (S. P. van Vuuren, personal communication). Strains of the Asian form have also been reported. In India different isolates showed varying degrees of virulence when grafted onto a common host: These isolates were grouped as mild, severe, and very severe (176). The mild strains offered only partial cross protection against the severe. Monoclonal antibodies raised against greening from Poona, India, failed to detect greening from China, Thailand, Malaysia, and other parts of India (86), but did react with an African greening sample (85).

VARIETAL SUSCEPTIBILITY AND HOST RANGE

Greening in South Africa is primarily a disease of sweet orange *(Citrus sinensis),* with valencias showing more pronounced leaf symptoms than navels (185). It is also particularly severe on mandarins *(C. reticulata)* and tangelos *(C. sinensis* × *C. reticulata),* but less so on lemon *(C. limon).* The least affected is the acid lime *(C. aurantifolia).* In Taiwan (162), India (181), and the Philippines (93) sweet orange and mandarin are the most susceptible, with lime, lemon, sour orange *(C. aurantium),* and grapefruit *(C. paradisi)* more tolerant. In India the rough lemon *(C. jambhiri),* sweet lime *(C. limettoides)* and pomelo *(C. grandis)* are tolerant and the trifoliate orange *(Poncirus trifoliata)* is fairly tolerant (181). As with African greening, some citrus selections are more affected by Asian greening than others—Blood Red sweet orange is more susceptible than Hamlin orange (72).

However, some differences have been reported. In India the widespread use of Kagzi lime *(C. aurantifolia)* as an indicator for greening (180) may have been effective because of co-infection with tristeza. In Australia the greening-like disease is most severe on grapefruit and sour orange (75) and in Taiwan the Wentan pomelo, once regarded as tolerant, has been suffering from a decline attributable to the likubin agent (106, 239).

Manicom & van Vuuren (140) have grouped the common citrus cultivars according to their general reaction to greening as follows: Severe (sweet orange, tangelo, mandarin), Moderate (grapefruit, lemon, sour orange) and Tolerant (lime, pomelo, trifoliate orange). That a particular type is not severely affected by greening does not necessarily reduce its significance—lemons probably serve as important reservoirs of infection because their more frequent flushes of new growth make them attractive to the insect vector (100).

In some cases the rootstock can affect symptom expression. Five out of 23 rough lemon rootstock selections in India induced a degree of tolerance in the sweet orange scion in greenhouse trials (59). In another study 100% of trees on rough lemon were infected, compared to only 25% of trees on "Blood Red" sweet orange rootstock (113). In South Africa the percentage of greening in valencias was higher on trifoliate orange rootstock than on Empress mandarin and Troyer citrange. Possibly the trifoliate rootstock causes an extension of the flushing period and thus extends the feeding time of the insect vector (263). However, no differences were found in a Chinese study on the effects of 13 rootstocks on symptoms in Ponkan mandarin (133).

Greening has been experimentally transmitted to many *Citrus* spp., the trifoliate orange (155) and to other rutaceous plants, causing leaf symptoms in kumquat (*Fortunella* sp.) (170), stunting (257), small leaves, and yellowing in *Murraya paniculata* (9), and stunting in *Atalantia missionis* and *Swinglea glutinosa* (240). These and other species may be natural hosts as the insect vectors of greening feed on them (18, 55, 173, 255). While the vector in Africa prefers citrus possibly because of its softer flush leaves (173), the vector in Asia displays a preference for *Murraya* spp. (55), which tend to flush all year round (188), but survives on *Clausena* and *Atalantia* (277). Moran (172) suggests that *Vepris undulata, Clausena anisata,* and *Zanthoxylum* (f. *Fagara*) *capense* are the original hosts of the vector in Africa. Efforts to transmit African greening to these plants by the vector were unsuccessful (J. V. da Graça and S. P van Vuuren, unpublished data). Other indigenous hosts may exist since the vector can feed on several nonrutaceous plants (250).

Greening has been experimentally transmitted by dodder to periwinkle *(Catharanthus roseus)* in which it induced marked yellowing (82, 116). As in its citrus hosts, African greening in periwinkle required temperatures below 27°C for symptom development, whereas the Asian form was more heat-tolerant. The dodder itself is a host for the greening pathogen with titers higher than has been observed in citrus (88, 116). Attempts in China to transmit the disease via dodder and the insect vector to cucumber, cabbage, tobacco and *Solanum nigrum* were unsuccessful (116).

TRANSMISSION

In 1943, Chen (64) suggested on the basis of graft inoculations that yellow shoot may be a viral disease. Similar opinions were soon expressed in South Africa (101, 150), strengthened by the finding in grafting trials that greening was inconsistently transmitted to healthy plants (184). Meanwhile, a report by Lin (132) in China confirmed that yellow shoot was indeed graft-transmissible. Graft-transmissibility of African greening was confirmed in 1965 by McClean & Oberholzer (154). The pathogen does not readily pass to progeny trees propagated by buds from infected trees (151), possibly because of necrosis of sieve tubes (154) and uneven distribution of the pathogen (105), but more transmission occurs if stem pieces are used. No infection could be obtained when material from apparently healthy sectors of diseased trees were used. Schwarz (218) reported a higher graft-transmissibility rate in winter.

In 1964 Schwarz (210) reported that seedlings exposed to insects in a greening-infected orchard developed yellowing symptoms similar to greening. McClean & Oberholzer (154) also noted that greening appeared to spread in the field. These investigators then placed insects from diseased trees on healthy seedlings, and found that only adults of the citrus psylla species, *Trioza erytreae*, transmitted greening (153). Schwarz et al (224) later showed a positive correlation between the degree of greening infection, the number of psylla, and the rate of transmission.

Research on Asian greening followed close behind. In 1966 Salibe & Cortez (199) demonstrated graft transmission in the Philippines, and reported on an insect vector, soon identified there (144, 145, 200) and in India (40) as another psylla, *Diaphorina citri*.

The number of adult psylla of either species in a population that carry the disease is relatively small (152, 278), but under experimental conditions, a single adult of either species can transmit greening (52, 192).

The pathogen can be acquired by *D. citri* in 15–30 min (184) with a latent period of 8–12 days (192). One hour or more is required for 100% transmission. *T. erytreae* acquires the organism after one day of feeding and transmits greening 7 days later (S. P. van Vuuren, personal communication), and can infect with an exposure time of less than 1 h (261). Longer feeds can render the psylla more infective. Long overwintering feeding on old leaves makes adults highly infective on young flush in spring (47). Psylla are strongly attracted by yellow green of wavelength 550 nm (202), making diseased trees attractive targets and thereby increasing the proportion of disease-carrying insects.

Although nymphs arc rcportedly unable to transmit greening (40, 154, 192), the fourth and fifth instars of *D. citri* can acquire the pathogen and adults from such nymphs transmit the disease (41, 278). Nymphs of *T. erytreae* can also acquire the greening agent (S. P. van Vuuren, personal communication).

Analysis of the spread of greening in orchards in Réunion, China, and the Philippines has shown that epidemics appear to follow a sigmoid curve (95–97). Clustering of diseased trees was observed, with a higher aggregation correlated in Réunion to analysis of rows running in the direction of prevailing winds, which presumably aid vector movement. In China, a north-south aggregation corresponded to general orchard traffic and closer tree spacing in that direction (96).

T. erytreae is sensitive to heat. In laboratory tests, high temperatures (a 32°C plateau) killed all stages, 27°C allowed rapid development of the insect but with a 52% mortality, whereas at 21°C 91% survived (174). In the field, populations were consistently higher in cool, moist upland areas of southern Africa, and always low in hot, drier lowlands, with eggs and first instars being particularly vulnerable to heat and desiccation (46). The prolonged flushing of trees in the cooler areas is more favorable to the insect (44, 49). The mortality of the developing stages was over 70% when the saturation deficit index (SDI), a value based on regression curves depicting the combined effects of temperature and humidity on egg and first instar mortality, is over 35 mb (98). These values were used to explain the distribution of *T. erytreae* in South Africa, connecting infestations of the insect with outbreaks of greening in past years (52). Since it is adult psylla that mainly transmit greening and are more tolerant to weather extremes, SDI is unsuitable, in Samways' view (204), for early detection of population upsurges. He recommends instead the use of sticky yellow traps to monitor numbers. Such traps, together with trap seedlings, were used to monitor psylla populations and greening transmission in an orchard for three years (262). High populations in 1976 were associated with high greening transmission, while the hotter, drier years that followed had both low populations and low transmission.

D. citri has a similar biology, but it is more resistant to extremes of temperature (49) and more sensitive to high rainfall and humidity (10). Since it prefers hot dry weather to cold wet conditions, high populations are found in early spring and summer, and dramatically lower numbers during high rainfall in spring (195). *D. citri* can withstand low temperatures for short periods, with a 45% survival at −3°C for 24 h under natural conditions and 39% survival at −5°C for 24 h under experimental conditions (274).

In Réunion and Mauritius both vectors occur, *T. erytreae* above 500 m and *D. citri* below 400 m where it is hotter (48, 51). Similarly in the Arabian peninsula, *D. citri* occurs in the low-lying citrus areas of Saudi Arabia and *T. erytreae* above 1000 m in neighboring Yemen (25). However, in Ethiopia only *T. erytreae* has been recorded (1).

There does not appear to be any specificity between the psylla species and the greening type. In Réunion and Mauritius the African form is spread by both vectors (48, 51). Indian greening has been experimentally transmitted by *T. erytreae* (147) and African greening by *D. citri* (127).

In the South African western Cape province greening disease does not spread despite the presence of the psylla (252). In the laboratory, trapped specimens from this area were able to transmit greening and breed with Transvaal specimens. The absence of greening transmission in this area may be influenced by climatic factors, this being a winter rainfall area with hot, dry summers.

In an ecological study, Samways & Manicom (206) found that adult *T. erytreae* invaded an orchard in exponentially increasing numbers; on an area basis 4% of new trees were infested per day. Mean population density is not significantly different for managed citrus orchards, neglected orchards, and natural bush (202).

Other psylla species have been noted on citrus, namely *T. eastopi (T. litseae)* and *Mesohomatoma lutheri* in Réunion (17), *D. communis* in India (10), *T. citroimpura, Psylla citrisuga,* and *P. citricola* in China (268a), *D. auberti* and *D. amoena* in Comoros, (10, 16) and *D. punctulata* and *D. zebrana* in Swaziland (52), but there is as yet no evidence that any transmit greening.

The only other means of transmission known is by dodder (*Cuscuta* spp.). *C. reflexa* was used to transmit the pathogen from citrus to citrus (190), and *C. campestris* for transmission from citrus to periwinkle (82, 116), but *C. japonica* failed to transmit (116).

NATURE OF THE CAUSAL AGENT

The demonstrations that greening is a graft- and insect-transmissible disease led to the conclusion that a virus was responsible. In China some researchers believed tristeza virus to be the cause (65, 131, 149), but Lin (135) publicly disagreed. In South Africa it was shown that tristeza and greening could readily be distinguished since the aphid *Toxoptera citricidus* transmitted tristeza but not greening, and psylla vice versa (153).

In 1970 Laflèche & Bové (125, 126) reported observing mycoplasma-like organisms (MLOs) in citrus phloem tissue infected with African and Indian greening. These organisms measured 100–200 nm in diameter, with filamentous forms up to 2 μ long. The phloem restriction of the greening pathogen was supported by the observation that girdling prevented spread within a plant (142).

Apparently identical organisms were soon observed in the hemolymph and salivary glands of *T. erytreae* (167), the phloem of yellow shoot (278) and likubin-infected citrus (61, 243), and in infective *D. citri* (63). Their identification as MLOs was soon questioned because, whereas MLOs are surrounded by a true unit membrane 10 nm thick, the greening organism has an outer envelope 20 nm thick (198). Comparative electron microscope studies on the structures of the greening organism and several other types of prokaryote clearly showed that it was not an MLO (84, 168). Garnier & Bové (81) suggested that it should be classified as a true bacterium, possibly belonging to the Gracilicute division of the prokaryotes (26). Although no distinct R-layer of peptidoglycan (PG) had been observed (168, 242), the inner layer of the outer membrane was somewhat thicker in places, suggesting the presence of an R-layer (81). Treatment of infected plants with penicillin G, which inhibits a late step in PG synthesis, caused symptom remission and the disappearance of organisms from the phloem (26). In a comparable cytochemical treatment study with the greening organism in periwinkle, and with gram negative and gram positive bacteria, *Escherichia coli,* and *Staphylococcus aureus,* respectively, Garnier et al (83) showed that papain caused a visible R-layer to separate from the outer membrane of both *E. coli* and the greening organism. The R-layer disappeared after lysozyme treatment. They concluded that the greening organism is gram negative.

The greening organism's resistance to culture on artificial media has made study of the organism difficult. Some isolates from Indian greening-infected citrus have been reported (89, 90, 179, 235), but all had the characteristics of true mycoplasmas and were therefore probably contaminants. Although some isolates induced foliar symptoms in inoculated citrus (179, 235), various microorganisms are known to cause transient leaf blotches (123, 139). Initial attempts to isolate the African greening organism were unsuccessful (164).

The presence of other bacteria in healthy (78) and greening-infected citrus vascular tissue, and in psylla (24) impedes attempts to isolate the greening organism.

In 1984 Garnett (79) reported the isolation of a long rod-shaped gram negative organism from African greening-infected citrus leaf mid-ribs. The ultrastructure of this organism was described as similar to that of the organisms observed in greening-infected citrus, periwinkle, and insect vectors (4). On a solid medium, it formed small round colonies with predominantly long rod-shaped cells near the edges, but rounder cells in the oldest parts. Antibodies raised against this isolate reportedly gave positive ELISA results only with greening-infected plant material (73). Gold-labeled antibodies reacted only with the cell wall of a bacterium resembling the observed greening organism in infected citrus phloem and extracted sap (5).

Garnett and others have established a collection of greening cultures in plants from South Africa, Réunion, China, Taiwan, India, and the Philippines for comparative studies in quarantine facilities in the USA (130). Isolates from these plants share common serological and protein profiles (80), and a high percentage of DNA homology (103). Inoculated plants develop foliar symptoms, but completion of Koch's postulates has not been reported.

Other research groups in South Africa have not been able to confirm the above results. Manicom (139) isolated several species but none resembled the greening organism. Chippindall (personal communication) has isolated a bacterium that on mechanical inoculation to tangelo induces foliar symptoms. However, this bacterium could not be graft- or psylla-transmitted to other plants, no gentiosyl-β-glucoside was present, and the symptoms later disappeared. This organism has been identified as a *Clavibacter* sp., and although it is not the greening organism, may, in the opinion of Chippindall & Whitlock (67), be a component of the overall disease syndrome. Labuschagne et al (123) isolated several bacterial species in South Africa, of which one, identified as *Acinetobacter lwoffii*, induced transient yellowing symptoms (234). Serological tests show it to be present in both healthy and African greening-infected trees, but of higher incidence in the latter, especially when foliar mottling occurs (208). It also reacted positively with infected material from the Philippines. *A. lwoffii* was one of the species isolated from citrus in Florida (78) where greening does not exist. Clearly, *A. lwofii* is not the greening organism, but it may play a role in the overall greening syndrome.

In an effort to confirm the isolation claim by Garnett (79), Garnier et al (85) found that antibodies against the cultured organism failed to react with greening-infected tissue in their laboratory, and that a chemically fixed culture failed to react with monoclonal antibodies raised against greening-infected phloem (268a). These researchers raised monoclonal antibodies against infected phloem tissues (85) and used them to purify the organism from infected periwinkle plants (86). Filamentous and round cells have been trapped in this way.

DETECTION

The use of indicator seedlings was the first diagnostic test developed. In South Africa either valencia sweet orange or Orlando tangelo are used (213). In Taiwan Ponkan mandarin is preferred (148), while in India the Kagzi lime (see above; 21, 179), has been replaced by the Mosambi sweet orange or the Darjeeling orange, a mandarin type (2). Graft sticks are grafted into the indicator stems, using about 20 seedlings per test: Symptoms appear in 3–4 months in a greenhouse at 21–23°C (213).

A quicker laboratory test is always desirable, and since the causal bacterium had not been isolated, attention turned to biochemical changes in infected plants. Schwarz (214) reported that a fluorescent substance, later identified as gentisoyl-β-glucoside (74), was detectable in ether extracts of greening-infected bark, but not in healthy ones. When separated by paper or thin layer chromatography using n-butanol:acetic acid:water (5 : 1 : 1) (214) or chloroform-methanol (223), a bright violet-blue spot fluoresced under ultraviolet light (365 nm). The same fluorescence can be observed in the albedo of greening-affected fruit (214), and was found to be reliable for early detection (217). The test is not absolutely specific for greening, however, as Californian stubborn-infected material also contained such a marker (214). The marker is detectable in all sweet oranges and mandarins, most tangelos and lemons, but only sporadically in grapefruit (111, 216).

In South Africa its concentration was reported to vary with seasons but not enough to affect its diagnostic use (218), whereas in India the concentrations were high all year (231). Flower buds were found to be as suitable as bark for extraction (57).

The test has been modified by using water extraction by centrifugation instead of ether (227), by reducing the volume of water to eliminate the lengthy evaporation step (260), and employing acid hydrolysis to liberate gentisic acid, which produces a clearer fluorescent spot (39). Gentisic acid has since been shown to be present in many healthy citrus species and cultivars, but its concentration increases significantly in some as greening symptoms develop (258).

Diagnostic tests using light microscopy have been developed in China. Sections of infected leaves are examined either by fluorescent microscopy, which reveals a yellow fluorescence in infected phloem that is absent from healthy, virus-infected and nutrient-deficient tissues (270), or by staining sections with safranin, which shows red patches in infected phloem (271).

Serological tests are being developed. Using infected plant tissues, French researchers have raised monoclonal antibodies against Indian and African greening that recognized greening isolates from India, the Philippines, and Réunion (85), but not ones from China (77), Thailand, Malaysia, or other parts of India (86), while monoclonals against a Chinese strain did not react with Indian greening (86). This group has now raised monoclonals against several Asian and African isolates (J. M. Bové, personal communication), which in combination may have potential universal use. In South Africa, monoclonals have also been prepared using extracts of greening-transmitting psylla that react positively with infected citrus (122), as does a polyclonal against infected citrus tissue (66).

Another diagnostic technique in France is the use of DNA probes prepared from cloned DNA from infected periwinkle (86), and from greening organism cells trapped by immunoaffinity chromatography (267). Research into developing probes from the cultured putative greening organism has begun in Australia (103).

CONTROL

Control of the Greening Pathogen

THERMOTHERAPY In 1964 Lin (134) reported eliminating yellow shoot disease by water-saturated hot air treatment of graftwood at 48–58°C with no loss of tissue viability. Other attempts to eliminate Asian greening with heat have had varying degrees of success. In India treatment of budwood at 47°C for 2 h reduced disease incidence (56) and longer treatments eliminated the pathogen (178). Treatment of infected young plants or seedlings budded with infected tissue at 38–40°C for three to four weeks also killed the pathogen (104, 178).

In South Africa budwood from infected trees heated over a hot water bath at

51°C for 1 h, 49°C for 2 h and 47°C for 4 h eliminated the disease although some tissue viability was lost at the higher temperatures (222). Infected trees covered for 2–5 months with polyethylene-covered fiberglass sheets showed a dramatic decrease in the number of diseased fruit. However this method is impractical for large-scale use.

Heat treatments have some application in the elimination of greening from horticulturally desirable trees, possibly in conjunction with shoot tip grafting (238).

CHEMOTHERAPY The association of prokaryotic organisms with greening prompted investigations into the use of antibiotics. Martinez et al (143) in the Philippines suppressed leaf symptoms with a foliar spray of tetracycline hydrochloride and apparently eliminated greening in budwood by immersion in the solution. In India immersion in a penicillin-carbendazin dip reportedly gave complete control (60). In South Africa Schwarz & van Vuuren (228) injected various tetracyclines into the trunks of greening-infected sweet orange trees and found that tetracycline hydrochloride gave the best results, reducing fruit symptoms in the next crop from 60 to 20%. The incidence of infection remained unchanged for three years when a severe psylla outbreak and a slight increase in greening occurred (169). The trees received a second treatment that reduced fruit symptom incidence to below 10% (171). Better results were achieved when two injections were given one month apart (259), while a 97% reduction of fruit symptoms was achieved when trees were injected continuously under pressure for 7 days (264). The best time for injection is spring (226), and distribution in the tree can be enhanced by the use of hyaluronidase (233). Trunk injections of tetracycline hydrochloride have also been successful in Taiwan (68), China (278), Réunion (13), and the Philippines (143).

Other antibiotics have been tested. Penicillin had a less suppressive effect than tetracycline hydrochloride in Asia and Réunion (13, 14, 161, 278), but was totally ineffective in South Africa (225). In India partial to total remission occurred following injection with a compound called BP-101 (Hindustan Antibiotics Ltd.) (42, 43, 190), ledermycin (demeclocycline hydrochloride) and a streptomycin-chlortetracycline mixture (178), while complete remission using foliar sprays of agrimycin and carbendazin has also been reported (58, 112).

However, tetracycline hydrochloride is phytotoxic. Buds immersed in concentrations of over 250 ppm did not survive after grafting (161). Van Vuuren (259) observed leaf narrowing, lamina yellowing, veinal browning, and occasional defoliation of individual twigs. A brown discoloration was found in the trunk in a zone extending some 300 mm from the hole, decreasing in intensity with distance. Leaf narrowing has also been recorded in

Taiwan with oxytetracycline, but not with tetracycline hydrochloride or chlortetracycline (68).

A derivative of tetracycline hydrochloride, N-pyrrolidinomethyl tetracycline (PMT) is more soluble in water, giving slightly better control of greening, and causing no foliar phytotoxicity and minimal wood discoloration (35). Midsummer treatments gave the best results (37). The solubility of PMT in water is 1250 mg/ml, compared to 10.8 mg/ml for tetracycline hydrochloride, so that smaller volumes can be injected from syringes without employing high pressures (99).

Moll (165) was unable to detect by fluorimetry any antibiotic residues in the juice of injected trees one month after treatment. Other studies using growth inhibition of *Bacillus megaterium* (36, 37), *B. cereus* var. *mycoides* (68) or *B. subtilis* (13), have confirmed that antibiotic residues drop rapidly.

BREEDING FOR RESISTANCE In South Africa various hybrids of sweet oranges and Tahiti lime have been produced, and are being tested for possible resistance (71). The use of callus tissue and protoplasts is also being evaluated. Peroxidase activity, which is higher in more tolerant varieties, may serve as a marker for tolerance to greening in screening new hybrids (256).

Control of the Vectors

USE OF INSECTICIDES Because psylla are sap feeders, systemic insecticides are the most effective. Those most commonly used in southern Africa are endosulfan sprays (187); monocrotophos applied as sprays (31), aerial sprays (109), trunk applications (38), and trunk injections (225); and dimethoate applied to the soil (158, 159, 269). Weekly sprays of monocrotophos did not prevent infection (32), while high volume and mist blower applications destroy natural enemies of other citrus pests (33). A single application of dimethoate by microject and drip irrigation gave control for over 7 weeks, and two applications gave season-long protection (159). Simultaneous injection of dimethoate and tetracycline hydrochloride controlled psylla without reducing antibiotic efficacy (160).

In India, *D. citri* has been controlled for up to 4 weeks with sprays of several insecticides, including dimethoate and monocrotophos (22, 117) and soil applications of dimethoate (193). In China field control with soil applications of O-methoate has been reported (272).

However insecticides can interfere with biological control of other pests. While dimethoate can reduce the efficacy of the biological of citrus red scale (102), at 0.01% a.i. it does not seriously affect the activity of psylla parasites (47). Another problem for small farmers is cost; one Indonesian farmer reportedly spent $2500 p.a. to control psylla (268).

BIOLOGICAL CONTROL The major parasite of *T. erytreae* in South Africa is a species of parasitoid wasp *Tetrastichus* that oviposits in the psylla nymphs (255), and appears to limit psylla populations (46).

A survey in Réunion in 1973 found no parasitized mummies of psylla (51), and it was suggested that parasitic wasps should be introduced (6). Three species, *Tetrastichus radiatus* (now called *Tamarixia radiata*) from India, and *Tetrastichus dryi* and *Psyllaephagus pulvinatus* from South Africa, were collected and mass reared (14). *T. dryi* was released in 1976, and had dramatically reduced the *T. erytreae* population by 1978 (17, 18), with complete eradication by 1982 (19). *P. pulvinatus* failed to establish itself. *T. radiata,* which is specific for *D. citri,* was introduced in 1978 (17) and by 1982 had virtually eliminated psylla from commercial orchards, surviving only on *M. paniculata* hedges (189). The incidence of greening has consequently been reduced (18).

Both species of wasp were subsequently introduced into Mauritius where they reduced psylla populations (13), but more slowly than in Réunion, possibly because they were not mass reared before release (199). *T. radiata* has also been introduced into Taiwan (237), the Philippines (87), Nepal (128), and Indonesia (183), but without much apparent success.

Other parasitic wasps of *D. citri* are *Diaphorencyrtus aligarhensis* (16, 69), *Psyllaephagus* sp. (16, 137), *Chartocerus walkeri,* and *Encarsia* (243a), but none appears to be a potential biological control agent.

The efficacy of parasitic wasps is limited by the activity of hyperparasitic wasps. In South Africa the effect of *Aphidencyrtus cassatus* varies according to its ability to maintain synchrony with its host (45). Another species is *Cheiloneurus cyanonotus* (157). In Taiwan *D. aligarhensis* has at least 10 hyperparasites, of which one, *Pachyneuron,* can also oviposit in *T. radiata* (69). In China a species of *Tetrastichus* attacks both *T. radiata* and *D. aligarhensis* (243a). No hyperparasites occur in Réunion (16), which may in part explain the success of biological control there.

A survey of predators of psylla in South Africa showed that none of them reduce densities low enough for greening control (251).

Fungi provide another means of biological control. *T. erytreae* is attacked by *Cladosporium oxysporium* (205) and *Capnodium citri* (10), but both are sensitive to desiccation and are density-dependent. *D. citri* is parasitized by *Beauveria* (87) and *Cephalosporium lacanii* (275), the latter somewhat effective at high density.

OTHER CONTROL MEASURES The production of greening-free trees is vital. The use of shoot-tip grafted material can ensure this goal (110, 237). Production can be facilitated by using sticky yellow traps to identify nursery and orchard sites that are less susceptible to psylla attack (20, 91). Such sites should also be distant from indigenous psylla hosts (20). Although psylla were considered not to have strong dispersal powers (50), recent investigation established that adult *T. erytreae* can be wind-dispersed in the absence of host plants 1.5 km or more, and can survive approximately 85 hr without food (249). Strong winds such as those associated with typhoons have transported *D. citri* over medium to long distances (8).

The removal of infected branches or trees (20, 34) and neglected trees (253) may also reduce inoculum sources (24, 38) although mean population densities in the latter are not always significantly higher (203). If winter irrigation is withheld from lemon trees to suppress new flush, such infection reservoirs (105) can be reduced (248). In China hedges of *M. paniculata* which serve as hosts for *D. citri,* are being removed (274).

Systems of internal quarantine in South Africa prohibit the movement of trees from the greening-affected Transvaal and Natal provinces to the Cape province (154). Similar restrictions apply in China (188).

In Shantou, China, the practice of high-density plantings aimed at generating faster economic returns through early bearing serves to keep greening in check provided that the trees are from a disease-free nursery, that tree canopies remain uniform, and that strict psylla control is enforced (12).

The use of SDI (98) to predict psylla population outbreaks or sticky yellow traps to detect a population threshold value (two or more per set of three traps during each of two consecutive weeks) (207) can assist insecticide application programs. If insecticides are applied from late winter to midsummer the natural enemies of psylla are not seriously affected. Because *D. citri* does not cause foliar damage traps must be used. They are now being employed in China (8, 274) and the Philippines (87). By themselves traps do not appreciably reduce populations (202), but trap trees, preferably heavily pruned valencia sweet orange, could help manage psylla populations (206).

The prevalence of greening in some of the cooler citrus-growing areas of South Africa has caused more citrus cultivation in the hotter areas where citrus blight is prevalent (166). This shift has therefore been problematical.

CONCLUDING REMARKS

The conclusive isolation of the causal organism of citrus greening remains the primary goal for many researchers. International cooperation between all laboratories currently involved would substantially increase the chance of success. Serious consideration should also be given to the possibility that the full greening syndrome may be caused by a combination of microorganisms such as the putative greening organism (79), *A. lwoffi* (234), *Clavibacter* (67) and possibly others, and/or environmental conditions. The so-called greening organism may not on its own produce all the greening symptoms.

For citrus industries control is paramount. Clearly, integrated control will be the most effective. In Réunion a combination of introducing parasites, treating trees with antibiotics, and raising new trees free of infection has had success (14), and in India the use of insecticides, injections of tetracycline hydrochloride, thermotherapy of budwood and the use of tolerant rootstocks has reduced losses (177). In China, where the use of antibiotics has not yet been widely adopted (278), a combination of quarantine, propagation of healthy plants, thermotherapy and shoot-tip grafting of infected budwood, removal of diseased trees, especially in areas of low psylla incidence (108), and the use of insecticides is recommended (110, 115). In foundation blocks in the Philippines a combination of tolerant varieties, eradication of diseased trees, replanting with healthy plants, and spraying against psylla showed that citrus production is still feasible in greening areas (92a). Shoot-tip grafting and the maintenance of disease-free trees in screenhouses is recommended in Taiwan (92). Experience in the Philippines where the country's most valuable budwood sources are to a large degree greening-infected (138) underscores the importance of disease-free stock.

Currently, vector control is emphasized in South Africa, combined with removal of infected trees and branches (34). Trees with 50–75% greening fruit symptoms should be removed, while removal of branches is recommended for lower infection levels. Injection with PMT is recommended only for trees over 10 years old with more than 40% greening. Alternative chemotherapeutic agents are needed, however, together with long-term breeding programs. The subject of integrated management of greening has recently been reviewed (11).

A concern for those countries that have so far escaped the ravages of greening is how to avoid it. For Brazil the danger is probably greatest because the vector, *D. citri,* is already present (10). If it should enter, the rest of South America and subsequently North America would be threatened. The northward migration of the citrus tristeza aphid vector *Toxoptera citricidus* (197) illustrates the danger. The presence of the vectors and the disease in Pakistan and the Arabian peninsula poses a similar danger for Mediterranean countries. The danger is real that the Asian vector, *D. citri,* may move either from Arabia or from Réunion and Mauritius, via Madagascar, to the African mainland, and become established in conjunction with the existing African form. Citrus in the hotter, drier areas would then also be threatened.

Acknowledgments

I thank M. J. Samways, S. P. van Vuuren, L. Korsten, B. Q. Manicom, J. M. Bové, M. Garnier, B. Aubert, S. Quilici, H. M. Garnett, and R-J. Chippindall, who all helped either by sending me publications, critically reviewing the manuscript, and/or holding informative discussions.

Literature Cited

1. Abate, T. 1988. The identity and bionomics of insect vectors of tristeza and greening diseases of citrus in Ethiopia. *Trop. Pest Manage.* 34:19–23
2. Ahlawat, Y. S., Raychaudhuri, S. P. 1988. Status of citrus tristeza and dieback diseases in India and their detection. *Proc. Int. Citrus Congr., 6th,* 2:871–79
3. Altamirano, D. M., Gonzales, C. I., Viñas, R. C. 1976. Analysis of the devastation of leaf-mottling (greening) disease of citrus and its control program in the Philippines. *Proc. Conf. Int. Org. Citrus Virol., 7th,* pp. 22–26
4. Ariovich, D., Garnett, H. M. 1984. Structure of the greening organism. *Citrus Subtrop. Fruit J.* 611:6–9
5. Ariovich, D., Garnett, H. M. 1989. The use of immuno-gold staining techniques for detection of a bacterium associated with greening diseased citrus. *Phytopathology* 79:382–84
6. Aubert, B. 1975. La lutte aménagée contre les ravageurs des agrumes en Afrique du sud et ses applications possible pour les Mascareignes. *Fruits* 30: 149–59
7. Aubert, B. 1979. Progrés accomplis dans la lutte contre le greening des citrus à la Réunion. *Rev. Agric. Sucr. Ile Maurice* 58:53–56
8. Aubert, B. 1987. Epidemiological aspects of the greening (huanglungbin) disease in Asia. *Proc. Workshop Citrus Greening Disease, Fuzhou, China.* 5 pp.
9. Aubert, B. 1987. *Le greening, une maladie infectieuse des agrumes d'origine bactérienne, transmise par des Homopteres psyllides.* IRFA/CIRAD, St. Pierre. 185 pp.
10. Aubert, B. 1987. *Trioza erytreae* Del Guercio and *Diaphorina citri* Kuwayama (Homoptera:Psyllidae), the two vectors of citrus greening disease: biological aspects and possible control strategies. *Fruits* 42:149–62
11. Aubert, B. 1988. Towards an integrated management of citrus greening disease. *Proc. Conf. Int. Org. Citrus Virol., 10th,* pp. 226–30
12. Aubert, B. 1990. High density planting (HDP) of Jiagan mandarine in the lowland area of Shantou (Guangdong, China). *Proc. Int. Asia-Pacific Conf. Citrus Rehabil. 4th,* pp. 149–57
13. Aubert, B., Bové, J. M. 1980. Effect of penicillin or tetracycline injections of citrus trees affected by greening disease under field conditions in Reunion Island. *Proc. Conf. Int. Org. Citrus Virol., 8th,* pp. 103–8
14. Aubert, B., Bové, J. M., Etienne, J. 1980. La lutte contre la maladie du "greening" des agrumes à l'île de la Réunion. Résultats et perspectives. *Fruits* 35:605–24
15. Aubert, B., Garnier, M., Cassim, J. C., Bertin, Y. 1988. Citrus greening disease in East and West African countries south of the Sahara. See Ref. 11, pp. 231–37
16. Aubert, B., Garnier, M., Guillaumin, D., Herbagyandodo, B., Setiobudi, L., Nurhadi, F. 1985. Greening, a serious threat for the citrus production of the Indonesian archipelago. Future prospects of integrated control. *Fruits* 40:549–63
17. Aubert, B., Quilici, S. 1984. Biological control of the African and Asian citrus psyllids (Homoptera:Psylloidea), through eulophid and encyrtid parasites (Hymenoptera:Chalcidoidea) in Reunion Island. *Proc. Conf. Int. Org. Citrus Virol. 9th,* pp. 100–8
18. Aubert, B., Quilici, S. 1988. Results of the monitoring of citrus psylla on yellow traps in Reunion Island. See Ref. 11, pp. 249–54
19. Aubert, B., Sabine, A., Geslin, P., Picard, L. 1984. Epidemiology of the greening disease in Reunion Island before and after the biological control of the African and Asian citrus psyllas. *Proc. Int. Soc. Citricult.* 1:440–42
20. Begemann, G. J. 1985. Die epidemiologie van vergroensiekte van sitrus by Zebediela, Noord-Transvaal. *Citrus Subtrop. Fruit J.* 619:12–14
21. Bhagabati, K. N., Nariani, T. K. 1980. Interaction of greening and tristeza pathogens in Kagzi lime (*Citrus aurantifolia* (Christm.) Swing.) and their effect on growth and development of disease symptoms. *Indian Phytopathol.* 33:292–95
22. Bhagabati, K. N., Nariani, T. K. 1983. Chemical control of citrus psylla, a vector of citrus greening disease. *J. Res. Assam. Agric. Univ.* 4:86–88 (*Rev. Appl. Entomol. Ser. A* 75:28)
23. Bhavakul, K., Intavimolsri, S., Vichitrananda, S., Kratureuk, C., Promminatara, M. 1981. The current citrus disease situation in Thailand with emphasis on citrus greening. *Proc. Int. Soc. Citricult.* 1:464–66
24. Botha, A. D., Nel, D. D., Labuschagne, N., Begeman, D., Kotzé, J. M. 1984. Research into the cause of citrus green-

ing disease. *Proc. Symp. Citrus Greening, Nelspruit, S. Afr.* pp. 1–10
25. Bové, J. M. 1986. Greening in the Arab Peninsula: Towards new techniques for its detection and control. *FAO Plant Prot. Bull.* 34:7–14
26. Bové, J. M., Bonnet, P., Garnier, M., Aubert, B. 1980. Penicillin and tetracycline treatment of greening disease-affected citrus plants in the glasshouse, and the bacterial nature of the procaryote associated with greening. See Ref. 13, pp. 91–102
27. Bové, J. M., Calavan, E. C., Capoor, S. P., Cortez, R. E., Schwarz, R. E. 1974. Influence of temperature on symptoms of California stubborn, South African greening, Indian citrus decline and Philippines leaf mottling diseases. *Proc. Conf. Int. Org. Citrus Virol., 6th,* pp. 12–15
28. Bové, J. M., Garnier, M. 1984. Citrus greening and psylla vectors of the disease in the Arabian Peninsula. See Ref. 17, pp. 109–14
29. Broadbent, P., Fraser, L. R., Beattie, A., Grylls, N., Duncan, J. 1977. Australian citrus dieback problem. *Proc. Int. Soc. Citricult.* 3:894–96
30. Broadbent, P., Fraser, L. R., McGechan, J. 1976. Australian citrus dieback. See Ref. 3, pp. 141–46
31. Buitendag, C. H. 1972. Die effek van Azodrin op die sitrus bladvlooi *Trioza erytreae* (del G.). *Citrus Grower Subtrop. Fruit J.* 465:15–18
32. Buitendag, C. H. 1976. Vergroening by sitrus—Is daar enige hoop om die oordraging van die siekte deur sitrusbladvlooi met behulp van insek middelaanwendings te voorkom? *Citrus Subtrop. Fruit J.* 510:15–18
33. Buitendag, C. H. 1986. *Vergroeningsiekte van sitrus: Verskeie beheermetodes van die vektor* Trioza erytreae *(Del G.) met Azodrin (monokrotophos) en 'n ondersoek na onderdrukking van vergroeningsiekte-vrugsimptome met staminspuitings.* PhD thesis, Univ. Natal, Piertermaritzburg, S. Afr. 265 pp.
34. Buitendag, C. H. 1988. Current trends in the control of greening disease in citrus orchards. *Citrus Subtrop. Fruit J.* 640:6–7, 10
35. Buitendag, C. H., Bronkhorst, G. J. 1983. Micro-injection of citrus trees with N-pyrrolidinomethyl tetracycline (PMT) for the control of greening disease. *Citrus Subtrop. Fruit J.* 592:8–10
36. Buitendag, C. H., Bronkhorst, G. J. 1984. An inhibition technique for monitoring the presence of tetracycline residues in citrus fruit. *Citrus Subtrop. Fruit J.* 605:5–7
37. Buitendag, C. H., Bronkhorst, G. J. 1984. Beheer van vergroening by sitrus met spesiale verwysings na PMT-inspuitings. See Ref. 24, pp. 49–58
38. Buitendag, C. H., Bronkhorst, G. J. 1986. Further aspects of trunk treatment of citrus with insecticides: Phytotoxicity, side effects on incidental pests and development of application apparatus. *Citrus Subtrop. Fruit J.* 623:7–10
39. Burger, W. P., van Vuuren, S. P., van Wyngaardt, W. 1984. Comparative evaluation of gentisic acid and gentisoyl-β-D-glucose as markers for the identification of citrus greening disease. See Ref. 24, pp. 183–94
40. Capoor, S. P., Rao, D. G., Viswanath, S. M. 1967. *Diaphorina citri* Kuway., a vector of the greening disease of citrus in India. *Indian J. Agric. Sci.* 37:572–76
41. Capoor, S. P., Rao, D. G., Viswanath, S. M. 1974. Greening disease of citrus in the Deccan Trap Country and its relationship with the vector *Diaphorina citri* Kuwayama. See Ref. 27, pp. 43–49
42. Capoor, S. P., Thirumalachar, M. J. 1973. Cure of greening affected plants by chemotherapeutic agents. *Plant Dis. Reptr.* 57:160–63
43. Capoor, S. P., Thirumalachar, M. J., Pandey, P. K., Chakraborty, N. K. 1974. Control of the greening disease of citrus by B.P.-101: a new chemotherapeutant. See Ref. 27, pp. 50–52
44. Catling, H. D. 1969. The bionomics of the South African citrus psylla, *Trioza erytreae* (Del Guercio) (Homoptera: Psyllidae) 1. The influence of the flushing rhythm of citrus and factors which regulate flushing. *J. Entomol. Soc. South. Afr.* 32:191–208
45. Catling, H. D. 1969. The bionomics of the South African citrus psylla, *Trioza erytreae* (Del Guercio) (Homoptera: Psyllidae) 2. The influence of parasites and notes on the species. *J. Entomol. Soc. South. Afr.* 32:209–33
46. Catling, H. D. 1969. The bionomics of the South African citrus psylla, *Trioza erytreae* (Del Guercio) (Homoptera: Psyllidae) 3. The influence of extremes of weather on survival. *J. Entomol. Soc. South. Afr.* 32:273–90
47. Catling, H. D. 1969. The control of citrus psylla *Trioza erytreae* (Del Guercio) (Homoptera:Psyllidae). *S. Afr. Citrus J.* 426:8–16
48. Catling, H. D. 1970. Distribution of the psylla vectors on greening disease, with notes on the biology and bionomics of

Diaphorina citri. *FAO Plant Prot. Bull.* 18:8–15

49. Catling, H. D. 1972. Factors regulating populations of psyllid vectors of greening. *Proc. Conf. Int. Org. Citrus Virol., 5th,* pp. 51–57
50. Catling, H. D. 1973. Notes on the biology of the South African citrus psylla *Trioza erytreae* (Del Guercio) (Homoptera:Psyllidae). *J. Entomol. Soc. South. Afr.* 36:299–306
51. Catling, H. D. 1973. Results of a survey for psyllid vectors of citrus greening disease in Reunion. *FAO Plant Prot. Bull.* 21:78–82
52. Catling, H. D., Atkinson, P. R. 1974. Spread of greening by *Trioza erytreae* (Del Guercio) in Swaziland. See Ref. 27, pp. 33–39
53. Catling, H. D., Garnier, M., Bové, J. M. 1978. Presence of citrus greening disease in Bangladesh and a new method for rapid diagnosis. *FAO Plant Prot. Bull.* 26:16–18
54. Catling, H. D., Green, G. C. 1972. The influence of weather on the survival and population fluctuations of *Trioza erytreae* (Del Guercio)—a vector of greening. See Ref. 49, pp. 58–64
55. Chakraborty, N. K., Pandey, P. K., Chatterjee, S. N., Singh, A. B. 1976. Host preference in *Diaphorina citri* Kuwayama, vector of greening disease in India. *Indian J. Entomol.* 38:196–97 (*Rev. Appl. Entomol. Ser. A* 67:232)
56. Cheema, S. S., Chohan, J. S., Kapur, S. P. 1982. Effect of moist hot air treatment on citrus greening-infected budwood. *J. Res. Punjab Agric. Univ.* 19:97–99 (*Rev. Plant Pathol.* 62:16)
57. Cheema, S. S., Dhillon, R. S., Kapur, S. P. 1982. Flowers as source material in chromatographic detection of citrus greening disease. *Curr. Sci.* 51:241 (*Rev. Plant Pathol.* 61:546)
58. Cheema, S. S., Kapur, S. P., Bansal, R. D. 1985. Efficacy of various therapeutic agents against greening disease of citrus. *J. Res. Punjab Agric. Univ.* 22:479–82 (*Rev. Plant Pathol.* 65:551)
59. Cheema, S. S., Kapur, S. P., Chohan, J. S. 1982/83. Evaluation of rough lemon strains and other rootstocks against greening disease of citrus. *Sci. Horticult.* 18:71–75
60. Cheema, S. S., Kapur, S. P., Sharma, O. P. 1986. Chemotherapeutic control of greening disease of citrus through bud dip treatment. *Indian J. Virol.* 2:104–7 (*Rev. Plant Pathol.* 68:648)
61. Chen, M-H., Miyakawa, T., Matsui, C. 1971. Mycoplasmalike bodies associated with likubin-diseased Ponkan citrus. *Phytopathology* 61:598
62. Chen, M-H., Miyakawa, T., Matsui, C. 1972. Simultaneous infections of citrus leaves with tristeza virus and mycoplasmalike organism. *Phytopathology* 62: 663–66
63. Chen, M-H., Miyakawa, T., Matsui, C. 1973. Citrus likubin pathogens in the salivary glands of *Diaphorina citri*. *Phytopathology* 63:194–95
64. Chen, Q. 1943. A report of a study on yellow shoot of citrus in Chaoshan. *New Agric. Q. Bull.* 3:142–75 (refs. 173, 343)
65. Chen, Y-H., Mei, J-H. 1965. A preliminary study of the citrus yellow shoot virus. *Acta Phytophylact. Sin.* 4:361–64 (*Rev. Appl. Mycol.* 45:439)
66. Chippindall, R-J., Whitlock, V. H. 1989. Development of an antiserum to detect greening disease of citrus. *Phytopathology* 79:1212 (Abstr.)
67. Chippindall, R-J., Whitlock, V. H. 1989. Transmission and reisolation of a bacterium isolated from greening-infected citrus in South Africa. *Phytopathology* 79:1182–3 (Abstr.)
68. Chiu, R-J., Tsai, M-Y., Huang, C-H. 1979. Distribution and retention of tetracyclines in healthy and likubin-infected citrus trees following trunk transfusion. *Proc. ROC-US Co-op. Sci. Sem. Mycoplasma Dis. Plants (Natl. Sci. Counc. Symp. Ser. No. 1),* pp. 143–52
69. Chiu, S-C., Aubert, B., Chin, C-C. 1988. Attempts to establish *Tetrastichus radiatus* Waterson (Hymenoptera, Chalcidoidea), a primary parasite of *Diaphorina citri* Kuwayama, in Taiwan. See Ref. 11, pp. 265–68
70. Chowdhury, A. R., Chaturvedi, H. C., Mitra, G. C. 1974. Quantitative changes in free amino acids in leaves and flower buds of healthy and greening-affected *Citrus sinensis*. *Indian J. Exp. Biol.* 12:461–62
71. de Lange, J. H., Vincent, A. P., Nel, M. 1985. Breeding for resistance to greening disease in citrus. *Citrus Subtrop. Fruit J.* 614:6–9
72. Dhillon, R. S., Cheema, S. S., Kapur, S. P., Deol, I. S. 1983. Reaction of nucellar lines of sweet orange varieties to viruses under field conditions. *J. Res. Punjab Agric. Univ.* 20:281–84. (*Rev. Plant Pathol.* 64:110)
73. Duncan, F., Garnett, H. M. 1984. Serological studies of the greening organism. *Citrus Subtrop. Fruit J.* 611:9–10
74. Feldman, A. W., Hanks, R. W. 1969.

The occurrence of a gentisic glucoside in the bark and albedo of virus-infected citrus trees. *Phytopathology* 59:603–6

75. Fraser, L. R. 1978. Recognition and control of citrus virus diseases in Australia. *Proc. Int. Soc. Citricult.*, pp. 178–81
76. Fraser, L. R., Singh, D., Capoor, S. P., Nariani, T. K. 1966. Greening virus, the likely cause of citrus dieback in India. *FAO Plant Prot. Bull.* 14:127–30
77. Gao, S. J. 1988. Production of monoclonal antibodies against the organism of Chinese greening disease (Progr. rept.). *Proc. FAO-UNDP Greening Workshop, Lipa, Philippines, 2nd,* pp. 82–83
78. Gardner, J. M., Feldman, A. W., Zablotowicz, R. M. 1982. Identity and behaviour of xylem-residing bacteria in rough lemon roots of Florida citrus trees. *Appl. Environ. Microbiol.* 43:1331–42
79. Garnett, H. M. 1984. Isolation and characterization of the greening organism. *Citrus Subtrop. Fruit J.* 611:4–6
80. Garnett, H. M., Mochaba, F., Civerolo, E. L., Lee, R. F., Brlansky, R. H., Bock, D. 1991. Serological comparison of putative greening isolates. *Proc. Conf. Int. Org. Citrus Virol., 11th.* In press
81. Garnier, M., Bové, J. M. 1978. The organism associated with citrus greening disease is probably a member of the Schizomycetes. *Zentralbl. Bakteriol. Parasitenkd. Infektions, Kra. Hyg. Reihe A* 241:221–22
82. Garnier, M., Bové, J. M. 1983. Transmission of the organism associated with the citrus greening disease from sweet orange to periwinkle by dodder. *Phytopathology* 73:1358–63
83. Garnier, M., Danel, N., Bové, J. M. 1984. The greening organism is a gram negative bacterium. See Ref. 17, pp. 115–24
84. Garnier, M., Latrille, J., Bové, J. M. 1976. *Spiroplasma citri* and the organism associated with likubin: comparison of their envelope systems. See Ref. 3, pp. 13–17
85. Garnier, M., Martin-Gros, G., Bové, J. M. 1987. Monoclonal antibodies against the bacterial-like organism associated with citrus greening disease. *Ann. Inst. Pasteur/Microbiol.* 138:639–50
86. Garnier, M., Villechanoux, S., Gao, S., He, Y., Bové, J. M. 1991. Study of the greening organism (GO) with monoclonal antibodies: Purification, serological identification, morphology and serotypes of the GO. See Ref. 80. In press
87. Gavarra, M. R., Mercado, B. G. 1988. Progress report on studies dealing with the psyllid vector (*Diaphorina citri* Kuwayama) of greening disease in the Philippines. See Ref. 77, pp. 23–28
88. Ghosh, S. K., Giannotti, J., Louis, C. 1977. Multiplication intense des procaryotes associés aux maladies de type "greening" des agrumes dans les cellules criblées de Cuscutes. *Ann. Phytopathol.* 9:525–30
89. Ghosh, S. K., Raychaudhuri, S. P., Chenula, V. V., Varma, A. 1975. Isolation, cultivation and characterization of mycoplasma-like organisms from plants. *Proc. Indian Natl. Sci. Acad. Ser. B* 41:362–66 (*Rev. Plant Pathol.* 56:593)
90. Ghosh, S. K., Raychaudhuri, S. P., Varma, A., Nariani, T. K. 1971. Isolation and culture of mycoplasma associated with citrus greening disease. *Curr. Sci.* 40:299–300
91. Gilbert, M. J. 1984. Trapping as a monitor for determining outbreaks of citrus psylla, *Trioza erytreae* (Del Guercio) (Hemiptera:Triozidae). See Ref. 24, pp. 84–89
92. Gonzales, C. I., Su, H. J. 1988. Recent concept on the establishment of disease free citrus foundation trees in southeast Asia. See Ref. 77, pp. 69–72

92a. Gonzales, C. I., Viñas, R. C. 1981. Field performance of citrus varieties and cultivars grown under control measures adopted against leaf mottling (greening) disease in the Philippines. *Proc. Int. Soc. Citricult.* 1:463–64

93. Gonzales, C. I., Viñas, R. C., Vergara, L. A. 1972. Observations on 110 citrus cultivars planted in an area severely infested by leaf mottling. See Ref. 49, pp. 38–40
94. Goswami, B. K., Raychaudhuri, S. P., Nariani, T. K. 1971. Free amino acid content of the greening-affected and healthy plants of sweet orange (*Citrus sinensis* Osbeck). *Curr. Sci.* 17:469–70
95. Gottwald, T. R., Aubert, B. 1991. Spatial patterns of citrus greening in Shantou, China. See Ref. 80
96. Gottwald, T. R., Aubert, B., Zhao, X-Y. 1989. Preliminary analysis of disease progress of citrus greening epidemics in the People's Republic of China and French Reunion Island. *Phytopathology* 79:687–93
97. Gottwald, T. R., Gonzales, C. I., Mercado, B. G. 1991. Analysis of the distribution of citrus greening in a heavily diseased grove in Philippines. See Ref. 80
98. Green, G. C., Catling, H. D. 1971. Weather-induced mortality of the citrus

psylla, *Trioza erytreae* (Del Guercio) (Homoptera: Psyllidae), a vector of greening virus, in some citrus producing areas of southern Africa. *Agric. Meteorol.* 8:305–17

99. Greening—what is it and how to treat it. Part 1. With special reference to the injection method and the use of PMT (Rolitetracycline). 1984. *Citrus Subtropical Fruit J.* 602:10–11
100. Greening. The symptoms, transmission and treatment for the control of greening—Part 2. 1984. *Citrus Subtropical Fruit J.* 603:11–12
101. Hofmeyr, J. D. J., Oberholzer, P. C. J. 1948. Genetic aspects associated with the propagation of citrus. *Farming S. Afr.* 23:201–8
102. Honiball, F. 1984. Probleme met die chemiese beheer van die sitrusbladvlooi, *Trioza erytreae* (Del G.) (Fam.:Triozidae), en die invloed daarvan op geintegreerde plaagbeheer. See Ref. 24, pp. 158–61
103. Hortelano Hap, G., Garnett, H. M. 1991. Analysis of DNA from putative greening isolates. See Ref. 80
104. Huang, C-H. 1978. Effect of hot air treatment on likubin, tristeza virus and exocortis viroid diseases of citrus. *J. Agric. Res. China* 27:193–97
105. Huang, C. H. 1979. Distribution of likubin pathogen in likubin-affected citrus plants. *J. Agric. Res. China* 28:29–33
106. Huang, C-H., Chang, C-A. 1980. Studies on the relation of mycoplasma-like organism with the decline of wentan pummelo in Taiwan. *J. Agric. Res. China* 29:13–19
107. Huang, C.-H., Chen, M.-J., Chiu, R.-J. 1980. Separation of a mycoplasma-like organism from the likubin complex in citrus. *Plant Dis.* 64:564–66
108. Huang, S. 1988. The relation between the distribution of *D. citri* and greening disease in Guangxi Zhuang Autonomous Region. See Ref. 77, pp. 78–79
109. Hughes, J. P., van Dyk, P. J. 1972. Evaluating aerial sprays for the control of citrus psylla. *Citrus Grow. Subtrop. Fruit J.* 468:13–15
110. Jiang, Y. H., Zhao, X. Y., Su, W. F., Huang, T. Y., Huang, Z. Q. 1987. Exclusion of citrus yellow shoot pathogen by shoot tip grafting. *Acta Phytophylact. Sin.* 14:184 (*Rev. Plant Pathol.* 67:54)
111. Kapur, S. P., Ananda, S. A., Cheema, S. S., Kapoor, S. K. 1977. Prevalence of greening marker substances in different species of citrus. *Indian J. Horticult.* 34:205–8 (*Rev. Plant Pathol.* 60:54)
112. Kapur, S. P., Cheema, S. S., Bansal, R. D., Singh, I. 1986. Chemotherapeutic control of citrus greening through foliar spray effect on chlorophyll and greening marker substances (GMS). *Indian J. Virol.* 2:68–72 (*Rev. Plant Pathol.* 68:648)
113. Kapur, S. P., Cheema, S. S., Dhillon, R. S. 1984. Reaction of certain citrus scionic combinations to viral/mycoplasmal diseases. *Indian J. Horticult.* 41:142–43 (*Rev. Plant Pathol.* 65:135)
114. Kapur, S. P., Kapoor, S. K., Cheema, S. S., Dhillon, R. S. 1978. Effect of greening disease on tree and fruit characters of Kinnow mandarin. *Punjab Horticult. J.* 18:176–79 (*Horticult. Abstr.* 50:470)
115. Ke, C., Lin, X. Z. 1987. Establishment of disease free foundation stock mother tree and nurseries for citrus in Fujian, China. See Ref. 8, 7 pp.
116. Ke, S., Li, K. B., Ke, C., Tsai, J. H. 1988. Transmission of the huanglungbin agent from citrus to periwinkle by dodder. See Ref. 11, pp. 258–64
117. Khangura, J. S., Singh, H. 1984. Effect of sprays on the effectiveness and persistence of LCV and EC formulations of dimethoate against *Diaphorina citri* Kuwayama on citrus. *J. Entomol. Res.* 8:31–35 (*Rev. Appl. Entomol. Ser. A* 73:479)
118. Knorr, L. C., Gupta, O. P., Saeed, A. 1970. Occurrence of greening virus disease of citrus in West Pakistan. *Plant Dis. Reptr.* 54:7
119. Knorr, L. C., Shah, M., Gupta, O. P. 1970. Greening disease of citrus in Nepal. *Plant Dis. Reptr.* 54:1092–95
120. Ko, W. W. 1988. Plant indexing to detect the greening disease in Malaysia. See Ref. 77, pp. 84–86
121. Koen, T. J., Langenegger, W. 1970. Effect of greening virus on the macroelement content of citrus leaves. *Farming S. Afr.* 45(12):65
122. Korsten, L., Labuschagne, N., Verschoor, J. A., de Bruyn, M., Kotzé, J. M. 1991. Progress with greening research in South Africa. See Ref. 80
123. Labuschagne, N., Botha, A. D., Kotzé, J. M. 1984. Study of greening symptoms on inoculated plants. See Ref. 24, pp. 28–33
124. Labuschagne, N., Kotzé, J. M. 1988. Effect of temperature on expression of greening disease symptoms and possible inactivation of the pathogen in Eureka lemon. *Phytophylactica* 20:177–78
125. Laflèche, D., Bové, J. M. 1970. Mycoplasmes dans les agrumes atteints de "greening", de stubborn, ou des maladies similaires. *Fruits* 25:455–65

126. Laflèche, D., Bové, J. M. 1970. Structures de type mycoplasme dans les feuilles d'orangers atteints de la maladie du "greening". *C. R. Acad. Sci. Ser. D* 270:1915–17
127. Lallemand, J., Fos, A., Bové, J. M. 1986. Transmission de la bacterie associé à la forme africaine de la maladie du "greening" par le psylle asiatique *Diaphorina citri* Kuwayama. *Fruits* 41: 341–43
128. Lama, T. K., Regmi, C., Aubert, B. 1988. Distribution of the citrus greening disease vector (*Diaphorina citri* Kuw.) in Nepal and attempt of establishing biological control against it. See Ref. 11, pp. 255–57
129. Lee, H. A. 1921. The relation of stocks to mottled leaf of citrus leaves. *Philipp. J. Sci.* 18:85–95
130. Lee, R. F., Civerolo, E. L., Brlansky, R. H., Hooker, M. E., Garnsey, S. M., Garnett, H. M. 1991. A U.S. collection of citrus greening and greening-like cultures *in planta*. See Ref. 80
131. Lin, C-K. 1963. Notes on citrus yellow shoot disease. *Acta Phytophylact. Sin.* 2:237–42 (*Rev. Appl. Mycol.* 43:136)
132. Lin, K-H. 1956. Observations on yellow shoot on citrus. Etiological studies of yellow shoot of citrus. *Acta Phytopathol. Sin.* 2:1–42
133. Lin, K-H. 1963. Further studies on citrus yellow shoot. *Acta Phytophylact. Sin.* 2:243–51 (*Rev. Appl. Mycol.* 43:136)
134. Lin, K-H. 1964. A preliminary study on the resistance of yellow shoot virus and citrus budwood to heat. *Acta Phytopathol. Sin.* 7:61–63
135. Lin, K-H. 1964. Remarks on "Notes on citrus yellow shoot disease." *Acta Phytophylact. Sin.* 3:165–71 (*Rev. Appl. Mycol.* 44:205)
136. Lin, K-H., Chu, M-L. 1957. The relation of *Fusarium* to yellow shoot of citrus. *Acta Phytopathol. Sin.* 3:169–76 (*Rev. Appl. Mycol.* 37:408)
137. Lin, S.-J., Tao, C. C. C. 1979. A new *Psyllaephagus* Ashmead parasitising *Diaphorina citri* Kuwayama in Taiwan. *Q. J. Taiwan Mus.* 32:117–21 (*Rev. App. Entomol. Ser. A* 69:512)
138. Magnaye, L. V., Olfato, R. B., Herradura, L., Dimailig, V. 1988. Indexing greening in the Philippines. See Ref. 77, p. 95
139. Manicom, B. Q. 1984. Attempts at culture, purification and fluorescent antibody tracking of the greening organism. See Ref. 24, pp. 66–69
140. Manicom, B. Q., van Vuuren, S. P. 1990. Symptoms of greening disease with special emphasis on African greening. See Ref. 12, pp. 127–31
141. Martinez, A. L. 1972. Combined effects of greening and seedling yellows pathogens in citrus. See Ref. 49, pp. 25–27
142. Martinez, A. L. 1972. On the transmission and translocation of the greening pathogen in citrus. See Ref. 49, pp. 22–24
143. Martinez, A. L., Nora, D. M., Armedilla, A. L. 1970. Suppression of symptoms of citrus greening disease in the Philippines with tetracycline antibiotics. *Plant Dis. Reptr.* 54:1007–9
144. Martinez, A. L., Wallace, J. M. 1967. Citrus leaf mottle-yellows disease in the Philippines and transmission of the causal virus by a psyllid, *Diaphorina citri*. *Plant Dis. Reptr.* 51:692–95
145. Martinez, A. L., Wallace, J. M. 1968. Studies on leaf mottle-yellows disease of citrus in the Philippines. *Proc. Conf. Int. Org. Citrus Virol., 4th* pp. 167–76
146. Martinez, A. L., Wallace, J. M. 1969. Citrus greening disease in the Philippines. *Proc. 1st Int. Citrus Symp.* 3:1427–31
147. Massonie, G., Garnier, M., Bové, J. M. 1976. Transmission of Indian citrus decline by *Trioza erytreae* (Del Guercio), the vector of South African greening. See Ref. 3, pp. 18–20
148. Matsumoto, T., Su, H. J., Lo, T. T. 1968. Likubin. In *Indexing Procedures for 15 Virus Diseases of Citrus. USDA Agric. Res. Serv. Agric. Handbk.* 333:63–67
149. Matsumoto, T., Wang, M. C., Su, H. J. 1961. Studies on likubin. *Proc. Conf. Int. Org. Citrus Virol., 2nd,* pp. 121–25
150. McClean, A. P. D. 1950. Virus infections of citrus in South Africa III. Stem-pitting disease of grapefruit. *Farming S. Afr.* 25:289–96
151. McClean, A. P. D. 1970. Greening disease of sweet orange: its transmission in propagative parts and distribution in partially diseased trees. *Phytophylactica* 2:263–68
152. McClean, A. P. D. 1974. The efficiency of citrus psylla, *Trioza erytreae* (del G.) as a vector of greening disease of citrus. *Phytophylactica* 6:45–54
153. McClean, A. P. D., Oberholzer, P. C. J. 1965. Citrus psylla, a vector of the greening disease of sweet orange. *S. Afr. J. Agric. Sci.* 8:297–98
154. McClean, A. P. D., Oberholzer, P. C. J. 1965. Greening disease of sweet orange: Evidence that it is caused by a transmissible virus. *S. Afr. J. Agric. Sci.* 8:253–76

155. McClean, A. P. D., Schwarz, R. E. 1970. Greening of blotchy-mottle disease of citrus. *Phytophylactica* 2:177–94
156. McClean, A. P. D., Schwarz, R. E., Oberholzer, P. C. J. 1969. Greening disease of citrus in South Africa. See Ref. 146, pp. 1421–25
157. McDaniel J. R., Moran, V. C. 1972. The parasitoid complex of the citrus psylla *Trioza erytreae* (Del Guercio) (Homoptera: Psyllidae). *Entomaphaga* 17:297–317
158. Milne, D. L. 1977. Control of citrus nursery pests by soil applications of dimethoate. *Citrus Subtrop. Fruit J.* 525:5–7
159. Milne, D. L., de Villiers, E. A. 1977. Dimethoate applied by microjet and drip irrigation for control of citrus psylla, *Trioza erytreae,* the vector of greening disease. *Citrus Subtrop. Fruit J.* 525: 15–17
160. Milne, D. L., van Vuuren, S. P., Moll, J. N. 1981. Simultaneous injection of greening-infected citrus trees with tetracycline hydrochloride and dimethoate. *Subtropica* 2(1):13–15
161. Miyakawa, T. 1979. Suppressive effect of penicillin and some other antibiotics on symptom development of citrus likubin (greening disease). *Ann. Phytopathol. Soc. Jpn.* 45:401–3
162. Miyakawa, T. 1980. Experimentally induced symptoms and host range of citrus likubin (greening disease). *Ann. Phytopathol. Soc. Jpn.* 46:224–30
163. Miyakawa, T., Tsuno, K. 1989. Occurrence of citrus greening in the southern islands of Japan. *Ann. Phytopathol. Soc. Jpn.* 66:667–70
164. Moll, J. N. 1974. *Citrus greening disease*. PhD thesis. Univ. Natal, Pietermaritzburg, S. Afr. 43 pp.
165. Moll, J. N. 1974. Fluorimetric determination of antibiotic residues in citrus trees injected with tetracycline hydrochloride. See Ref. 27, pp. 198–201
166. Moll, J. N. 1987. The interaction of greening, tristeza and blight in citrus production. *Phytophylactica* 19:124 (Abstr.)
167. Moll, J. N., Martin, M. M. 1973. Electron microscope evidence that citrus psylla *(Trioza erytreae)* is a vector of greening disease in South Africa. *Phytophylactica* 5:41–44
168. Moll, J. N., Martin, M. M. 1974. Comparison of the organism causing greening disease with several plant pathogenic gram-negative bacteria, rickettsia-like organisms and mycoplasma-like organisms. *Coll. INSERM* 33:89–96
169. Moll, J. N., van Vuuren, S. P. 1977. Greening disease in Africa. *Prot. Int. Soc. Citricult.* 3:903–12
170. Moll, J N., van Vuuren, S. P. 1982. Observations on the occurrence and annual variations of greening in the citrus. *Subtropica* 3(10):18–20
171. Moll, J. N., van Vuuren, S. P., Milne, D. L. 1980. Greening disease, the South African situation. See Ref. 13, pp. 109–17
172. Moran, V. C. 1968. Preliminary observations on the choice of host plants by adults of the citrus psylla, *Trioza erytreae* (Del Guercio) (Homoptera: Psyllidae). *J. Entomol. Soc. South. Afr.* 31:401–10
173. Moran, V. C. 1968. The development of the citrus psylla, *Trioza erytreae* (Del Guercio) (Homoptera: Psyllidae) on *Citrus limon* and four indigenous host plants. *J. Entomol. Soc. South. Afr.* 31:45–54
174. Moran, V. C., Blowers, J. R. 1967. On the biology of the South African citrus psylla, *Trioza erytreae* (Del Guercio) (Homoptera:Psyllidae). *J. Entomol. Soc. South. Afr.* 30:96–106
175. Moreira, S. 1967. Mauritius and Reunion. Survey of citrus diseases. *FAO Plant Prot. Bull.* 15:59–60
176. Naidu, R., Govindu, H. C. 1981. Strains and strain interactions of the citrus greening pathogen in India. In *Mycoplasma Diseases of Trees and Shrubs,* ed. K. Maramarosch, S. P. Raychaudhuri, pp. 299–313. New York: Academic
177. Nariani, T. K. 1981. Integrated approach to control citrus greening disease in India. *Proc. Int. Soc. Citricult.* 1:471–72
178. Nariani, T. K., Bhagabati, K. N. 1980. Studies on the therapeutic and insect vector control of the greening disease of citrus in India. See Ref. 13, pp. 122–28
179. Nariani, T. K., Ghosh, S. K., Kumar, D., Raychaudhuri, S. P., Viswanath, S. M. 1975. Detection and possibilities of therapeutic control of the greening disease of citrus caused by mycoplasmas. *Proc. Indian Nat. Sci. Acad. Ser. B* 41:334–39
180. Nariani, T. K., Raychaudhuri, S. P., Bhalla, R. B. 1967. Greening virus of citrus in India. *Indian Phytopathol.* 20:146–50
181. Nariani, T. K., Raychaudhuri, S. P., Viswanath, S. M. 1973. Tolerance to greening disease in certain citrus species. *Curr. Sci.* 42:513–14
182. Nariani, T. K., Singh, G. R. 1971. Epidemiological studies on the citrus die back complex in India. *Proc.*

Indian Natl. Sci. Acad. Ser. B 37:365–71
183. Nurhadi, F. 1987. Records of important parasites attacking *Diaphorina citri* in East Java, Indonesia. See Ref. 8, (Abstr.)
184. Oberholzer, P. C. J., Hofmeyr, J. D. J. 1955. The nature and control of clonal senility in commercial varieties of citrus in South Africa. *Bull. Fac. Agric. Univ. Pretoria,* pp. 1–46
185. Oberholzer, P. C. J., von Staden, D. F. A., Basson, W. J. 1965. Greening disease of sweet orange in South Africa. *Proc. Conf. Int. Org. Citrus Virol., 3rd* pp. 213–19
186. Ōtake, A. 1990. Bibliography of citrus greening disease and its vectors attached with indices, and a critical review on the ecology of the vectors and their control. *Jpn. Int. Coop. Agency,* 161 pp.
187. Pyle, K. R. 1977. Control of citrus psylla *Trioza erytreae* (Del G.) with sprays of endosulfan on Mazoe Citrus Estates. *Citrus Subtrop. Fruit J.* 520:7–11
188. Qian, J. 1988. Quarantine and control strategies of the citrus greening disease in China. See Ref. 77, pp. 57–62
189. Quilici, S. 1988. Biological control of citrus psylla in Reunion Island. See Ref. 77. pp. 39–42
190. Raychaudhuri, S. P., Nariani, T. K., Ghosh, S. K., Viswanath, S. M., Kumar, D. 1974. Recent studies on citrus greening in India. See Ref. 27, pp. 53–57
191. Raychaudhuri, S. P., Nariani, T. K., Lele, V. C. 1969. Citrus die-back problem in India. *Proc. Int. Citrus Symp., 1st,* 3:1433–37
192. Raychaudhuri, S. P., Nariani, T. K., Lele, V.C., Singh, G. R. 1972. Greening and citrus decline in India. See Ref. 49, pp. 35–37
193. Raychaudhuri, S. P., Sharma, D. C. 1983. Role of disease in citrus in India and their management. *Proc. Int. Congr. Plant Prot., 10th,* 3:1019
194. Reddy, D. B., ed. 1971. Outbreaks of pests and diseases and new records. *FAO Q. Newsl., Plant Prot. Comm. SE Asia Pac. Reg.* 14:5–15
195. Regmi, C., Lama, T. K. 1988. Greening incidence and greening vector population dynamics in Pokhara. See Ref. 11, pp. 238–42
196. Reinking, O. A. 1919. Diseases of economic plants in southern China. *Philipp. Agric.* 8:109–35
197. Roistacher, C. N., Bar-Joseph, M. 1987. Aphid transmission of citrus tristeza virus: A review. *Phytophylactica* 19:163–67
198. Saglio, P., Laflèche, D., Bonissol, C., Bové, J. M. 1971. Isolement, culture et observations au microscope électronique des structures de type mycoplasme associés à la maladie du stubborn des agrumes et leur comparison avec les structures observées dans le cas de la maladie Greening des agrumes. *Physiol. Veg.* 9:569–82
199. Salibe, A. A., Cortez, R. E. 1966. Studies on the leaf mottling disease of citrus in the Philippines. *FAO Plant Prot. Bull.* 14:141–44
200. Salibe, A. A., Cortez, R. E. 1968. Leaf mottling—a serious virus disease of citrus in the Philippines. See Ref. 145, pp. 131–36
201. Salibe, A. A., Tirtawidjaja, S. 1984. Incidencia da doença "greening" em variedades citricos na Indonesia. *Summa Phytopathol.* 10:35 (*Rev. Plant Pathol.* 65:196)
202. Samways, M. J. 1987. Phototactic response of *Trioza erytreae* (Del Guercio) (Hemiptera: Triozidae) to yellow-coloured surfaces, and an attempt at commercial suppression using yellow barriers and trap trees. *Bull. Entomol. Res.* 77:91–98
203. Samways, M. J. 1987. Prediction and upsurges in populations of the insect vector (*Trioza erytreae,* Hemiptera: Triozidae) of citrus greening disease using low-cost trapping. *J. Appl. Ecol.* 24:881–91
204. Samways, M. J. 1987. Weather and monitoring the abundance of the adult citrus psylla, *Trioza erytreae* (Del Guercio) (Hom., Triozidae). *J. Appl. Entomol.* 103:502–8
205. Samways, M. J., Grech, N. M. 1986. Assessment of the fungus *Cladosporium oxysporum* as a biological control agent of certain Homoptera. *Agric. Ecosyst. Environ.* 15:231–39
206. Samways, M. J., Manicom, B. Q. 1983. Immigration, frequency distributions and dispersal patterns of the psyllid *Trioza erytreae* (Del Guercio) in a citrus orchard. *J. Appl. Ecol.* 20:463–72
207. Samways, M. J., Tate, B. A., Murdoch, E. 1986. Monitoring the citrus thrips and psylla using fluorescent yellow sticky traps—a practical guide. *Citrus Subtrop. Fruit J.* 629:9–15
208. Sanders, G., Korsten, L., Kotzé, J. M. 1990. Assessing the incidence of *Acinetobacter lwoffii* in greening infected and symptomless citrus material by means of monoclonal antibodies. *Phytophylactica* 22:156 (Abstr.)
209. Schneider, H. 1968. Anatomy of green-

ing diseased sweet orange shoots. *Phytopathology* 58:1155–60

210. Schwarz, R. E. 1964. An insect-transmissible virus trapped on sweet orange seedlings in orchards where greening disease is common. *S. Afr. J. Agric. Sci.* 7:885–89
211. Schwarz, R.E. 1965. A fluorescent substance present in tissues of greening-infected sweet orange. *S. Afr. J. Agric. Sci.* 8:1177–79
212. Schwarz, R. E. 1967. Results of a greening survey on sweet orange in the major citrus growing areas of the Republic of South Africa. *S. Afr. J. Agric. Sci.* 10:471–76
213. Schwarz, R. E. 1968. Greening disease. See Ref. 148, pp. 87–90
214. Schwarz, R. E. 1968. Indexing of greening and exocortis through fluorescent marker substances. See Ref. 145, pp. 118–24
215. Schwarz, R. E. 1968. The distribution of greening in citrus areas of South Africa. See Ref. 145, pp. 124–27
216. Schwarz, R. E. 1968. Thin layer chromatographical studies on phenol markers of the greening virus in various citrus species. *S. Afr. J. Agric. Sci.* 11:797–801
217. Schwarz, R. E. 1970. Comparative indexing of the annual and seasonal incidence of greening in sweet orange fruits by external symptoms and by the albedo fluorescence test. *Phytophylactica* 2:1–16
218. Schwarz, R. E. 1970. Seaonal graft-transmissibility and quantification of gentisoyl glucoside marker of citrus greening in the bark of infected trees. *Phytophylactica* 2:115–20
219. Schwarz, R. E. 1972. Strains of the greening pathogen. See Ref. 49, pp. 40–44
220. Schwarz, R. E., Bové, J. M. 1981. Greening. In *Description and Illustration of Virus and Virus-like Diseases of Citrus* 2nd Ed., ed. J. M. Bové, R. Vogel, 1: 10 pp. Paris:IRFA
221. Schwarz, R. E., Green, G. C. 1970. Das "Citrus-Greening" und der Citrusblattfloh *Trioza erytreae*—ein temperaturabhängiger Erreger-Überträgerkomplex. *Z. Pflanzenkrankh. Pflanzenschutz* 77:490–93
222. Schwarz, R. E., Green, G. C. 1972. Heat requirements for symptom expression and inactivation of the greening pathogen. See Ref. 49, pp. 44–51
223. Schwarz, R. E., Knorr, L. C., Prommintara, M. 1973. Presence of citrus greening and its psylla vector in Thailand. *FAO Plant Prot. Bull.* 21:132–38
224. Schwarz, R. E., McClean, A. P. D., Catling, H. D. 1970. The spread of greening disease by citrus psylla in South Africa. *Phytophylactica* 2:45–54
225. Schwarz, R. E., Moll, J. N., van Vuuren, S. P. 1974. Control of citrus greening and its psylla vector by trunk injections of tetracyclines and insecticides. See Ref. 27, pp. 26–29
226. Schwarz, R. E., Moll, J. N., van Vuuren, S. P. 1974. Incidence of fruit greening on individual citrus trees in South Africa. See Ref. 27, pp. 30–32
227. Schwarz, R. E., van Vuuren, S. P. 1970. Centrifugal extraction of phenolic markers for indexing citrus greening and avocado sun-blotch diseases. *Phytophylactica* 2:65–68
228. Schwarz, R. E., van Vuuren, S. P. 1970. Decreases in fruit greening of sweet orange by trunk injections with tetracyclines. *Plant Dis. Reptr.* 55:747–50
229. Sec. Agric., Rhodesia, 1967. Plant pathology report, 1 Oct. 1965–30 Sept 1966:27–28 (*Rev. Appl. Biol.* 47:9)
230. Seif, A. A., Whittle, A. M. 1984. Diseases of citrus in Kenya. *FAO Plant Prot. Bull.* 32:122–27
231. Sharma, R. C., Bakshi, J. C., Jeyarajam, R. 1974. Periodic changes in the concentration of fluorescent marker substance in relation to leaf chlorosis of sweet orange. *Indian J. Agric. Sci.* 44:18–21 (*Rev. Plant Pathol.* 54:837)
232. Singh, G. R., Raychaudhuri, S. P., Kapoor, I. J. 1971. Interaction of fungi, pathogen in producing citrus die-back. *Citrograph* 57:27–28
233. Smith, J. H. 1984. Enhancement of tetracycline distribution in citrus trees with enzymes, imazalil and nitrate or urea. See Ref. 24, pp. 44–48
234. Smith, L., Korsten, L., Verschoor, J. A., Kotze, J. M. 1988. Development of monoclonal antibodies against an organism associated with greening of citrus. *Phytophylactica* 20:95 (Abstr.)
235. Sodhi, S. S., Dhillon, S. S., Jeyarajam, R., Cheema, S. S. 1973. Mycoplasma-like organism associated with citrus greening. *Proc. Indian Natl. Acad. Sci. Ser. B* 41:386–89
236. Su, H. J., Chang, S. C. 1976. The responses of the likubin pathogen to antibiotics and heat therapy. See Ref. 3, pp. 27–34
237. Su, H. J., Cheon, J. U., Tsai, M. J. 1986. Citrus greening (likubin) and some viruses and their control trials. In *Plant Virus Diseases of Horticultural Crops in the Tropics and Subtropics, FFTC Book Ser.* 33:142–47
238. Su, H. J., Chu, J. Y. 1984. Modified

technique of citrus shoot tip grafting and rapid propagation method to obtain citrus budwood free of citrus viruses and likubin organism. *Proc. Int. Soc. Citricult.* 1:332–34

239. Su, H. J., Wu, R. Y. 1979. Preliminary study on the etiology of Wentan pomelo decline. See Ref. 68, pp. 45–57
240. Suryanarayana, D., Upadhyay, R., Chona, B. L. 1968. Studies on the amino acid status of dieback-affected citrus trees in India. *Indian Phytopathol.* 21:118–20
241. Swai, I. S. 1988. Citrus diseases in Tanzania. *Acta Horticult.* 218:329–32
242. Tanaka, S., Doi, Y. 1974. Studies on mycoplasma-like organisms suspected cause of citrus likubin and leaf mottling. *Bull Fac. Agric. Tamagawa Univ.* 14:64–70
243. Tanaka, S., Doi, Y. 1976. Further investigations of likubin of Ponkan mandarin. See Ref. 3, pp. 35–37

243a. Tang, Y. Q. 1988. Preliminary survey on the parasite complex of *Diaphorina citri* Kuwayama (Homoptera:Psyllidae) in Fujian. See Ref. 77, pp. 10–15

244. Tirtawidjaja, S. 1980. Citrus virus research in Indonesia. See Ref. 13, pp. 129–32
245. Tirtawidjaja, S. 1981. Insect, dodder and seed transmission of citrus vein phloem degeneration (CVPD). *Proc. Int. Soc. Citricult.* 1:469–71
246. Tirtawidjaja, S., Hadewidjaja, T., Lasheen, A. M. 1965. Citrus vein phloem degeneration virus, a possible cause of citrus chlorosis in Java. *Proc. Am. Soc. Horticult. Sci.* 86:235–43
247. van Bruggen, A. H. C., Yilma, A. 1985. Virus and virus-like diseases of citrus in Ethiopia. *FAO Plant Prot. Bull.* 33:2–12
248. van den Berg, M. A. 1986. Effects of citrus cultivars and adapted irrigation on availability of new growth for the breeding of citrus psylla, *Trioza erytreae* (Del Guercio). *Fruits* 41:597–604
249. van den Berg, M. A., Deacon, V. E. 1988. Dispersal of the citrus psylla, *Trioza erytreae* (Del Guercio) (Hemiptera:Triozidae), in the absence of its hosts plants. *Phytophylactica* 20:361–68
250. van den Berg, M. A., Deacon, V. E. 1989. Flight activities of the citrus psylla, *Trioza erytreae* (Hemiptera:Triozidae). *Phytophylactica* 21:391–95
251. van den Berg, M. A., Deacon, V. E., Fourie, C. J., Anderson, S. H. 1987. Predators of the citrus psylla, *Trioza erytreae* (Hemiptera:Triozidae), in the lowveld and Rustenburg areas of Transvaal. *Phytophylactica* 19:285–89
252. van den Berg, M. A., van Vuuren, S. P., Deacon, V. E. 1987. Cross breeding and greening disease transmission of different populations of the citrus psylla, *Trioza erytreae* (Hemiptera:Triozidae). *Phytophylactica* 19:353–54
253. van den Berg, M. A., Vercuil, S. W. 1985. Neglected and dry-land trees as a source for the breeding of citrus psylla, *Trioza erytreae* (Del Guercio) (Hemiptera:Triozidae). *Subtropica* 6(12):12–15
254. van der Merwe, A. J., Andersen, F. G. 1937. Chromium and manganese toxicity. Is it important in Transvaal citrus greening? *Farming S. Afr.* 12:439–40
255. van der Merwe, C. P. 1942. The citrus psylla (*Spanioza erytreae,* del G.). *Sci. Bull. Dept. Agric. For. Union S. Afr.* No. 233, 12 pp.
256. van Lelyveld, L. J., van Vuuren, S. P. 1988. Peroxidase activity as a marker in greening disease of citrus for assessment of tolerance and susceptibility. *J. Phytopathol.* 121:357–62
257. van Lelyveld, L. J., van Vuuren, S. P. 1988. The effect of gentisic acid on activity of peroxidases from *Citrus aurantifolia*. *J. Phytopathol.* 121:363–65
258. van Lelyveld, L. J., van Vuuren, S. P., Visser, G. 1988. Gentisic acid concentration in healthy and greening-infected fruit albedo and leaves of citrus. *S. Afr. J. Plant Soil* 5:209–11
259. van Vuuren, S. P. 1977. The determination of optimal concentration and pH of tetracycline hydrochloride for trunk injection of greening-infected citrus trees. *Phytophylactica* 9:77–81
260. van Vuuren, S. P., da Graça, J. V. 1977. Comparison of the thin layer chromatographic methods for indexing citrus greening disease. *Phytophylactica* 9:91–96
261. van Vuuren, S. P., da Graça, J. V. 1977. The effects of exposure time and monocrotophos on citrus greening transmission by psylla *(Trioza erytreae). Citrus Subtrop. Fruit J.* 536:13–14
262. van Vuuren, S. P., Moll, J. N. 1984. Population dynamics and greening transmission by field psylla, *Trioza erytreae* (Del Guercio). See Ref. 24, pp. 90–94
263. van Vuuren, S. P., Moll, J. N. 1985. Influence of the rootstock on greening fruit symptoms. *Citrus Subtrop. Fruit J.* 612:7
264. van Vuuren, S. P., Moll, J. N., da Graça, J. V. 1977. Preliminary report on extended treatment of citrus greening with tetracycline hydrochloride by trunk injection. *Plant Dis. Reptr.* 61:358–59
265. Verma, A. K., Singh, B. P. 1977. In-

fluence of tristeza, greening and their complex on respiration in Kagzi lime leaves. *Indian Phytopathol.* 31:79

266. Verma, A. K., Singh, B. P. 1977. Metabolic changes induced by greening, tristeza and complex form in Kagzi lime, *Citrus aurantifolia*. *Indian J. Exp. Biol.* 15:811–14

267. Villechanoux, S., Garnier, M., Bové, J. M. 1990. Purification of the bacterium-like organism associated with greening disease of citrus by immunoaffinity chromatography and monoclonal antibodies. *Curr. Microbiol.* 21:175–800

268. Whittle, A. M., Muharam, A. 1988. Control measures against vein phloem degeneration in Indonesia. See Ref. 77, pp. 63–65

268a. Whittle, A. M., Nurhadi, F., Muharam, A. 1987. Report on greening workshop, Fuzhou, P. R. China, 6–12 Dec. 1987. Comments of the Indonesian delegation. See Ref. 8, 6 pp.

269. Wortmann, G. B., Schafer, E. 1977. The control of sucking insects with particular reference to citrus psylla (*Trioza erytreae* Del G.) by means of a soil applied insecticide. *Citrus Subtrop. Fruit J.* 520:14–16

270. Wu, S. P. 1987. Direct fluorescence detection for diagnosing citrus yellow shoot disease. See Ref. 8, 3 pp.

271. Wu, S. P., Faan, H. C. 1987. A microscopic method for rapid diagnosis of the citrus yellow shoot disease. See Ref. 8, 1 p.

272. Wu, S. P., Faan, H. C. 1988. Recent research on citrus yellow shoot in Guangdong Province. See Ref. 77, pp. 66–68

273. Xia, Y., Xu, C. 1988. First results of monitoring *Diaphorina citri* (Homoptera:Chermidae), the vector of huanglungbin, with yellow traps. See Ref. 77, pp. 32–34

274. Xia, Y., Xu, C., Chen, J. 1987. Population dynamics of *D. citri* in Fuzhou on *Murraya paniculata*. See Ref. 8, 2 pp.

275. Xie, P., Su, C., Lin, Z. 1988. A preliminary study on the parasite fungus of citrus psyllid *Cephalosporium lecanii* Zimm. See Ref. 77, pp. 18–22

276. Xie, P., Su, C., Lin, Z. 1988. A study on the cold endurance of the Asian citrus psyllid in Zhejiang. See Ref. 77, pp. 35–38

277. Xu, C. F., Xia, Y. H., Li, K. B., Ke, C. 1988. Further study of the transmission of citrus huanglungbin by a psyllid, *Diaphorina citri* Kuwayama. See Ref. 11, pp. 243–48

278. Zhao, X. Y. 1981. Citrus yellow shoot disease (Huanglongbin)—a review. *Proc. Int. Soc. Citricult.* 1:466–69

Annu. Rev. Phytopathol. 1991. 29:137–48

GENETICS OF SMALL-GRAIN SMUTS[1]

P. L. Thomas

Agriculture Canada Research Station, 195 Dafoe Road, Winnipeg, Manitoba, Canada R3T 2M9

KEY WORDS: Tilletia, ustilago, wheat, barley, oat

INTRODUCTION

A review of the genetics of the small-grain smuts and bunts would not be complete without reference to the classic book *Biology and Control of the Smut Fungi* (12), which contains a comprehensive survey of research done before 1957. Reviews published in 1968 (20), 1983 (37), and 1987 (4) have included information on the genetics of these fungi. Three recent reviews, in 1988, dealt specifically with the genetics of *Tilletia* spp. affecting wheat (32) and *Ustilago* spp. affecting wheat, barley, and oat (36, 41). The primary objective of this review is to bring the reader up to date by covering genetic research published after these earlier reviews. The following diseases (and causal species of fungi) have been used in such genetic studies and thus are of interest in this context: the common bunts of wheat (*Tilletia tritici* Bjerk. R. Wolff, syn. *T. caries* [DC] Tul., and *T. laevis* Kühn, syn. *T. foetida* [Wallr.] Liro); dwarf bunt of wheat (*T. controversa* Kühn); Karnal bunt of wheat (*T. indica* Mitra, syn. *Neovossia indica* [Mitra] Mundkur); the floral infecting smuts of wheat and barley (*Ustilago tritici* [Pers.] Rostr. and *U. nuda* [Jens.] Rostr.); and the seedling infecting smuts of barley and oat (*U. hordei* [Pers.] Lagerh., *U. nigra* Tapke, *U. avenae* [Pers.] Rostr., and *U. kolleri* Wille).

This whole group of causal organisms have life cycles involving a diploid teliospore that germinates to produce a promycelium (metabasidium) in which meiosis occurs, leading to a tetrad of haploid nuclei. The nuclei divide and are parcelled into gametic cells (basidiospores). Different types of gametes are present among the smut and bunt species. Those most suitable for genetic studies are from species in which the gametes are self-replicating, yeast-like cells called sporidia, which can be easily isolated and cultured in the laboratory. It is then relatively easy to produce infective, dikaryotic inoculum by mixing compatible sporidial isolates. For other species, haploid cultures can be established (35), albeit with difficulty, or hybridization can be accomplished by placing germinating teliospores in close proximity and subsequently detecting hybrids among the progeny by the presence or absence of known genetic markers, including virulence genes.

All members of this group of fungi are seedborne in that infective inoculum may be present on the exterior of the seed, under the hull, or may have already invaded the embryo during its development. The inoculum of some of the species (e.g. some bunts) can also survive in soil and cause infection in subsequent crops.

Under natural conditions, the dikaryon invades the host either at the flowering stage or through the cotyledon of the germinating seedling, depending on the fungal species. Those species that infect seedlings produce sori in the spikes and, in some instances, the upper foliage of the host plant. The species that infect at the flowering stage are of two types, *Ustilago* spp. that grow into the embryo and do not sporulate until the adult plant stage, and *T. indica,* which sporulates locally in the seed developing from the infected spike.

Artificial means may be used to induce infection at stages of plant growth other than those which are receptive in nature. Inoculum of the floral-infecting loose smuts of barley and wheat can be forced into a developing seedling (21–23), thereby producing an infected plant and sporulation in one generation (normally, infection in these species takes place in the flower of the preceding generation). Injection of bunt inoculum just above the nodes in primary stems of susceptible wheat results in the production of adventitious shoots, almost all of which subsequently developed sori containing teliospores (24). A similar technique, in which the uppermost node is inoculated at a later growth stage so as to infect the primary spike, has been used in genetic studies of bunt (11, 44).

Completion of the life cycle of the smut and bunt fungi in culture would accelerate genetic research but the techniques for most species are not reliable. Recent work with *Tilletia* spp. (45, 46) may allow this approach to be used with the bunt fungi.

GENERAL GENETICS

Most of the genetic work done on the small-grain smuts has involved covered smut of barley (*U. hordei*). This is not surprising, considering that the organism is economically important and that it has easily cultured sporidia amenable to controlled hybridization. Studies of the host-parasite genetics of *U. hordei/Hordeum vulgare* L. are facilitated by the relatively extensive genetic knowledge of the 14-chromosome host.

Mutation and Linkage

Mutants can be isolated after treating haploid sporidia of *U. hordei* with either ultraviolet light or a chemical mutagen, followed by an antibiotic/enzyme enrichment technique (16). The auxotrophs produced have requirements like those found in earlier studies with this organism (41). Morphological and temperature-sensitive mutants were also found. Complementation tests were used to obtain a preliminary identification of the allelic groups that the auxotrophs belonged to. Data from random sporidia isolated from the progeny of 10 crosses indicated a loose linkage between two genes and a recombination value of 12% between *pan-1* and *pro-2*.

Mating-type Loci

A much lower frequency of recombination was found between genes governing production of pantothenic acid (*PAN1*) and proline (*PRO1*) in another study of *U. hordei* (P. Thomas, unpublished data). A cross involving the two auxotrophs and mating type (*pan1*T481 *PRO1*T15 *MAT1-2* × *PAN1*T481 *pro1*T15 *MAT1-1*) was subjected to tetrad analysis. A total of 361 tetrads were isolated. Only eight of these exhibited recombination between the three markers, i.e. produced sporidia with nonparental associations of the markers (Table 1). The ratio for at least one of the markers within each of these tetrads was 1:3 or 0:4 rather than the expected 2:2; such abnormal ratios are typically produced by gene conversion. The data indicate that the genes responsible for these genotypes are very closely linked, such that single recombination events occurring at the complex locus lead to tetrads with abnormal segregation ratios. The discrepancy between the recombination frequency in these data and that noted above (16) will have to be resolved by further study.

The mating-type locus is known to be tightly linked to, if not the same as, a locus conferring proline auxotrophy in isolates of wild-type strains of another barley smut, *U. nuda* (35). All of the *a (MAT1-2)* mating-type lines that were isolated from a worldwide collection of *U. nuda* were proline auxotrophs (35). Several hundred teliospores from Canadian field isolates have subsequently been tested for proline auxotrophy and mating type with the same result (P. Thomas, unpublished data). In addition, sensitivity of haploid cultures to temperatures above 20°C was always found associated with the other known allele at the *MAT1* locus (35).

Considerably more work at the fine structure level will be required to determine the true nature of this seemingly complex, mating-type locus in *Ustilago,* especially as it appears to behave differently in different species,

Table 1 Nonparental associations of mating type and requirements for pantothenic acid and proline among 361 ordered tetrads of *U. hordei*

Recombinant tetrad no.	Auxotrophy (−), prototrophy (+) and mating-type alleles observed among the four sporidial lines in each ordered tetrad												Ratio of auxotrophy to prototrophy and of *MAT* alleles		
	1			2			3			4					
	PAN	*PRO*	*MAT*	*PAN*	*PRO*	*MAT*	*PAN*	*PRO*	*MAT*	*PAN*	*PRO*	*MAT*	*PAN*	*PRO*	*MAT*
1	−	+	2	−	−	2	−	−	1	+	−	1	1:3	1:3	2:2
2	−	−	1	+	−	1	−	−	2	−	−	2	1:3	0:4	2:2
3	−	−	2	−	+	2	−	−	1	+	−	1	1:3	1:3	2:2
4	−	−	1	−	+	2	−	+	2	−	−	2	0:4	2:2	1:3
5	−	+	2	+	+	1	−	−	1	−	−	1	1:3	2:2	3:1
6	−	−	1	+	+	2	−	−	1	−	+	2	1:3	2:2	2:2
7	−	−	2	+	−	1	−	−	2	+	−	1	2:2	0:4	2:2
8	−	−	2	+	−	1	−	+	2	−	−	1	1:3	1:3	2:2

i.e. until now, no species other than *U. nuda* has been found that is auxotrophic for proline in nature.

Genetic studies in *Tilletia indica* are relatively difficult because the mating compatibility between haploid lines cannot be determined by an in vitro test, such as the Bauch test that is used for sporidia-forming *Ustilago* spp. (12). Therefore, paired monosporidial lines must be inoculated into a susceptible host to determine compatibility or incompatibility. Compatibility in other *Tilletia* species can be determined by microscopic observation of sporidial fusion (18). Most investigators have reported a single mating-type locus (i.e. a bipolar mating system) in *Tilletia* spp., with evidence for multiple alleles in *T. controversa* (18) and at least four alleles in *T. indica* (9, 39). Progeny of a hybrid between *T. controversa* and *T. tritici* segregated as if mating type were controlled by two alleles at one locus (44). Resistance to an antibiotic segregated independently from mating type in this study.

The genetic designations that are currently the most used to denote alleles at bipolar mating-type loci in these smuts are "*A*" and "*a*". Yoder et al (47) pointed out that this type of usage is clumsy and confusing because it implies dominance and is different from the more commonly used gene designations in the fungi. They suggested that the symbols *MAT1* and *MAT1-x* be used for the first (identified) locus and its alleles, respectively, in all plant pathogenic fungi. Informally, the abbreviations *MAT-1* and *MAT-2* could then be used for the alleles in diallelic, bipolar systems. They further gave the suggested example of investigators agreeing to change "*A*" to *MAT1-1* and "*a*" to *MAT1-2*. Since the "*A*" and "*a*" designations in the smuts have generally been arbitrarily assigned, with different investigators not comparing their strains to a standard, I suggest that the "*a*", proline-requiring strains of *U. nuda* would be a convenient, naturally occurring standard for *MAT1-2*, against which any *Ustilago* species that will mate with *U. nuda* could be tested. Strains of *U. hordei, U. nigra, U. tritici, U. kolleri,* and *U. avenae* have already been standardized with the "*a*" strain of *U. nuda* in this laboratory. Note that all letters in the proposed gene symbol are uppercase since both alleles are needed for the activity of forming the dikaryon. This should not be confused with the linked function of control of proline production that is associated with the mating-type locus of *U. nuda* and *U. hordei*.

Echinulation

The genetic control of the ornamentation on teliospores, as observed in the light microscope, is well established as being due to two genes in the species *U. hordei, U. nigra, U. kolleri,* and *U. avenae* (36, 41). The scanning electron microscope was used to examine teliospores from test crosses of progeny from a *U. hordei* × *U. nigra* hybrid (42). At least one gene (in addition to the two mentioned above) was shown to be modifying the height of the rounded, wart-like echinulations.

Some wild-type isolates of *U. kolleri,* and progeny of crosses between these isolates and *U. hordei* (both nonechinulate species), possessed surface ornamentation that was approximately 30% of the diameter of normal echinulations and thus only visible in the scanning electron microscope (31). Teliospores with this ornamentation were termed "stippled". However, more study is required to determine the heritability of this character.

Germination Temperature

Teliospores of *T. controversa* germinate only at temperatures that are low relative to the requirements of *T. tritici*. This response appears to be under the control of an incompletely dominant gene(s) because hybrids germinate at intermediate temperatures (44).

VIRULENCE

Experiments designed to elucidate the inheritance of virulence are often adversely affected by the seemingly capricious nature of the most easily quantifiable symptom of infection by smuts and bunts, the proportion of plants with sporulation. The proportion observed is the result of a complex interaction between virulence genes, genes that modify the expression of virulence genes, the fitness of the pathogen, and the environment. Environment is a compound complication because it involves a host, perhaps with a complement of resistance genes. Among-plant and within-plant disease severity is also an important consideration (14, 15). Experiments repeated in different years are affected not only by changes in the environment in fields and greenhouses, but also by changes that take place in cultures during storage. A recent series of studies on the inheritance of virulence in *U. hordei* illustrates many of the complications due to the variables affecting the expression of this factor (6–8, 38).

Polygenic Modification

In the initial study in the series, a tetrad of sporidia from a race avirulent on cv. Trebi was hybridized, in all possible combinations, with a tetrad from a race virulent on this cultivar (38). Twenty-seven teliospores from these crosses were selfed and backcrossed to the avirulent parent. The results showed that virulence on Trebi was under the control of a single, dominant gene. This was contrary to the results of other investigators (41), who had shown that virulence on Trebi was due to a single recessive gene. Twenty-four sporidia were selected from the 27 tetrads according to the level of infection that they produced in the self- and back-crosses. The results from a diallele cross of these sporidia indicated that several genes, each producing a small incremental change in effect, modified the effect of the major gene. As well, certain combinations of polygenes interacted specifically to enhance or depress the effects of the major gene.

Reversal of Dominance

The parental lines used above were hybridized on Trebi and the universal suscept Odessa in the next experiments (7). As before, F_1 tetrads were selfed and backcrossed while additional data was produced by crossing F_2 tetrads deemed to be heterozygous for virulence with tetrads that were homozygous. There was a nonspecific interaction of the polygenes affecting virulence in that infection levels on Odessa were higher than, but correlated with, those on Trebi. The data also indicate that Odessa may have a previously undetected resistance gene that is interacting with the pathogen genotypes. Genetic analysis of the host would be required to confirm this. The most interesting aspect of the interpretation of data from these experiments is that virulence on Trebi was attributed to a recessive gene, contrary to the results of the earlier study with the same parental strains (38). This reversal in dominance was not understood, but several factors may have been responsible. In particular, the over-all amount of infection on Trebi was approximately one tenth of that observed in the earlier experiments (38), possibly due to changes in the parental cultures during storage between the experiments but probably due to environment.

It is interesting, despite the complications for researchers, to speculate that the virulence gene may be dominant under one environment and recessive under another. The expression of virulence of *U. hordei* on cv. Excelsior changed from recessive to partially dominant with changes in the environment in an earlier study (41).

Selection

Another aspect of the inheritance of virulence was addressed in studies that were concerned with virulence genes in fungal populations exposed to selection by the host. Field collections of the four-seedling infecting smuts of barley and oat (*U. hordei, U. nigra, U. kolleri,* and *U. avenae*) were subjected to selection by differential hosts for as many as 11 generations (5). Some stable races were produced and some of these were considerably more virulent than the original collection, indicating selection for virulence alleles. A more refined experiment using *U. hordei* indicates that the increased virulence was a consequence of the accumulation of genes modifying the expression of virulence (6). Three populations of five dikaryons each were formed by random selection from a group of 15 dikaryons that were known to be heterozygous for virulence on Trebi. These populations were exposed to three generations of selection on the cultivars Trebi and Odessa. Random changes were observed over the generations, with no significant difference between selection on the two cultivars. This indicates that selection did not result in genetic stability in the population, nor did it lead to loss of the gene for virulence on Trebi when selection took place on the susceptible host, Odessa. The variability was attributed to the complex nature of the interaction between varying genotypes, possibly compounded by environmental interaction (6).

It is often assumed that tetrad-forming fungi are homozygous in nature due to their tendency to inbreed because most infections probably result from one teliospore, as has been shown for species such as *U. nuda* (36). However, an examination of the theoretical implications of intratetrad selfing showed that heterozygosity will tend to be maintained or preserved (29). Therefore, variability seen in the offspring of heterozygous teliospores after several generations of selection should not be surprising.

Fitness

The impact of the parasitic fitness of the fungus on studies of virulence is clearly shown by the data from one of the three populations of dikaryons mentioned above (6). This population was unable to produce spores in the host cultivars after the second generation of selection (8). The results were attributed to a mutation modifying chromosomal behavior during meiosis (8). Why this mutation occurred in all five dikaryons in this population, but only in one of the 10 dikaryons in the other two populations, was not addressed. If an infective agent were responsible, one would have expected the population with one abnormal dikaryotic line to succumb.

FUNGICIDE RESISTANCE

The most cost-effective, efficient, and environmentally safe method that can be used to control the smuts and bunts is to grow resistant cultivars. However, when cultivars that are both agronomically desirable and resistant are not available to the farmers, or when resistance is overcome by the pathogen, the use of seed-treatment fungicides becomes a consideration. In an era of environmental consciousness, this class of fungicides has an advantage over other classes because relatively small quantities are required to control the fungus in its restricted location on or in the seed (soilborne bunts require larger quantities). Effective use of these fungicides requires knowledge of the genetics of any strains of fungi that may express resistance.

Mutants of *U. hordei* with resistance to benomyl and carboxin were produced by ultraviolet irradiation. The genetic control of resistance was dominant in tests of the fungus in both culture and the host (1). Resistance of other strains was monogenic at one rate of carboxin but polygenic over a range of concentrations (2). Spontaneous and induced mutants of *U. hordei* were obtained that were resistant to four fungicides (17). A limited genetic study indicated that the resistance could be due to more than one locus, with genes that were dominant, semidominant or recessive, and stable or unstable.

Two field isolates of *U. nuda* from Europe were shown to be resistant to carboxin when germinating on media containing the fungicide (30). Since the fungicide normally acts on dikaryotic hyphae rather than the germinating teliospore, a laboratory test was developed that allowed screening after germination (33). Some field isolates of *U. nuda* obtained from Europe were shown to be resistant in this laboratory test and also in tests involving inoculated host plants (34). This resistance has been determined, by tetrad analysis, to be due to a single dominant gene that is not linked to the *MAT1* locus (G. Newcombe & P. Thomas, unpublished data).

BIOCHEMICAL TECHNIQUES

Electrophoresis

Two-dimensional, isoelectric-focusing polyacrylamide-gel electrophoresis was used to compare detergent-soluble, polypeptide extracts from teliospores of several *Ustilago* species, including *U. hordei* and *U. nigra* (27). The same technique was then applied to isolates of *U. hordei* with known genes for virulence (43). An initial correlation of a dominant allele for avirulence with a prominent polypeptide was proven wrong when tetrads from a heterozygote were backcrossed to the virulent parent. The prominent polypeptide clearly segregated independently from avirulence. This study provided an experimental example of the importance, as stressed by Ellingboe (10), of conducting genetic tests for the independence of phenomena involving host-parasite systems before drawing causal relationships. However, the prominent polypeptide was under the control of a single gene. Since it should be possible to biochemically characterize such polypeptides, this type of study could contribute toward an understanding of the basic biology of *U. hordei* through identification of gene products. Auxotrophic mutants would also be good subjects for this type of study.

The inheritance of a total of 10 isozymic alleles at four polymorphic loci of *T. indica* was examined in the progeny of 15 crosses (3). Monoteliospore cultures generally exhibited the enzyme patterns that were expected. However, the basidiospores from single teliospores frequently produced non-1:1 ratios for pairs of alleles at single loci. These abnormal ratios could have been caused by postmeiotic segregation. Basidiospores frequently inherited both alleles from teliospores that were heterozygous at isoenzyme loci. Diploid sporidia were ruled out, based on an earlier cytological study (13). Aneuploids could not be ruled out. Heterokaryotic basidia were also deemed possible, especially given a germination biology in which up to 180 basidiospores are produced by each teliospore (3), and in which nuclei appear to move out of basidiospores after dividing and then into basidiospores that have not yet received nuclei (3).

Transformation

U. hordei and *U. nigra* sporidia were genetically transformed to resistance to hygromycin B by exposing sphaeroplasts to a plasmid containing a bacterial gene encoding hygromycin B phosphotransferase (19). Cells from one *U. hordei* transformant retained resistance through 25 mitotic divisions while those from a *U. nigra* transformant were unstable. The transformation rates were 1/10th of those generally obtained for *Ustilago maydis* (DC) Cda, probably due to relatively poor regeneration of sphaeroplasts with the techniques used.

Plasmids

A promising genetic tool has been identified in *Tilletia* spp. Homologous linear plasmids with no homology with mitochondrial, ribosomal, or nuclear DNA were found (26). They are clearly of mitochondrial origin but their functional role has not been ascertained (28).

Karyotypes

Biochemical karyotyping techniques will also see use in the future. In an initial study, chromosomal DNA of sporidia of *U. hordei* was obtained after extracting from protoplasts and subsequent separation by orthogonal-field, alternation-gel electrophoresis (25). Electrophoretic karyotyping of *U. hordei, U. tritici,* and *T. controversa* is also possible without generating protoplasts (31a). Hopefully, such studies will improve our knowledge of chromosome complement, "probably the least known aspect of cytology in the smut fungi" (12). This observation, published 34 years ago, still holds true today.

SUMMARY

Despite being the subject of study since the early days of plant pathology, the amount of genetic work done with the smuts and bunts affecting small grains is relatively small. This is because they have not been used in large-scale, basic genetic research as has a smut like *U. maydis*. This situation is beginning to change with many laboratories selecting organisms like *U. hordei* for molecular studies.

However, even on the practical side of research related to agronomy, we are still in the infancy of the study of the genetics of virulence in the smuts and bunts. We do not know how any virulence gene functions, nor have any been mapped. The complex, host-parasite interaction leading to the expression of the diseases is poorly understood, especially with regard to factors controlling the level of infection. Perhaps it would be more informative, in the future, to first study the inheritance of all known virulence and resistance genes in pathogens bred and selected for high levels of virulence and in hosts bred and selected for low levels of resistance. Modifying genes could then be added in stages through hybridization to determine their effects.

Literature Cited

1. Ben-Yephet, Y., Henis, Y., Dinoor, A. 1974. Genetic studies of tolerance of carboxin and benomyl at the asexual phase of *Ustilago hordei*. *Phytopathology* 64:51–56
2. Ben-Yephet, Y., Henis, Y., Dinoor, A. 1975. Inheritance of tolerance to carboxin and benomyl in *Ustilago hordei*. *Phytopathology* 65:563–67
3. Bonde, M. R., Peterson, G. L., Royer, M. H. 1988. Inheritance of isozymes in the smut pathogen *Tilletia indica*. *Phytopathology* 78:1277–79
4. Caten, C. E. 1987. The concept of race in plant pathology. In *Populations of Plant Pathogens: Their Dynamics and Genetics*, ed. M. S. Wolfe, C. E. Caten, pp. 21–37. Oxford: Blackwell Sci. Publ.
5. Cherewick, W. J. 1958. Cereal smut races and their variability. *Can. J. Plant Sci.* 38:481–89
6. Christ, B. J., Person, C. O. 1987. Effects of selection by host cultivars on populations of *Ustilago hordei*. *Can. J. Bot.* 65:1379–83
7. Christ, B. J., Person, C. O. 1987. Nonspecific action of polygenes in *Ustilago hordei*. *Can. J. Bot.* 65:1390–95
8. Christ, B. J., Person, C. O. 1988. Relationship between abnormal germination of *Ustilago hordei* teliospores and reduction in smut. *Can. J. Plant Pathol.* 10:105–9
9. Duran, R., Cromarty, R. 1977. *Tilletia indica:* A heterothallic wheat bunt fungus with multiple alleles controlling incompatibility. *Phytopathology* 67:812–15
10. Ellingboe, A. H. 1976. Genetics of host parasite interactions. In *Physiological Plant Pathology, Encyclopedia of Plant Pathology*, ed. R. Heitefuss, P. H. Williams, pp. 760–78. New York: Springer Verlag
11. Fernandez, J. A., Duran, R. 1978. Hypodermic inoculation, a rapid technique for producing dwarf bunt in wheat. *Plant Dis. Rep.* 62:336–37
12. Fischer, G. W., Holton, C. S. 1957. *Biology and control of the smut fungi*. New York: The Ronald Press Co. 622 pp.
13. Goates, B. J., Hoffmann, J. A. 1987. Nuclear behaviour during teliospore germination and sporidial development in *Tilletia caries, T. foetida,* and *T. controversa*. *Can. J. Bot.* 65:512–17
14. Groth, J. V., Person, C. O., Ebba, T. 1976. Relation between two measures of disease expression in barley-*Ustilago hordei* interactions. *Phytopathology* 66: 1342–47
15. Groth, J. V., Person, C. O., Ebba, T. 1977. The measurement of disease severity in cereal smut. In *Genetic Diversity in Plants*, ed. A. R. Muhammed, R. Aksel, R. C. von Borstel, pp. 193–204. New York: Plenum
16. Henry, C. E., Bullock, B., Smith, V., Steward-Clark, E. 1988. Genetics of *Ustilago hordei:* mutagenesis and linkage tests. *Bot. Gaz.* 149:101–6
17. Henry, C. E., Gaines, V., Bullock, B., Schaefer, R. W. 1987. Genetics of *Ustilago hordei:* fungicide resistant mutants. *Bot. Gaz.* 148:501–6
18. Hoffmann, J. A., Kendrick, E. L. 1969. Genetic control of compatibility in *Tilletia controversa*. *Phytopathology* 59:79–83
19. Holden, D. W., Wang, J., Leong, S. A. 1988. DNA-mediated transformation of *Ustilago hordei* and *Ustilago nigra*. *Physiol. Mol. Plant Pathol.* 33:235–39
20. Holton, C. S., Hoffmann, J. A., Duran, R. 1968. Variation in the smut fungi. *Annu. Rev. Phytopathol.* 6:213–42
21. Kavanagh, T. 1961. Inoculating barley seedlings with *Ustilago nuda* and wheat seedlings with *U. tritici*. *Phytopathology* 51:175–77
22. Kavanagh, T. 1963. Seedling inocula-

tion with *Ustilago nuda* (Jens.) Rostr. *Nature* 200:1021
23. Kavanagh, T. 1964. Factors influencing seedling infection of barley by *Ustilago nuda* (Jens.) Rostr. *Ann. Appl. Biol.* 54:225–30
24. Kawchuck, L. M., Nielsen, J. 1987. Improved bunt inoculation of wheat utilizing adventitious shoots. *Can. J. Bot.* 65:1284–85
25. Kim, W. K. 1988. Analysis of fungal genome—Gene analysis technique using orthogonal field alternation gel electrophoresis. *Kor. J. Mycol.* 16:101–5
26. Kim, W. K., McNabb, S. A., Klassen, G. R. 1988. A linear plasmid in *Tilletia controversa,* a fungal pathogen of wheat. *Can. J. Bot.* 66:1098–100
27. Kim, W. K., Rohringer, R., Nielsen, J. 1984. Comparison of polypeptides in *Ustilago* spp. pathogenic on wheat, barley, and oats: A chemotaxonomic study. *Can. J. Bot.* 62:1431–37
28. Kim, W. K., Whitmore, E., Klassen, G. R. 1990. Homologous linear plasmids in mitochondria of three species of wheat bunt fungi, *Tilletia caries, T. laevis* and *T. controversa*. *Curr. Genet.* 17:229–33
29. Kirby, G. C. 1984. Breeding systems and heterozygosity in populations of tetrad forming fungi. *Heredity* 52:35–41
30. Leroux, P., Berthier, G. 1988. Resistance to carboxin and fenfuram in *Ustilago nuda* (Jens.) Rostr., the causal agent of barley loose smut. *Crop Protection* 7:16–19
31. Maksymetz, L. C. M. 1989. *Pathogenicity and spore morphology in hybrids between* Ustilago hordei and U. kolleri. MSc thesis. Univ. Manitoba, Winnipeg. 84 pp.
31a. McCluskey, K., Russell, B. W., Mills, D. 1990. Electrophoretic karyotyping without the need for generating protoplasts. *Curr. Genet.* 18:385–86
32. Mills, D., Churchill, A. C. L. 1988. *Tilletia* spp., bunt fungi of the Gramineae. See Ref. 40, pp. 401–14
33. Newcombe, G., Thomas, P. L. 1990. Fungicidal and fungistatic effects of carboxin on *Ustilago nuda*. *Phytopathology* 80:509–12
34. Newcombe, G., Thomas, P. L. 1991. Incidence of carboxin resistance in *Ustilago nuda*. *Phytopathology*. 81:247–50
35. Nielsen, J. 1968. Isolation and culture of monokaryotic haplonts of *Ustilago nuda*, the role of proline in their metabolism, and the inoculation of barley with resynthesized dikaryons. *Can. J. Bot.* 46:1193–200
36. Nielsen, J. 1988. *Ustilago* spp., smuts. See Ref. 40, pp. 483–90
37. Person, C., Fleming, R., Cargeeg, L., Christ, B. 1983. Present knowledge and theories concerning durable resistance. In *Durable Resistance in Crops,* ed. F. Lamberti, J. M. Waller, N. A. Vander, pp. 27–40. New York: Plenum
38. Person, C. O., Christ, B. J., Ebba, T. 1987. Polygenic modification of a major virulence gene in *Ustilago hordei*. *Can. J. Bot.* 65:1384–89
39. Royer, M. H., Rytter, J. 1985. Artificial inoculation of wheat with *Tilletia indica* from Mexico and India. *Plant Disease* 69:317–19
40. Sidhu, G. S., Williams, P. H., Ingram, D. S. 1988. *The Genetics of Phytopathogenic Fungi*. Vol. 6, *Advances in Plant Pathology*. London: Academic. 566 pp
41. Thomas, P. L. 1988. *Ustilago hordei,* covered smut of barley and *Ustilago nigra,* false loose smut of barley. See Ref. 40, pp. 415–25
42. Thomas, P. L. 1989. Genetic modification of echinulation on teliospores of *Ustilago hordei* X *U. nigra* hybrids. *Bot. Gaz.* 150:319–22
43. Thomas, P. L., Kim, W. K., Howes, N. K. 1987. Independent inheritance of genes governing virulence and the production of a polypeptide in *Ustilago hordei*. *J. Phytopathology* 120:69–74
44. Trail, F., Mills, D. 1990. Growth of haploid *Tilletia* strains in planta and genetic analysis of a cross of *Tilletia caries* X *Tilletia controversa*. *Phytopathology* 80:367–70
45. Trione, E. J., Hess, W. M., Stockwell, V. O. 1989. Growth and sporulation of the dikaryons of the dwarf bunt fungus in wheat plants and in culture. *Can. J. Bot.* 67:1671–80
46. Trione, E. J., Stockwell, V. O., Latham, C. J. 1989. Floret development and teliospore production in bunt-infected wheat in planta and in cultured spikelets *Phytopathology* 79:999–1002
47. Yoder, O. C., Valent, B., Chumley, F. 1986. Genetic nomenclature and practice for plant pathogenic fungi. *Phytopathology* 76:383–85

Annu. Rev. Phytopathol. 1991. 29:149–66

FACTORS AFFECTING THE EFFICACY OF NATURAL ENEMIES OF NEMATODES*

R. M. Sayre

Nematology Laboratory, USDA-ARS, Beltsville, Maryland 20705

D. E. Walter

Department of Ecology and Evolutionary Biology, Monash University, Clayton, Melbourne, Victoria 3168, Australia

KEY WORDS: root-knot nematodes, cyst nematodes, soil antagonists of nematodes, nematode-trapping fungi, nematophagous soil invertebrates

INTRODUCTION

Nematologists generally agree that the many field plot studies on nematicides initiated and carried out during the late 1940s and early 1950s were responsible in part for the growth of nematology in the United States (43, 54, 87). These demonstration plots conducted in several states showed convincingly the efficacy of broad spectrum nematicides. Beneficial results coming from the plots were not overlooked by state and federal funding agencies. Agencies recognized the economic gains that were being fostered by the burgeoning discipline of nematology. Early and continued monetary support by agencies at sustained levels (i.e. formula funding) helped underwrite the growth of the science (2). Conversely, recent removal of several effective nematicides from the marketplace resulting from federal deregulation (31) eventually may have the reverse effect. Support of nematological research might ebb if replacement materials and practices no longer show the dramatic results once associated with broad spectrum nematicides.

In searching for and reporting on alternatives to nematicides, investigators and reviewers alike have listed many natural soil enemies of plant-parasitic nematodes, and considered some as potential biocontrol agents (18–20, 30, 32, 34, 36, 46, 47, 53, 60, 62, 63, 67, 68, 77, 83, 84, 88, 89). However, results of field studies of natural enemies rarely approach the degree of control that was obtained by the use of nematicides. Nevertheless, reviewers almost without exception concluded that future replacements for nematicides probably will be found among the natural soil enemies of nematodes.

NATURAL CONTROL OF PHYTONEMATODES (SUPPRESSIVE SOILS)

Some modern "demonstration plots" were neither designed nor executed by nematologists, but were fortuitous field observations of abrupt population declines of phytonematodes with little or no subsequent resurgence. Rovira (64) reviewed two examples of chance observation of nematode suppressive soils: the first by Gair et al (16), the second by Ferris et al (13). In both soils an unexpected and sustained population decline of pest nematodes had developed under systems of crop monocultures. Since these two initial discoveries, other investigators (5, 27, 51, 58, 81) have observed similar declines. These latter observations of pest nematode declines were also in crop monocultures where plant nematode diseases were no longer economically significant.

These modern demonstration plots serve to initiate our discussion on the efficacy of natural enemies in controlling nematodes. Knowledge of the edaphic factors inherent in suppressive soils that favor populations of natural enemies over pest nematodes should provide some clues for using these potential biological control agents.

The term "biological control" has engendered conflicting viewpoints (15). We accept the definition offered by Baker & Cook (1) that served as the basis for Stirling's (77) definition: "Biological control is the reduction in nematode damage by organisms antagonistic to nematodes through the regulation of nematode populations and/or a reduction in the capacity of nematodes to cause damage, which occurs naturally and is accomplished through the manipulation of environment or by the mass introduction of the antagonist." In the past when nematologists observed spontaneous declines in field populations of plant nematodes, they dismissed the reductions as resulting from changes in cultural or cropping practices. Recently, investigators have recognized the significance of the natural control phenomenon and have initiated studies to understand these inexplicable declines. Investigations of suppressive soils, we feel, marked a historical turning point from an unquestioned reliance on nematicides to the search for alternative control strategies. This new research initiative in nematology paralleled the ongoing investigations in progress in allied disciplines of entomology, plant pathology, and weed sciences.

In this review we examine factors affecting the efficacy of natural enemies of nematodes (i.e. biological control agents) and, in particular, factors amenable to manipulation by growers. We use an organismal approach. Interactions between plant-parasitic nematodes and fungal (Figure 1 *a–c*), bacterial (Figure 2 *a–c*) and invertebrate antagonists (Figure 3 *a–c*) are each influenced differently by edaphic factors. Nematicidal efficacy of a biological control agent is likely to be affected by environment conditions, and control will probably be more difficult to achieve than when using traditional chemical control methods. Generally, organisms that grow actively in soil are more

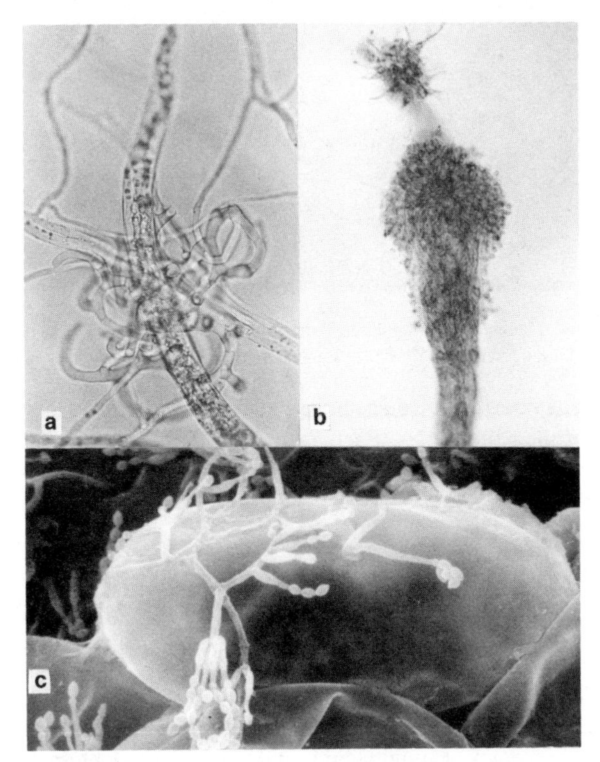

Figure 1 Fungal antagonists: (*a*) Nematode in the net of an *Arthrobotrys* sp. (450 X); (*b*) A nematode colonized by *Catenaria* sp. (550 X); (*c*) Nematode egg colonized by *Paecilomyces* sp. (3000 X).

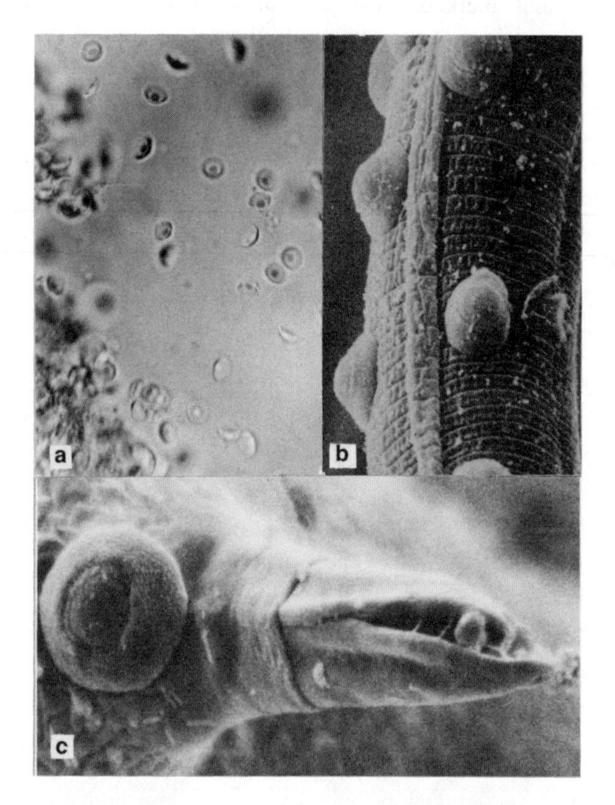

Figure 2 Bacterial antagonists: (*a*) Endospores of *Pasteuria* sp. (875 X); (*b*) Endospores attached to root-knot nematode (450 X); (*c*) Endospore on male of soybean cyst nematode (6000 X).

affected by edaphic factors than those that remain dormant in soils for the greater part of their life cycles. Each agent has its temperature, moisture, and pH requirements for growth, infection, or predacious activity so it is not possible to generalize about what soil conditions may favor the control of any particular nematode group (36).

Fungal Biocontrol Agents

The first documented study of a nematode suppressive soil was a decline in population of cereal cyst nematode (CCN), *Heterodera avena*. Gair et al (16) observed that continuous cropping of oats did not, as would be expected, result in high populations of CCN in the oat and barley fields of southern England. Three species of fungi were found that parasitized females and cysts of CCN (34). Apparently, these fungal populations kept the numbers of CCN below threshold densities that caused economic loss (39). Also, Kerry et al

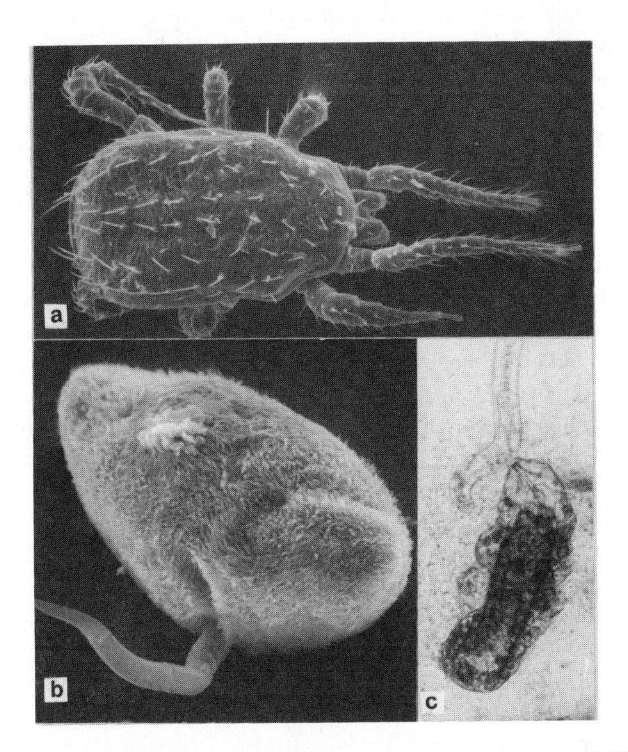

Figure 3 Invertebrate antagonists: (*a*) Nematophagous mite, *Lasioseius* sp. (150 X); (*b*) Turbellarian, *Adenoplea* sp. ingesting a nematode (500 X); (*c*) Tardigrade, *Hypsibius* sp., feeding on a nematode (375 X).

(38) observed that development and parasitism of fungal antagonists was greater on CCN during prolonged periods of high rainfall or in poorly drained soils than in the dry periods or in well-drained soils. They concluded that high soil moisture was a key factor contributing to the efficacy of these fungi as biocontrol agents. Their finding was confirmed in part with experiments in arid soils in Australia, where some of the same fungi were present as in England, but limited soil moisture prevented the fungi from functioning as biocontrol agents of the CCN (65).

In their examination of soils from old peach orchards, Ferris et al (13) noted that previously high populations of the root-knot nematode *Meloidogyne javanica* had declined to low concentrations. In ensuing studies Stirling et al (79) found the egg masses of *M. javanica* on peach roots were colonized by the fungus *Dactylella oviparasitica*. They postulated that the fungus was responsible for reduction in the root-knot nematode populations, and was the primary factor responsible for the nematode-suppressive soil. However, this fungus was unable to control the same root-knot species when it occurred on roots of tomato and grape. On roots of the latter hosts, egg masses of *M. javanica* were larger and more abundant. Apparently, the fungus was incapable of keeping pace with the higher reproductive capacity of *M. javanica* in its more favorable host. They postulated that a delicate balance existed between the fungus and eggs of the nematode. *Dactylella oviparasitica* appeared to be only a weak fungal parasite and was more successful in regulating the nematode populations that were under stress on the less than favorable host, peach.

Similarly, Jaffee & Muldoon (27) found that the nematophagous fungus, *Hirsutella rhossilienses,* was responsible for the natural suppression of *Heterodera schachtii* on cabbage roots. In loamy soils, 40–63% of the second-stage juveniles (J2) of the sugar beet cyst nematode had spores of the fungus adhering to their cuticles. Of those with spores, 82–92% were infected with the fungus.

Bacterial Biocontrol Agents

In vineyards near the towns of Loxton and Cooltong in South Australia, Stirling & White (81) conducted a survey and found that root-knot nematodes infected with *Pasteuria penetrans* occurred more frequently on grapes over 25 years old and were rarely found in vineyards under 10 years in age. In older vineyards 33–40% of females were parasitized, suggesting that the bacterium might be useful in the biological control of the nematode. In ensuing investigations of these same vineyards, Bird & Brisbane (5) examined seven soils for their suppressiveness to the reproduction of the root-knot nematode, *M. javanica*. Soils from Cooltong significantly reduced the numbers of egg masses produced by the root-knot nematodes. Suppressiveness of these soils was biological in origin because it was removed by autoclaving. Where inhibition of nematode reproduction occurred, the endospores of *P. penetrans* replaced all egg production of the adult females. Other bacterial isolates taken from these same soils showed no inhibitory effect on *M. javanica*. Except for the presence of the *Pasteuria* sp., they found no other soil factors associated with the incidence of parasitism caused by the bacterium.

Similarly, Nishizawa (58) followed populations of soybean and rice cyst nematodes in newly established experimental plots in monocultures of soybean and upland rice. Periodically, the plots were sampled for the rice cyst nematode, *Heterodera elachista*. During the six-year period of the experiment populations of the nematode reached their maximum then declined rapidly (Figure 4). Greenhouse studies showed that the observed decline of cyst nematode populations was due to the presence of a biocontrol agent—an unnamed species of *Pasteuria* (Figure 2c).

The "decline phenomenon," which Nishizawa (58) observed, parallels

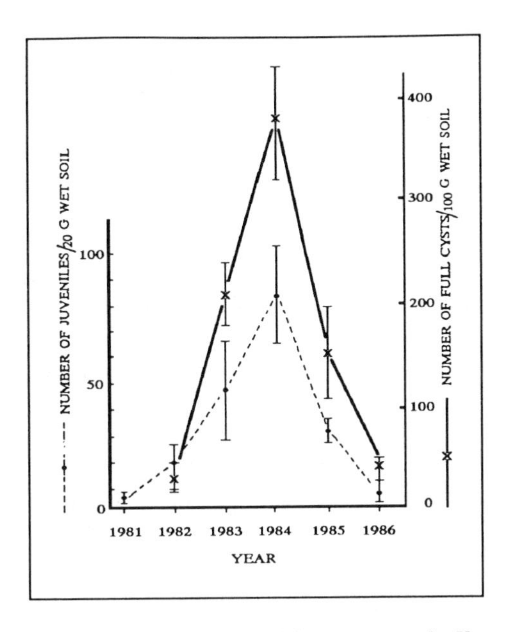

Figure 4 Annual population changes of the upland rice cyst nematode, *Heterodera elachista*, in plots where upland rice was cultivated successively for 6 years and associated with the occurrence of a *Pasteuria* sp. (59). Vertical bars denote LSD (P = 0.05).

the observations of Gair et al (16). In graphic form the decline phenomenon showed a sharp rise in the two nematode populations during the first years. This rise was followed by a rapid and sustained population decline in the ensuing years and, consequently, relatively little economic crop loss occurred. Such results are typical of biological control phenomena as discussed by Schneider (71) and Kerry (36). A typical population response begins with an induction period when the antagonist builds up in soil in response to the presence of the nematode host populations. This period may vary from a few to many years. It is then followed by the suppressive phase during which nematode populations fall to densities below the threshold of economic damage. Kerry (36) postulated that long periods of association between antagonists and nematodes are necessary for the development of suppressive soils. Consequently, natural control of nematodes is more likely to be expressed in monocultures and under perennial cropping systems (36).

Minton & Sayre (51) also found an association between a low population of *M. arenaria* and the occurrence of *P. penetrans* in soils in Georgia. For two decades, field plots of about two hectares at Tifton, Georgia, were used for research on plant-parasitic nematodes. Despite the continuous cropping of the plots to hosts of *M. arenaria*, populations of that nematode in the latter years dropped to densities that did not seriously affect peanut yields. These nematode-suppressive soils were bioassayed using the migratory second-stage juveniles (J2) of *M. arenaria*. Most J2 emerging from the suppressive soil were encumbered with >25 endospores of a bacterium, *P. penetrans* (Figure 2*b*). Endospore loads of this magnitude were considered sufficient to prevent some J2 from penetrating plant roots. While lesser numbers of endospores (i.e. 10–24) did not prevent J2 from penetrating roots, most developing females within the host plants were found parasitized by the bacterium. In further greenhouse experiments, incidence of galling caused by *M. arenaria* and its reproduction were inversely related to numbers of endospores found in soil. These investigators concluded that a naturally occurring population of *P. penetrans* acted as a biological control agent of *M. arenaria* (51).

Invertebrate Biocontrol Agents

Interest in invertebrate nematophages has a long history (6, 82), but rarely have invertebrates been considered as potential antagonists in suppressive soils. Even weakly pathogenic fungi are often credited with reducing nematode numbers, but any nematophagous arthropods (Figure 3*a*), turbellarians (Figure 3*b*), tardigrads (Figure 3*c*), protozoa, annelids, and nematodes (52, 61, 63, 70, 73, 74, 91) that may be present are generally ignored. Below we consider what factors may influence the control potential of invertebrate nematophages, adapting McMurtry's (48) approach.

VORACITY, PREY SPECIFICITY, AND SURVIVAL AT LOW PREY DENSITIES Nelmes (56) estimated that the predatory nematode *Prionchulus punctatus* consumed between one and six times its body volume in nematodes per day. Similarly high consumption rates with little indication of satiation or a numerical response to prey density were characteristic of most predatory nematodes that have been studied (7, 23, 93). Nematophagous mites consumed between 22 and 109% of their body weight in nematode biomass (dry weight) per day (91) but often stopped feeding, presumably for digestion to occur.

Nematophagous mites had faster development times, lower mortality, and higher egg-laying rates on nematodes in comparison to arthropod prey (88, 91, 92). There was little indication of feeding specificity in either predatory nematodes or arthropods. Walter (89) recognized three categories of nematophagous arthropods: omnivores feeding on microbes and on nematodes; general predators of soil invertebrates; and nematode specialists. High population densities of nematode specialists are sometimes reported (22, 32), but even specialized nematophages attacked a variety of nematode prey (22, 55, 70, 73, 90).

Generalist predators (21) can be effective biological control agents. Omnivorous arthropods (collembolans) have been shown to reduce numbers of plant-parasitic nemtodes in pot studies (17, 85). Perhaps the ability to use alternative foods allows for survival or even population growth during periods of low pest populations (49).

Certain nematode species, life stages, or feeding types do appear to be more resistant to predation than others. In general, all vermiform nematodes tested were attacked by the predatory nematodes discussed above, but attacks against large dorylaims *(Xiphinema, Longidorus)* and *Helicotylenchus* species were not successful. Feeding by the earthworm *Lampito mauritii* (Kinberg) reduced populations of predatory and microbivorous nematodes by about 90%, compared to only a 47% reduction in plant-parasitic species (10). Adult females and eggs of root-knot and cyst nematodes were often immune to predation by mites and nematodes that fed on their vermiform stages (23, 25, 55). The mesostigmatic mite *Hypoaspis* nr. *aculeifer* (Canestrini) and the acarid mite *Caloglyphus* sp., however, have both been reported to prey on the eggs of root-knot nematodes (26, 72), and the acarid mite *Tyrophagus similis* Volgin has been reported to feed on females of *H. avenae* (78).

POWER OF DISPERSAL AND DISTRIBUTION IN RELATION TO PREY Microcosm studies have shown that soil pore structure, soil moisture, and trophic interactions can be complexly interconnected. Protozoa, for example, are restricted in distribution and in predatory ability by low soil moisture conditions (24, 86). Nematodes, including predatory species, are generally assumed to be restricted to existing soil passageways and to require a film of water for movement (8). Elliott et al (12) found that the omnivorous nematode *Mesodiplogaster lheritieri* grew better in microcosms with bacteria and protozoa than in those with only bacteria. They suggested that protozoa were able to graze on bacteria from pores too small to admit nematodes, but that the protozoa were then eaten by *M. lheritieri*.

The assumption that the larger nematophages are restricted to existing soil pores (74), however, must be questioned. *Digamasellus quadrisetus* Berlese, a nematophagous mite associated with bark beetles, has been observed to mine a passageway through packed bark beetle frass in its search for beetle and nematode prey (40). Collembola are strongly associated with plant roots where they may help suppress plant disease (9). An ability to burrow through soil is a likely explanation for the association between nematophagous mites and plant roots (26, 72).

REPRODUCTIVE POTENTIAL Generation time and reproductive output are important determinants of reproductive potential. Plant-parasitic nematodes generally have relatively long generation times in relation to their invertebrate antagonists, usually on the order of 3 to 6 weeks (11). In contrast, the amoeba *Theratromyxa weberi* Zwillenberg can complete its life cycle in 23 hr at 21°C

(66). Many predatory nematodes also have long generation times, but the aphelenchoid *Seinura tenuicaudata* (De Man) is a notable exception. At 28°C generation time is 5.5 to 6 days, and about 12 eggs are laid each day, 1–2 eggs being produced for each *Aphelenchus avenae* eaten (23). Many nematophagous mites also have rapid developmental times and high reproductive outputs (91).

MANIPULATING NATURAL ENEMIES OF PHYTONEMATODES

Pioneering research by Linford (41) on the influence of soil organic amendments on root-knot nematodes kindled an interest in many investigators that persists to the present. Investigators still examine organic materials and suggest reasons for nematode control that occurs during decomposition of various amendments. Besides using soil organic amendments in biocontrol, as was suggested by Linford et al (42), other cultural practices (i.e. measurement of soil moisture, use of chemical fertilization of plants, regulation of soil pH, etc) have been investigated. Recently, the influence of these practices on fungal antagonists of nematodes has been reviewed (18–20). Several practices can be manipulated, monitored, and shown to have a positive or negative correlation with the activity of natural enemies and with the decline of nematode populations. The exploitation of beneficial effects of natural enemies by altering soil environment is a major and current objective of nematologists who are interested in biocontrol, and the subject of this review. There are three types of biological control agents: (*a*) Natural, where an agent has increased to concentrations that suppress nematode populations without having been specifically introduced, (*b*) the augmentation or inundation of a soil with an agent(s) that is already present in soils and recognized as containing a pest nematode population, and (*c*) the introduction of an exotic enemy with the hope that it may become established and increase to densities that have an economic impact on a pest nematode. Already, we have looked at natural enemies of nematodes. Now, we examine instances of biological control through the augmentation of a soil with resident nematode antagonists.

Exploiting Resident Soil Antagonists

FUNGAL ANTAGONISTS Over 200 species of nematophagous fungi have been mentioned in reviews (3, 4, 33, 35). Nematophagous fungi form the single largest group of natural soil enemies of nematodes and have received the greatest share of research attention. Based on modes of parasitism the nematophagous fungi have been divided into the following three groups: (*a*) the predacious or nematode-trapping fungi (Figure 1*a*), (*b*) the endoparasitic fungi, and (*c*) the opportunistic fungi that have the capability of colonizing females, cysts, and eggs in such genera as *Heterodera, Globodera,* and *Meloidogyne* (37, 53; Figure 1*c*).

NEMATODE-TRAPPING FUNGI Species of trapping fungi vary in their abilities to capture nematodes; those producing network traps (Figure 1*a*) are good soil saprophytes and grow well in vitro, but are less effective at trapping nematodes than predacious fungi that form adhesive knobs and branches or constricting rings (29). Generally, the trapping activity of all species, regardless of their trapping mechanisms, are short-lived, and restricted to vermiform nematodes. As a result of limited trapping activity and their lack of specificity for preying on phytonematodes, these fungi have not been exploited extensively as control agents. Where biological control has been achieved through the use of trapping fungi, the results may have been confounded by the abundance of organic material that was added as a food base for the fungi. Degradation products of organic carriers, even without living predacious fungi, may be sufficiently nematicidal to have an impact on phytonematode populations (53).

ENDOPARASITIC FUNGI Obligate endoparasites occur in the Chytridiomycetes, Oomycetes, Zygomycetes, Deuteromycetes (including some opportunistic species), and Basidiomycetes (3). The three most studied of this group are *Catenaria auxiliaris* (Kuhn) Tribe, *Nematophthora gynophila* Kerry and Crump, and a lagenidiaceous fungus found in cyst nematodes. These three fungi were not influenced by organic amendments. Their survival is not dependent on saprophagous growth, but is directly related to the density of host nematodes. Most of these fungi produce a thick-walled resting spore for their survival in soils. Consequently, organic amendments have little influence on these fungi (37). Other edaphic factors such as soil moisture, pH, metals, and nematode density do influence the obligate parasitic fungi of nematodes (19). Cropping practices largely determine population density of a phytonematode species and consequently determine populations of their obligate parasites. As Kerry (39) pointed out, high population densities of nematodes are important initially in the establishment of nematode-suppressive soils. To induce a suppressive soil, some crop losses would necessarily be incurred. High initial population of nematodes might be achieved by planting tolerant crops or double cropping with cover crops that are susceptible to nematode parasitism, but are of little economic value.

Irrigation can provide high moisture potentials that favor the zoosporous endoparasites of nematodes. *Nematophthora gynophila,* a parasite of *H. avenae* (38), was active from May to July, when female nematodes were exposed on roots during a period of ample moisture in the soil. But during dry summers in well-drained coarse soils this fungus was less active. The decrease in parasitism has been attributed to the possible inhibition of zoospore movement in the drier soil.

During a survey at sites near Woburn, Tidworth, and Sydmonton in southern England, Kerry et al (39) observed that populations of *H. avenae* decreased on spring barley at two sites and increased only slightly at the other. Only 40% of females forming cysts contained any eggs. This reduction in fecundity was coupled with an increase in numbers of encysted eggs containing fungi. *Verticillium chlamydosporium* was the main parasite occurring in encysted eggs at all sites. Other saprophytic fungi isolated from eggs were considered weak parasites. *Nematophthora gynophila* was present in the soil at all sites but was most active at Woburn where there was irrigation. The latter factor, high soil moisture, increased the effectiveness of *N. gynophila* and *V. chlamydosporiun* as control agents of CCN.

In water-saturated soils, locomotion of the zoosporangia of phycomycetous fungi may be considerable and may allow for the diffusion of nematode waste products that are attractive to the motile fungal stages (69; Figure 2*b*).

OPPORTUNISTIC FUNGAL PARASITES About 150 species of opportunistic fungi have been found that parasitize eggs (Figure 1*c*) and colonize developing females of cyst nematodes (37). The edaphic factors affecting these fungal antagonists of nematodes have recently been reviewed by Gray (18, 19).

Altering soil pH with lime, anhydrous ammonia, or fertilizer treatments may stress nematodes, making them more susceptible to parasitism. More specifically, Jaffee & Zehr (28) observed that *Criconemella xenoplax* is more susceptible to infection by *Hirsutella rhossiliensis* in solutions of certain salts at tensions of 30 to 300 kPa than in distilled or tap water, or solutions of polyethylene glycol at the same moisture potentials. If the degree of parasitism is influenced by osmotic tension, parasitism might be increased by application of fertilizers at appropriate times.

BACTERIAL ANTAGONISTS When plant roots decay each root-knot nematode parasitized by *Pasteuria penetrans* releases over two million endospores into the soil to continue the bacterium's life cycle (45). The influence of edaphic factors on the fate of these endospores has not been investigated. Only temperature relationships were studied by Stirling, (75, 77). Generally, this obligate parasite appeared to be little influenced by the addition of organic amendments. Perhaps endospore degradation might be accelerated by an increase in microbial activity.

INVERTEBRATE ANTAGONISTS The introduction of agents that may facilitate predation on nematodes is an unexplored facet of nematode biological control. Most studies of soil amendments have been concerned with their effects on microbial antagonists, although the resulting changes in soil structure will undoubtedly affect invertebrate antagonists as well. Van de Bund (85) found that the addition of stable manure (110 tons/ha) to fields caused a 63% decline in plant-parasitic nematode numbers at the same time that predatory and omnivorous arthropods increased four- to sevenfold. Yeates (94) found that earthworms introduced into pastures in New Zealand caused marked reductions in nematode densities, but attributed the effect to the resulting changes in soil structure, and not to predation on nematodes by earthworms per se. Relationships between soil invertebrates and microbes can be complex, and may include both grazing on and vectoring of fungi antagonistic to plant parasites (9).

Introducing Exotic Antagonists into Soil

Consideration has already been given to natural and augmentative/inundative releases of biological control agents into soils. The introduction of exotic agents into soils as an approach imposes new requirements on the biological control agent—their competitiveness and survival in hostile soils.

FUNGAL ANTAGONISTS Several methods for adding fungal antagonists of nematodes to soil have been suggested. One such method is, of course, the addition of small amounts of the suppressive soil to a soil conducive to nematode (44) development.

Another method of adding inocula to soils was devised by Fravel et al (14). Life stages of antagonistic fungi are embedded in a matrix of calcium alginate using the sodium alginate method, with bran as a food base. Other constituents also may be added to enhance the survival of the fungus or make the pellet more attractive to the target nematode. More recently, Meyer (50) developed a method specifically for a fungal antagonist in controlling soybean cyst nematodes.

Selective soil treatments were sometimes used prior to the mass transfer of an antagonist from a suppressive to a conducive soil. The high temperatures of solarization rid soils of competitors of nematode antagonist and allowed the antagonist an opportunity to become established.

The zoosporangia of the fungus, *Catenaria* sp., are very susceptible to copper ions. Any cupric fungicide or fertilizer would suppress this fungus. Conversely, the utilization of another pesticide was shown to have the reverse effect and favor the fungal species. Ethoprop nematicide in sublethal doses increased susceptibility of the nematode *Meloidogyne incognita* to attack from the zoosporangia of *Catenaria anguillulae* (65). High moisture reduces soil oxygen. Under these conditions, fungi tolerant to low oxygen may become antagonists of phytonematodes. Generally, under oxygen stress nematodes are more susceptible to fungal attack, particularly from opportunistic fungi.

BACTERIAL ANTAGONISTS Using microplots, Mankau (44) tested the ability of *P. penetrans* to control root-knot nematodes. He employed the following three treatments: (*a*) quantities of air-dried soil infested with endospores of the bacterium were used to fill holes 3 inches (76.2mm) in diameter and 6 inches (152.4mm) deep; (*b*) 1-month old tomato seedlings infected with root-knot and parasitized by the bacterium were transplanted into the plots; and (*c*) 240,000 *M. incognita* J2 free of the bacterium were inoculated 4 inches (101.6 mm) deep into soil. Each treatment was replicated 24 times. These microplots were planted to tomatoes and 11 months later samples were taken from nematode-infested areas and assayed by the addition of healthy J2: 98% of those from treatment (*a*) were encumbered with 20–50 spores, while 53% of those from treatment (*b*) and only 7% of those from treatment (*c*) were found with spores. Thus, infested soil appeared to be an effective and convenient method for introducing the parasite into the field.

Stirling & Wachtel (80) devised yet another method for the introduction of *P. penetrans* into soils. Tomato roots containing females of the root-knot nematode *M. javanica* infected with *P. penetrans* were air-dried and finely ground to a powdery material heavily laden with bacterial endospores. When this material was incorporated into root-knot nematode-infested soils at the rate of 212 to 600 mg/kg of soil, galling of tomato roots and the numbers of J2 at harvest time were reduced significantly. Nematode control was similar to that usually obtained with nematicides. Before the tomatoes were planted, treated soils were bioassayed for the bacterium by adding J2 and counting those encumbered with endospore that were extracted 24 hr later. Significant root-knot nematode control was obtained in the field when at least 80% of J2 were encumbered with 10 or more endospores (76). Laboratory experiments showed the spore-encumbered J2 were less able to penetrate plant roots than were those free of spores. Numbers penetrating the roots decreased with the increasing spore concentrations and with the distance they moved through soil. Some spores did not always germinate, consequently five or more spores per nematode were required to ensure that infection occurred.

In Nishizawa's method (57), suspensions of endospores of a *Pasteuria* sp. were obtained by leaching 1kg of soil suppressive to the soybean cyst nematode with 4 liters of water. The four liters of water were then added to microplots that were infested with the susceptible nematodes. Using this method, suppressive activity was transferred to the microplots and resulted in significant decreases in cyst populations (57).

CONCLUSIONS AND FUTURE PROSPECTS

Clearly, the future is bright for the biological control of phytonematodes. The availability of enumerable natural enemies of nematodes, coupled with the clear demonstration of nematode suppressiveness of certain soils provide not only good sources of effective antagonists, but also clear evidence that biological control of phytonematodes does occur. In examining the efficacy of soil factors influencing the antagonist-nematode relationship, there is clear demonstration that management practices can be used effectively to manipulate the natural enemies of nematodes. Eventually, healthy root growth, little impaired by nematode pests, may be achieved by the introduction of populations or individual strains responsible for suppression.

One imperative remains: The demonstration of the successful biological control of nematodes in tangible field plots. As nematologists compete for the available research dollars, there is a continuing need to demonstrate the economic worth of the research, not only to garner the continued funding from governmental agencies but maintain the general support of those who build upon the research or ultimately put it into practice for control of nematode damage on crop plants.

Literature Cited

1. Baker, K. F., Cook, R. J. 1974. *Biological Control of Plant Pathogens*. San Francisco, CA: W. H. Freeman
2. Barnes, J. M. 1987. Impact of formula funding on the science of nematology. See Ref. 87, pp. 15–21
3. Barron, G. L. 1977. *The Nematode-Destroying Fungi. Topics in Mycobiol.* 1. Guelph, Ontario: Canadian Biol. Publ. Ltd. 140 pp.
4. Barron, G. L., 1981. Predators and parasites of microscopic animals. In *Biology of Conidial Fungi*, ed. G. F. Cole, E. Kendrick, 1:167–200. New York: Academic
5. Bird, A. F., Brisbane, P. G. 1988. The influence of *Pasteuria penetrans* in field soils on the reproduction of root-knot nematodes. *Rev. Nematol.* 11(1):75–81
6. Cobb, N. A. 1917. The mononchs, a genus of free-living predatory nematodes. Contributions to a science of nematology. *Soil Sci.* 3:431–86
7. Cohn, E., Mordechai, M. 1974. Experiments in suppressing nematode populations by use of a marigold and a predacious nematode. *Nematolog. Mediterr.* 1:43–53
8. Croll, N. A. 1970. *The Behavior of Nematodes*. New York, NY: St. Martin's Press 117 pp.
9. Curl, E. A., Lartey, R., Peterson, C. M. 1988. Interactions between root pathogens and soil microarthropods. *Agric. Ecosyst. Environ.* 24:249–61
10. Dash, M. C., Senapati, B. K., Mishra, C. C. 1980. Nematode feeding by tropical earthworms. *Oikos* 34:322–25
11. Dropkin, V. H. 1989. *Introduction to Plant Nematology*. New York: Wiley. 304 pp. 2nd. ed.
12. Elliott, E. T., Anderson, R. V., Coleman, D. C., Cole, C. V. 1980. Habitable pore space and microbial trophic interactions. *Oikos* 35:327–35
13. Ferris, H., McKenry, M. V., McKinney, H. E. 1976. Spatial distribution of nematodes in peach orchards. *Plant Dis. Rep.* 60:18–22
14. Fravel, D. R., Marios, J. J., Lumsden, R. D., Connick, W. J. Jr. 1985. Encapsulation of potential biocontrol agents in an alginate-clay matrix. *Phytopathology* 75:774–77
15. Gabriel, C. J., Cook, R. J. 1990. Biological control—the need for a new scientific framework. *Bioscience* 40(3): 204–7

16. Gair, R., Mathias, P. L., Harvey, P. N. 1969. Studies of cereal cyst-nematode populations and cereal yields under continuous or intensive culture. *Ann. Appl. Biol.* 63:503–12
17. Gilmore, S. K. 1970. Collembola predation on nematodes. *Search Agric.* 1:1–12
18. Gray, N. F. 1985. Ecology of nematophagous fungi: Effect of soil moisture, organic matter, pH and nematode density on distribution. *Soil Biol. Biochem.* 17:499–507
19. Gray, N. F. 1988. Ecology of nematophagous fungi: Effect of the soil nutrients N, P and K, and seven major metals on distribution. *Plant Soil* 108: 286–90
20. Gray, N. F. 1988. Fungi attacking vermiform Nematodes. See Ref. 62, 2:3–38
21. Greenstone, M. H. 1989. Foreign exploration for predators: A proposed new methodology. *Environ. Entomol.* 18: 195–200
22. Haabersaat, U. 1989. *The importance of predatory soil mites as predators of agricultural pests, with special reference to* Hypoaspis angusta *Karg, 1965 (Acari: Gamasina)*. Doctoral diss. Fed. Inst. Technol. Zurich
23. Hechler, H. C. 1963. Description, developmental biology, and feeding habits of *Seinura tenuicaudata* (De Man) J. B. Goodey, 1960 (Nematoda: Aphelenchoididae), a nematode predator. *Proc. Helminth. Soc. Wash.* 30:183–95
24. Homma, Y., Cook, R. J. 1985. Influence of matric and osmotic water potentials and soil pH on the activity of giant vampyrellid amoebae. *Phytopathology* 75:243–46
25. Imbriani, J. L., Mankau, R. 1983. Studies on *Lasioseius scapulatus,* a mesostigmatid mite predaceous on nematodes. *J. Nematol.* 15:523–28
26. Inserra, R. N., Davis, D. W. 1983. *Hypoaspis* nr. *aculeifer:* a mite predacious on root-knot and cyst nematodes. *J. Nematol.* 15:324–25
27. Jaffee, B. A., Muldoon, A. E. 1990. Suppression of cyst nematode by natural infestation of a nematophagous fungus. *J. Nematol.* 21:505–10
28. Jaffee, B., Zehr, E. I. 1983. Suppression of *Criconemella xenoplax* by the fungus *Hirsutella rhossiliensis. Phytopathology* 72:1378–81
29. Jansson, H. B., Norbring-Hertz, B. 1980. Interactions between nematophagous fungi and plant-parasitic nematodes: attraction, induction of trap formation and capture. *Nematologica* 26:383–89
30. Jatala, P. 1986. Biological control of plant-parasitic nematodes. *Annu. Rev. Phytopathol.* 24:452–89
31. Johnson, A. W., Feldmesser, J. 1987. Nematicides—A historical review. See Ref. 87, pp. 448–54
32. Karg, W. von. 1983. Verbreitung und Bedeutung von Raubmilben der Cohors Gamasina als Antagonisten von Nematoden. *Pedobiologia* 25:419–32
33. Kerry, B. 1978. Natural control of cereal cyst nematode by parasitic fungi. *ARC Res. Rev.* 4:17–20
34. Kerry, B. R. 1980. Biocontrol: Fungal parasites of female cyst nematodes. *J. Nematol.* 12:253–59
35. Kerry, B. R. 1984. Nematophagous fungi and the regulation of nematode populations in soil. *Helminth. Abstr. Ser. B. Plant Nematol.* 53:(1):1–14
36. Kerry, B. R. 1987. Biological control. In *Principles and Practices of Nematode Control in Crops,* ed., H. R. Brown, B. R. Kerry, pp. 233–63. Sydney: Academic. 434 pp.
37. Kerry, B. 1988. Fungal parasites of cyst nematodes. *Agric. Ecosyst. Environ.* 24:293–305
38. Kerry, B. R., Crump, D. H., Mullen, L. A. 1980. Parasitic fungi, soil moisture and multiplication of the cereal cyst nematode *Heterodera avenae. Nematologica* 26:57–68
39. Kerry, B. R., Crump, D. H., Mullen, L. A. 1982. Natural control of the cereal cyst nematode *Heterodera avenae* Woll. by soil fungi at sites. *Crop Prot.* 1:99–109
40. Kinn, D. N. 1967. Notes on the life cycle and habits of *Digamasellus quadrisetus* (Mesostigmata: Digamasellidae). *Ann. Entomol. Soc. Am.* 60:862–65
41. Linford, M. B. 1937. Stimulated activity of natural enemies of nematodes. *Science* 85:123–24
42. Linford, M. B., Yap, F., Oliveira, J. M. 1938. Reduction of soil populations of the root-knot nematode during decomposition of organic matter. *Soil Sci.* 45:127–41
43. Mai, W. F., Motsinger, R. E., 1987. History of the Society of Nematologists. See Ref. 87, pp. 1–6.
44. Mankau, R. 1973. Utilization of parasites and predators in nematode pest management ecology. *Proc. Tall Timbers Conf. Ecolog. Anim. Control by Habitat Manag.* 4:129–43
45. Mankau, R. 1975. *Bacillus penetrans* n. comb. causing a virulent disease of plant-parasitic nematodes. *J. Invertebr. Pathol.* 26:333–39
46. Mankau, R. 1980. Biological control of

nematode pests by natural enemies. *Annu. Rev. Phytopathol.* 18:415–40
47. Mankau, R. 1981. Microbial control of nematodes. In *Plant Parasitic Nematodes* ed. B. M. Zuckerman, R. A. Rhode, 3:475–94. New York: Academic
48. McMurtry, J. A. 1982. The use of phytoseiids for biological control: Progress and future prospects. In *Recent Advances in Knowledge of the Phytoseiidae*, ed. M. A. Hoy, pp. 23–48. Div. Agric. Sci., Univ. Calif. Publ. 3284, Berkeley
49. McMurtry, J. A. 1984. A consideration of the role of predators in the control of acarine pests. In *Acarology VI*, ed. D. A. Griffiths, C. E. Bowman, 1:109–21 Chichester: Ellis Horwood Ltd.
50. Meyer, S. L. F. 1990. Evaluation of potential biocontrol agents of soybean cyst nematode. *Mycolog. Soc. Am. Newsl.* 41(1):29(Abstr.)
51. Minton, N. A., Sayre, R. M. 1989. The suppressive influence and *Pasteuria penetrans* discovered in Georgia soils on the reproduction of the peanut root-knot nematodes, *Meloidogyne arenaria*. *J. Nematol.* 24(4):574–75
52. Moore, J. C., Walter, D. E., Hunt, H. W., 1988. Arthropod regulation of micro- and mesobiota in below ground detrital food webs. *Annu. Rev. Entomol.* 33:419–39
53. Morgan-Jones, G., Rodriguez-Kabana, R. 1987. Fungal biocontrol for the management of nematodes. See Ref. 87, pp. 94–99
54. Morton, H. V. 1987. Industry perspectives in nematology. See Ref. 87, pp. 47–51
55. Muraoka, M., Ishibashi, N. 1976. Nematode-feeding mites and their feeding behavior. *Appl. Entomol. Zool.* 11:1–7
56. Nelmes, A. J. 1974. Evaluation of the feeding behavior of *Prionchulus punctatus* (Cobb), a nematode predator. *J. Anim. Ecol.* 43:553–65
57. Nishizawa, T. 1986. On a strain of *Pasteuria penetrans* parasitic to cyst nematodes. *Rev. Nématol.* 9:303–4 (Abstr.)
58. Nishizawa, T. 1987. A decline phenomenon in a population of upland cyst nematode, *Heterodera elachista*, caused by bacterial parasite, *Pasteuria penetrans*. *J. Nematol.* 19:546 (Abstr.)
59. Nishizawa, T. 1988. Suppression of cyst nematode populations caused by bacterial hyperparasite, *Pasteuria penetrans*. *Proc. Int. Congr. Plant Pathol.* 5th, Kyoto, 20–27 July
60. Norton, D. 1978. *Ecology of Plant-Parasitic Nematodes*. New York: Wiley. 268 pp.
61. Poinar, G. O. Jr., Hess, R. 1988. Protozoan diseases. See Ref. 62, 1:103–32
62. Poinar, G. O. Jr., Jansson, H.-B. eds. 1988. *Diseases of Nematodes*. Boca Raton: CRC Press. Vol. 1, 149 pp. Vol. 2, 150 pp.
63. Rodríguez-Kábana, R. 1986. Organic and inorganic nitrogen amendments to soil as nematode suppressants. *J. Nematol.* 18:129–35
64. Rovira, A. D. 1982. Organisms and mechanisms involved in some soils suppressive to soilborne plant diseases. See Ref. 71, pp. 23–33
65. Roy, A. K. 1982. Effect of ethoprop on the parasitism of *Catenaria anguillulae* on *Meloidogyne incognita*. *Rev. Nematol.* 5:335–36
66. Sayre, R. M., 1973. *Theratromyxa weberi*, an amoeba predatory on plant-parasitic nematodes. *J. Nematol.* 5:258–64
67. Sayre, R. M. 1986. Pathogens for the biological control of nematodes. *Crop Prot.* 5:268–76
68. Sayre, R. M. 1988. Bacterial diseases of nematodes and their role in controlling nematode populations. *Agric. Ecosyst. Environ.* 24:262–79
69. Sayre, R. M., Keeley, L. S. 1969. Factors influencing *Catenaria anguillulae* infections in a free-living and plant-parasitic nematode. *Nematologica* 15: 492–502
70. Sayre, R. M., Powers, E. M. 1966. A predacious turbellarian that feeds on free-living and plant-parasitic nematodes. *Nematologica* 12:619–29
71. Schneider, R. W., ed. 1982. *Suppressive Soils and Plant Diseases*. St Paul, MN. *Am. Phytopathol. Soc.* 88 pp.
72. Sell, P. 1988. *Caloglyphus* sp. (Acarina: Acaridae), an effective nematophagous mite on root-knot nematodes (*Meloidogyne* spp.). *Nematologica* 34:246–48
73. Small, R. W. 1987. A review of the prey of predatory soil nematodes. *Pedobiologia* 30:179–206
74. Small, R. W. 1988. Invertebrate predators. See Ref. 62, pp. 74–92
75. Stirling, G. R. 1981. Effect of temperature on infection of *Meloidogyne javanica* by *Bacillus penetrans*. *Nematologica* 27:458–62
76. Stirling, G. R. 1984. Biological control of *Meloidogyne javanica* with *Bacillus penetrans*. *Phytopathology* 74:55–60
77. Stirling, G. R. 1988. Biological control of plant-parasitic nematodes, See Ref. 62, 2:93–139
78. Stirling, G. R., Kerry B. R. 1983. An-

tagonists of the cereal cyst nematode *Heterodera avenae* Woll. in Australian soils. *Aust. J. Exp. Agric. Anim. Husb.* 23:318–24

79. Stirling, G. R., McKenry, M. V., Mankau, R. 1979. Biological control of root-knots (*Meloidogyne* spp.) on peach. *Phytopathology* 69:806–9
80. Stirling, G. R., Wachtel, M. F. 1980. Mass production of *Bacillus penetrans* for the biological control of root-knot nematodes. *Nematologica* 26:308–12
81. Stirling, G. R., White, A. M. 1982. Distribution of a parasite of root-knot nematodes in south Australian vineyards. *Plant Dis.* 66:52–53
82. Thorne, G. 1927. The life history, habits and economic importance of some mononchs. *J. Agric. Res.* 34(3):265–86
83. Tribe, H. T. 1977. Pathology of cyst-nematodes. *Biol. Rev. Cambridge Philos. Soc.* 52:477–508
84. Tribe, H. T. 1980. Prospects for the biological control of plant-parasitic nematodes. *Parasitology* 81:619–39
85. Van de Bund, C. F. 1972. Some observations on predatory action of mites on nematodes. *Zesz. Probl. Postepow Nauk Roln.* 129:103–10
86. Vargas, R., Hattori, T. 1986. Protozoan predation of bacterial cells in soil aggregates. *FEMS Microbiol. Ecol.* 38:233–42
87. Veech, J. A., Dickson, D. W., eds. 1987. *Vistas on Nematology.* Hyattsville, MD: Soc. Nematol. 509 pp.
88. Walter, D. E. 1988. Predation and mycophagy by endeostigmatid mites (Acariformes: Prostigmata). *Exper. Appl. Acarol.* 4:159–66
89. Walter, D. E. 1988. Nematophagy by soil arthropods from the short grass steppe, Chihuahuan desert, and Rocky Mountains of the central United States. *Agric. Ecosyst. Environ.* 24:307–16
90. Walter, D. E., Hunt, H. W., Elliott, E. T., 1987. The influence of prey type on the development and reproduction of some predatory soil mites. *Pedobiologia* 30:419–24
91. Walter, D. E., Ikonen, E. K. 1989. Species, guilds and functional groups: Taxonomy and behavior in nematophagous arthropods. *J. Nematol.* 21:315–27
92. Walter, D. E., Oliver, J. H. Jr. 1990. *Geolaelaps oreithyiae,* n. sp. (Acari: Laelapidae), a thelytokous predator of arthropods and nematodes, and a discussion of clonal reproduction in the Mesostigmata. *Acarologia* 30:293–303
93. Yeates, G. W. 1969. Predation by *Mononchoides potohikus* (Nematoda: Diplogasteridae) in laboratory culture. *Nematologica* 15:1–9
94. Yeates, G. W. 1981. Soil nematode populations depressed in the presence of earthworms. *Pedobiologia* 22:191–201

Annu. Rev. Phytopathol. 1991. 29:167–192

RESISTANCE TO AND TOLERANCE OF PLANT PARASITIC NEMATODES IN PLANTS

D. L. Trudgill

Zoology Department, Scottish Crop Research Institute, Invergowrie, Dundee, Scotland DD2 5DA

KEY WORDS: population dynamics, modeling, pathotypes and races, effects on yield, terminology

---

INTRODUCTION

Resistant cultivars have several advantages over other methods of reducing nematode populations: their use requires little or no technology and is cost effective; they allow rotations to be shortened and best use to be made of the land; and they do not leave toxic residues. In contrast, nematicides are uneconomic on many lower-value crops and when used on high-value crops are applied at relatively high rates with the consequential risk of toxic residues (16). Resistant cultivars also need to be tolerant; those that are intolerant will suffer extreme damage if grown in heavily infested soil. Equally, tolerant cultivars that are not resistant tend to increase nematode population densities to damagingly high numbers.

The purpose of this review is to outline the mechanisms and principles governing nematode/plant interactions and to consider some of the useful parallels between nematodes and other plant pathogens, especially fungi. Terminology, which often differs between branches of plant pathology (51), is briefly discussed. These differences arise because many plant pathologists are concerned with the expression of disease in whole crops in the field and the epidemiology of mobile pathogens capable of explosive population in-

0066-4286/91/0901-0167$02:00

creases. Plant parasitic nematodes are relatively immobile and have only one or, at most, a few generations per year so that proper study of their epidemiology could take many years. Consequently, the attention of most nematologists has focused on the dynamics of the interaction between the host and single generations of a nematode.

Many reviews of resistance to nematodes have been published, including ones that have concentrated on active incompatibility mechanisms (9, 45, 136), the role of phytoalexins (136), biochemical interactions (47), the genetics of resistance (110), and the concept of race and physiological specialization (26, 110). Tolerance of potato and soybean cyst nematodes (31, 55) and resistance to cyst and root-knot nematodes (15) have been recently reviewed and Cook & Evans (16) have comprehensively reviewed resistance and tolerance of nematodes in potato, tomato, tobacco, soybean, cereals, alfalfa, vines, *Prunus,* and citrus. Even so, their review is largely restricted to cyst (*Heterodera* and *Globodera*), root-knot (*Meloidogyne*), stem and bulb (*Ditylenchus*), dagger (*Xiphinema*), citrus (*Tylenchulus*), and burrowing (*Radopholus*) species, reflecting the relatively narrow range of nematodes on which most research effort is being concentrated.

NEMATODE BIOLOGY

Based on their feeding behavior, three broad groups of nematodes can be distinguished. The least specialized, generally with wide host ranges, are the migratory ectoparasites. More specialized, but still often polyphagous, are the migratory endoparasites. The third, and most important group are the sedentary endoparasites, containing both polyphagous species (e.g. *Meloidogyne* spp.) and more host-specific species (e.g. *Globodera* and *Heterodera* spp.). However, in morphology and behavior these three groups are very similar. All have a hollow mouth stylet (except for trichodorid nematodes where the stylet is solid) that they use during feeding or invasion to pierce host cells. Ingestion is usually preceded by the injection of "saliva" coming from esophageal gland cells (53, 54) and plant cell contents are removed by an esophageal pump (142, 143). The relationship with the plant of most migratory ectoparasites (e.g. *Trichodorus, Tylenchorhynchus, Tylenchus* spp.) is, in mycological terms, close to necrotrophic. In the few studies made (143), the injected saliva liquefies the cytoplasm that accumulates around the stylet tip and is rapidly ingested, usually killing the host cell. The nematode then moves to a new cell and repeats the process. Such behavior limits the possibilities for effective induced resistance (110).

Migratory endoparasites (e.g. *Pratylenchus, Radopholus* spp.) invade the plant, moving and feeding inter- or intracellularly. Most attack roots but some specialize in stem and leaf tissues (e.g. *Ditylenchus, Aphelenchoides, An-*

guina spp.) and yet others inhabit vessels (e.g. *Bursaphelenchus* spp.). Frequently, species that attack plant aerial parts have a specific host range and high reproduction rates and are extremely damaging. Feeding may be necrotrophic, but in some species it is partly biotrophic as it involves the induction of favorable changes in cells adjacent to the feeding site (9). Resistant cultivars are available to species within this group, especially the stem and leaf parasites (110).

The sedentary endoparasites are, in mycological terms, obligate biotrophs. The invasive juveniles lose their mobility and, if they are to develop into adult females, require the cells on which they feed to remain alive and change in ways that improve the supply of food. These responses range from the induction of a single, uninucleate giant cell (e.g. *Nacobbus*) to several multinucleate giant cells (e.g. *Meloidogyne*), a multinucleate syncytium (e.g. *Globodera* and *Heterodera*) or a group of discrete nurse cells (e.g. *Tylenchulus*) (29, 63, 142). To feed on such cells, which have enhanced concentrations of RNA, polyploid nuclei, and are metabolically highly active (48), the nematode extrudes a "feeding tube" from its stylet tip. This "tube" is formed from dorsal gland cell secretions and probably acts as a sieve, enabling the cytosol and the nutrients it contains to be removed without damage to the plant cell (29, 102, 143). As a consequence of this close association of the nematode with its host, the developing females have no need to retain their mobility and therefore can become much enlarged, with increased reproductive rates. Nematodes within this group are among the most damaging to plants and resistance to them is widespread (9, 34, 110).

A subgroup of migratory ectoparasites with long stylets (*Hemicycliophora, Longidorus, Paralongidorus,* and *Xiphinema*) is also biotrophic, at least for some species during parts of their feeding. They induce root-tip galls containing highly modified cells (10, 143, 144) and may produce structures analogous to feeding tubes (132). Resistance is also available to some species within this group.

TERMINOLOGY

Resistance and Tolerance

For terminology to be logical and unambiguous, there should be an opposite term (e.g. tolerant and intolerant) for each term relating to host or parasite and corresponding terms relating to host and pathogen (e.g. resistance and virulence).

Resistance describes the effects of host genes that restrict or prevent nematode multiplication in a host species (120). *Tolerance* of damage is independent of resistance (14, 87) and relates to the ability of a host genotype to withstand or recover from the damaging effects of nematode attack and

Table 1 Effect of a nematicide treatment (aldicarb, 3.36 kg a.1./ha) on the yield of two potato genotypes of different resistance and tolerance in plots heavily infested with *Globodera pallida**

	Cultivar/clone		
Nematicide	Maris Piper (nonresistant/intermediate tolerance)		12380 ac2 (> 90% resistant/intolerant)
Yield (kg/plant)			
−	1.25		0.26
+	2.45		1.84
L.S.D P=5%		0.4	
G. pallida multiplication			
−	2.9		0.4
+	0.2		0.1
L.S.R.** P=5%		2.0	

*Data from ref. (1a)

**Least significant ratio. The nematode multiplications were converted to logarithms for statistical analysis. The detransformed means are shown here. The L.S.R. is used multiplicatively, two means differing significantly if one is twice the value of the other.

yield well. A similar distinction has been proposed in the use of resistance and tolerance in relation to plant viruses (17). Both terms are frequently used comparatively, and these comparisons have to be made in the same environment and *with the same inoculum pressure*. Resistant plants are frequently attacked by similar numbers of nematodes as nonresistant plants, i.e. resistance to nematodes does not usually protect plants from invasion damage. Indeed, some resistant genotypes of cereals, potato, and soybean are relatively intolerant (i.e. more damage than susceptible genotypes) of their respective cyst nematodes (37, 55, 124). An example is given for two genotypes of potato in plots heavily infested with *G. pallida* (Table 1). The extremes of resistance and tolerance can be represented in a four-celled chart (Table 2).

These terms can apply to whole crops or single plants, and their expression in a particular genotype is usually measured over a single or, at most, a few generations of the nematode. Yield losses and nematode reproduction can be mathematically modeled (Figures 1, 2).

Table 2 Extremes of resistance and tolerance

		Host growth	
		Good	Poor
Nematode reproduction	Good	Tolerant/ nonresistant	Intolerant/ nonresistant
	Poor	Tolerant/ resistant	Intolerant/ resistant

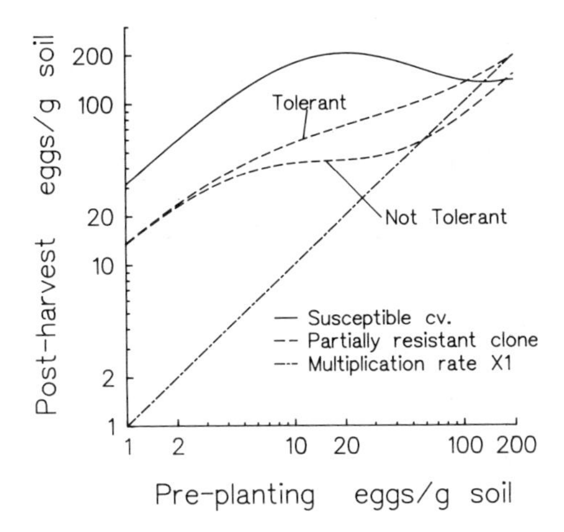

Figure 1 The effect of numbers of eggs of potato cyst nematode (*Globodera pallida*) per g of soil at planting on their numbers after harvest (both on a logarithmic scale) for a nontolerant susceptible (solid line) and tolerant and nontolerant partially (60%) resistant potato (dotted line) genotype. These lines have been modeled using a modified Jones & Perry (63) equation.

Pathogenicity and Virulence

The corresponding terms that apply to the nematode can also be represented in a four-celled chart (Table 3). Pathogenicity relates to the capacity to cause disease or damage (usually measured in yield reduction) and virulence to the capacity to overcome/circumvent/ suppress resistance genes.

Again, these terms are frequently used on a relative basis but, as with resistance and tolerance, mathematical models exist by which they can be compared (28, 90).

Resistance, Susceptibility, and Aggressiveness

Vanderplank (134) subdivided resistance into vertical and horizontal, depending upon whether or not in an analysis of variation the interaction between a

Table 3 Extremes of pathogenicity and virulence

		Pathogen reproduction	
		Good	Poor
Host growth	Good	Nonpathogenic/ virulent	Nonpathogenic/ avirulent
	Poor	Pathogenic/ virulent	Pathogenic/ avirulent

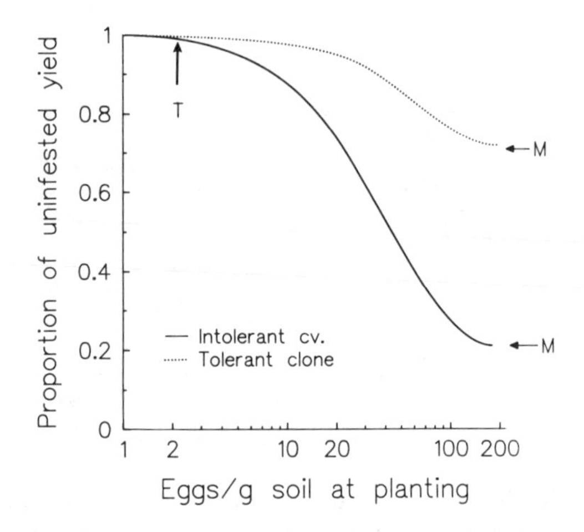

Figure 2 The effect of increasing numbers of potato cyst nematode (on a logarithmic scale) on the proportional yield of potato. The lines have been modelled using the Seinhorst (106) equation $Y = m + (1 - m)z^{P_i - T}$ with $T = 2.0$, $z = 0.975$ and two different values of m (0.7 and 0.2, respectively).

range of host and pathogen genotypes was significant. Vertical resistance usually relates to major genes against which large differences in pathogen virulence have been identified, a gene-for-gene interaction (39, 40). Gene-for-gene interactions have been identified in many plant-pathogen interactions (108), including nematodes (60, 61). During incompatible (resistant) associations there is usually a positive resistant response, termed active resistance (65), which often involves a so-called hypersensitive reaction. It has been argued that horizontal resistance differs from vertical only in that it involves several genes with small effects (85). Although this may sometimes be so, experience with quantitative resistance to potato cyst nematodes suggests that, in part, resistance could take a negative (passive) form involving reduced susceptibility.

The term "susceptibility" is often used imprecisely, either to indicate lack of resistance or lack of tolerance, or both. It actually relates to host status and is defined as the sum total of the qualities that make a plant a fit (good) host for the pathogen (101). However, apparently completely susceptible plants may contain resistance genes when probed with appropriate avirulent populations (72). The opposite of a fully susceptible host is a non-(immune) host (i.e. no recognition of, invasion of, or feeding on the host). Many authors assume the existence of susceptiblility genes and it has been shown that successful infection by microbial pathogens involves over 100 genes, including host-recognition genes (25). Differences in susceptibility could take several forms, some of which might affect the quantity or quality of nutrients the pathogen can obtain to support its development and reproduction (65).

Resistance involving reduced susceptibility was proposed in relation to viruses (44). Resistance to nematodes based on reduced susceptibility can be expected to differ in several ways from resistance involving vertically, positively acting genes. Resistance is likely to be sensitive to environmental factors (other than increased temperature), with expression being reduced in environments favorable to pathogen development. It will be nonspecific, as will virulence differences between populations, and may be more likely to be recessively inherited (44). Selection for increased virulence, where it occurs, will be slow and largely nonspecific, reflecting a general increase in overall fitness.

Examples relating to potato cyst nematodes (*Globodera* spp.) can be used to illustrate the above. Only a small proportion of juveniles that invade the thin lateral roots of susceptible potato plants become female, probably because the syncytia they induce and depend upon for nutrients are physically restricted in size (80, 115, 116). Such juveniles usually become male (sex determination in many cyst and root-knot species occurs after the juveniles enter the root), which require less nutrition for development than females. In contrast, a large proportion of juveniles entering main roots become female. Hence, any host plant with abnormally thin main roots will support the development of few females and, compared to standard cultivars, will appear to be partially resistant. Similarly, genotypes of potato with quantitative (taken here to partially equate with horizontal) resistance to *G. pallida* derived from *Solanum vernei* (94) are comparatively much less resistant when tested in petri dishes, an environment favorable to female development, than in plastic canisters, a less favorable environment (79). Such an interaction with test method distinguishes this type of resistance from the qualitative resistance to pathotype Ro1 of *G. rostochiensis* that is conferred by the *H1* gene (derived from *Solanum tuberosum* ssp. *andigena* CPC 1673), which is not influenced by environment. Also, selection of potato cyst nematodes for increased reproduction on genotypes of potato with quantitative resistance derived from *S. vernei* is erratic and slow (41) but, for some populations, is accompanied by increased reproduction on other genotypes, including nonresistant (13) and those with resistance derived from a different wild source (92, 133). However, this increase is sometimes proportionally much less than on the genotype used to select the population, implying either that some of the same vertically acting genes are present in all genotypes or that selection has been to a mixture of vertically (positively?) and horizontally (negatively?) acting genes. Equally, when populations of *G. pallida* selected over 9 generations for increased virulence on an ex *vernei* clone were tested on the *S. vernei* source of that resistance, they showed only a small increase in virulence over the unselected population (133).

Vanderplank (134) used "aggressiveness" to describe differences in the reproduction of pathogen populations on plants with horizontal resistance. In the context discussed here, aggressiveness largely equates to general fitness (i.e. the capacity to reproduce well on all genotypes) and is therefore the corresponding term in the area of susceptibility/horizontal resistance. Because cyst nematodes are very heterogeneous (6, 36), some species (e.g. potato cyst nematodes) suffer from inbreeding depression (M. S. Phillips, unpublished results) and general fitness may, in part, relate to the degree of heterogeneity of populations.

Biotype, Pathotype, and Race

Mode (78) proposed dual systems of balanced polymorphism in pathogens and their hosts as a necessary condition for coevolution. Pathogens produce selection pressures in their hosts and this leads to changes in host resistance, and subsequently in pathogen virulence alleles. Hence, variations in pathogen virulence are generated. The term "race" has been used to identify such variations, although some definitions imply a restriction in gene flow between populations. Dropkin (26) favored a phenotypic rather than a genotypic basis for race identification and suggested that races/pathotypes could be reliably distinguished in improved, standard tests. Other nematologists suggest that this identification is impractical where genotypes with quantitative resistance are used to separate races/pathotypes because environmental interactions will affect reproductive rates (79, 88, 118). Consequently, these researchers favor pathotype or race differentiation only in relation to major genes and where there is a demonstrated or inferred gene-for-gene relationship; this is the view favored here.

Nematode reproduction/host plant status can be recognized by three distinguishing criteria: variations in relation to (*a*) major gene (qualitative) resistance, (*b*) polygenic (quantitative) resistance, and (*c*) host range (for polyphagous nematodes). There are also apparent nonhosts (immune hosts), especially for very polyphagous species such as some *Meloidogyne* spp., that, rather than lacking specific susceptibility genes, may possess unrecognized resistance genes.

Race has been used to distinguish differences in virulence in the soybean cyst nematode *Heterodera glycines* (46) where the genetics of resistance is only partially resolved (1), for host range differences within the stem and bulb nematode *Ditylenchus dipsaci* (52) and for polyphagous, parthenogenetic root-knot nematodes (*Meloidogyne* spp.), (103) where differences occur in both host range and in virulence against major resistance genes. Race has also been used to determine both physiological and pathogenicity differences

(112). It is suggested here that the term race should be used with a prefix that specifies the basis for differentiation, e.g. host race.

Pathotype was initially used to differentiate populations of cyst nematode species that differ in virulence in relation to major gene resistance. Subsequently, for potato cyst nematodes this criterion was extended to pathotypes with quantitative resistance (69), but later this extension was strongly criticized (118). Occasionally, genotype has also been used for populations of noncyst nematodes differing in host range (12). It is suggested here, and elsewhere (2, 118), that pathotype should be used only in relation to gene-for-gene interactions. Biotype is used to delineate virulence differences in aphids and other insects, and rarely in relation to nematode virulence differences. However, for nematodes with the same genes for virulence, biotype is a more logical term than pathotype, which implies differences in pathogenicity.

GENETICS OF RESISTANCE AND VIRULENCE

The genetics of resistance to nematodes has been reviewed by Fassuliotes (34) and Sidhu & Webster (110). However, because the current breeding effort is largely restricted to a few major crops, the sources of nematode resistance so far identified probably represent only a small proportion of that available, e.g. recent studies (104) have revealed widespread resistance to several races of five *Meloidogyne* spp. in a range of crops. Almost all the genetically identified resistance used in breeding is against endoparasitic *Ditylenchus, Meloidogyne, Heterodera* or *Globodera* spp. and is conferred by dominant major genes (110), except for resistance in cotton to *Meloidogyne,* in soybean to *H. glycines* and in potato to *G. pallida* where recessive or polygenic resistance is used. Race/pathotype schemes have been proposed for species within all these genera and for species of *Pratylenchus, Radopholus* and *Tylenchulus* (110). Gene-for-gene relationships have been demonstrated for *Globodera* spp. (57, 60, 61) and assumed for several other interactions involving major gene resistance. Some genes confer resistance to more than one species of nematode. Examples include the *Mi* gene that gives resistance to *M. incognita, M. javanica* and *M. arenaria,* polygenic resistance from. *S. vernei* that gives resistance to *G. rostochiensis* and *G. pallida,* and common resistance in soybean to both *H. glycines* and *Rotylenchulus reniformis* (15).

Selection for virulence from within heterogenous populations has been demonstrated but, within a species such as *G. pallida,* takes a number of generations (133). Because nematodes are relatively immobile, virulence does not spread rapidly. Where populations are homozygously avirulent extended periods of selection have often failed to increase virulence. Such is the situation in north-east USA and in Britain (130) where potatoes with the *H1* gene for resistance to pathotype Rol of *G. rostochiensis* have been grown widely for more than 20 years. Populations of *G. rostochiensis* pathotypes Ro2, 3, and 5 (69), which are virulent against this gene, are widespread in South America and were introduced to other parts of Europe. Virulence is conferred by the homozygous recessive condition (57, 86) and it appears that the initial introductions of *G. rostochiensis* to the USA and Britain were homozygously avirulent.

Selection for increased virulence is a problem in the use of resistance to soybean cyst nematode *H. glycines* (146). The reason appears to be the heterogenous nature of most populations (33, 72). Similarly, although the root-knot nematode *M. incognita* reproduces by mitotic parthenogenesis and has almost no variation in isozyme phenotypes, progressive selection for virulence against the *Mi* gene has been reported. Initially, this was thought to involve selection from within a heterogenous population, but selection from within the progeny from a single female has recently been reported (58). The genetic basis of this change, which may involve regulatory genes as has been proposed for parthenogenetic species of aphids (135), is of considerable interest. The production of otherwise isogenic avirulent and virulent lines offers possibilities for identifying avirulence gene products (21).

MEASURING RESISTANCE AND VIRULENCE

Except on genotypes with qualitative resistance, nematode multiplication rates are density-dependent (Figure 1). Various equations have been produced to model single generation population changes (62, 90, 107). As multiplication rates are density-dependent they vary with the inoculum and the plant-root densities. This basic principle has important implications for the conduct of statutory and breeder-pot tests designed to assess levels of quantitative resistance and for the field performance of genotypes with such resistance. For the same inoculum density of *G. pallida,* multiplication rates in the field were only one half to one third of those obtained in pots (92). Qualitatively (major, usually vertically) acting genes can be readily identified in a pot test. All that is required is the necessary expertise and experience, a largely avirulent nematode population, and a susceptible control. Good tests are those where multiplication on the susceptible control is high and little (sometimes there are a few "escapes") or none on those with the resistance gene. Such tests are widespread in breeding for resistance to potato, soybean, and cereal cyst nematodes and several simplified tests have been developed (49, 89, 94).

Measuring quantitative (polygenic and possibly horizontal) resistance is fraught with problems. There is usually a continuous range in its expression in progeny derived from crosses between resistant and susceptible parents. Rather than recognize this variation in expression, the tendency in nematology has been to fix arbitary limits for resistance. For potato cyst nematodes, plants were classified as resistant if the multiplication rate was < 1 in a standard pot test, irrespective of the multiplication rate on the nonresistant controls which may be as low as $\times 10$ or as great as $\times 50$-fold. For soybean cyst nematode a multiplication rate of 10% of the susceptible controls was used to delineate resistance. When resistant/susceptible is classified in such arbitrary ways the problem of variation in the expression of resistance due to tolerance differences (Figure 1) or to environmental influences has been discussed (88). For potato cyst nematode it is proposed to use instead a 1–9 classification system based on the expression of resistance in the test clones in relation to certain well-tested, standard clones that represent specific points on the 1–9 scale (M. S. Phillips & I. Clayden, personal communication).

If the test nematode population is heterogenous for virulence, plants with major, vertically acting genes may appear partially resistant. If virulence and hence pathotype distinctions have a genetic base, then such populations should be regarded as pathotype mixtures (81). The lack of pure avirulent/virulent test populations has restricted analysis of the genetics of resistance and virulence in relation to soybean cyst nematode (*H. glycines*), which is being overcome by producing inbred lines (73).

Routine estimates of the virulences of heterogenous populations on genotypes with either qualitative or quantitative resistance can only be done by comparing population reproductive rates on the test genotypes with the mean of those on one or more nonresistant genotypes. Even so, experience suggests that, because of environmental influences, the apparent proportion of virulent nematodes will differ between tests (91) and comparison with a standard population, of well-tested virulence, is also recommended (81).

MECHANISMS OF RESISTANCE

Veech (136) suggested that most plants are resistant to most nematodes. This is debatable as many ectoparasitic and migratory endoparasitic and some sedentary endoparasitic (e.g. parthogenetic *Meloidogyne*) species have wide host ranges. However, certain crops, such as raspberry, appear to be poor or nonhosts for many nematodes, including ectoparasitic species. The mechanisms involved in such "resistance" are unknown, but there is a growing literature on the use of plant wastes and plant extracts for nematode control (8).

Certain plants contain toxins effective against endoparasitic species. Foremost are some plants of the genus *Tagetes* that contain α terthienyl, a compound toxic to *Pratylenchus* and *Meloidogyne* (9, 136). Some plant defense chemicals may actively repulse invading nematodes (59) and evidence is growing that in some potato hybrids resistance to potato cyst nematodes involves a loss of hatching-factor activity. However, the best information on resistance mechanisms comes from positive responses mediated by major resistance genes.

Vanderplank (134) proposed that vertically acting resistance genes are modified susceptibility genes and that they, and pathogen avirulence genes, serve other useful purposes in the plant and pathogen. In compatible associations the protein products of the avirulence and susceptibility genes copolymerize to provide the nutrients necessary for pathogen development. This process makes available the specific host proteins required for pathogen development, and their production is increased through a feedback system (70).

The process of copolymerization also removes the pathogen proteins from the cell, thereby preventing them from acting as elicitors of the resistant response. Vanderplank hypothesized that in resistant plants the pathogen and host proteins do not copolymerize because a susceptibility gene, and hence the protein it codes for, has been modified. Copolymerization is an endothermic process, and Vanderplank suggested this may account for the breakdown of some resistance genes at high temperatures. Nematode resistance genes of this type include the *Mi* gene (148) and certain genes in bean (84).

Plant mycologists and bacteriologists have identified many potential elicitors of resistance. These derive either directly from the pathogen, or are fragments of host or pathogen cell walls. They include various polysaccharides, oligosaccharides, proteins, glycoproteins, and fatty acids (25). Although most elicitors identified so far have been non race-specific, in fungi a glycoprotein produced by the α race of *Colletotrichum lindemuthianum* specifically induces phytoalexin accumulation in resistant but not in nonresistant beans (114) and a peptide has been isolated as the race-specific elicitor of the resistant response of tomato to *Cladosporium fulvum* (105).

The concept of elicitors assumes a recognition process (68) involving binding sites within the host cell. Binding of an elicitor to such a site initiates a signal leading to a relatively nonspecific resistant response based on the activation of defense genes coding for the synthesis of phytoalexins, hydroxyproline-rich proteins, PR proteins, chitinases, etc. There is relatively little direct evidence for such elicitor binding sites, but they are thought to be located in cell membranes (25).

The fact that genes for qualitative resistance are usually dominant and the corresponding virulence genes are recessive, together with the elicitor/receptor hypothesis, implies that resistance relates to constitutive changes in the plant involving the recognition process and that virulence involves avoidance of recognition by changes to, or nonproduction of, the elicitor.

The mechanisms of resistance to nematodes have been comprehensively reviewed by Kaplan & Davis (65). No nematode-derived elicitors of resistance have been identified, but potential sources include the saliva injected via the stylet, excretory products, and various surface components. For endoparasites all three could be involved, but for resistance in grape to the migratory ectoparasitic *Xiphinema index* (140) the elicitor must be in the secretions injected via the stylet.

Glycoproteins are involved in many cell-to-cell recognition systems. Lectins, which bind specifically to particular carbohydrate moieties, have been used to characterize carbohydrate distributions on the surfaces of nematodes (66). Various carbohydrates have been identified, emanating from cuticular pores, the excretory system, or from the paired amphidial chemoreceptors situated on the nematode head (100). The role of surface glycoproteins in plant-pathogen interactions remains controversial but there is indirect evidence, through their blocking by certain lectin treatments, for their involvement in host finding by nematodes (149). In addition, treatment of invasive juveniles of a root-knot nematode (*M. incognita,* Race 3) with lectins and their complementary sugars increased the tendency for a hypersensitive response to be induced in resistant soybeans and for an incompatible response in a susceptible variety (22); incubating juveniles in distilled water decreased the proportion of juveniles that initiated a hypersensitive response in the resistant cultivar. As these effects were relatively nonspecific, an indirect effect is implied, perhaps by influencing the production of the elicitor.

In biotrophic species the esophageal gland cell secretions, and particularly those from the dorsal gland cell, produce the feeding tube (143) and initiate the susceptible response. Much attention has focused on their role in the susceptible response because it can involve such profound changes in cell development (29, 143). Equally, however, these secretions are a potential source of elicitors in incompatible interactions. Hussey (53) has reviewed recent research on the function and nature of nematode gland cell secretions and monoclonal antibodies are now being used to localize and extract secretory granules from cyst and root-knot species (5, 56). Various studies have been made on the gland cell secretions of root-knot nematodes, either in situ or in stylet secretions (53). Basic proteins and peroxidase and acid phosphatase activity were detected, but several other enzymes, including cellulase, were not. Staining was also obtained for nucleic acids. *Ditylenchus dipsaci,* a migratory endoparasitic species inhabiting spaces created between cells in leaf and stem tissues, produces cellulases and pectinases, as do several other migratory endoparasites (24). Differences between host races in pectin methylesterase were also reported. A wide range of enzymes has been detected in nematode homogenates, but they have not been localized. Recently, one of the secretory components from the dorsal esophageal gland cell of *M. incognita* was immunoaffinity purified and shown to be a large molecular weight glycoprotein (56).

THE RESISTANT RESPONSE

A range of responses in resistant plants to invasion/feeding by incompatible nematodes has been reported (66). These range from nonspecific tissue necrosis, which may not impair nematode development, through delayed localized necrosis around the nematode or its feeding site, which prevents the development of females but not of males, to a more rapid "hypersensitive" response that prevents development of the nematode and its feeding site. It is uncertain whether the hypersensitive response induced by nematodes is always comparable with that classically induced by incompatible associations involving fungi. Because of their endoparasitic nature, timing of the initiation of the hypersensitive response is uncertain. With *Tylenchulus semipenetrans* the hypersensitive response in citrus took up to two weeks to develop (64). In contrast, a necrotic response to *Meloidogyne* in resistant tomatoes was visible within 2 days of infection and some changes could be observed within 8–12 h (87). The reaction was localized around the nematode and its feeding site and initially involved a loss of electron-dense inclusions in the vacuoles, cell membrane disruption, and a rapid increase in the electron density of the cytoplasm. Membrane-bound organelles disappeared but the endoplasmic reticulum became extended, suggesting changes in cell wall and membrane permeability and the synthesis of enzymes leading to general disorganization (87).

Resistance to *Rotylenchulus,* cyst nematodes and, in some crops, to *Meloidogyne* generally involves a slower response, with syncytial or giant cell feeding sites being initiated (11, 67). Some juvenile development occurs, often to adult males but not females (98, 131, 145). Cells surrounding the nematode and the developing feeding site become necrotic, limiting feeding site development. Cells incorporated in the feeding site show an increase in the amount of rough endoplasmic reticulum and fail to develop the dense cytoplasm characteristic of the susceptible response (98, 145). Eventually the cytoplasmic contents degenerate and nematodes not already developing as males leave the roots or die (42).

The mechanisms involved in the positive resistant response have been the subject of much speculation (65). Growth regulators have been implicated in the induction of the susceptible but not in the resistant response (136). Cytokinin treatment was reported to reduce the resistance of tomato to *M. incognita* (27). Arrigoni (4) suggested that early, increased induction of cyanide-resistant respiration leading to an increase in hydroxyproline-rich proteins is a feature of nematode-resistant mechanisms. Hydrogen peroxide, and an increase in superoxides due to increased peroxidase activity, will also be generated. Superoxides are extremely toxic unless suppressed by superoxide dismutase (SOD). Differences in SOD activity consistent with this general mechanism have been detected (147).

Phytoalexins, which may be synthesized in response to superoxide generation, have been implicated in the resistant responses of cotton and soybean to *M. incognita* (2a,136). In soybean, glyceollin accumulated in the stele within three days of infection in resistant but not in nonresistant cultivars (67). Similarly, toxic accumulations of gossypol and related terpenoid aldehydes occurred in the stele and endodermis of *M. incognita*-infected resistant cotton, but decreased in the susceptible controls. Lima beans are hypersensitively resistant to *Pratylenchus scribneri* whereas snapbeans are susceptible, but only in lima beans does infection induce toxic accumulations of commestrol and other coumestans (99).

Phenolic compounds have been implicated in resistance to several nematodes (9) and Giebel (45) suggested that potato cyst nematodes inject a β-glycosidase that hydrolyzes a glycoside releasing phenolic aglycones, present in the resistant but not the susceptible plant, that directly or indirectly causes necrosis. In addition, the glycosidase affects the ratio of mono- to polyphenols in such a way that IAA oxidase activity is inhibited in susceptible but not resistant plants.

TOLERANCE

Cultivars of a crop are regarded as differing in their tolerance if the proportional decrease in their growth and/or yield due to nematode damage differs significantly when they are grown in uniformly infested soil. Where cultivars of different yield potentials are being compared, the proportional effect on yield must be compared in adjacent, uniformly uninfested and infested pots/plots. Recent reviews on resistance and tolerance are by Cook & Evans (16) and on tolerance of cyst nematodes are by Evans & Haydock (31). Resistance and tolerance are independent attributes of the plant (43, 55) but resistance may confer tolerance, especially if it decreases the incidence of nematode attack or parasitism (123). Equally, mechanisms of tolerance exist that are independent of resistance and there are many examples of hypersensitive resistant responses decreasing tolerance. Although tolerance and resistance can be independent attributes of the plant, differences in tolerance will affect nematode multiplication rates on susceptible and partially resistant plants (Figure 1), especially at high initial population densities (90).

Tolerance is probably widespread and important in wild plants and landrace varieties but, unless associated with other desirable characters being selected for, is likely to be lost during crop breeding (38). Selection for tolerance is usually secondary to other factors and, if done, is usually on a comparative basis. Also, routine selection is only possible against nematodes whose population densities can be so manipulated that they cause obvious yield losses. Many plant parasitic nematodes, especially ectoparasitic species, do not readily achieve such densities, either because they are only weakly pathogenic or because their maximum population density is generally kept low by natural factors (e.g. parasitism and predation).

Because tolerance reflects the capacity of the plant to withstand or overcome nematode damage, an understanding of the mechanisms of plant damage and of modeling yield-loss relationships is helpful to understanding the mechanisms of tolerance.

MODELING YIELD LOSS AND TOLERANCE

Nematode effects in the field on plant growth and yield are generally proportional to the numbers of infective nematodes per unit of soil at planting (P_i, Figure 2). Several equations have been developed (28) to model yield losses due to nematodes. Most are based on a progressively diminishing effect as P_i increases, with P_i often presented on a logarithmic scale. The equation used here was developed by Seinhorst (106) and is based on Nicholson's competition model (82). In its simplest form it states $y = m + (1 - m)z^{P_i - T}$ where y at any value of P_i is the total dry matter yield on a proportional basis (i.e. $y =$ unity in the absence of nematodes), m is the minimum yield achieved at high values of P_i (again on a proportional basis), z is a constant that relates to the proportion of the plant undamaged by the attack of a single nematode, and T (the tolerance limit) is the P_i below which plant growth is not decreased because (*a*) the plant has surplus root, (*b*) very low densities of nematodes can stimulate root growth, or (*c*) there is rapid compensation for any damage. Where T is greater than the P_i then $y = 1.0$. The sigmoidal nature of the relationship when P_i is on a log scale is shown in Figure 2.

Clearly, any differences between cultivars in the values of z or m will affect the slopes of their regressions (119) and, together with variations in T, the overall tolerance of the cultivars. Tolerance differences can only be assessed at P_is between T and that when m is first reached and will increase with increasing P_i. Also, if T and z are kept constant, then the minimum yield (m) will be reached at the same P_i, but if T or z but not m are changed, then m will be reached at a different P_i. Views differ as to whether tolerance differences are largely due to differences in the values of z or m (129). Where a crop supports several generations of a nematode, a progressive increase in P_i, which will usually be greater on tolerant than intolerant genotypes, may eventually swamp any initial tolerance differences.

MECHANISMS OF DAMAGE

Possible Components of Damage

Factors that decrease the rate of growth of aerial parts of plants will reduce total light interception and hence rates of dry matter productivity. The possible mechanisms by which root-feeding nematodes might reduce top growth are: (*a*) direct parasitic effects involving withdrawal of nutrients; (*b*) mechanical or physiological damage that reduce root growth and function; and (*c*) systemic physiological effects that directly affect top growth or change assimilation, respiration, dry matter partitioning, or other growth and activity characteristics (123). However, nematode attack and damage occurs over a period of time and interacts dynamically with plant growth, as well as with the environment in which the plant is grown.

PARASITIC EFFECTS Feeding nematodes withdraw nutrients and it has been estimated that for potato plants heavily infested with potato cyst nematodes up to 5% of the total nitrogen is in the developing females (126). The proportion extracted is likely to be considerably more. Similarly, a moderate infestation of vines (*Vitis vinifera*) by *M. incognita* was estimated to divert 15% of the total energy assimilated (75). Resistant plants largely avoid these parasitic drains on their food reserves and, everything else being equal, should be more tolerant than nonresistant genotypes (123). Hence, the potato cultivar Maris Piper is comparatively more tolerant of pathotype Ro1 of *G. rostochiensis* to which it is resistant than of *G. pallida* to which it is susceptible (31, 128).

IMPAIRED ROOT FUNCTION AND PHYSIOLOGICAL EFFECTS For endoparasitic species such as potato cyst nematodes, the tunnelling involved in invasion and/or feeding is a major cause of root damage. This damage directly reduces the rate of root extension (32, 95, 116), and tends to decrease rates of uptake and leaf content of macro-nutrients (117, 127) and water. Leaf concentrations of calcium tend to increase (7, 30, 96, 127), probably because the endodermis is bridged by nematode feeding sites allowing apoplastic movement (30).

Meloidogyne juveniles usually invade intercellularly (29) and hence usually cause less direct damage than cyst nematodes. However, the galls they induce considerably disrupt the xylem (76) and divert assimilates, and they can have an equally depressing effect on water relations (141) and root and top growth.

Nematode damage interacts with a number of soil factors and the effects of nutrient or water stress are proportionally greater for infected than uninfected plants (83, 122). Interactions with pathogenic fungi are also important and can greatly increase the level of damage and apparent intolerance (31, 74).

The invasion damage caused by cyst nematodes increases the formation of abscisic acid (35) and substantial reductions in photosynthetic rate per unit area of leaf have been reported in cyst and root-knot infected plants (3, 71). The mechanisms by which these systemic effects on photosynthesis are induced are unclear (138, 139), although reductions in leaf N status are probably involved. Other systemic changes include the induction of pathogenesis related (PR) protein in the leaves of cyst-nematode infected plants

(50) and systemic suppression of resistance to *Fusarium* wilt in tomato infected by *Meloidogyne* (109).

MECHANISMS OF YIELD REDUCTION AND OF TOLERANCE

The most comprehensive studies have been on cyst nematodes. Various mechanisms of tolerance have been suggested, including differences in numbers of roots, compensatory root growth, delayed plant senescence, and enhanced water and nutrient uptake (30, 37, 77, 97, 125). Increased chlorophyll content was suggested as a mechanism of tolerance of rice to *M. graminicola* (113). Tolerance is independent of resistance (38, 55) and appears to be largely nonspecific (139). A common feature seems to be a capacity for enhanced root growth after the initial period of nematode attack (23, 139) that may also confer tolerance to other stresses such as drought (18). Whatever the mechanism, the total dry matter productivity of a crop is determined by the efficiency with which the available light is intercepted and converted into dry matter. Harvestable yield (the farmer's yield) is further influenced by partitioning and dry matter content. Nematode damage to roots reduces the rate of top growth, as already discussed. With heavy infestations of potato cyst nematodes, the primary mechanism appears to be a reduction in the rate of nutrient uptake (122). The rate of top growth and leaf expansion is reduced, thereby preventing intolerant potato cultivars such as Pentland Dell from achieving 100% groundcover, and hence 100% light interception. In contrast, although top growth is initially reduced by proportionally the same amount, a heavy infestation on the tolerant cultivar Cara only delays the time taken for it to achieve 100% ground cover. The key to this difference in response was the much greater growth rate of Cara, which at all stages of growth and intensities of nematode infestation had tops twice as large as those of the corresponding Pentland Dell plants. Analysis of leaf area/dry matter relationships showed that per unit of dry matter Cara consistently produced 25% more leaf area than Pentland Dell (129). Cara's vigorous top growth supported vigorous root growth that allowed adequate nutrient uptake and hence further top growth. Grafting experiments between Pentland Dell and Cara showed that the scion contributed most to overall growth and partitioning, but that both scion and stock contributed to tolerance of potato cyst nematode (121).

OCCURRENCE AND USE OF TOLERANCE

Nematological Abstracts consider tolerance a form of resistance, and most of their abstracts so classified relate to cyst or root-knot nematode species.

However, in recent reports useful tolerance of migratory endoparasitic genera including *Radopholus* (coconut) and *Pratylenchus* (*Prunus*, sugar-cane, alfalfa) and of ectoparasitic genera including *Criconemella* (*Prunus*), *Hoplolaimus* (soybean), *Paralongidorus* (rice) and *Tylenchorhynchus* (rice), is indicated. These reports point to the value of tolerance in many crops or situations where alternative control measures are not available. However, tolerance of cyst and root-knot nematodes generally needs to be combined with a degree of resistance, otherwise populations will be increased to densities where even tolerant cultivars are damaged. Susceptible, tolerant genotypes may, however, be useful in rotation following resistant genotypes as a means of reducing the rate of selection for virulence (55). Experience in The Netherlands has already shown that in rotations alternating nematicide-treated susceptible and untreated resistant potato cultivars, tolerance of potato cyst nematode is an important adjunct to resistance.

SCREENING FOR TOLERANCE

Tolerance differences are usually determined by comparing the relative yield losses of a range of genotypes growing in uniformly infested soil. Experience has shown that tolerance differences generally cannot be assessed in pots (16) and have to be assessed in field trials. Initially, to assess tolerance of cyst nematodes split plot trials were favored in which yields in nematicide-treated and untreated subplots were compared. However, it was observed that tolerance differences were positively correlated with yields in the untreated subplots (20) and this led to trial designs that used replicated, single plants growing in untreated, heavily infested soil (93). Top growth differences also tend to be correlated with yield and can be used to separate genotypes of potato that are tolerant and intolerant of potato cyst nematodes (19). Tolerance in wheat of *H. avenae* is closely correlated with early growth (111), especially the maximum length of leaf 4 (the first leaf to be initiated outside the embryo). An alternative approach to assessing tolerance differences, particularly used to investigate the mechanisms of tolerance and how tolerance should be modeled, is to compare the performance of different cultivars in plots previously prepared to contain a range of initial population densities (as in Figure 2) (31, 129).

SUMMARY AND CONCLUSIONS

Nematologists work with relative immobile pathogens with comparatively low reproductive rates. This has enabled them to separate clearly resistance and tolerance as independent characteristics of the host. Koch's postulates cannot be applied to most nematode problems because damage is proportional to the inoculum density at planting. However, nematologists have been able to develop equations modeling both yield losses and population dynamics within a single crop and within rotations.

Much resistance to nematodes is conferred by major, dominant genes and their use has usually revealed complementary virulence in the nematode. In the examples studied, this has been shown to be on a gene-for-gene basis. However, quantitative resistance to obligate, biotrophic nematodes may also include a passive component involving reduced susceptibility. Monoclonal antibodies are being used to identify the nematode gland cell products that, when injected into susceptible host cells, transform them into multinucleate, polyploid and metabolically highly active "nurse" cells. These gland cell products may be the elicitors of the resistant response; isogenic avirulent and virulent lines of a root-knot nematode have been developed and comparisons between them may lead to identification of the avirulence gene and its product. Other possible elicitors include surface glycoproteins and excretory products.

Tolerance of nematode damage is a desirable character in resistant cultivars, many of which have an unacceptable degree of intolerance. Equally, as tolerant, susceptible cultivars have the potential to increase nematode population densities to extremely damaging levels, tolerance should be combined with resistance. Several mechanisms are involved in tolerance, but these are usually nonspecific and include vigor of root and top growth and their response to nematode attack. A dynamic relationship exists between the effects of nematode attack (which usually lasts several weeks) on nutrient and water uptake and root and top growth. Damage reduces uptake, which reduces the rate of top growth and hence the rate at which carbohydrate synthesis increases. This, in turn, affects the rate of root growth and hence uptake. In the field, nematode damage to the host roots also interacts with the availability of nutrients and water and the effects of nematode damage on final yields are strongly influenced by their effects on crop canopy development and light interception.

Overall, for crops to perform well in nematode-infested soil, it has been found that tolerance is a useful and necessary adjunct to resistance; until recently it has been largely ignored. Even without resistance, tolerance is an important attribute of many low-value crops, especially in regions where nematicides and resistant cultivars are not available.

ACKNOWLEDGMENTS

I am grateful to my colleagues Dr. Jock Forrest and Mr. Mark Phillips, to the staff of the *Annual Reviews of Phytopathology,* and especially to Dr. Ken Evans for their helpful comments.

Literature Cited

1a. Alphey, T. J. W., Phillips, M. S., Trudgill, D. L. 1988. Integrated control of potato cyst nematodes using small amounts of nematicides and potatoes with partial resistance. *Ann. Appl. Biol.* 113:545–52

1. Anand, S. C., Luedders, V. D. 1989. Use of soybean cyst nematode inbreds to determine genetic diversity among resistant soybeans. *Crop Prot.* 8:380–82

2. Andersen, S., Andersen, K. 1982. Suggestion for determination and terminology of pathotypes and genes for resistance in cyst-forming nematode, especially *Heterodera avenae*. *EPPO Bull.* 12:379–86

2a. Apostol, I., Heinstein, P. F., Low, P. S. 1990. Rapid stimulation of an oxidative burst during elicitation of cultured plant cells. *Plant Physiol.* 90:109–16

3. Arntzen, F. K., Schans, J. 1988. The effect of density of potato cyst nematodes and plant genotype on photosynthesis and biomass production. *Nematologica* 34:255 (Abstr.)

4. Arrigoni, 0. 1979. A biological defence mechanism in plants. In *Root-Knot Nematodes* (Meloidogyne *species) Systematics, Biology and Control*, ed. F. Lamberti, C.E. Taylor, pp. 457–67. London: Academic. 477 pp.

5. Atkinson, H. J., Harris, P. D., Halk, E. J., Novitsk, C., Leighton-Sands, J., Nolan, P., Fox, P. C. 1988. Monoclonal antibodies to the soya bean cyst nematode *Heterodera glycines*. *Ann. Appl. Biol.* 112:459–69

6. Bakker, J., Bouwman-Smits, L. 1988. Genetic variation in polypeptide maps of two *Globodera rostochiensis* pathotype. *Phytopathology* 78:894–904

7. Been, T. H., Schomaker, C. H. 1986. Quantitative analysis of growth, mineral composition and ion balance of the potato cultivar Irene infested with *Globodera pallida* (Stone). *Nematologica* 32:339–55

8. Bhattacharya, D., Goswami, B. K. 1987. A study on the comparative efficiency of neem and ground nut oil cakes against root-knot nematode, *Meloidogyne incognita* as influenced by microorganisms on sterilised and unsterilised soil. *Ind. J. Nematol.* 17:81–83

9. Bingefors, S. 1982. Nature of inherited nematode resistance in plants. In *Pathogens, Vectors and Plant Diseases: Approaches to Control*, ed. K. F. Harris, K. Maramorosch, pp. 187–218. New York: Academic. 310 pp.

10. Bleve-Zacheo, T., Lamberti, F., Chinappen, M. 1987. Root cell response in rice attacked by *Hemicycliophora typica*. *Nematol. Mediterr.* 15:129–38

11. Bleve-Zacheo, T., Melillo, M. T., Zacheo, G. 1990. Ultrastructural response of potato root resistant to cyst nematode *Globodera rostochiensis* pathotype Ro1. *Rev. Nematol.* 13:29–36

12. Bolla, R. I., Weaver, C., Koslowski, P., Fitzsimmons, K., Winter, R. E. K. 1987. Characterisation of a non-parasitic isolate of *Bursaphelenchus xylophilus*. *J. Nematol.* 19:304–10

13. Caligari, P. D. S., Phillips, M. S. 1984. A re-examination of apparent selection in *Globodera pallida* on *Solanum vernei* hybrids. *Euphytica* 33:583–86

14. Cook, R. 1974. Nature and inheritance of nematode resistance in cereals. *J. Nematol.* 6:165–74

15. Cook, R. 1991. Resistance in plants to cyst and root-knot nematodes. *Agric. Zool. Rev.* 4. In press

16. Cook, R., Evans, K. 1987. Resistance and tolerance. In *Principles and Practice of Nematode Control in Crops*, ed. R. H. Brown, B. R. Kerry, pp. 179–231. Sydney: Academic. 447 pp.

17. Cooper, J. I., Jones, A. T. 1983. Responses of plants to viruses: proposals for the use of terms. *Phytopathology* 73:127–28

18. Dale, M. F. B. 1991. Breeding for tolerance to potato cyst nematode. *Aspects Appl. Biol.* 17:95–101

19. Dale, M. F. B., Brown, J. 1989. The use of foliage assessment to improve the identification of tolerance to damage by nematodes (*Globodera pallida)* in potatoes. *Ann. Appl. Biol.* 115:313–19

20. Dale, M. F. B., Phillips, M. S., Ayres, R. M., Hancock, M., Holliday, M., et al. 1988. The assessment of the tolerance of partially resistant potato clones to damage by the potato cyst nematode *Globodera pallida* at different sites and in different years. *Ann. Appl. Biol.* 113: 79–88

21. Dalmasso, A., Castagnone-Sereno, P., Bingiovanni, M., de Jonck, A. 1991. Acquired virulence in the phytophagous nematode *Meloidogyne incognita* II. Bidimensional analysis of isogenic isolates. *Rev. Nematol.* 14:305–8

22. Davis, E. L., Kaplan, D. T., Dickson, D. W., Mitchell, D. J. 1989. Root tissue response of two related soybean cultivars to infection by lectin-treated *Meloidogyne* spp. *J. Nematol.* 21:219–28

23. Davy de Virville, J., Person-Dedryver, F. 1989. Growth and respiratory activity of roots of various Triticeae tolerant or resistant to *Heterodera avenae* Woll. with or without infection by the nematode. *Rev. Nematol.* 12:379–86
24. Deubert, K. H., Rohde, R. A. 1971. Nematode enzymes. In *Plant Parasitic Nematodes*, ed. B. M. Zuckerman, W. F. Mai, R. A. Rohde, 2:73–90. New York/London: Academic. 347 pp.
25. Dixon, R. A., Lamb, C. J. 1990. Molecular communication in interactions between plants and microbial pathogens. *Annu. Rev. Plant Physiol. Plant Mol. Biol.* 41:339–67
26. Dropkin, V. H. 1988. The concept of race in phytonematology. *Annu. Rev. Phytopathol.* 26:145–61
27. Dropkin, V. H., Helgeson, J. P., Upper, C. D. 1969. The hypersensitive reaction of tomatoes resistant to *Meloidogyne incognita:* reversal by cytokinins. *J. Nematol.* 1:55–61
28. Elston, D. A., Phillips, M. S., Trudgill, D.L. 1991. The relationship between initial population density of potato cyst nematode *Globodera pallida* and the yield of partially resistant potatoes. *Rev. Nematol.* 14:213–20
29. Endo, B. Y. 1987. Histopathology and ultrastructure of crops invaded by certain sedentary endoparasitic nematodes. See Ref. 137, pp. 196–210
30. Evans, K., Franco, J. 1979. Tolerance to cyst-nematode attack in commercial potato cultivars and some possible mechanisms for its operation. *Nematologica* 25:153–62
31. Evans, K., Haydock, P. P. J. 1990. A review of tolerance by potato plants of cyst nematode attack with consideration of what factors may confer tolerance and methods of assaying tolerance and improving it in crops. *Ann. Appl. Biol.* 117:703–40
32. Evans, K., Trudgill, D. L., Brown, N. J. 1977. Effects of potato cyst nematodes on potato plants. V. Root system development on lightly- and heavily-infested susceptible and resistant varieties, and its importance in nutrient and water uptake. *Nematologica* 23:153–64
33. Faghihi, J., Ferris, J. M., Ferris, V. R. 1986. *Heterodera glycines* in Indiana: 1. Reproduction of geographical isolates on soybean differentials. *J. Nematol.* 18:169–72
34. Fassuliotis, G. 1987. Genetic basis of plant resistance to nematodes. See Ref. 137, pp. 364–71
35. Fatemy, F., Trinder, P. K. E., Wingfield, J. N., Evans, K. 1985. Effects of *Globodera rostochiensis,* water stress and exogenous abscisic acid on stomatal function and water use in Cara and Pentland Dell potato plants. *Rev. Nematol.* 8:249–55
36. Ferris, V. R., Faghihi, J., Ireholm, A., Ferris, J. M. 1989. Two-dimensional protein patterns of cereal cyst nematodes. *Phytopathology* 79:927–33
37. Fisher, J. M. 1982. Problems with the use of resistant wheat to the Australian pathotype of *Heterodera avenae*. *EPPO Bull.* 12:417–21
38. Fisher, J. M., Rathjen, A. J., Dube, A. J. 1981. Tolerance of commercial cultivars and breeders' lines of wheat to *Heterodera avenae* Woll. *Aust. J. Agric. Res.* 32:545–51
39. Flor, H. H. 1942. Inheritance of pathogenicity of *Melampsora lini*. *Phytopathology* 32:653–69
40. Flor, H. H. 1971. Current status of the gene for gene concept. *Annu. Rev. Phytopathol.* 9:275–96
41. Forrest, J. M. S., Phillips, M. S. 1984. The effect of continuous rearing of a population of *Globodera pallida* (Pa2) on susceptible or partially resistant potatoes. *Plant Pathol.* 33:53–56
42. Forrest, J. M. S., Trudgill, D. L., Cotes, L. M. 1986. The fate of juveniles of *Globodera rostochiensis* pathotype Rol in roots of susceptible and resistant potato cultivars with gene Hl. *Nematologica* 32:106–14
43. Fox, J. A., Spasoff, L. 1976. Resistance and tolerance of tobacco to *Heterodera solanacearum*. *J. Nematol.* 8:284–85 (Abstr.)
44. Fraser, R. S. S. 1987. Genetics of plant resistance to viruses. In *Plant Resistance to Viruses* pp. 6–22. Chichester: Wiley
45. Giebel, J. 1982. Mechanisms of resistance to plant nematodes. *Annu. Rev. Phytopathol.* 20:275–79
46. Golden, A. M., Epps, J. M., Riggs, R. D., Duclos, L. A., Fox, J. A., Bernard, R. L. 1970. Terminology and identity of intraspecific forms of the soybean cyst nematode (*Heterodera glycines*). *Plant Dis. Reptr.* 54:544–46
47. Gommers, F. J. 1981. Biochemical interactions between nematodes and plants and their relevance to control. *Helminth. Abstr. Ser. B* 50:9–24
48. Griffiths, B., Robertson, W. M. 1989. A quantitative study of changes induced by *Xiphinema diversicaudatum* in root-tip galls of strawberry and ryegrass. *Nematologica* 34:198–207
49. Halbrendt, J. M., Dropkin, V. H. 1986. *Heterodera glycines*-soybean associ-

tion: A rapid assay using pruned seedlings. *J. Nematol.* 18:370–74
50. Hammond-Kosack, K. E., Atkinson, H. J., Bowles, D. J. 1989. Systemic accumulation of novel proteins in the apoplast of the leaves of potato plants following root invasion by the cyst nematode *Globodera rostochiensis*. *Physiol. Mol. Plant Pathol.* 35:495–506
51. Harris, M. K., Frederiksen, R. A. 1984. Concepts and methods regarding host plant resistance to arthropods and pathogens. *Annu. Rev. Phytopathol.* 22:247–72
52. Hesling, J. J. 1965. Biological races of stem eelworm. *Rep. Glasshouse Crops Res. Inst.* pp. 132–41
53. Hussey, R. S.1987. Secretions of esophageal glands of tylenchid nematodes. See Ref. 137, pp. 221–28
54. Hussey, R. S. 1989. Disease-inducing secretions of plant-parasitic nematodes. *Annu. Rev. Phytopathol.* 27:123–41
55. Hussey, R. S., Boerma, H. R. 1991. Tolerance to *Heterodera glycines*. In *Biology and Management of Soybean Cyst Nematode*, ed. J. A. Wrather, R. D. Riggs, Minneapolis: APS Press. In press
56. Hussey, R., Paguio, O. R., Seabury, F. 1990. Localization and purification of a secretory protein from the esophageal glands of *Meloidogyne incognita* with a monoclonal antibody. *Phytopathology* 80:709–14
57. Janssen, R. 1990. *Genetics of virulence in potato cyst nematodes*. PhD thesis, Agric. Univ., Wageningen, The Netherlands, 71 pp.
58. Jarquin-Barberena, H., Dalmasso, A., de Guiran, G., Cardin, M. C. 1991. Acquired virulence in.the phytophagous nematode *Meloidogyne incognita* I. Biological analysis of the phenomenon. *Rev. Nematol.* 14:299–303
59. Jatala, P., Russell, C. C. 1972. Nature of sweet potato resistance to *Meloidogyne incognita* and the effects of temperature on parasitism. *J. Nematol.* 4:1–7
60. Jones, F. G. W. 1975. Host parasite relationship of potato cyst nematodes: A speculation arising from the gene for gene hypothesis. *Nematologica* 20:437–43
61. Jones, F. G. W., Parrott, D. M., Perry, J. N. 1981. The gene-for-gene relationship and its significance for potato cyst nematodes and their Solanaceous hosts. See Ref. 150, pp. 23–36
62. Jones, F. G. W., Perry, J. N. 1978. Modelling populations of cyst nematodes (Nematoda : Heterodoridae). *J. Appl. Ecol.* 15:349–71
63. Jones, M. G. K. 1981. The development and function of plant cells modified by endoparasitic nematodes. See Ref. 150, pp. 255–78
64. Kaplan, D. T. 1981. Characterisation of citrus rootstock responses to *Tylenchulus semipenetrans* (Cobb). *J. Nematol.* 13: 492–98
65. Kaplan, D. T., Davis, E. L. 1987. Mechanisms of plant incompatibility with nematodes. See Ref. 137, pp. 267–76
66. Kaplan, D. T., Keen, N. T. 1980. Mechanisms conferring plant incompatibility to nematodes. *Rev. Nematol.* 3:123–34
67. Kaplan, D. T., Keen, N. T., Thomason, I. J. 1980. Association of glyceollin with the incompatible response of soybean roots to *Meloidogyne incognita*. *Physiol. Plant Pathol.* 16:309–18
68. Keen, N. T., Bruegger, B. 1977. Phytoalexins and chemicals that elicit their production in plants. In *Host Plant Resistance to Pests*, ed. P. Hedin. *Am. Symp. Chem. Soc. Ser.* 62:1–26
69. Kort, J., Ross, H., Rumpenhorst, H. J., Stone, A. R. 1977. An international scheme for identifying and classifying pathotypes of potato cyst nematodes *Globodera rostochiensis* and *G. pallida*. *Nematologica* 23:333–39
70. Lobell, R. B., Schleif, R. F. 1990. DNA looping and unlooping by AraC protein. *Science* 250:528–32
71. Loveys, B. R., Bird, A. F. 1973. The influence of nematodes on photosynthesis in tomato plants. *Physiol. Plant Pathol.* 3:525–29
72. Luedders, V. D. 1989. Genes for resistance to soybean cyst nematode populations. *Crop Sci.* 29:259–62
73. Luedders, V. D. 1990. Linkage of soybean cyst nematode alleles for incompatibility with soybean. *Ann. Appl. Biol.* 116:313–19
74. Mai, W. F., Abawi, G. S. 1987. Interactions among root-knot nematodes and Fusarium wilt fungi on host plants. *Annu. Rev. Phytopathol.* 25:317–38
75. Melakeberhan, H., Ferris, H. 1989. Impact of *Meloidogyne incognita* on physiological efficiency of *Vitis vinifera*. *J. Nematol.* 21:74–80
76. Meon, S., Wallace, H. R., Fisher, J. M. 1978. Water relations of tomato (*Lycopersicon esculentum* Mill. cv. Early Dwarf) infected with *Meloidogyne javanica* (Treub), Chitwood. *Physiol. Plant Pathol.* 13:275–81
77. Miltner, E. D., Karnok, K. J., Hussey, R. S. 1991. Root responses of tolerance and intolerant soybean cultivars to soybean cyst nematodes. *Agron. J.* In press

78. Mode, C. J. 1958. A mathematical model for the co-evolution of obligate parasites and their hosts. *Evolution* 12:158–65
79. Mugniery, D., Balandras, C. 1989. Influence du type de test et de la population de *Globodera pallida* Stone sur l'evaluation de la resistance des varietes de pommes de terre en inscription. *Potato Res.* 32:311–20
80. Mugniery, D., Fayet, G. 1981. Determination de sexe chez *Globodera pallida* Stone. *Rev. Nematol.* 4:41–45
81. Mugniery, D., Phillips, M. S., Rumpenhorst, H. R., Stone, A. R., Treur, A., Trudgill, D. L. 1989. Assessment of partial resistance of potato to, and pathotypes and virulence differences in, potato cyst nematodes. *EPPO Bull.* 19:7–25
82. Nicholson, A. J. 1933. The balance of animal populations. *J. Anim. Ecol.* 2: 132–78
83. O'Bannon, J. H., Reynolds, H. W. 1965. Water consumption and growth of root-knot nematode infested and uninfested cotton plants. *Soil Sci.* 99:251–55
84. Omwega, C. O., Thomason, I. J., Roberts, P. A. 1990. Effect of temperature on expression of resistance to *Meloidogyne* spp. in common bean. *J. Nematol.* 22:446–51
85. Parlevliet, J. E., Zadocks, J. L. 1977. The integrated concept of disease resistance; a new view including horizontal and vertical resistance in plants. *Euphytica* 26:5–21
86. Parrott, D. M. 1981. Evidence for a gene-for-gene relationship between resistance gene *HI* from *Solanum tuberosum* ssp. *andigena* and a gene in *Globodera rostochiensis,* and between *H2* from . *multidissectum* and a gene in *G. pallida*. *Nematologica* 27:372–84
87. Paulson, R. E., Webster, J. M. 1972. Ultrastructure of the hypersensitive reaction in roots of tomato *Lycopersicon esculentum* L., to infection by the root-knot nematode, *Meloidogyne incognita*. *Physiol. Plant Pathol.* 2:227–34
88. Phillips, M. S. 1985. Environmental differences and their effect on the assessment of quantitative resistance to potato cyst nematodes. *EPPO Bull.* 15:179–83
89. Phillips, M. S., Forrest, J. M. S., Wilson, L. A. 1980. Screening for resistance to potato cyst nematode using closed containers. *Ann. Appl. Biol.* 96:317–22
90. Phillips, M. S., Hackett, C. A., Trudgill, D. L. 1991. The relationship between the initial and final population densities of the potato cyst nematode *Globodera pallida* for partially resistant potatoes. *Appl. Ecol.* 27. In press
91. Phillips, M. S., Rumpenhorst, H. J., Trudgill, D. L., Evans, K., Gurr, G., et al. 1988. Environmental interactions in the assessment of partial resistance to potato cyst nematodes. I. Interactions with centres. *Nematologica* 35:187–96
92. Phillips, M. S., Trudgill, D. L. 1986. A comparison of partially resistant potato clones to different populations and densities of *Globodera pallida* in a pot test and field trial. *Nematologica* 31:433–42
93. Phillips, M. S., Trudgill, D. L. Evans, K. 1988. The use of single plants to assess their tolerance by potato cyst nematodes. *Potato Res.* 31:469–75
94. Phillips, M. S., Wilson, L. A., Forrest, J. M. S. 1979. General and specific combining ability of potato parents for resistance to the white potato cyst nematode (*Globodera pallida*). *J. Agric. Sci.* 92:255–56
95. Price, N. S., Clarkson, D. T., Hague, N. G. 1983. Effect of invasion by cereal cyst nematode (*Heterodera avenae*) on the growth and development of the seminal roots of oats and barley. *Plant Pathol.* 32:377–83
96. Price, N. S., Sanderson, J. 1984. The translocation of calcium from oat roots infected by the cereal cyst nematode *Heterodera avenae* (Woll.). *Rev. Nematol.* 7:239–43
97. Radcliffe, D. E., Hussey, R. S., McClendon, R. W. 1990. Cyst nematode vs. tolerant and intolerant soybean cultivars. *Agron. J.* 82:855–60
98. Rice, S. L., Leadbeater, B. S., Stone, A. R. 1985. Changes in cell structure in roots of resistant potatoes parasitized by potato cyst-nematodes. 1. Potatoes with resistance gene H_1 derived from *Solanum tuberosum* ssp. *andigena*. *Physiol. Plant Pathol.* 27:219–34
99. Rich, J. R., Keen, N. T., Thomason, I. J. 1977. Association of coumestans with the hypersensitivity of Lima bean roots to *Pratylenchus scribneri*. *Physiol. Plant Pathol.* 10:105–16
100. Robertson, W. M., Spiegel, Y., Yansson, H. B., Marban-Mendoza, N., Zuckerman, B. M. 1989. Surface carbohydrates on plant parasitic nematodes. *Nematologica* 35:180–86
101. Robinson, R. A. 1969. Disease resistance terminology. *Rev. Appl. Mycol.* 48:593–606
102. Rumpenhorst, H. J. 1984. Intracellular feeding tubes associated with sedentary plant parasitic nematodes. *Nematologica* 30:77–85

103. Sasser, J. N. 1979. Pathogenicity, host ranges and variability in *Meloidogyne* species. In *Root-Knot Nematodes* (Meloidogyne *species*) *Systematics, Biology and Control,* ed. F. Lamberti, C. E. Taylor, pp. 257–68. London/New York/San Francisco: Academic. 477 pp.
104. Sasser, J. N., Hartman, K. M., Carter, C. E. 1987. Summary of preliminary crop germplasm evaluation for resistance to root-knot nematodes. NC State Univ./US Agency Int. Dev., Raleigh, NC. 88 pp.
105. Schottens-Toma, I. M. J., De Wit, P. J. G. M. 1988. Purification and primary strucuture of a necrosis-inducing peptide from the apoplastic fluids of tomato infected with *Cladosporium fulvum* (syn. *Fulvia fulva*). *Physiol. Mol. Plant Pathol.* 33:59–67
106. Seinhorst, J. W. 1965. The relation between nematode density and damage to plants. *Nematologica* 11:137–54
107. Seinhorst, J. W. 1970. Dynamics of populations of plant parasitic nematodes. *Annu. Rev. Phytopathol.* 8:131–56
108. Sidhu, G. S. 1975. Gene-for-gene relationships in plant parasitic systems. *Sci. Prog. Oxf.* 62:467–85
109. Sidhu, G., Webster, J. M. 1977. Predisposition of tomato to the wilt fungus (*Fusarium oxysporum lycopersici*) by the root-knot nematode (*Meloidogyne incognita*). *Nematologica* 23:436–42
110. Sidhu, G. S., Webster, J. M. 1981. Genetics of plant-nematode interactions. See Ref. 150, pp. 61–87
111. Stanton, J. M., Fisher, J. M. 1987. Field assessment of factors associated with tolerance of wheat to *Heterodera avenae. Nematologica* 33:357–60
112. Sturhan, D. 1971. Biological races. In *Plant Parasitic Nematodes*, ed. B. M. Zuckerman, W. F. Mai, R. A. Rohde, 2:51–71. New York/London: Academic. 347 pp.
113. Swain, B., Prasad, J. S. 1988. Chlorophyll content in rice as influenced by the root-knot nematode, *Meloidogyne graminicola,* infection. *Curr. Sci.* 57:85–96
114. Tepper, C. S., Anderson, A. J. 1986. Two cultivars of bean display a differential response to extracellular components of *Colletotrichum lindemuthianum. Physiol. Mol. Plant Pathol.* 29:411–20
115. Trudgill, D. L. 1967. The effect of environment on sex determination in *Heterodera rostochiensis. Nematologica* 13:263–72
116. Trudgill, D. L. 1968. *The effect of the environment on sex determination in* Heterodera rostochiensis. PhD thesis. Univ. London. 187 pp.
117. Trudgill, D. L. 1980. Effects of *Globodera rostochiensis* and fertilisers on the mineral nutrient content and yield of potato plants. *Nematologica* 26:243–54
118. Trudgill, D. L. 1985. Potato cyst nematodes: a critical review of the current pathotype scheme. *EPPO Bull.* 15:273–79
119. Trudgill, D. L. 1986. Concepts of resistance, tolerance and susceptibility in relation to cyst nematodes. In *Cyst Nematodes,* ed. F. Lamberti, C. E. Taylor, pp. 179–89. New York/London: Plenum. 467 pp.
120. Trudgill, D. L. 1986. Yield losses caused by potato cyst nematodes: a review of the current position in Britain and future prospect for improvements. *Ann. Appl. Biol.* 108:189–98
121. Trudgill, D. L. 1987. Effect of *Globodera rostochiensis* on the growth of reciprocal grafts between two potato cultivars of different tolerance in a field trial. *Ann. Appl. Biol.* 110:97–103
122. Trudgill, D. L. 1987. Effects of rates of a nematicide and of fertiliser on the growth and yield of cultivars of potato which differ in their tolerance of damage by potato cyst nematodes (*Globodera rostochiensis* and *G. pallida*). *Plant Soil* 104:185–93
123. Trudgill, D. L. 1991. Mechanisms of damage and of tolerance in nematode infested plants. Presented at 2nd Int. Nematol. Congr., Veldhoven, The Netherlands, ed. F. Gommers, P. Maas. In press
124. Trudgill, D. L., Cotes, L. M. 1983. Differences in the tolerance of potato cultivars to potato cyst nematodes (*Globodera rostochiensis* and *G. pallida*) in field trials with and without nematicides. *Ann. Appl. Biol.* 102:373–84
125. Trudgill, D. L., Cotes, L. M. 1983. Tolerance of potato to potato cyst nematodes (*Globodera rostochiensis* and *G. pallida*) in relation to the growth and efficiency of the root system. *Ann. Appl. Biol.* 102:385–97
126. Trudgill, D. L., Evans, K., Parrott, D. M. 1975. Effects of potato cyst nematodes on potato plants: 1. Effects in trials with irrigation and fumigation on the growth and nitrogen and potassium contents of a resistant and susceptible variety. *Nematologica* 21:169–82
127. Trudgill, D. L., Evans, K., Parrott, D. M. 1975. Effects of potato cyst nematodes on potato plants. II. Effects on haulm size, concentration of nutrients in haulm tissue and tuber yield of a nema-

tode resistant and susceptible potato variety. *Nematologica* 21:183–91
128. Trudgill, D. L., Holliday, J. M., Mathias, P. L., French, M., MacKintosh, G. M., Tones, S. J. 1983. Effects of the nematicide oxamyl on the multiplication of *Globodera rostochiensis* and *G. pallida* and on the haulm growth and yield of six potato cultivars with different levels of resistance and tolerance. *Ann. Appl. Biol.* 103:477–84
129. Trudgill, D. L., Marshall, B., Phillips, M. S. 1990. A field study of the relationship between preplanting density of *Globodera pallida* and the growth and yield of two potato cultivars of differing tolerance. *Ann. Appl. Biol.* 117:107–18
130. Trudgill, D. L., Parrott, D. M. 1972. Effects of growing resistant potatoes with gene HI from *Solanum tuberosum* ssp. *andigena* on populations of *Heterodera rostochiensis* pathotype A. *Ann. Appl. Biol.* 73:67–75
131. Trudgill, D. L., Parrott, D. M., Stone, A. R. 1970. Morphometrics of males and larvae of ten *Heterodera rostochiensis* populations and the influence of resistant hosts. *Nematologica* 16:410–16
132. Trudgill, D. L., Robertson, W. M., Wyss, U. 1991. Analysis of the feeding of *Xiphinema diversicaudatum*. *Rev. Nematol.* 14:107–12
133. Turner, S. J. 1990. The identification and fitness of virulent potato cyst nematode populations (*Globodera pallida*) selected on resistant *Solanum vernei* hybrids for up to eleven generations. *Ann. Appl. Biol.* 117:385–97
134. Van der Plank, J. E. 1975. *Principles of plant infection*. New York: Academic. 216 pp.
135. Van Emden, H. F. 1990. Aphid–plant genotype interactions—a perspective. In *Aphid-Plant Genotype Interactions*, ed. R. K. Campbell, R. D. E. Kenbary, pp. 1–6. Amsterdam: Elsevier. 378 pp.
136. Veech, J. A. 1981. Plant resistance to nematodes. See Ref. 150, pp. 377–403
137. Veech, J. A., Dickson, D. W., eds. 1987. *Vistas on Nematology*. Hyattsville, Maryland: Soc. Nematol. 509 pp.
138. Wallace, H. R. 1987. Effects of nematode parasites on photosynthesis. See Ref. 137, pp. 253–59
139. Wallace, H. R. 1988. A perception of tolerance. *Nematologica* 33:419–32
140. Weischer, B. 1982. Histopathological studies on grape species with different degrees of resistance to *Xiphinema index*. *Nematologica* 28:178–79 (Abstr.)
141. Wilcox-Lee, D., Loria, R. 1987. Effects of nematode parasitisim on plant-water relations. See Ref. 137, pp. 280–66
142. Wyss, U. 1981. Ectoparasitic root nematodes: Feeding behaviour and plant cell responses. See Ref. 150, pp. 325–51
143. Wyss, U. 1987. Video assessment of root cell responses to dorylaimid and tylenchid nematodes. See Ref. 137, pp. 211–20
144. Wyss, U., Lehmann, H., Jank-Landwig, R. 1980. Ultrastructure of modified root-tip cells in *Ficus carica,* induced by the ectoparasitic nematode *Xiphinema index*. *J. Cell Sci.* 41:193–208
145. Wyss, U., Stender, C., Lehmann, H. 1984. Ultrastructure of feeding sites of the cyst nematode *Heterodera schachtii* Schmidt in roots ofsusceptible and resistant *Raphanus sativus* L. var. *oleiformis* Pers. cultivars. *Physiol. Plant Pathol.* 25:21–37
146. Young, L. D. 1984. Changes in reproduction of *Heterodera glycines* on different lines of *Glycine max*. *J. Nematol.* 16:304–9
147. Zacheo, G., Arrigoni-Liso, R., Bleve-Zacheo, T., Lamberti, F., Arrigoni, O. 1983. Mitochondrial peroxidase and superoxide dismutase activities during the infection by *Meloidogyne incognita* of susceptible and resistant tomato plants. *Nematol. Mediterr.* 11:107–14
148. Zacheo, G., Pricolo, G., Bleve-Zacheo, T. 1988. Effect of temperature on resistance and biochemical changes in tomato inoculated with *Meloidogyne incognita*. *Nematol. Mediterr.* 16:107–12
149. Zuckerman, B. M., Jannson, H. B. 1984. Nematode chemotaxis and mechanisms of host/prey recognition. *Annu. Rev. Phytopathol.* 22:95–113
150. Zuckerman, B. M., Rohde, R. A., eds. 1981. *Plant Parasitic Nematodes*, Vol. 3. New York: Academic. 508 pp.

Annu. Rev. Phytopathol. 1991. 29:193–217

VIRUS-HOST INTERACTIONS: Induction of Chlorotic and Necrotic Responses in Plants by Tobamoviruses

J. N. Culver, Alwyn G. C. Lindbeck, and W. O. Dawson

Department of Plant Pathology, University of California Riverside, California 92521

KEY WORDS: virus hypersensitivity, virus systemic symptoms

INTRODUCTION

Viruses cause chronic diseases of plants, resulting in substantial reductions in the production of food and fiber throughout the world. Although the characteristics of these viruses have been comprehensively described, much less has been published on virus-host interactions that result in disease or resistance to disease. Two basic and unresolved questions of plant virology have been: "How do viruses cause diseases?" and, "How do plants recognize and resist viruses?" Until recently, most of the available information was limited to morphological descriptions of the development of symptomatic tissues and surveys of host enzymatic processes. Currently, the techniques of molecular biology have substantially advanced our understanding of virus-host interactions.

The relatively small genome sizes of plant viruses along with recent developments in recombinant DNA technologies, have greatly expanded our understanding of these pathogens. The elucidation of the complete genomic sequences of numerous viruses with diverse characteristics has led to the creation of taxonomic groupings and an understanding of the modular evolution of viruses from hosts spanning different kingdoms. The ability to make defined mutations in viral genomes at the DNA plasmid level and then to examine the effects of the mutations on the biology of the virus has allowed

0066-4286/91/0901-0193$02:00

precise mapping of viral sequences involved in replication, regulation, and gene expression. These technologies are now significantly facilitating our understanding of virus-host interactions. In fact, it is becoming relatively simple to define precise viral sequences involved in virus-host interactions, although the host processes affected by these sequences are considerably more difficult to define.

Only some of the various types of virus-host interactions result in visible symptoms. Several interactions, often referred to as "nonhost" interactions, do not allow the virus to replicate or move within the plant. However, interactions in which visual symptoms develop have received the most attention. Two of the most common symptoms associated with viral infections are chlorosis and necrosis.

Chlorosis is usually associated with susceptible interactions in which the virus replicates and moves throughout the plant. Chlorotic symptoms usually appear as yellowed areas in expanded leaves that developed before infection, or as a mosaic pattern of dark green mixed with light green, yellow, or white areas in leaves that developed after infection. This symptom is usually associated with a disruption of chloroplast structure, function, and/or development.

A number of different necrotic responses in plants are associated with viral infections. In a necrotic hypersensitive response (HR), the plant specifically recognizes the pathogen at the onset of infection and mounts an active defense response that usually restricts further movement of the virus. This process often provides resistance to virus diseases. However, sometimes the virus induces necrosis but is not confined to the initially infected area and is able to move to other parts of the plant. This movement often results in the development of necrosis in systemically infected tissue and/or the intermixing of sporadic necrosis with mosaic symptoms. When necrosis occurs in vascular tissues, the result can be plant death. One of the more economically important types of virus-induced necrotic reactions is graft incompatibility, which results in the decline or death of grafted trees. It is not clear whether all these necrotic reactions are genetically and physiologically similar.

There are excellent reviews describing host genetics and physiological responses to virus infections (18, 54, 80, 83). To complement those works, we summarize recent progress in defining the genetics of virus-host interactions. Advances have been made with several virus groups, including caulimoviruses (3, 9), hordeiviruses (53), geminiviruses (72), cucumoviruses (60, 70), satellite RNAs (30, 71), and viroids (50). It is becoming evident that virus-host interactions may be relatively complex, with numerous factors able to alter the homeostatic balance of the host plant and cause restrictions in plant growth, development, and output. Faced with the vast amount of available information, we have, for simplicity's sake, focused primarily on one group of viruses, the tobamoviruses.

The tobamoviruses generally have wide host ranges and are distributed throughout the world. Tobamovirus virions are rigid rods of about 300 X 18 nm. Tobacco mosaic virus (TMV) is the type strain for the tobamovirus group. The genome of TMV resides on one molecule of plus-sense RNA of approximately 6400 nucleotides and encodes at least four proteins (21). The 5' proximal open reading frame (ORF) encodes a protein of approximately 126kd. Read-through of the amber stop codon results in the production of a protein of approximately 183kd (52). Both 126kd and 183kd proteins are required for efficient replication (29). The 3' proximal genes encode a cell-to-cell movement protein of approximately 30kd and a coat protein of 17.5kd (43). Both 5' proximal genes are translated from the genomic RNA, while the 3' proximal genes are translated from subgenomic mRNAs (5, 28). (See ref. 11 for a detailed review of the genomic organization and gene expression of tobamoviruses.)

INDUCTION OF CHLOROTIC SYMPTOMS

Chlorotic symptoms generally develop from a decrease in chlorophyll content and are usually associated with alterations of chloroplast structure and function. The development of chlorotic symptoms by TMV involves at least two distinct processes: (*a*) the appearance of chlorosis associated with mosaic symptoms in systemically infected leaves involves the prevention of normal chloroplast development in nascent leaves where chloroplast division and development and cell division are occurring; (*b*) the appearance of chlorotic symptoms in directly inoculated leaves, where both chloroplast and cell division have ceased, involves the degradation of fully developed chloroplasts. In this section we discuss each system separately and show how each differs.

Chlorosis Associated with Systemic Infections

In systemically infected developing leaves, the typical symptom induced by TMV is a dark green and light green mosaic, caused by alterations in the development of chloroplasts in viral-infected cells (37, 38). The mosaic pattern that develops in a leaf appears not to depend upon its position, but rather on the stage of leaf development at the time of viral infection (2, 73, 75). Mosaic symptoms develop in tobamovirus-infected tobacco plants only in nascent leaves that are less than 1.5 cm in length at the time of the initial infection (2, 20, 47). All subsequently developing leaves display the mosaic symptom. With most tobamoviruses, the inoculated leaves and leaves greater than approximately 5–6 cm in length at the time of infection develop no visible symptoms, even though they support high levels of viral replication (47).

When a virus infects a cell in the leaf primordium, it is theoretically capable of replicating and moving to neighboring cells, giving rise to uniform light green, chlorotic tissue throughout the developing leaf. However, leaves with mosaic symptoms consist of apparently normal dark green tissue (green islands), containing little or no virus, surrounded by light green tissue containing high levels of virus. These dark green islands appear to be resistant to viral infection. A proportion of plants regenerated from protoplasts isolated from this tissue were virus-free (46). These plants were only susceptible to TMV infection after considerable growth; as small plants they maintained resistance. Similar results were obtained with potato virus X-infected tobacco plants (68). Apparently some cells in the leaf primordium are resistant to the virus. Division of these cells results in clusters of noninfected cells that appear as the dark green tissue of the mosaic pattern. The nature of this resistance is not known. Several lines of evidence suggest a correlation between the presence of TMV virions or coat protein within chloroplasts and the development of mosaic symptoms. Although TMV replicates and accumulates primarily in the cytoplasm, virions can be found in chloroplasts (14). Additionally, pseudovirions (75), consisting of viral coat protein encapsulating chloroplast RNA (59), have been found inside chloroplasts. More recently, both TMV coat protein (57, 58) and TMV genomic RNA (63) have been isolated from Percoll-purified chloroplasts. In chloroplasts isolated from spinach infected with TMV (24), the coat protein is associated with photosystem II complexes in the thylakoids. Ubiquitinated coat protein can accumulate in chloroplasts (12) and Reinero & Beachy (58) found a positive correlation between the amount of coat protein found in chloroplasts and the degree of visible symptoms. These observations have led to the suggestion that coat protein plays a role in the development of mosaic symptoms.

While some evidence suggests that the presence of coat protein somehow contributes to mosaic symptoms, more compelling evidence suggests that the coat protein is not involved in the induction of mosaic symptoms in developing leaves. Although virions can be found in chloroplasts, various strains of TMV differ greatly with regard to the levels of virions found in tobacco chloroplasts. However, the severity of the symptoms were not always correlated with the levels of virions or coat protein found within chloroplasts. For example, chloroplasts from tissue infected with the U5 strain of TMV often contain many virions whereas chloroplasts from tissue infected with the U1 strain of TMV contain few (19, 66). Yet the U1 strain of TMV induces more severe symptoms than those induced by the U5 strain. In fact, immunogold labeling of thin sections of tobacco leaves systemically infected with U1 suggests that very little coat protein accumulates in chloroplasts (23). Additionally, transgenic tobacco plants expressing the TMV coat protein (56; J. N. Culver & W. O. Dawson, in preparation) do not show symptoms normally associated with a TMV infection.

TMV mutants that produce no coat protein and yet induce normal mosaic symptoms in *Nicotiana sylvestris* Speg. & Comes and *N. tabacum L.* have recently been described (8; Lindbeck et al, in preparation). Although coat protein-less mutants move poorly from leaf to leaf, these mutants do move long distances sporadically. In *N. sylvestris,* which has short internodes, coat protein-less mutants can move into the apex of the plants within a few weeks of infection. When this movement occurs, a normal mosaic symptom appears. Western immunoblots demonstrated that normal amounts of viral 126kd protein were present in the mosaic tissues but that coat protein was absent. Thus, in developing leaves the abnormal development of chloroplasts was induced by viral infection without the production of coat protein.

The genomes of two attenuated or masked mutants of TMV that produce greatly reduced symptoms in tobacco plants have been sequenced. Comparisons of the tomato TMV mutant $L_{11}A$ sequence, which produces attenuated symptoms in both tomato and tobacco plants, to the wild-type virus sequence have identified 10 nucleotide alterations that map within the TMV 126/183kd ORF (48). The Holmes' masked strain (25) was recently sequenced and compared to the parental U1 strain (27). Although there were 55 nucleotide differences, substitution experiments that examined hybrids of the two viruses demonstrated that the alterations responsible for symptom attenuation were within the 126/183kd ORF.

These results demonstrate that while TMV is responsible for the appearance of mosaic symptoms in systemically infected leaves, the correlations between the presence of coat protein or virions in chloroplasts appear to be coincidental rather than causative. The coat protein clearly is not required for the development of mosaic symptoms. In fact, it now appears that the 126/183kd gene is somehow involved in the prevention of normal chloroplast development in systemically infected leaves.

Development of Chlorosis in Mature Leaves

Chloroplasts in leaves allowed to expand before inoculation are already fully developed before the time of infection. Upon inoculation of these leaves, TMV U1 replicates to high levels but causes no visible changes in the chloroplasts and the leaves remain virtually symptomless. However, several strains and mutants of TMV cause visible chlorotic symptoms on directly inoculated leaves (10, 40, 62). These symptoms are due to alterations of mature chloroplasts rather than the prevention of normal chloroplast development that occurs during the appearance of mosaic symptoms.

A variety of symptoms (10) occurred when expanded tobacco leaves containing fully developed chloroplasts were infected with TMV mutants that had insertions or deletions in the coat protein gene. The symptoms induced by the different mutants ranged from none to severe yellowing, transient yellowing, and necrosis. Lindbeck et al (36) showed that the chlorosis in these leaves resulted from significant rearrangements of the photosynthetic membranes in the chloroplasts of inoculated leaves. These changes were due to the degradation of already existing photosynthetic membranes present in fully developed chloroplasts. Chloroplasts from leaves inoculated with coat protein mutants cp +2, cp 4, cp 10, cp 27, cp 28, cp 35, and cp 35-5 were examined (Table 1). Although wild-type TMV (Figure 1A) and the coat protein mutants that caused weak or transient yellowing (cp +2, cp 28, and cp 35) (Figure 1B,C,D) did not induce significant ultrastructural changes in fully developed chloroplasts, mutants that caused significant yellowing on expanded leaves (cp 4, cp 10, cp 27, and cp 35-5) also caused significant degradation of the chloroplasts in infected cells. Chloroplasts from these tissues (Figure 2A) contained interconnected tubular networks that appeared to be derived from rearrangements of existing photosynthetic membranes. Chloroplasts from tissue infected with the yellowing mutant cp 4 contained grana that appeared to be unstacking (Figure 2B), forming convex membrane structures at four days post inoculation. In some chloroplasts, the photosynthetic membranes appeared to be completely unstacked and arranged in circular structures (Figure 2C). At seven days post inoculation, tubules were formed in these chloroplasts. Disruption of the chloroplast membrane structure appeared soon after infection and paralleled the development of visible yellow areas.

Coat protein mutants were created with a *Xho*I linker at the insertion or deletion sites in the cDNAs (10). Thus, any 5' coat-protein gene sequences of one mutant could be ligated to any 3' coat-protein gene sequences of another mutant. The symptomatology of all 5' to 3' coat-protein gene combinations of the original mutants was examined (A. G. C. Lindbeck et al, in preparation). In general, a correlation exists between the amount of defective coat protein produced and the degree of yellowing. Mutants that produced small amounts of defective coat protein were symptomless, whereas mutants that produced large amounts of the defective coat protein induced yellowing symptoms.

Saito et al (62) also produced a series of coat-protein deletion and insertion mutants of TMV-L. Three of these mutants (D104, D101, and D63/77) caused chlorotic spots on inoculated leaves. No common deletion or insertion sequence has been identified among the coat-protein mutants that induce chlorotic symptoms on inoculated leaves. Thus, although the wild-type coat-protein gene did not induce chlorotic symptoms in directly inoculated leaves, a number of specific mutations in the coat-protein gene affect the structure and/or function of fully developed chloroplasts.

The correlation between the production of defective coat protein and chloroplast degradation suggested that defective proteins somehow entered the chloroplast and destabilized its structure. However, immunocytochemical localization of these defective coat proteins showed that very little of this

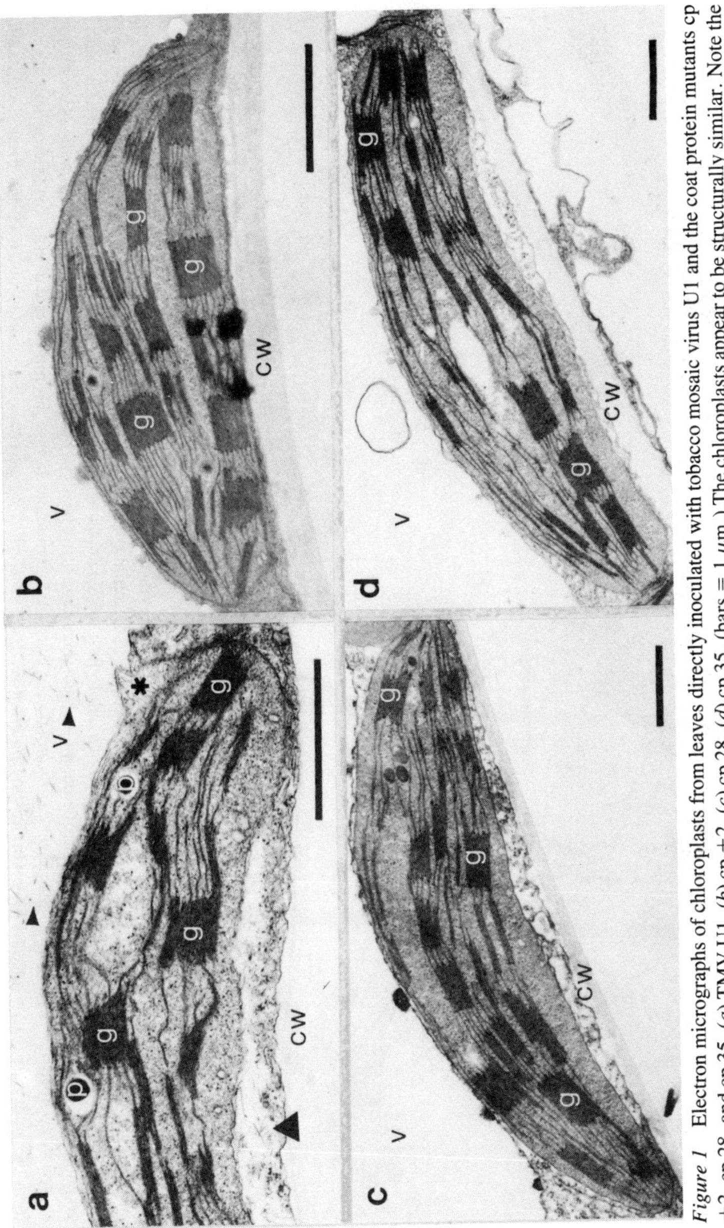

Figure 1 Electron micrographs of chloroplasts from leaves directly inoculated with tobacco mosaic virus U1 and the coat protein mutants cp +2, cp 28, and cp 35. (*a*) TMV U1. (*b*) cp +2. (*c*) cp 28. (*d*) cp 35. (bars = 1 μm.) The chloroplasts appear to be structurally similar. Note the presence of a small virion inclusion in the chloroplast from the TMV U1-infected leaf (*); such inclusions were rare. Virions can also be seen in the cytoplasm (large arrowhead) and the vacuole (small arrowheads). (cw = cell wall; g = grana; p = plastoglobuli; v = vacuole). Reprinted from Lindbeck et al (36).

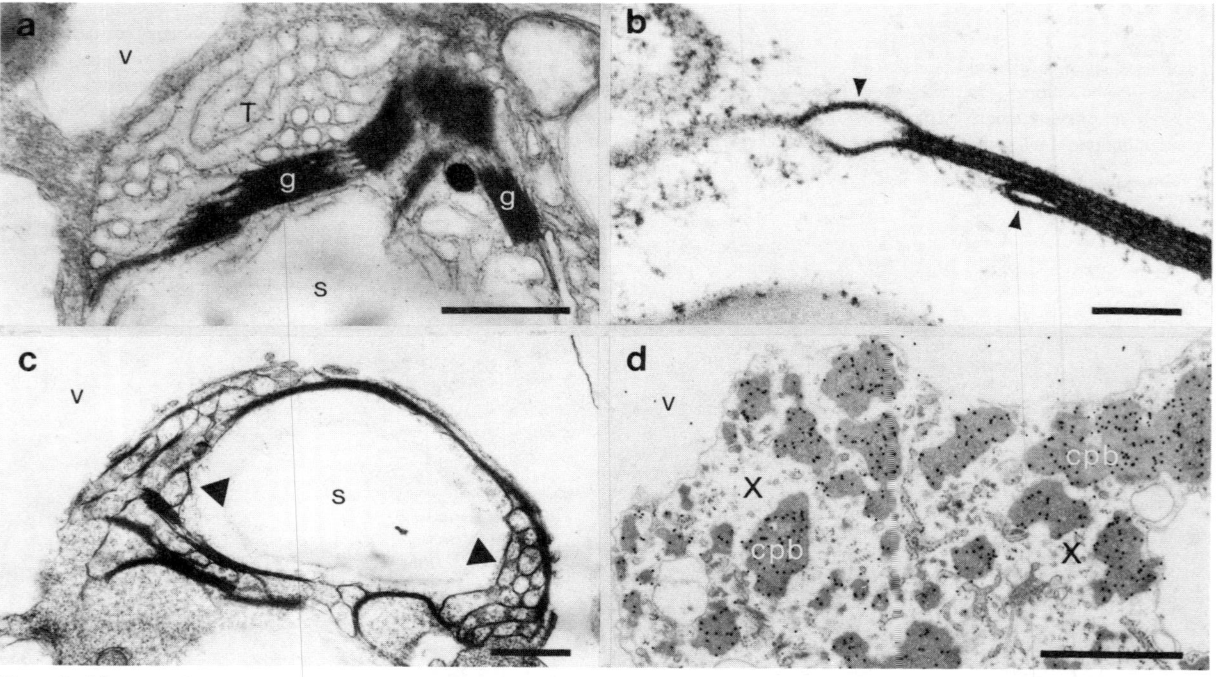

Figure 2 Electron micrographs of chloroplasts from chlorotic leaf areas resulting from the replication of the coat-protein mutants cp 4, cp 10, and cp 35-5. (*a*) Micrograph of a tubular complex (T) in a chloroplast from the center or the chlorotic area caused by cp 35-5. (bar = 1 μm). (*b*) Convex membrane structures (small arrowheads) in a chloroplast from a cp 4-infected leaf that appear to be formed by the unstacking of a granum. (bar = 0.25 μm). (*c*) Micrograph of the putative precursors (large arrowheads) of the tubules in a chloroplast from the chlorotic area caused by cp 4. (bar = 1 μm). Note the presence of normal photosynthetic membranes in this chloroplast and in 2(*a*). (*d*) Micrograph of tissue from the outer half (early stage of infection) of the thin chlorotic band adjacent to the necrotic area caused by cp 10. Discrete dark staining bodies, labeled with gold particles, can be seen embedded in an X-body (X). These bodies have been called coat-protein bodies(cpb). (bar = 1 μm). (g = grana; s = starch granule; v = vacuole). Reprinted from Lindbeck et al (36).

Table 1 Summary of coat protein mutants

Mutants	Nucleotides[a] deleted in coat protein gene	Amino acids deleted in coat protein	Symptoms on inoculated leaves
TMV	0	0	none
cp +2	+6[b]	+2	mild yellowing
cp 4	6020–6153	102–147	yellowing
cp 10	6043–6084	110–158	necrosis
cp 27	5949–6079	79–122	necrosis
cp 28	5874–6191	54–158	mild yellowing
cp 35	5841–6219	43–158	none
cp 35-5	5841–6055	43–114	yellowing

[a]Nucleotide numbering as described by Goelet et al (21).
[b]Number of nucleotides added at nucleotide 6056.

protein accumulated in the chloroplasts of infected cells (36). The bulk of these defective coat proteins was localized to discrete dark-staining coat-protein bodies located in the cytoplasm (Figure 2D). This observation suggests that the mutant coat proteins exert their influence on chloroplasts from the cytoplasm instead of from inside of the chloroplasts.

One simple mechanism by which mutant coat proteins might cause chloroplast degradation involves an interference with the synthesis or transport of cytoplasmic proteins bound for the chloroplasts. A positive correlation existed between mutants that produced cytoplasmic coat-protein bodies and mutants that caused the degradation of chloroplasts. Other coat-protein mutants and strains of TMV that form insoluble coat protein-containing aggregates also induce yellowing (32). Mutant coat proteins, which fail to assemble into virions and form aggregates of coat-protein in the cytoplasm, may possibly capture host proteins needed for chloroplasts maintenance. This would interfere with the transport of these proteins into the chloroplasts and result in the eventual degradation of the mature chloroplasts.

This mechanism of coat protein-induced chloroplast disorganization that occurs in expanded leaves may also occur to a certain extent in developing leaves. The systemic movement of mutants and strains of TMV that cause yellowing in inoculated leaves should result in a mosaic symptom. However, these mutants should also cause chloroplast degeneration in systemically infected leaves, inducing a bright yellow mosaic symptom. Such a symptom would result from the combination of two different mechanisms: (*a*) the process that prevents the normal development of chloroplasts and does not involve the coat protein; *b*) the process that causes mature chloroplast disorganization and can be related to coat protein. In fact, TMV strains that form intact virions and induce yellowing in inoculated leaves tend to cause a brighter yellow mosaic pattern than strains that do not induce yellowing in inoculated leaves. The PV230 strain of TMV, used by Reinero & Beachy (57, 58) and Hodgson et al (24) to examine the affect of TMV on chloroplast function in systemically infected leaves, does induce yellowing in inoculated, expanded leaves. Their results probably reflect both chloroplast degradation and prevention of normal chloroplast development in systemically infected leaves.

INDUCTION OF NECROTIC SYMPTOMS

A number of necrotic responses are induced by tobamoviruses in different hosts. These include the hypersensitive response, which is usually controlled by a single dominant host gene, a slower-forming local necrosis that is not related to a single host gene, and various degrees of systemic necrosis. It is not clear whether these necrotic responses are distinctly different or only variations of the same process. In many virus-host systems, confinement of the virus by the necrotic response is a race between how fast the virus replicates and moves and how fast the HR occurs. Manipulations that alter either the rate of viral replication or the kinetics of HR affect the outcome. Certainly, in the same plant, different viral mutants and strains induce the HR at different rates. Also, different tissues respond differently to the same virus, usually with juvenile tissues responding more slowly or, in some cases, not at all to HR-inducing viruses.

Viral-induced Hypersensitive Reactions

Viral-induced HR parallels that induced by other pathogens. It has become evident in the last few years that the HR of plants is a generalized response to different pathogens. The major difference in this defense mechanism toward different pathogens appears not to be the cascade of induced events that limit the pathogens, but the mechanisms of induction of that cascade of events. Viral-induced HRs appear to differ from bacterial- and fungal-induced HRs relative to their sites of induction. These extracellular pathogens are thought to interact with the host cell surface to induce HR (33). In contrast, for viral-induced HRs, the plant appears to recognize and respond to an intracellular viral gene product. In addition, mutation of bacterial and fungal genes involved in the induction of the HR tends to change these genes from avirulence to virulence (33), whereas in the best-defined tobamovirus system (7), mutations more frequently change the phenotypes from virulence to avirulence.

Most information concerning tobamovirus-induced necrosis comes from TMV interactions with plants in the genus *Nicotiana*. Plants with the N gene from *Nicotiana glutinosa* L. (also bred into *N. tabacum* cvs. Xanthi nc and Samsun NN) respond to essentially all tobamoviruses with the HR, resulting in formation of necrotic lesions on inoculated leaves and prevention of further virus spread (25, 26). The plant is thus resistant to this viral disease. Other species of *Nicotiana,* for example *N. sylvestris* and some varieties of tobacco (Java and Burley), contain the N' gene that localizes some tobamoviruses and TMV mutants but not others (81). Both these genes segregate as single dominant genes (81). Although it has been assumed that these genes are allelic (13, 79, 81), they are more likely to be different, closely linked genes. Plants with both N and N' genes have an additive HR. These plants produce faster-forming and smaller lesions when induced by a virus with an elicitor to both genes than by a virus with an elicitor to only the N gene (81). Also, as will be described here, each different gene responds to a different viral elicitor. Plants with the corresponding recessive alleles (nn and n'n' genotype) respond to most tobamoviruses with systemic infection and disease.

N' Gene of Nicotiana sylvestris

The N' gene, originating from *Nicotiana sylvestris,* controls an HR directed against most strains of tobamoviruses. The tobacco strains of TMV, U1 (vulgare) and OM, are two tobamoviruses that do not elicit this resistance response. Both strains move systemically and produce a mosaic symptom in N' genotype plants (35, 61). A precise interaction between TMV and plants with the N' gene is required to induce this HR. Small changes in the genome of either will alter this reaction. Mutants of TMV that induce necrosis can easily be isolated from viral strains that cause systemic infections (1, 65). A spontaneous mutation in *N. tabacum* var. Samsun, which has the nn, n'n' genotype, resulted in the necrotic localization of specific strains of TMV, similar to the HR phenotype of the N' gene (41). Since one of the progenitors of *N. tabacum* is *N. sylvestris,* it has been suggested (13) that this Samsun HR phenotype is a result of a back-mutation of the n' gene to the N' gene.

Saito et al (61) first demonstrated the involvement of coat protein gene sequences in the induction of the N' gene HR. They substituted the coat-protein gene of the TMV-OM strain, which does not induce the N' gene HR, into the genome of the HR-inducing TMV-L strain. This hybrid virus no longer induced the development of necrotic lesions in N' gene hosts. Thus, all TMV-L sequences except those in the coat-protein ORF were eliminated from direct responsibility for the induction of the HR. Therefore, the induction of the N' gene HR was localized to the coat-protein ORF.

Knorr & Dawson (35) demonstrated that only the alteration of a single nucleotide in the coat-protein ORF was required for the induction of the N' gene HR. This alteration was accomplished by cloning portions of the genomes of mutants of the U1 strain of TMV that induced the HR in *N.*

sylvestris. Hybrids between this mutant and the wild-type U1 strain were then used to identify sequences of the mutant responsible for the induction of the HR. This substitution procedure localized the HR-inducing mutation to the coat-protein ORF. Sequence analysis of this region showed a single nucleotide difference (nucleotide 6157, cytosine to uracil) between the mutant and wild-type virus. Additionally, this nucleotide alteration led to an amino acid substitution of serine to phenylalanine at position 148 in the coat protein. Thus, a single nucleotide alteration can dramatically alter this plant-virus interaction from a susceptible to a resistance response.

Six additional HR-inducing mutants were also partially sequenced (35). Of these mutants five contained the same mutation at nucleotide 6157. However, it was possible that other single-nucleotide changes in the coat-protein ORF would induce the HR. Earlier work that sequenced the coat proteins of TMV mutants identified a number of different amino acid substitutions from mutants that induce the HR in *N. sylvestris* (16, 45, 82). However, no direct evidence eliminated other possible mutations in the rest of the genomes of these mutants from being responsible for the induction of the HR. To establish whether additional mutations within the coat-protein ORF could also induce the HR, four specific nucleotide alterations, resulting in four different amino acid substitutions, were individually made in the coat-protein ORF of the U1 strain of TMV (7). Each amino acid substitution recreated one of the previously described coat protein alterations that had been associated with the induction of HR in *N. sylvestris* (16, 82). However, these substitutions were the only alterations in the viral genomes. Inoculation of each of these four mutants onto leaves of *N. sylvestris* resulted in the formation of necrotic local lesions, demonstrating that changes throughout the coat protein ORF could induce the HR.

Interestingly, each mutant gave a distinctly different pattern of lesion development. These patterns ranged from a rapid necrosis resulting in small lesions that confined the infection to that area within two to three days, to a more slowly developing necrosis that required five to seven days to appear (7). The slower-developing lesions continually expanded from the initial site of infection and occasionally spread systemically, causing necrosis as they moved. Mutants that induced rapid lesion formation were defined as "strong elicitors" and mutants that induced slower-forming lesions were defined as "weak elicitors". All gradations of mutants with intermediate phenotypes have been found. Coat-protein mutants that induce the rapid development of HR may have a stronger affinity for the N' gene product than mutants that induce the slower development of necrosis.

Each nucleotide alteration responsible for the induction of the N' gene HR resulted in amino acid substitutions in the coat protein. Thus, the active determinant required for the induction of HR could have resided in either the altered RNA or the altered protein. To determine which of the two molecules, RNA or protein, was required for the induction of the HR, the coat protein translational start codon was changed from AUG to AGA for one of the previously identified HR-inducing mutants (8). This alteration did not affect the infectivity of the virus or its ability to move from cell to cell in the plant. Other than the start codon, this mutation left the coat-protein ORF intact, including the original nucleotide substitution associated with the induction of the HR. However, no coat protein was produced. Inoculation of this mutant onto leaves of *N. sylvestris* failed to induce the HR, even though the mutant replicated efficiently. Therefore, the production of altered coat protein during viral replication was required for induction of the HR.

It was possible that the altered coat protein alone was the elicitor of the HR, but it also was possible that it interacted in concert with other viral gene products or functions to induce the resistant reaction. Mutations in the coat protein may have created new HR elicitor sites or interfered with an existing HR suppressor site. The latter possibility was supported by the high frequency of mutation from virulence to avirulence. Starting with virus from a cDNA clone of TMV U1, which should begin as a uniform population, mutants with the ability to induce local lesions constitute about 0.1% of the progeny population after a single passage (1). Reversion back to virulence occurs at a much lower frequency, estimated at less than 10^{-5} (1). This finding suggests that perhaps only a few specific sequences of the coat protein will result in the systemic infection of plants with the N' gene, whereas numerous alterations in those sequences will result in coat proteins that retain the ability to assemble, but now induce the HR. Compared to most other genetic systems, this would suggest that the wild-type coat-protein sequences that do not induce the HR are active in avoiding that response. However, the absence of wild-type coat-protein production during viral infection does not result in the induction of the HR (8, 10). Thus, the coat protein could possibly have both elicitor and suppressor functions.

To determine if any viral components or replication processes other than the coat protein were involved in the induction of the N' gene HR, transgenic *N. sylvestris* plants were produced that expressed only the coat-protein ORF of either HR-inducing mutants or wild-type virus that did not induce the HR (J. N. Culver & W. O. Dawson, in preparation). Coat-protein ORFs originating from either mutant TMV 11, a weak elicitor of the HR, or from mutant TMV 25, a strong elicitor of the HR (7), were selected for integration into the genome of *N. sylvestris*. Each coat-protein ORF was expressed constitutively behind the 35S promoter of cauliflower mosaic virus. Transgenic *N. sylvestris* expressing the wild-type coat protein developed normally when compared to nontransgenic control plants. In contrast, transgenic *N. sylvestris* plants expressing either of the two HR-inducing coat proteins displayed various degrees of growth reduction and necrosis on expanded leaves. In general, chlorosis and small necrotic areas became visible after two months of growth. Necrotic areas then slowly expanded for several weeks, eventually leading to collapse of entire leaves. These phenotypes were similar to that observed in *N. sylvestris* plants regenerated from callus tissue infected with a high titer of an HR-inducing strain of TMV (34). These data demonstrated that the coat protein by itself was sufficient to induce the HR.

Accumulations of coat protein in transgenic plants were many times lower than in plants infected with TMV. Also, a longer time was required for the development of a necrotic response in transgenic plants expressing elicitor coat proteins than for a response to an HR-inducing viral infection. This suggests that the rapid induction of the HR, leading to resistance, requires the accumulation of elicitor protein at higher thresholds than that produced in many cells of the transgenic plants. Interestingly, protoplasts of *N. sylvestris* infected with an HR-inducing strain of TMV do not display any visible form of necrosis or cell death (J. L. Sherwood, personal communication). However, necrosis associated with a HR has been observed in clumps of cultured NN genotype tobacco cells (4). These data may indicate that the expression of the HR necrosis is dependent upon clusters of cells containing threshold levels of elicitor. It is also possible that only specific cells in the cluster interact with the elicitor and these cells subsequently induce the HR in the surrounding cells.

Comparisons between transgenic plants expressing strong or weak elicitors of the N' gene HR revealed that reductions in growth and development of necrosis appeared earlier and more severely in plants expressing the strong elicitor coat protein than in plants expressing the weak elicitor coat protein. In addition, lower accumulations of the strong elicitor coat protein produced a more severe response than higher accumulations of weak elicitor coat protein. Thus, the phenotypic differences between the two HR-inducing mutants were maintained through the expression of their respective coat proteins in N' gene plants.

One possible hypothesis for the mechanism of induction of the HR in N' gene plants is that these plants recognize specific three-dimensional configurations of the TMV coat protein. N' gene plants may fail to recognize the wild-type coat-protein configuration (that does not induce the HR), but may recognize specific three-dimensional configuration changes in that structure (mutations in the coat protein that induce the HR). It is possible that the resistance gene product acts as a receptor, recognizing a specific coat-protein structure. Such a receptor could recognize specific tertiary and/or quaternary structures dependent upon the proper folding of an entire coat-protein subunit or it could recognize only a segment of the coat-protein sequence. Data presented by Saito et al (62) suggest that the entire coat protein structural configuration is required to actively elicit the HR. Deletions made in an elicitor coat protein resulted in loss of activity, but substitution of the deletion with the corresponding region of a nonelicitor coat protein retained elicitor activity. Only small deletions at the carboxy-terminus were tolerated in the maintenance of HR induction. These data suggest that the tertiary and/or quaternary configuration of the entire coat protein is recognized by some component of an N' gene plant to induce the HR.

The apparent importance of the three-dimensional coat protein structure in eliciting the N' gene HR suggests that structural alterations produced by HR-inducing amino acid substitutions are being directly recognized by the plant. However, each amino acid substitution that elicits the HR is uniquely different, with no apparent similarity in the type of amino acid or location on a single coat protein subunit. In addition, these alterations have not been related to any known glycosylation signals. G. Stubbs and J. N. Culver et al (in preparation), in collaboration, have defined a possible elicitor region on the surface of the coat protein that may play a role in the induction of the N' gene HR. Amino acid substitutions that induce HR primarily were located in a surface region between two coat-protein subunits positioned in a protohelix or disk formation (Figure 3). Substitutions that did not induce the HR were located outside this region. In addition, Mundry et al (45) have mapped a number of possible HR-inducing amino acid substitutions to three "hot" regions in the coat protein. These hot regions fall into the same general elicitor region we have identified from single amino acid substitutions. Thus, it is possible that this region of the coat protein interacts with the host to induce the HR.

Coat protein monomers may be the active molecules required to induce this response. Evidence from transgenic plants and from coat-protein mutants deficient in assembly (62) has eliminated virions as being required for the induction of HR. The question is whether monomers or aggregates of coat protein are the active elicitor units. Under the normal physiological conditions of the cell, individual coat-protein subunits preferentially aggregate into protohelical disks and few monomers are thought to be present (55). Powell-Abel et al (56) demonstrated that multimers of coat protein could be isolated from transgenic plants expressing the protein. The HR-inducing amino acid substitutions may act to shift the equilibrium from aggregates to monomers of coat protein. Siegel (64) found that mutants with the ability to induce the HR in N' gene plants were more susceptible to UV inactivation, possibly indicating a more relaxed or open virion. Thus, the induction of the N' gene HR might depend upon the presence of higher accumulations of monomer coat-protein subunits or weaker subunit-to-subunit interactions. However, Fraser (17) demonstrated a positive correlation between TMV mutants that produced large lesions in N' gene plants and the thermal instability of their coat

Figure 3 Schematic representation of two tobacco mosaic virus coat-protein subunits (G. J. Stubbs, personal communication). Shaded area indicates the general location of amino acid substitutions that induce the hypersensitive response in *Nicotiana sylvestris*.

proteins. Therefore, if production of large lesions is characteristic of a weak elicitor, and one assumes that thermal stability represents stronger subunit-to-subunit interactions, then these results would indicate that alterations in subunit equilibrium from multimers to monomers are not essential for the induction of the HR. This hypothesis should be testable by making specific amino acid substitutions that prevent or interfere with coat-protein subunit-to-subunit interactions. It will be interesting to determine if these mutants induce an HR response.

The N gene of Nicotiana glutinosa

The N gene, originating from *Nicotiana glutinosa,* is a single dominant gene that controls an HR against almost all strains of tobamoviruses (25, 81). A detailed analysis of this gene has previously been presented (13). To date, a precise viral component or process responsible for eliciting this response has not been identified. However, several studies have identified viral components that are not involved in the induction of the N gene HR.

The coat protein clearly is not involved in induction of the HR in plants with the N gene. TMV mutants with a deletion of the entire coat-protein ORF induce normal necrotic lesions that are indistinguishable from those induced by the wild-type virus (10, 62). In addition, a mutant with an alteration in the coat protein translational start codon, preventing the production of coat protein but maintaining an intact coat-protein ORF, was also found to induce the HR in N gene plants (8). Replacement of the coat-protein ORF with the bacterial chloramphenicol acetyltransferase ORF also did not affect the induction of necrotic lesions in N gene plants (10, 76). Thus, the coat protein and its ORF are not directly involved in the induction of the N gene HR.

The N gene has proven to be one of the most resilient resistance genes, with only one strain of tobamovirus (ToMV-Ob) identified as virulent (6). Although this virulent strain induces the development of lesions on inoculated leaves, the virus is not confined to these lesions and moves systemically, inducing latent or chlorotic symptoms that are temperature dependent. Interestingly, ToMV-Ob does not induce necrosis in upper systemically infected leaves. Thus, this strain may be able to suppress or avoid the induction of the N gene HR in systemically infected tissues. Further studies, looking at hybrids between the non-HR-inducing ToMV-Ob strain and HR-inducing strains of tobamoviruses, may lead to the discovery of the viral components responsible for the elicitation of this resistance gene.

Nicotiana Plants with the nn, n'n' Genotype

Tobacco plants with the nn, n'n' genotype are susceptible to essentially all strains of TMV with production of systemic mosaic symptoms. A series of deletions in the coat-protein gene were made to examine secondary functions of the coat protein in this systemic host (10). A single phenotype was expected, similar to that of free RNA mutants previously described (65). Instead, each mutant had a distinct symptomatology as described earlier in this chapter. Most surprisingly, two mutants, cp 10 and cp 27, induced local lesions in Xanthi tobacco, which has the nn, n'n' genotype and normally does not respond with localized necrosis to wild-type TMV (10).

Perhaps most noteworthy is that the induction of necrosis by these mutants was much less specific than those previously described. Mutants cp 10 and cp 27 also induced local lesions in plants with the N' gene as well as almost all other plants tested, including *N. glauca* Graham, *Solanum nigrum* L., *Lycopersicon esculentum* Mill, *Petunia hybrida* Vilm., *Physalis floridana* Rydb., and *Nicandra physalodes* (L.) Gaertn. Specificity for the induction of this necrotic response was observed for mutant cp 10, which induced necrosis in *N. glauca* and *Physalis floridana,* whereas mutant cp 27 did not. In contrast, mutant cp 27 induced necrosis in Ace tomato and mutant cp 10 did not. In other varieties of tomato, both mutants replicated without inducing necrosis. These mutants may provide a system comparable to general elicitors reported for fungal systems.

Other coat-protein mutants of TMV give a similar phenotype in nn, n'n' genotype tobacco. Saito et al (62) recently demonstrated this phenomenon with the tomato strain of TMV where a deletion in the coat-protein gene that was 5' of the sequence deleted in mutants cp 10 and cp 27 induced necrosis. Several other coat-protein deletion mutants did not induce necrosis (62). Additionally, adding 6 amino acids to the C terminus (Met-Tyr-Gly-Gly-Phe-Leu) produced a free RNA virus with a similar phenotype (77). Nondeletion mutants of TMV and other tobamoviruses have been reported to have a similar phenotype (26, 31, 62).

From a comparison of coat-protein deletion mutants, it is difficult to identify precise coat-protein sequences that result in necrosis when deleted. Both mutants cp 10 and cp 27 have deletions in the region where the coat protein binds to the RNA in virions (amino acids 90–123). Mutant cp 10 and cp 27 were deleted 14 and 108 nucleotides in the 5' direction of this region, respectively (Table 1). However, additional mutants with intermediate deletions, such as cp 4 and cp 25 (37 and 38 nucleotides in the 5' direction of this region, respectively), did not induce necrosis. Also, mutants with much less or much more deleted in the 5' direction of this region did not induce necrosis. All possible hybrids between mutants cp 10 and cp 27 and other deletion mutants that did not induce necrosis were examined. Hybrids with either the cp 10 or cp 27 5' coat protein gene sequences induced necrosis regardless of their 3' sequence or whether the native carboxy-terminus was present (A. G. C. Lindbeck et al, in preparation). None of the hybrid combinations with a 5' sequence from mutants other than cp 10 or cp 27 induced necrosis. Thus, only specific deletions in the coat protein induce necrosis in nn, n'n' genotype plants.

We do not know whether necrosis induced by these mutants is a typical HR, a partial HR, or a necrotic reaction different from the HR. Pathogenesis-related proteins associated with the HR are produced before this necrosis. Lesions in nn, n'n' genotype plants developed more slowly than in NN or N'N' genotype plants (5 days compared to 3 days) and yellowing frequently preceded the development of necrosis. This necrotic reaction was temperature sensitive, similar to the HR in plants with the N or N' gene (10). However, the temperature that prevented necrosis in nn, n'n' genotype plants was approximately 5°C higher (35°C) than in NN genotype plants. Also, the induction of this necrosis was related to plant development. These mutants induced less necrosis and were not confined in younger tissues. However, in expanded leaves these mutants induced a strong necrotic response and were confined to the infection sites. The result is a resistance response similar to the HR.

Induction of necrosis, leading to resistance in hosts that are normally susceptible to TMV, suggests that this form of necrotic response is nonspecific and does not involve an interaction with a host resistance gene. The truncated coat proteins produced by these mutants may act as nonspecific "general" elicitors of necrosis. However, the ability of these normally susceptible plants to mount a necrotic resistance response suggests the presence of the mechanisms and components necessary for resistance. Only the components required for pathogen recognition of this response are missing.

Tm-2 Gene of Tomato

Two HR resistance factors, Tm-2, and Tm-2^2, occur in the genus *Lycopersicon* (22, 51). Upon infection by certain strains of tobamoviruses, the expression of these resistance factors can be displayed as either localized necrosis, resulting in resistance, or systemic necrosis, resulting in severe disease. The type of necrosis displayed is dependent upon the strain of infecting virus, the genotype of the plant, and environmental conditions, especially temperature. Tm-2 and Tm-2^2 resistance factors are thought to be allelic (22, 51). Interestingly, no strains of tobamoviruses have yet been reported to fully overcome the resistance of the Tm-2^2 locus (18).

Several lines of evidence have implicated the involvement of the viral 30kd movement protein in overcoming both Tm-2 and Tm-2^2 resistance. Tm-2 resistance can be overcome by coinoculation with potato virus X, which complements the TMV movement function (78). Also, both Tm-2 and Tm-2^2 resistance responses are expressed in whole-leaf tissues, not in protoplasts (44). Finally, restrictions in the movement of TMV have been observed in initially infected tissue of Tm-2 and Tm-2^2 plants (49).

Meshi et al (42) examined the nucleotide sequence of a TMV mutant (Ltb1) capable of overcoming Tm-2 resistance in tomato. Sequence comparison of this resistance-breaking mutant to that of the wild-type virus revealed two amino acid differences located in the 30kd movement protein. Both substitutions (Cys to Phe at position 68 and Glu to Lys at position 133) were required to overcome Tm-2 resistance. Analysis of a second independently isolated resistance-breaking mutant (C32) also revealed similar amino acid substitutions in the 30kd protein (Glu to Lys at position 52 and Glu to Lys at position 133). These data demonstrate that the ability to overcome the Tm-2 resistance response is specific to the 30kd movement protein. These specific amino acid substitutions in the 30kd protein may act to prevent or avoid the induction of the HR that normally occurs in the presence of the wild-type virus. However, it is not known whether the wild-type 30kd protein acts as an elicitor of this necrotic response.

SYSTEMIC NECROSIS

The systemic spread of viral-induced necrosis can result in severe disease. Several tobamovirus strains and mutants induce different types of systemic necrosis in different host plants. The expression of systemic necrosis can be displayed in a variety of ways, including necrosis in a few areas of the upper leaves, sporadic necrotic spots intermixed with mosaic symptoms, and/or widespread necrosis throughout the plant, resulting in death. The display of these systemic necrotic symptoms is dependent upon host genotype, strain of infecting virus, and environmental conditions (39). This section focuses on specific changes in the genome of TMV that are known to induce this response.

One type of systemic necrosis results from "weak elicitor" mutants infecting plants with the N' gene. As previously described, specific amino acid substitutions in the TMV coat protein induce the N' gene HR (7). Many of these mutants weakly elicit the HR and are often able to escape the developing HR and move systemically in the plant. Movement of these weak elicitor mutants into the apical axis results in the development of mosaic symptoms on small, newly expanding leaves. However, as the leaf expands, necrosis develops and eventually the entire leaf collapses. Interestingly, these mutants do not travel systemically as fast as the wild-type virus. This pattern of necrotic development is apparently due to the virus escaping the resistance response and moving systemically into leaves that are too immature to express the HR, resulting in the display of mosaic symptoms. Necrosis then develops after these leaves mature and are capable of expressing the HR.

Some TMV mutants, having single amino acid substitutions in the coat protein of TMV and capable of inducing the N' gene HR, can induce the development of sporadic patches of necrosis in systemically infected leaves of nn, n'n' genotype plants (*N. tabacum* cv. Xanthi), which are normally susceptible to most strains of TMV. These mutants produce distinctly different mosaic symptoms, intermixed with patches of necrosis in upper uninoculated leaves (7). No correlation between the ability of the mutants to elicit the N' gene HR (strong vs weak elicitors) and the development of this type of systemic necrosis was found. This necrotic response appears to be independent of the mechanisms responsible for the induction of the N' gene HR.

CONCLUSIONS AND FUTURE DIRECTIONS

We are just beginning to understand the mechanisms by which viruses cause diseases of plants. Studies described in this review demonstrate the involvement of several different tobamovirus genes in the induction of disease symptoms or resistance responses. Virus-host interactions differ widely in the mechanisms involved in the display of symptoms. This complexity is reinforced by the variety of viral mechanisms found in other plant virus systems. Examples include: involvement of the cauliflower mosaic virus gene VI protein in the display of mosaic and necrotic symptoms (3, 74); changes in the regulation of the barley stripe mosaic virus gamma-a protein associated with a necrotic HR (A. O. Jackson, personal communication); coinfections with satellite viruses resulting in increases or deceases in the systemic display of symptoms (30, 71); and changes in the nucleotide sequences of viroids resulting in milder systemic symptoms (72). Clearly we must examine numerous other virus-host interactions before we can formulate generalizations on how viruses cause diseases.

Plants appear to have evolved complex processes for recognizing infecting viruses. From examinations of tobamovirus-host interactions, we know that tobacco plants with the N' gene recognize the coat protein, whereas tobacco plants with the N gene recognize something other than the coat protein, and tomato plants with the Tm-2 gene recognize the movement protein. Within the host range of tobamoviruses, numerous other plants recognize the viral infection and respond with the HR. Which components of TMV do these plants recognize? There is no reason for us to suspect any one virus gene product more than others. It is likely that plants have evolved to recognize any viral specific product that was presented to them. Evolution of viruses might be closely tied to the manner in which resistance genes in plants evolved.

While it is important to understand the role of viral genes in causing disease, it is equally important to understand how plants respond to viral infections. Unfortunately, most of our understanding has been limited to the viral contribution of these interactions. For efficient progress in this field, plant pathologists need to collaborate more with plant physiologists and plant molecular biologists to determine at what level plant cell homeostasis is affected by viral infections and what genes are involved. Without access to the plant contributions to these interactions, it will be difficult to understand how viruses cause diseases or to develop new approaches to controlling the diseases. For example, it should be particularly enlightening to identify and locate the plant receptors that recognize plant viruses, to understand how the genes associated with these receptors evolve, and how the recognition event is able to turn on a cascade of defense genes. Current molecular technologies are available to define how viruses cause diseases and how plants resist viral diseases, and rapid progress is being made. As these mechanisms are revealed, it will be of interest to see how this information will be utilized to improve crop production.

ACKNOWLEDGMENTS

This work was supported in part by grants 90–37262-5384 from UDSA Competitive Grants and DMB–9005225 from NSF.

Literature Cited

1. Aldaoud, R. 1987. *Biological characteristics, variations, and factors affecting variation in cloned tobacco mosaic virus*. PhD thesis, Univ. Calif., Riverside, 85 pp.
2. Atkinson, P. H., Matthews, R. E. F. 1970. On the origin of dark green tissue in tobacco leaves infected with tobacco mosaic virus. *Virology* 40: 344–56
3. Baughman, G. A., Jacobs, J. D., Howell, S. H. 1988. Cauliflower mosaic gene VI produces a symptomatic phenotype in transgenic tobacco plants. *Proc. Natl. Acad. Sci. USA* 85:733–37
4. Beachy, R. N., Murakishi, H. H. 1971. Local lesion formation in tobacco tissue culture. *Phytopathology* 61:887–78
5. Beachy, R. N., Zaitlin, M. 1977. Characterization and in vitro translation of the RNAs from less than full-length, virus related, nucleoprotein rods present in tobacco mosaic virus preparations. *Virology* 81:160–69
6. Csillery, G., Tobias, I., Rusko, J. 1983. A new pepper strain of tomato mosaic virus. *Acta Phytopathol. Acad. Sci. Hung*. 18:195–200
7. Culver, J. N., Dawson, W. O. 1989. Point mutations in the coat protein gene of tobacco mosaic virus induce hypersensitivity in *Nicotiana sylvestris*. *Mol. Plant-Microb. Interact*. 2:209–13
8. Culver, J. N., Dawson, W. O. 1989. Tobacco mosaic virus coat protein: An elicitor of the hypersensitive reaction but not required for the development of mosaic symptoms in *Nicotiana sylvestris*. *Virology* 173:755–58
9. Daubert, S., Routh, G. 1990. Point mutations in cauliflower mosaic virus gene VI confer host-specific symptom changes. *Mol. Plant-Microb. Interact*. 3:341–45
10. Dawson, W. O., Bubrick, P., Grantham, G. L. 1988. Modifications of the tobacco mosaic virus coat protein gene affect replication, movement, and symptomatology. *Phytopathology* 78:783–89
11. Dawson, W. O., Lehto, K. M. 1990. Regulation of tobamovirus gene regulation. *Adv. Virus Res*. 38:307–42
12. Dunigan, D. D., Dietzgen, R. G., Schoelz, J. E., Zaitlin, M. 1988. Tobacco mosaic virus particles contain ubiquitinated coat protein subunits. *Virology*. 165:310–12
13. Dunigan, D. D., Golemboski, D. B., Zaitlin, M. 1987. Analysis of the N gene of *Nicotiana*. In *Plant Resistance to Viruses*, ed. B. D. Harrison, 133:120–35. Chichester: Wiley
14. Esau, K. 1968. *Viruses in Plant Hosts: Form, Distribution, and Pathologic Effects*, pp. 130–31. Madison: Univ. Wisconsin Press. 225 pp.
15. Deleted in proof
16. Funatsu, G., Fraenkel-Conrat, H. 1964. Location of amino acid exchanges in chemically evoked mutants of tobacco mosaic virus. *Biochemistry* 3:1356–61
17. Fraser, R. S. S. 1983. Varying effectiveness of the N' gene for resistance to tobacco mosaic virus in tobacco infected with virus strains differing in coat protein properties. *Physiol. Plant Pathol*. 22:109–19
18. Fraser, R. S. S. 1990. The genetics of resistance to plant viruses. *Annu. Rev. Phytopathol*. 28:179–200
19. Ganett, A. L., Shalla, T. A. 1970. Discrepancies in the intracellular behavior of three strains of tobacco mosaic virus, two of which are serologically indistinguishable. *Phytopathology* 60: 419–25
20. Gianinazzi, S., Deshayes, A., Martin, C., Vernoy, R. 1977. Differential reactions to tobacco mosaic virus infection in Samsun 'nn' tobacco plants. I. Necrosis, mosaic symptoms and symptomless leaves following the ontogenic gradient. *Phytopathol. Z*. 88:347–54
21. Goelet, P., Lomonossoff, G. P., Butler, P. J. G., Akam, M. E., Gait, M. J., Karn, J. 1982. Nucleotide sequence of tobacco mosaic virus RNA. *Proc. Natl. Acad. Sci. USA* 79:5818–22
22. Hall, T. J., Resistance at the TM-2 locus in the tomato to tomato mosaic virus. *Euphytica* 29:189–97
23. Hills, G. J., Plaskitt, K. A., Young, N. D., Dunigan, D. D., Watts, J. W., Wilson, T. M. A. Zaitlin, M. 1987. Immunogold localization of the intracellular sites of structural and non-

structural tobacco mosaic virus proteins. *Virology* 161:488–96
24. Hodgson, R. A. J., Beachy, R. N., Pakrasi, H. B. 1989. Selective inhibition of photosystem II in spinach by tobacco mosaic virus: an effect of the viral coat protein. *FEBS Lett.* 245:267–70
25. Holmes, F. O. 1938. Inheritance of resistance to tobacco-mosaic disease in tobacco. *Phytopathology* 28:553–61
26. Holmes, F. O. 1952. Use of primary lesions to determine relative susceptibility of individual plants. *Phytopathology* 42:113
27. Holt, C. A., Hodgson, R. A. J., Coker, F. A., Beachy, R. N., Nelson, R. S. 1990. Characterization of the masked strain of tobacco mosaic virus: identification of the region responsible for symptom attenuation by analysis of an infectious cDNA clone. *Mol. Plant-Microb. Interact.* 3:417–43
28. Hunter, T., Hunt, T., Knowland, J., Zimmern, D. 1976. Messenger RNA for the coat protein of tobacco mosaic virus. *Nature* 260:759–64
29. Ishikawa, M., Meshi, T., Motoyoshi, F., Takamatsu, N., Okada, Y. 1986. In vitro mutagenesis of the putative replicase genes of tobacco mosaic virus. *Nucleic Acids Res.* 14:8291–305
30. Jaegle, M., Devic, M., Longstaff, M., Baulcombe, D. 1990. Cucumber mosaic virus satellite RNA (Y strain): analysis of sequences which affect mosaic symptoms on tobacco. *J. Gen. Virol.* 71:1905–12
31. Jensen, J. H. 1937. Studies on representative strains of tobacco mosaic virus. *Phytopathology* 27:69–84
32. Jockusch, H., Jockusch, B. 1968. Early cell death caused by TMV mutants with defective coat proteins. *Mol. Gen. Genet.* 102:204–09
33. Keen, N. T., Staskawicz, B. 1988. Host range determinants in plant pathogens and symbionts. *Annu. Rev. Microbiol.* 42:421–40
34. Khan, I. A., Jones, G. E. 1988. Accumulation of necrotic lesion inducing variants in TMV-infected plantlets derived from leaf disks of Nicotiana sylvestris. *Can. J. Bot.* 67:984–89
35. Knorr, D. A., Dawson, W. O. 1988. A point mutation in the tobacco mosaic capsid protein gene induces hypersensitivity in *Nicotiana sylvestris*. *Proc. Natl. Acad. Sci. USA* 85:170–74
36. Lindbeck, A. G. C., Dawson, W. O., Thomson, W. W. 1991. Coat protein-related polypeptides from in vitro tobacco mosaic virus mutants do not accumulate in the chloroplasts of directly inoculated leaves. *Mol. Plant-Microb. Interact.* 4:89–94
37. Martelli, G. P., Russo, M. 1985. Virus host relationships: Symptomatological and ultrastructural aspects. In *Plant Viruses, Polyhedral Virions with Tripartite Genomes,* ed. R. I. B. Francki, 1:163–205, New York: Plenum
38. Matthews, R. E. F. 1973. Induction of disease by viruses, with special reference to turnip yellow mosaic virus. *Annu. Rev. Phytopathol.* 11:147–70
39. Matthews, R. E. F. 1981. *Plant Virology*, pp. 218–25, 345–48, 405–07, 430–32. New York: Academic. 897 pp. 2nd ed.
40. McKinney, H. H. 1935. Evidence of virus mutation in the common mosaic of tobacco. *J. Agric. Res.* 51:951–59
41. Melchers, G., Jockusch, H., Sengbusch, P. V. 1966. A tobacco mutant with a dominant allele for hypersensitivity against some TMV-strains. *Phytopathol. Z* 55:86–88
42. Meshi, T., Motoyoshi, F., Maeda, T., Yoshiwoka, S., Watanabe, H., Okada, Y. 1989. Mutations in the tobacco mosaic virus 30-kd protein gene overcome Tm-2 resistance in tomato. *Plant Cell* 1:515–22
43. Meshi, T., Watanabe, Y., Saito, T., Sugimoto, A., Maeda, T., Okada, Y. 1987. Function of the 30kd protein of tobacco mosaic virus: Involvement in cell-to-cell movement and dispensability for replication. *EMBO J.* 6:2557–63
44. Motoyoshi, F., Oshima, N. 1975. Infection with tobacco mosaic virus of leaf mesophyll protoplasts from susceptible and resistant lines of tomato. *J. Gen. Virol.* 29:81–91
45. Mundry, K. W., Schaible, W., Ellwart-Tschurtz, M., Nitschko, H., Hapke, C. 1990. Hypersensitivity to tobacco mosaic virus in N'-gene hosts: which viral genes are involved. In *Recognition and Response in Plant-Virus Interactions*, ed. R. S. S. Fraser, 41:345–59. Berlin/Heidelberg: Springer-Verlag. 467 pp.
46. Murakishi, H. H., Carlson, P. S. 1976. Regeneration of virus-free plants from dark green islands of tobacco mosaic virus-infected leaves. *Phytopathology* 66:931–32
47. Nilsson-Tillgren, T., Kolehmainen-Seveus, L., Von Wettstein, D. 1969. Studies on the biosynthesis of TMV I. A system approaching a synchronized virus synthesis in a tobacco leaf. *Mol. Gen. Genet.* 104:124–41
48. Nishiguchi, M., Kikuchi, S., Kiho, Y., Ohno, T., Meshi, T., Okada, Y. 1985. Molecular basis of plant viral virulence;

the complete nucleotide sequence of an attenuated strain of tobacco mosaic virus. *Nucleic Acids Res.* 13:5585–90

49. Nishiguchi, M., Motoyoshi, F. 1987. Resistance mechanisms of tobacco mosaic virus strains in tomato and tobacco. In *Plant Resistance to Viruses,* ed. D. Evered, S. Harnett, pp. 38–46. Chichester/Wiley
50. Owens, R. A., Candresse, T., Diener, T. O. 1990. Construction of novel viroid chimeras containing portions of tomato apical stunt and citrus exocortis viroids. *Virology* 175:238–46
51. Pelham, J. 1966. Resistance in tomato to tobacco mosaic virus. *Euphytica* 15: 258–67
52. Pelham, H. R. B. 1978. Leaky UAG termination codon in tobacco mosaic virus RNA. *Nature* 272:469–71
53. Petty, I. T. D., Edwards, M. C., Jackson, A. O. 1990. Systemic movement of an RNA plant virus determined by a point substitution in a 5' leader sequence. *Proc. Natl. Acad. Sci. USA* 87:8894–97
54. Ponz, F., Bruening, G. 1986. Mechanisms of resistance to plant viruses. *Annu. Rev. Phytopathol.* 24:355–81
55. Potschka, M., Koch, M. H. J., Adams, M. L., Schuster, T. M. 1988. Time-resolved solution X-ray scattering of tobacco mosaic virus coat protein: kinetics and structure of intermediates. *Biochemistry* 27:8481–91
56. Powell-Abel, P., Nelson, R. S., De, B., Hoffmann, N., Rogers, S. G., et al. 1986. Delay of disease development in transgenic plants that express the tobacco mosaic virus coat protein gene. *Science* 232: 738–43
57. Reinero, A., Beachy, R. N. 1986. Association of TMV coat protein with chloroplast membranes in virus-infected leaves. *Plant Mol. Biol.* 6:291–301
58. Reinero, A., Beachy, R. N. 1989. Reduced photosystem II activity and accumulation of viral coat protein in chloroplasts of leaves infected with tobacco mosaic virus. *Plant Physiol.* 89:111–16
59. Rochon, D., Siegel, A. 1984. Chloroplast DNA transcripts are encapsidated by tobacco mosaic virus coat protein. *Proc. Natl. Acad. Sci. USA* 81:1719–23
60. Roosnick, M., Palukaitis, P. 1990. Genetic mapping of symptom timing and severity of cucumber mosaic virus. *Phytopathology* 80:969
61. Saito, T., Meshi, T., Takamatsu, N., Okada, Y. 1987. Coat gene sequence of tobacco mosaic virus encodes host response determinant. *Proc. Natl. Acad. Sci. USA* 84:6074–77
62. Saito, T., Yamanaka, K., Watanabe, Y., Takamatsu, N., Meshi, T., Okada, Y. 1989. Mutational analysis of the coat protein gene of tobacco mosaic virus in relation to hypersensitive response in tobacco plants with the N' gene. *Virology* 173:11–20
63. Schoelz, J. E, Zaitlin, M. 1989. Tobacco mosaic virus RNA enters chloroplasts in vivo. *Proc. Natl. Acad. Sci. USA* 86:4496–500
64. Siegal, A. 1959. Mutual exclusion of strains of tobacco mosaic virus. *Virology* 8:470–77
65. Siegal, A., Zaitlin, M., Sehgal, O. P. 1962. The isolation of defective tobacco mosaic virus strains. *Proc. Natl. Acad. Sci. USA* 48:1845–51
66. Shalla, T. A. 1968. Virus particles in chloroplasts of plants infected with the U5 strain of tobacco mosaic virus. *Virology* 35:194–203
67. Shalla, T. A., Petersen, L. J., Guichedi, L. 1975. Partial characterization of virus-like particles in chloroplasts infected with the U5 strain of TMV. *Virology.* 66:94–105
68. Shepard, J. F. 1975. Regeneration of plants from protoplasts of potato virus X-infected tobacco leaves. *Virology* 66:492–501
69. Deleted in proof
70. Shintaku, M. H., Palukaitis, P. 1990. Mapping determinants of pathogenicity and transmission of cucumber mosaic virus. *Phytopathology* 80:1035
71. Sleat, D. E., Palukaitis, P. 1990. Site-directed mutagenesis of a plant viral satellite RNA changes its phenotype from ameliorative to necrogenic. *Proc. Natl. Acad. Sci. USA* 87:2946–50
72. Stanley, J., Frischmuth, T., Ellwood, S. 1990. Defective viral DNA ameliorates symptoms of geminivirus infection in transgenic plants. *Proc. Natl. Acad. Sci. USA* 87:6291–95
73. Suwa, M., Takahashi, T. 1975. Studies on viral pathogenesis in plant hosts. VII. Symptoms and external morphology of 'Samsun' tobacco plants following systemic infection with tobacco mosaic virus. *Phytopathol. Z.* 83: 348–59
74. Takahashi, H., Shimamoto, K., Ehara, Y. 1989. Cauliflower mosaic virus gene VI causes growth suppression, development of necrotic spots and expression of defence-related genes in transgenic tobacco plants. *Mol. Gen. Genet.* 216: 188–94
75. Takahashi, T. 1971. Studies on viral pathogenesis in plant hosts. 1. Relation

between host leaf age and the formation of systemic symptoms induced by tobacco mosaic virus. *Phytopathol. Z.* 71: 275–84

76. Takamatsu, N., Ishikawa, M., Meshi, T., Okada, Y. 1987. Expression of bacterial chloramphenicol acetyltransferase gene in tobacco plants mediated by TMV-RNA. *EMBO J.* 6:307–11
77. Takamatsu, N., Watanabe, Y., Yanagi, H., Meshi, T., Shiba, T., Okada, Y. 1990. Production of enkephalin in tobacco protoplasts using tobacco mosaic virus RNA vector. *FEBS Lett.* 269:73–76
78. Taliansky, M. E., Malyshenko, S. I., Pshennikova, E. S., Atabekov, J. G. 1982. Plant virus-specific transport function. II. A factor controlling virus host range. *Virology* 122:327–31
79. Valleau, W. D. 1952. The evolution of susceptibility to tobacco mosaic in *Nicotiana* and the origin of the tobacco mosaic virus. *Phytopathology* 42:40–42
80. Van Loon, L. C. 1987. Disease induction by plant viruses. *Adv. Virus Res.* 33:205–55
81. Weber, P. V. 1951. Inheritance of a necrotic-lesion reaction to a mild strain of tobacco mosaic virus. *Phytopathology* 41:593–609
82. Wittmann, H. G., Wittmann-Liebold, B. 1966. Protein chemical studies of two RNA viruses and their mutants. *Cold Spring Harbor Symp. Quant. Biol.* 31:163–72
83. Zaitlin, M., Hull, R. 1987. Plant virus–host interactions. *Annu. Rev. Plant Physiol.* 38:291-315

Annu. Rev. Phytopathol. 1991. 29:219–46

EXCLUSION AS A PLANT DISEASE CONTROL STRATEGY[1]

R. P. Kahn

14104 Flint Rock Terrace, Rockville, Maryland 20853

KEY WORDS: plant quarantine, regulatory control, plant pests, pest and pathogen risk analysis, plant protection

INTRODUCTION

Plant protection and quarantine (PPQ) programs, or the plant health or quarantine services in most countries, including the United States, usually have three components:

1. Exclusion of pests and pathogens of quarantine and economic significance that might inadvertently be moved along man-made pathways when articles are imported; or reduction of the risk of introducing such hazardous organisms to an acceptable level.
2. Containment, suppression, and eradication of exotic pests and pathogens *recently* introduced along natural or man-made pathways.
3. Assistance to exporters of plant products such as fruits, vegetables, plants, cut flowers, commodities, etc, in meeting the quarantine or exclusion requirements of importing countries and, therefore, based on plant health, biologically facilitating the acceptance of exports.

The first component falls under the aegis of plant quarantine or regulatory control. The second belongs to plant protection but may include some regulatory control. The third relates to the first, but the thrust is in the opposite direction. An exporting country assists the importing country in the latter's

[1]The views expressed herein are the author's and do not necessarily represent those of the Animal and Plant Health Inspection Service, USDA, from which he retired, or any other agency or quarantine service with which he is currently associated on short-term assignments.

0066-4286/91/0901-0219$02:00

practice of exclusion—a treaty obligation for 94 signatory nations under the International Plant Protection Convention of 1951 (5, 11, 43, 48); but also adhered to by many countries, particularly exporting countries, that are not yet signatories.

This chapter is concerned with the first component, i.e. exclusion or regulatory control—not because it is necessarily the most important component, but because its effectiveness as a disease control strategy is threatened and impeded by (*a*) a negative image among many, but not all, scientists as well as much of the traveling public and some importers of agricultural products, and (*b*) a shortage of supportive research in plant pathology that can be applied to exclusion.

The objectives of this paper are (*a*) to illustrate the negative image and its possible origins, and to expose problems and misconceptions that obstruct a positive image from developing; and (*b*) to suggest areas of research that can advance exclusion as a disease control methodology[2]

DEFINITIONS

Quarantine Significance

As used in this paper, a pathogen species that does not occur in a given country, or an exotic strain of domestic species, is of quarantine significance (QS) to that country if the pathogen is known to cause economic damage elsewhere; or has a life cycle or host/pathogen interaction that shows a potential to cause economic damage under favorable host, inoculum, and environmental conditions. An importation of a pathogen that already occurs in a given country is also of QS if an on-going regional or national containment, suppression, or eradication program is directed against that pathogen species.

A more general definition is that an exotic organism is of QS if its exclusion is perceived as important enough to the agriculture or natural vegetation of a country for a government to allocate resources to prevent or delay its entry along man-made pathways.

[2]The review is based on the author's survey of the scientific and regulatory literature and contacts with quarantine services in the United States and other countries. The term "quarantine services" is used to reflect the general results of this survey; but it is beyond the scope of this review to cite references to specific countries. Reference, to illustrate concepts or to provide examples, is sometimes made to procedures used in the United States; but it is also beyond the scope of this report to serve as an information source about any program activity or the entry status of any article. Persons interested in the entry status of any article into the United States or other countries should contact an office of the Animal and Plant Health Inspection Service (APHIS), USDA, whose telephone number and address may be found in the government telephone listings in many cities where international airports and major seaports are located.

Pest Risk

Pest risk (PR) defines the chances that a pest or pathogen of QS will enter along a man-made pathway. Risk is often expressed, in the absence of a generally accepted quantitative methodology, as low-, medium- or intermediate-, or high-risk. Low-risk means that there is little chance that the pathogen will enter; high-risk means that the chances are high the pathogen could enter. An acceptable level of risk, as related to risk/benefit factors, means that the benefits to be derived after taking the risk are high enough to justify taking the risk with safeguards in place.

Safeguards

Safeguards are actions taken to lower the risk (chances) of introducing pests and pathogens of QS along man-made pathways.

Entry Status

Entry status is determined by the gamut of rules, regulations, policy statements, guidelines, procedures, and decisions of exclusion or quarantine officers that govern whether plants or plant parts, and certain other specified imported articles are enterable, and if so, under what conditions or safeguards.

A Favorable Risk/Benefit Consideration

The benefits to be derived from accepting a risk should exceed the potential costs should a pest or pathogen of QS enter and become established a result of the importation. An example of a favorable risk/benefit consideration for a high-risk importation is the entry under approved safeguards of certain accessions of germplasm of plant species that are prohibited because of high-risk host/pathogen interactions (13, 18, 45, 53). The benefits gained by crop improvement exceed the risk that safeguards will not prevent pathogen entry. The importation of the same accession in large quantities for commercial purposes would not have a favorable risk/benefit ratio.

CONCEPTS

Regulatory Principles and Philosophy

A detailed discussion of general quarantine principles and philosophy, except as they pertain to exclusion, is beyond the scope of this review: moreover the topics have already been extensively reviewed (8, 15, 18, 28, 31, 32, 34–42, 44, 52).

Exclusion as a Plant Disease Control Strategy

A survey of 15 English-language plant pathology textbooks published between 1935 and 1989 shows that exclusion (also known as regulatory control, or quarantine) is considered to be one of the major categories of plant disease control (Author's unpublished data). The breakdown of these control categories from different sources is shown in the following columns:

1	2	3	4
Exclusion	Regulatory	Exclusion	Sanitation
Eradication	Protection	Chemical	Cultural
Chemical	Resistance	Physical	Physical
Adjustment to the environment	Eradication	Cultural	Chemical
Resistance	Therapy	Biological	Resistance
	Avoidance		Legislation

5	6	7
Chemical	Exclusion	Prophylaxis (Avoidance)
Resistance	Protection	Immunization (resistance)
Cultural	Resistance	Eradication
Quarantine	Eradication	Direct protection (chemical)
		Exclusion

8	9	10	11
Cultural	Cultural	Chemical	Exclusion
Biological	Quarantine	Cultural	Avoidance
Resistance	Eradication	Biological	Resistance
	Resistance	Quarantine	Therapy

The Legal Basis of Exclusion

In most countries, actions taken in the practice of exclusion as a disease control strategy are authorized by governmental regulations to reduce the chances of inadvertent introduction of exotic pests and pathogens of quarantine significance on imported articles. The legal basis of such regulatory actions is either (*a*) legislation passed by national and, sometimes, State or Provincial governments as acts, regulations, etc; or (*b*) enabling legislation that authorizes the Secretary or Minister of Agriculture to issue orders, rules, directives, etc, that have the force of law. In almost all such legal decrees the term "pest" is defined to include pathogens.

In the United States, the principal legislation authorizing exclusion activities is: (*a*) The Plant Quarantine Act (PQA) of 1912, as amended, which provides the basis of regulatory actions to exclude pests and pathogens that might enter on imported nursery stock and other plants and plant products, and provides authority for domestic and foreign plant quarantines; (*b*) The

Organic Act of 1944, which authorizes the Secretary of Agriculture to cooperate with the States, organizations, and individuals to detect, eradicate, suppress, control, and prevent or retard the spread of plant pests; and (*c*) The Federal Plant Pest Act (FPPA) of 1957, as amended, which regulates the movement (by persons) of plant pests into the United States, or interstate, and authorizes emergency actions to prevent the introduction and domestic movement of plant pests not covered by the PQA.

The International Plant Protection Convention (IPPC) of 1951 (5, 11, 43, 48), signed by 94 nations and conformed to by most other countries, established a basis for international cooperation in regulatory actions (for review see (8, 15, 18, 22, 29, 53)). The UN Food and Agriculture Organization (FAO) is the repository for the treaty; and the phytosanitary certificate conforming to the model FAO certificate is an instrument of that treaty. Disputes between nations may be brought to the attention of an international court of law—an event that has not yet occurred. In addition, international plant protection organizations (6, 32–34, 58) (e.g. the European and Mediterranean Plant Protection Organization (EPPO) and the North American Plant Protection Organization (NAPPO)), suggest standards or regulations with the thrust towards harmonizing regulations among member countries. Also, some countries have banded together into unions with parliaments that promulgate binding regulations on members, e.g. the European Economic Community.

Geographic Basis of Exclusion

The distribution of exotic pests and pathogens in various countries, regions, or continents is the rationale for promulgating rules and regulations, and for taking regulatory actions. Of critical concern is the known ecological range of the pathogen compared to the ecological range of its host(s). If the known ecological range of the pathogen is the same ecological range as its hosts, i.e. the pathogen is widespread, the pathogen species is usually not considered of QS (See definition). If the pathogen does not occur in the country taking regulatory action, and has an ecological range similar to the ecological range of its host(s) in that country, that country is likely to take regulatory actions if its perception of risk is high.

From an exclusion point of view, if the status quo of pathogen distribution were maintained, problems of regulatory officers would be simplified. But the status quo changes since pathogens move, or are moved, along natural and/or man-made pathways (2, 20, 21, 29, 35, 36, 38, 41).

Many pathogens (e. g. most viruses, bacteria, and nematodes, and some obligately parasitic fungi) have life cycles and disease cycles that do not foster long-distance spread by natural means. Such pests and pathogens can move, or be moved, by the cumulative effect of a series of short-distance natural moves until a barrier, including absence of a host, is reached. While some of these pathogens are moved by vectors (e.g. leafhoppers and aphids), it is unlikely that a pathogen retained by a vector for short periods of time will be spread by that vector long distances. A viruliferous aphid transmitting the plum pox virus in a nonpersistent manner could survive a trip from Europe to the east coast of the United States. If the aphid survived the journey with still infectious virus, it is more likely that it would feed first on a nonhost than a host in a susceptible condition (22). Similarly, if birds were to transport pathogens, it is also not likely that potential inoculum, if not lost in flight, would be fit into circumstances that support the disease triangle (See discussion under Biological Basis of Exclusion). Viruses could be easily moved in budwood, which could be grafted on to a U.S. root stock, and then the plant could serve as a source of inoculum for virus spread by aphids already present in the USA.

The problem of insects moving pathogens over long distances on man-made pathways has been addressed (12). Insects that carry the pathogens throughout their life cycles are more likely to spread pathogens over long distances.

Data about the geographical occurrence of such organisms are found in the scientific and regulatory literature. Reports (e.g. 46,50) of pests intercepted at ports of entry provide information about the detection of exotic pests, but the information is useful only if the country of origin of the imported article can be ascertained. Countries that are signatory nations to the IPPC are obligated to report in the *FAO Plant Protection Bulletin* the occurrence of recently introduced pests and pathogens. Some countries with strong import/export relationships often cable reports of pest or pathogen occurrence to each other.

Unfortunately, the published information is not always up-to-date. There may be a lag between the first report and the confirmation of the identity of the causal agent (e.g. a few years for viruses of woody hosts). Furthermore, information about occurrence is often scanty from countries that are understaffed in entomology, plant pathology, etc. Little information is available about pest and pathogen occurrence in remote regions or the wilds where plant germplasm is often collected.

Biological Basis of Exclusion

The biological foundation of exclusion rests on knowledge about: (*a*) the identity of pathogens as well as their geographic distribution; (*b*) pathogen life cycles as influenced by the environment, including weather, climate, vectors, and farm practices; (*c*) natural and man-made pathways; and (*d*) host range. The understanding of the interaction of these factors on entry, infection (or colonization), and establishment and spread is the basis of PR analysis and the implementation of regulatory actions.

The goals of regulatory actions are (*a*) to prevent or delay entry of pathogens along *man-made* pathways; (*b*) if entry succeeds, to prevent infection; (*c*) if infection succeeds, to prevent establishment of pathogens entering on man-made pathways, but also natural pathways; and (*d*) if establishment succeeds, to minimize or retard spread. Goal (*c*) is implemented by eradication, and goal (*d*) by containment or suppression procedures that are beyond the scope of this paper.

THE DISEASE TRIANGLE Goal (*a*) and to a limited extent goal (*b*) are addressed by regulatory actions that prevent the synchronization of three events by converting at least one to a limiting or nonoperating factor. The events are: the presence of (*a*) susceptible hosts in a susceptible condition; (*b*) sufficient inoculum to initiate infection; and (*c*) favorable environmental conditions. These factors have been depicted as the disease triangle (Figure 1), or when combined with the other factors (e.g. time, economic considerations) as pyramids, cones, or polygons (1, 22, 31).

The most widely employed regulatory approach is to reduce or eliminate the inoculum. Examples include: (*a*) not issuing permits for the importation of high-risk plant material from countries where pathogens of QS, which cannot be detected by inspection at ports of entry, are known to occur; (*b*) requiring treatment as a condition of entry; (*c*) denying entry to an entire consignment, or to infected plants or plant parts if an exotic pathogen is detected at a port of entry; (*d*) prohibiting, by some countries, alternate hosts of rust fungi that infect conifers; (*e*) requiring an official verification by the exporting country

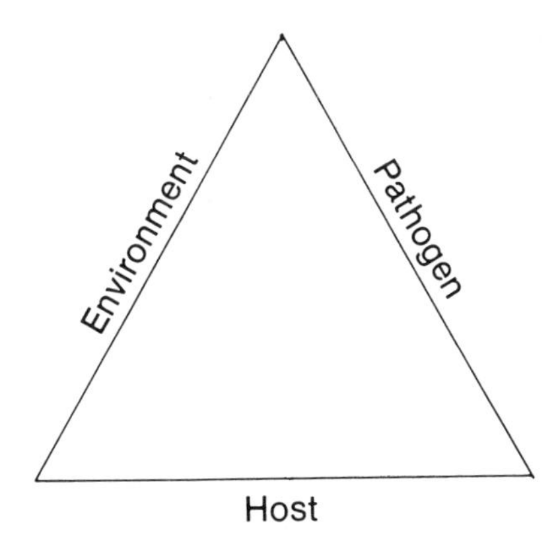

Figure 1 Diagram of the disease triangle whose area shows the interaction of three factors (a host in a susceptible condition, sufficient inoculum, and a favorable environment) on the establishment of a disease agent in a new area, or the development of a disease. If one factor becomes limiting, the side of the triangle would be shortened and its area reduced, illustrating a failure of disease development due to the lack of synchronization of the factors.

that a named pathogen (e.g. *Erwinia amylovora,* the bacterium causing fire blight of pome fruits) did not occur "at the place of production or within 5 km of the place of production since the beginning of the last two cycles of vegetation"; and (*f*) growing plants aseptically in tissue culture to deny access to pathogens.

Another approach is to insure that a susceptible host is not present under favorable environmental conditions. One example is the importation into the United States, for scientific purposes only, of maize from countries where exotic downy mildew fungi occur. The entry of maize is prohibited from such countries unless an exemption is approved under a Departmental Permit issued for exemptions. The permit specifies that as a condition of entry the maize must be grown in greenhouses located in northern states during November through March when outside temperatures are too low to support the growth of hosts should the pathogen be introduced into the greenhouse and escape to the outside. Another example is the use of third-country quarantine (3, 18) by which plants, usually germplasm, of a tropical crop, e.g. cacao, banana, rubber, are grown in quarantine greenhouses in temperate countries where hosts are not present on the outside throughout the year. During the detention period (incubation period) the plants are inspected and sometimes tested for pathogens.

Still another approach is to insure that the environment is not favorable. An example is the one cited above since if fungi were present and spores escaped to the outside, the low temperature would prevent germination during the period of longevity of the spores (furthermore, a host is not present). Plants could also be grown under surface irrigation in arid areas. If the pathogen of QS were favored by a cool or moist environment, the imported host could be grown in the greenhouse at high temperatures and low relative humidities, provided precautions were taken not to allow the plants to become wet.

THE INTERACTION OF PERCEIVED PEST RISK WITH EXCLUSION The perception of and reaction to PR by quarantine services determines the entry status of imported plants, plant products, and other articles that could serve as hosts or carriers of pathogens. If the entry status is matched to the known or perceived risk, (low-risk with liberal entry status, and high-risk with conservative entry status), the actions taken may be considered biologically sound (19), as shown by the "biological line" in Figure 2. To the contrary, if the risk is perceived to be low, and the entry status is conservative, or vice versa, then matching does not fall on or near the biological line. It may be inferred from these observations that such a matching may be economically or politically based (Figure 2), but it cannot be proved so because the mismatching may also be due either to biological misjudgments or differences in perception of risk.

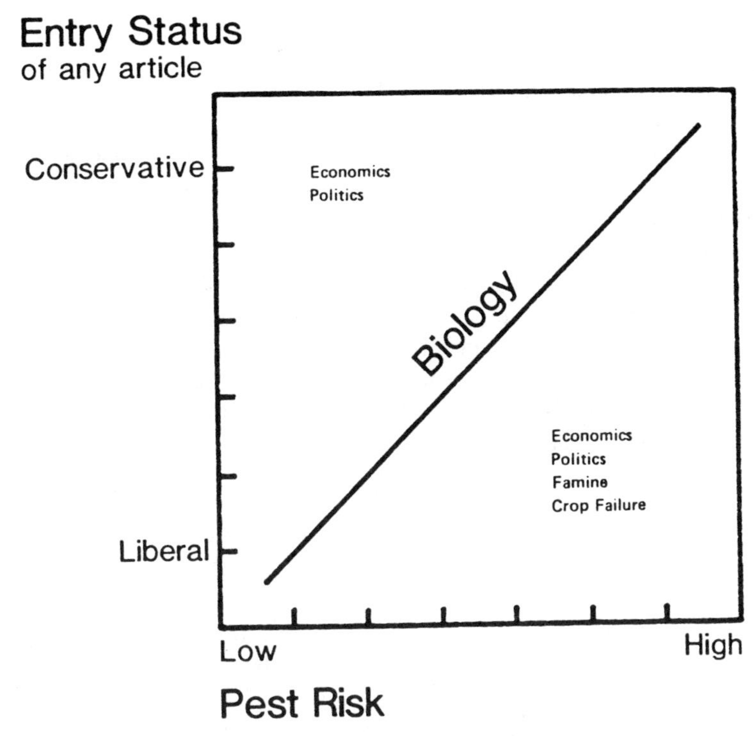

Figure 2 A graphic representation showing the interaction of the entry status of imported articles with known or perceived pest risk. Points that fall on or near the biological line show a matching of risk with entry status, or vice versa; e.g. high risk with conservative entry status. Points that are distant from the biological line represent nonbiological interactions, e.g. low risk with conservative entry status.

If the risk is low, or perceived to be low, entry status is liberal and the article may enter with few or no restrictions. Safeguards implemented in the United States in response to a very low level of risk have been reviewed (56). If the risk is intermediate, or perceived so, the entry status is often more conservative, and the action taken is moderate, e.g. entry is authorized subject to inspection and treat ment, if necessary. Examples of safeguards taken for medium to high levels of risk perception have been reviewed (7, 13–15, 18, 21, 30, 39–42, 45, 52, 53).

If the risk is high, or perceived to be high, the entry status is likely to be even more conservative, and the action taken is likely to be drastic, e.g. prohibition. In addition to outright prohibition of the host as specified in the regulations, nine other methods have been used (20) in the worldwide quarantine regulations (49) by which an otherwise enterable host is, nevertheless, in effect prohibited. Examples include: (*a*) the inability for the exporting country to meet conditions of entry specified in the permit; (*b*) the government's desire to protect the genetic integrity of local varieties by denying permits for named varieties of an otherwise enterable species; and (*c*) the policy of denying entry of enterable material if the consignment contains other material that is prohibited.

When an importing country prohibits an importation of a plant or plant product because of the threat of entry of a pathogen of QS, it is assumed that, based on PR perception, the host/pathogen interaction is judged to be high-risk. It is assumed that quarantine officials of the importing country have concluded that the level of risk cannot be dealt with by the safeguard of inspection and/or treatment at origin (exporting country) or at a port of entry (importing country). Consequently, the importing country prohibits the host (e.g. commercial quantities of plants, fruits, grains for consumption, agri cultural raw materials, seeds for planting, etc), with exception of approved entry of small amounts for scientific purposes.

The Economic, Social, and /or Political Bases of Exclusion

Within quarantine circles, it is axiomatic that regulations and regulatory actions should be biologically based, and should not be used as an economic embargo (e.g. 19, 34, 36, 40). The consensus is that if economic embargoes are to be implemented by a country other methods such as duties or tariffs should be used—a concept that may be in conflict with the General Agreement on Trade and Tariffs. Another biological axiom is that the regulatory action taken should be the least drastic response that can provide the safeguard necessary to deal with the perceived level of risk.

Notwithstanding, there is an undercurrent of suspicion that some exclusion regulations may cover up economic motivations. The doubts arise, for a given host/pathogen interaction, when the importing country perceives the risk to be high, and the exporting country perceives the risk to be low. Often, the quarantine service of an exporting country questions its counterparts in the importing country, and differences are resolved by either side presenting data, seeking expert "second" opinions, and negotiating biological safeguards.

If an impasse is reached and the status quo of the importing country prevails, the quarantine service of an exporting country may suspect an economic barrier. Is the importing country protecting its own growers or agribusinesses by not making available to its consuming public foreign products, thus thwarting market forces? Are commodity- or crop-oriented groups (i.e. lobbies) exerting pressure on the government to exclude foreign plants, produce, or commodities? Examples of possible economic problems, which overflow into local political problems, have been reviewed (14, 36).

A government may be faced with the dilemma of pressure from opposing special interest groups; one advocating a liberal entry status, and the other a conservative one. The lot of the senior quarantine official, to paraphrase Gilbert and Sullivan, is not always a happy one.

Exclusionary actions can be overridden for social reasons in some countries during years of famines, crop failures, or food shortages. Under such extenuating circumstances, a country may change its announced position on the importation of agricultural commodities (e.g. potato, wheat) from conservative (i.e. prohibited) to liberal, (i.e. enterable) to provide food. When social problems are involved, an importation, which previously received an unfavorable risk/benefit analysis, suddenly becomes enterable because risk/benefit considerations have been reversed from unfavorable to favorable. (For review of other social aspects, the reader is referred to (4, 36)).

There is sometimes a political aspect to PPQ activities in general, but in exclusion the political component is minimal. The dictionary definition of "politics" used herein includes the following concept: government policies, actions, strategies, maneuvers, etc, carried out to obtain a position of influence or control. This definition does not include the politics of pressure groups discussed earlier.

In countries with strong but separate Federal and State or Provincial quarantine services, politics may play a role in cooperative programs within the country. In the United States, the Federal (The Organic Act) and State Governments (55) conduct cooperative pest and pathogen control activities (51; See Introduction, Component 2), and export certification programs (See Introduction, Component 3). For the most part, politics, as defined, do not constitute a problem, except in some cooperative Federal/State eradication, suppression, and/or containment programs where problems of jurisdiction, procedure, and strategy are apparent. As an undertow in the tide of some Federal/State interactions in PPQ, as in many other Federal/State relations, some States are less than pleased with a strong Federal position—a thrust that runs countercurrent to effective cooperative regulatory action. Notable examples include the citrus canker bacteria and fruit fly eradication programs, which are beyond the scope of this paper's concerns with Component 1.

In the United States, exclusion is practiced as a disease control strategy by the Federal and State governments, with negligible political rivalries. The jurisdictional aspects are settled by Federal laws and regulations that take precedence over State laws. The Federal government takes regulatory actions to exclude *exotic* pathogens of QS. States take regulatory actions to exclude exotic and US pathogens of economic or QS that do not occur in the state, but do occur in other states. Most commonly the states regulate the entry of nursery stock and seeds. The Federal government may take regulatory actions to assist the states in their regulatory programs. Problems in exclusion policy or procedures between Federal and state governments are more likely to be biological, i.e. risk perception, than political.

The Administrative Basis of Exclusion

Administrative aspects relating to organizational framework and funding influence PPQ program activities in general and exclusion practices in particular. In any country, there is always competition for funds among government agencies. Within any agency, such as a quarantine service, organizational units compete with sister units for agency funds for staffing, facilities, equipment, supplies, and program activities.

Program managers usually attempt to allocate scarce resources to program activities that deal with the highest perceived risk although there is, in many countries, an alleged interaction with other nonbiological factors. The allocation of funds according to risk is compounded by the need not only to deal with program activities for pathogens, which are the subject of this paper, but also with exotic insects, mites, snails, slugs, and weeds of QS. The allocation of funds is confounded by the absence of a generally accepted quantitative method to assess risk in order to set priorities, and by nonbiological pressures. Since risk, like beauty, lies in the eyes of the perceiver, the lot of a senior manager is, again, not always a happy one.

In countries where the organizational framework separates the "line" (program or enforcement activities) from the "staff" (technical and administrative support activities), the line often takes precedence over the staff, and controls funding for technical support activities. Often, the line and the staff compete for available resources.

Most countries are divided into PPQ regions with a central headquarters. The line is organized into hierarchical supervisory layers for each geographical area. The staff, which is usually located at or near headquarters, is often organized into units according to technical function. Line officers, as a result of knowledge of program activities, experience, and demonstrated managerial skills, may advance up the line-career ladder. Scientists, as a result of experience, scientific skills and knowledge, reputation, and/or publications may advance up a much shorter scientific career ladder. Except for those few who breach the ceiling by promotions to supervisory scientist positions, most other scientists wishing to advance often switch to the line career ladder where more supervisory or managerial positions are available at higher grade levels. Some scientists elect to stay at the ceiling to pursue scientific interests, some transfer to other agencies, and others become supervisory line officers.

The author has belabored this topic to describe an administrative policy that results in a shortage of scientists engaged in technical support. The reservoir of experienced staff scientists representing entomology, plant pathology, virology, microbiology, weed science, nematology, etc, available to advise the line on PR for nonprecedented events, as well as to biologically update regulations and procedures, is sometimes reduced by these administrative policies. Most activities of line officers are routine or precedented, and covered by manuals that officers may consult in the performance of their duties. However, when a line officer or official is faced with an unprecedented event, the staff is usually called upon to develop a biological position or risk assessment. It is not that a line officer is not capable of making risk assessments, but rather that an employee is usually trained in one, or sometimes two, scientific disciplines. Once such an officer climbs the supervisory and managerial ladder, burgeoning administrative and supervisory responsibilities prevent the officer from keeping up with recent professional developments. A well-rounded staff with specialists in the various sciences is required to deal with PR and support the line.

A negative image may result from the dilution of scientific expertise caused by administrative policies. It is not well appreciated among the general, scientific, and commercial publics that a quarantine service must deal with viroids, viruses, mycoplasma-like organisms, spiroplasma, bacteria, fungi, protozoa, nematodes, a wide spectrum of diverse insect groups, mites, snails, slugs, parasitic plants, and named weeds. A quarantine service, even in an economically well-developed country, is unlikely to be able to afford the depth of scientific expertise to provide technical support. Any administrative action that dilutes scientific support detracts from the ability of that service to provide in-house technical support. As a back-up, most quarantine services develop a network of cooperating scientists in other government agencies and universities, but often these scientists are not familiar with exotic pathogens.

THE NEGATIVE IMAGE OF PLANT QUARANTINE

A positive image for plant quarantine services enhances voluntary compliance with regulations and guidelines. For widespread voluntary compliance to succeed, people need to understand that the regulations and restrictions are biologically sound and not just bureaucratic impediments. Some of the factors that contribute to the negative image problems and potential solutions are discussed in this section.

The absence of data for administrators who allocate government funding is perhaps the single most important factor contributing to the negative image of the quarantine service. The scientific and commercial publics would also like to see data supporting the expenditure of funds for exclusion and justifying the inconvenience of quarantine restrictions such as the requirement for permits, delays due to inspection, or prohibitions. Data are available for low- and intermediate-risk activities (e.g. 10, 46, 50), but corresponding data for high-risk events are more difficult to obtain.

If the host is prohibited, how do we know how many times the pathogen would have entered but for the prohibition? The efficacy of preventive measures are notoriously difficult to substantiate: controlled experiments are not feasible.

In the United States, some data exist on the incidence of pathogens detected in prohibited plant species imported for scientific purposes under Departmental Permits, and passed through detention and testing in a quarantine station (25, 27). For example, between 1968–78, the number of accessions tested and the percentage found to be infected with one or more viruses were, respectively: *Ipomoea batatas* (sweet potato) 50, 48%; *Solanum* spp. (potato) 424, 56%; *Vitis* spp. (grapevine) 409, 65%; stone fruits 62, 45%; and pome fruits 373, 55%. Data are also available for plants collected in the wild (24, 26). For example, of 106 *Solanum* spp. tested, 29% were virus infected.

However, there are no data for would-be commercial importations of these same species since commercial importations of prohibited materials are not allowed; consequently, tests could not have been conducted.

However, particularly for vegetatively propagated crops species, if the mother plant from which the imports were derived was infected with a systemic virus or other pathogen, almost all daughter plants would be similarly infected. Furthermore, if the pathogen were latent at the time vegetative propagations were harvested in the exporting country, the pathogen could not be detected at ports of entry using the methodology and equipment available at inspection stations.

Contribution of Public Attitude

The term "plant quarantine" has not received the universal acceptance in plant science that the term "animal quarantine" has in animal science. Similarly, animal quarantine is more likely to be accepted by the general public than plant quarantine. For example, animal breeders not only accept animal quarantine but welcome it, even on a voluntary basis. A breeder of thoroughbred racehorses or other pedigree animals would not add a new acquisition to his or her stock without isolation or quarantine for a suitable incubation period. Similarly, zoo directors isolate new accessions. Tropical fish hobbyists are advised to place new acquisitions in a holding aquarium to allow time for completion of an incubation period should the fish be infected and symptomless at the time of purchase. In human medicine, quarantine is relied upon to stop the spread of highly infectious disease. In some parts of the world, newborn babies are placed in quarantine that excludes visitors.

By contrast, some but, by no means all, plant breeders regard plant quarantine as an impediment to their breeding programs. Some commercial nurserymen and plant hobbyists regard regulatory activities as government red tape that can be circumvented. This attitude also extends to the other members of the traveling public.

Contribution of The Words "Quarantine" and "Embargo"

"Plant Quarantine" is not a meaningful term to describe exclusion as a disease control strategy because its literal and historic connotations do not reflect the thrust of the three components. Historically, "quarantine" was derived from the Latin word *quadraginata* and the Italian word *quaranta,* both meaning "forty." The Italian word *quarantina* was applied in 14th century Venice and other European cities to the 40-day period of isolation (incubation period) required for ships, passengers, and cargo arriving from a country where the Black Death was raging (22, 34, 35).

It follows then that "plant quarantine" should mean the detention of plants for an incubation period, but in actual practice it means the activities described by components 1 and 3. While component 1 does include some detention of plants, plant quarantine as a disease control strategy has very little to do with the quarantine of plants. The number of employees involved and the funding required for the detention of plants represents an insignificant amount of the total required for exclusion or other regulatory activities.

The term "embargo" also has negative implications when used in connection with plant health. In some older scientific and regulatory literature, embargo was equivalent to exclusion. However, a survey of dictionary definitions revealed the following concepts: (*a*) a government order prohibiting the movement of merchant ships into or out of its ports, (*b*) a suspension of trade, especially of a particular commodity, (*c*) a constraint, hindrance, or prohibition, and (*d*) any legal restriction imposed on commerce. The connotation is clearly an economic restriction of trade. Although an exclusion may restrict the international movement of certain plants or plant products, the rationale is biological rather than commercial or economic.

The negative attitude conjured up by the traveling, commercial and scientific publics when faced with the words "quarantine," or "embargo" saddles exclusion or regulatory programs with an unjustified burden. The load would be lightened if "exclusion," "regulatory control," or "plant health" were substituted. For example, within the USDA, the agency responsible for exclusion as a national disease control strategy is APHIS, but activities are conducted by Plant Protection and Quarantine. Why not "Plant Protection and Exclusion" since the word "quarantine" is not an asset?

The author's survey (22) of the names of 164 national government agencies listed in the FAO directory (47) showed that only 32 used the words "Quarantine" or "Plant Protection and Quarantine" in the title of the agency responsible for components 1 and 3 (as discussed in the Introduction). The others used the following: Plant Protection, Plant Health, Ministry of Agriculture, Plant Inspection, Phytosanitary, Plant Control, or Plant Pathology and Entomology—any one of which creates a more positive image than "quarantine."

Contribution of Plant Pathology Textbooks

Fifteen English-language texts, which discussed plant disease control methods, were reviewed to assess their coverage of exclusion or regulatory control (author's unpublished data). The number of pages devoted to these topics ranged from 0–11, with an average of 4.2. Ten out of 16 devoted 3 pages or fewer. With the exception of 3 texts with 8–11 pages about regulatory control, the treatment was too general to explain how exclusion as a disease control strategy works. Quarantine or regulatory concepts or principles were usually not discussed, nor were references given to the regulatory literature. Similarly, in discussing regulatory control, authors usually did not make any cross reference to general plant pathology principles or concepts discussed elsewhere in the text. In textbooks currently in print, regulatory control has not fared well in the competition with other topics for space.

Contribution of the University Courses in Plant Pathology

The author does not have first-hand information about what is taught currently in the first course of plant pathology concerning exclusion as a disease control strategy. Nevertheless, he hypothesizes, admittedly without the support of a survey of the course outlines but based on conversations with a small sample of instructors and students, that short shrift has been given to regulatory control. The coverage may have improved in recent years due to the publication of relevant review articles (1–3, 6, 12, 22, 23, 30–33).

Does the first course in plant pathology impart a negative image, which persists throughout graduate training, to regulatory control in general and exclusion in particular? If so, plant pathology graduates may complete their graduate training without any positive exposure to the principles and concepts of regulatory control or exclusion.

Is there any specific course in regulatory control at universities? A course could be justified by either: (*a*) job opportunities since many Federal and State Agencies, and international organizations employ plant pathologists in regulatory control or support programs, or (*b*) the need for scientists to have an understanding of how regulations and regulatory control may impact on their research programs.

Contribution of Review Articles and Dissertations

The Annual Review of Phytopathology has published two articles on regulatory control in the first 27 volumes (not counting two articles on the history of plant pathology in Ireland and Australia, which discuss quarantine activities in those countries). Only 30 out of more than 13,000 pages can be reached through the index by entries of exclusion, quarantine, or regulatory control (again excluding the historical reviews). Does the observation that editors have not assigned reviews (except very recently) to regulatory control contribute to negative image?

Apparently, no doctoral dissertations have been written on exclusion as a disease control strategy. A computer-assisted survey of the abstracts of dissertations (author's unpublished data) revealed that from 1862 to date there were no dissertations in which the words "exclusion" or "regulatory control" (in the quarantine sense rather than regulation of chemical, physiological or genetic processes or pathways) were linked in dissertation titles or descriptors with the words "plant pests" or "plant pathogens." This observation supports the hypothesis that graduate student advisors and/or graduates are not involved in principles and concepts of exclusion as a disease control strategy—perhaps contributing to a negative image.

Fortunately, research in other countries on exotic plant pathogens provides information and data for countries where such organisms do not occur on life cycles, geographic distribution, host range, dissemination, overwintering, detection, identification, and chemical and physical treatments, etc, for pathogens of QS.

Contribution of Integrated Pest Management Programs

Most integrated pest management programs apparently disregard exclusion as a component since they are geared toward the present pest/pathogen status quo. Do groups of crop-oriented scientists working on integrated pest management look into regulatory control as a means of excluding yet another organism whose establishment would add yet another task to pest management strategies? Do they use their collective expertise for a given crop to examine government regulatory control policies and procedures and, having done so, suggest to State or Federal governments deletions or additions that would make regulatory practices more biologically sound?

Contribution of the Plant Quarantine Act

Another contribution to the negative image of the exclusion as a disease control strategy is the Plant Quarantine Act (PQA) itself. Again, the word quarantine comes into play. The Act is designed to exclude pathogens and other pests, or to reduce chances of their entry along man-made pathways by eliminating or reducing the inoculum by treatments or regulating a potentially infected host. The negative image comes from a critic (e.g. see ref. 9) not understanding that when the government prohibits a plant or plant product, the agency is really excluding inoculum of a pathogen of QS, particularly one that cannot be detected by inspection using equipment and methods available at a plant inspection station.

In the United States, the only plants that are usually placed in quarantine, i.e. detention, are the relatively small number of plant genera that are prohibited under the PQA (e.g. plants or plant parts of species of stone fruits, pome fruits, citrus, potatoes, grapes, etc). Prohibited means that the importation of the named plants is not allowed unless a special Departmental Permit is issued that provides an exception to the prohibition for plants used for scientific purposes and specifies safeguards required as a condition of entry. An example of such a safeguard is the requirement that the importation be grown in detention for testing.

A positive image comes from the realization that the PQA allows for the entry of *most* plant species, subject to detection of pathogens of QS at ports of entry, while prohibiting only a relatively few species to the general public. The negative image generated by the requirement for a permit to import most species can be developed into a positive one by the understanding that permit is a tool to restrict importation of a plant species from a country where exotic pathogens of QS are known to occur. The phytosanitary certificate that accompanies the shipment serves as a indication of the geographic origin of the importation as well as its general plant health status.

Contribution of the Federal Plant Pest Act

The Federal Plant Pest Act itself (FPPA) bears perhaps the most responsibility for the negative image of exclusion. The Act requires that a person have a permit to knowingly move a plant pest across a state line. Similar state legislation controls the entry of pathogens into many states. Consequently, a permit must be approved by APHIS and the government of the state to which the pathogen is to be moved. Plant pathologists are aware that application of the FPPA to the movement of any and all pathogens is less than biologically sound. Despite the regulation of human activities, many pathogens can move naturally across state lines to adjacent states, and beyond.Furthermore, many domestic pathogens, at least at the species level, but perhaps not always at the strain level, are either ubiquitous or occur throughout the ecological range of their hosts in the United States.

It is alleged that scientists sometimes defiantly move plant pathogens across state lines without permits, in violation of the FPPA and state Plant Pest Acts. Plant pathologists who knowingly do so may be aware of the geographical distribution of the pathogen, including its strains, life cycle, and natural means of spread, and correctly make a judgment that no risk to US agriculture is involved; they then use this fox-watching-the-chickens justification to violate the law. Some plant pathologists and other scientists move plant pathogens in violation of the FPPA, and may not be aware of these factors but nevertheless, rightly *or wrongly,* reach the same no-risk conclusion. Some pathologists and other scientists may move pathogens in ignorance of the law.

Every time a scientist knowingly moves a pathogen across a state line without a permit, however safe or dangerous it might be, the negative image of exclusion as disease control strategy is bolstered. Quarantine services would serve their own best interests by reducing the incidence of such violations though better communication and updating biological procedures. Quarantine officials should either change the procedures wherever their application is not biologically meaningful, or Department heads should establish policies about the responsibilities of scientists under existing regulations.

Since the artificial movement across state lines of *any and all* domestic-spread plant pathogens is regulated under the FPPA, and since many of these organisms are already widely spread and/or can be readily moved along natural pathways, then from a biological point of view, *scientists are being overregulated*. If the number of pathogens subject to regulation were reduced through an updating of the biological regulations or other methodology *without an increase in risk to US agriculture*, the administrative burden of scientists would be lightened, time saved, and the workload of regulatory officials reduced. A workload reduction should speed up the approval process for those organisms that must be regulated, such as exotic pathogens or those not widely distributed in the United States.

Should scientists petition the US Government to exempt the movement (by scientists) across state lines of named unbiquitous pathogens (and pests), and exempt such organisms from the Federal permit requirements under the FPPA? Or could the Act remain unchanged, but with the authority to approve permits for named ubquitous pathogens (and pests) delegated to the state quarantine offices, or even further delegated by the state to the Heads of Departments of Plant Pathology (Entomology) or a committee of scientists?

Whether or not deregulation or delegation of authority takes place, it would be biologically prudent for the Heads of Departments of Plant Pathology (Entomology) to implement a strong Departmental-wide policy to ensure that faculty members and graduate students comply with the permit requirements as they now are, or as they may be if changed in the future.The Federal and State governments are not in a position to monitor the activities of each scientist. There must be a willingness to comply, reinforced by Department Head policy. As discussed earlier, the content of the first course in plant pathology would have to be revised to include principles and concepts and the biological basis of the PPQ and FPPA, including the permit system.

RESEARCH TO SUPPORT EXCLUSION AS A DISEASE CONTROL STRATEGY

Developing Pest Risk Assessment Methods

A limiting factor in the cost-effective and biologically sound use of exclusion as a disease control strategy is the absence of a general purpose, ready-to-use *quantitative* method for pathogen risk assessment, including a biologically sound means for determining the QS of exotic organisms. Any method should be generally accepted by the scientific community. If quantitative methods were available to ascertain risk, quarantine services might be in a better position to move from a conservative to a more liberal position on entry status of certain imported plants and commodities. In so doing, less drastic action could be taken, and the quarantine service might gain a positive image without sacrificing safeguards.

Current methods, which are mostly qualitative, sometimes do not serve managers well in allocating resources, deploying officers, and setting priorities to areas of the highest potential risk to agriculture. Some qualitative methods include: (*a*) "when in doubt, keep it out;" (*b*) "eye-of-the perceiver;" (*c*) "eyes-of-perceivers;" (*d*) definitions; or (*e*) official decrees. Method (*a*) may be used in some countries on a case-by-case basis when information or data upon which to base a risk perception or entry status determination are not available in the scientific or regulatory literature, particularly when an organism has only recently been discovered and described. Method (*b*), which is the opinion of an individual, and method (*c*), which is a consensus, are used when data are available but a quantitative method is not, and lead to nothing more definite than low-, intermediate-, or high-risk assessment perceptions.Methods (*d*) and (*e*), which often list names of organisms, are official pronouncements that may have been biologically sound at the time of publication, but usually require biological updating by deletion or addition of names. The 50 most frequently cited names of pests and pathogens, as of 1982, in the quarantine regulations of 124 countries have been listed (20, 22). The names of pests and pathogens that appear in worldwide quarantine regulations have been listed as of 1985 (16) and in digests of quarantine regulations (e.g. 49).

One approach to developing quantitative methods for risk assessment or determining QS is the following: (*a*) compile lists of exotic pathogens for selected crops (e.g. 23, 37, 49, 57); (*b*) develop criteria (21, 22, 36, 39, 54, 60; (*c*) collect and store biological data (17); (*d*) evaluate the contribution of biological factors by weighting criteria to emphasize the most important ones (22); (*e*) use (*b*), (*c*), and (*d*) to develop a model (22, 23, 37, 38, 54, 59); and (*f*) use well-characterized pathogens and some pathogens listed in (*a*) to test the model (22, 54). Models for pest and pathogen applications to exclusion are under development or planned in several countries, e.g. Canada, United Kingdom, Australia, United States, and some European countries (M. H. Royer, personal communication).

A coordinated research effort is required to develop the data and evaluate modeling systems for PPQ applications. Unfortunately, the Ministries or Departments of Agriculture in most countries have separated research from regulatory activities. In the past, quarantine services usually were funded, and sometimes minimally so, to conduct methods development research such as in taxonomy, detection, and treatments. Recently, some quarantine services and international plant protection organizations have initiated or planned research and development projects on PR analysis.

Quarantine agencies interested in taking risk perception out of the eyes of perceivers into a quantitative approach should: (*a*) increase their staffing and funding to develop methodology for this purpose; (*b*) cooperate, by pooling resources with adjacent countries, which face the same exotic PR; (*c*) support regional plant protection organizations that are promoting research in this area; and/or (*d*) set up grant programs and/or temporarily assign scientists to centers of excellence where the specific modeling can be developed.

Research and Development in Computer-Assisted Information Storage and Retrieval Systems

Quarantine agencies are interested in accumulating data about exotic pathogens of QS. Although an agency may send a scientist to study a few exotic pathogens where they occur, more often agencies depend on published information and data, and on manual information storage and retrieval systems. Where funding is available, agencies use computerized information storage and retrieval systems (17). EPPO has a pest and pathogen data base, and FAO is in the final stages of developing one oriented towards the needs of quarantine services. International cooperation is an imperative to avoid the danger of developing data bases whose formats may include code numbers, abbreviations, or acronyms for names of organisms that are not compatible.

Zero Tolerance and Threshold Requirement Research

Another limiting factor is that not enough information or data are available for a wide spectrum of pathogens (or pests) about how much inoculum (or how many individuals) are required to initiate an infection (or infestation), i.e. threshold requirements. The role of inoculum in disease development or the establishment of a pathogen is discussed earlier in this paper (Figure 1).

Quarantine or exclusion officials often have a zero tolerance for pathogens and pests, which cannot be detected by inspection at a port of entry, if they can enter along man-made but not along natural pathways. Under a zero tolerance policy, the host or carrier of such organisms is generally assigned a conservative entry status and prohibited. If officials could find a biological justification to be less conservative, regulatory actions in general and exclusion in particular would benefit from a more positive image.

If officials were more willing to set a tolerance and took less drastic action, more articles would be admitted and treatments that are less than eradicative might be accepted. The willingness to set tolerances could be based on publication of relevant threshold data. Since quarantine scientists of importing countries cannot conduct the necessary research without taking the risk of importing exotic pathogens, officials may ask counterparts in the exporting country where the pathogen occurs to supply available data, or to conduct, or to arrange for supporting research.

Officials may also have a zero tolerance for certain exotic pathogens that can be detected at a port of entry, and may require 100% inspection of the consignment. When a designated pest or pathogen is detected, inspection may be halted, and a decision about the entry status of the consignment is made. If a practical eradicant treatment is available, the consignment may be treated, otherwise it may be destroyed or sent back to its origin. If statistical sampling methods were used and the detection represented a small number of individuals, and a threshold were known, perhaps a tolerance could be set by the officials of the importing country.

Research on threshold requirements is required for many pathogens of QS (Also see section on the disease triangle). Supporting research is needed also for pathogens that are carried in or on seeds. Not all seedlings are infected by the pathogens they carry. What is the relationship between seedborne inoculum and the establishment of a new pathogen in the soil, in annual crops, or in spread to perennial plants?

Eradicant Treatments

For host/pathogen interactions, which have been assigned a zero tolerance, officials usually require that physical (e.g. hot water dips) and chemical treatments (e.g. formaldehyde, 0.5 to 5.0% sodium hypochlorite with a wetting agent) be eradicants. Even though a treatment that is 80–90% effective may provide control at the farmer's level, the same pesticide would be unacceptable as an exclusion treatment. The farmer's situation has a tolerance with a goal towards suppression of a pathogen that is already established, and is geared towards the farmer's (or groups of farmers) holdings. The regulatory situation has a stated zero tolerance with a goal of eradication to prevent the entry and subsequent establishment of a pathogen that does not occur in the country or in certain regions of large countries.

Research is needed to develop practical eradicant treatments. If tolerances can be set, through knowledge of threshold requirements, perhaps some chemicals that are not eradicants and therefore not currently eligible for zero tolerance situations, might become acceptable.

Research on Molecular Probes and Other Advanced Detection Methodology

The detection, characterization, and identification of pathogens has been facilitated by recent advances in laboratory procedures (22). Among the techniques recently developed or improved are electron microscopy, enzyme immunosorbent assay (ELISA, EIA), radioimmunosorbent assay, immunoflorescence microscopy, immunosorbentelectron microscopy, molecular probes and hydridization, etc, for a wide spectrum of pathogens (22). These procedures are conducted in well-equipped laboratories with trained personnel rather than in inspection stations that are equipped mostly for visual examination and treatment.

Methods development research is required to render the advanced techniques more practical for inspection station conditions. Inspection stations, which process samples of large volumes of commercial imports, have laboratory facilities or work stations to facilitate inspection and treatment activities for low- to intermediate-risk importations. Due to the high cost of holding or delaying imports, or the perishability of some commodities, consignments can be held only for short periods pending a decision of entry status. Research is needed to develop kits and simple equipment that can be used in upgraded laboratories at inspection stations to detect exotic pathogens.

Personnel at quarantine stations have been using some of these new procedures, or cooperating with scientists in other nearby agencies or universities, to detect latent pathogens in intermediate- to high-risk plant importations, which are prohibited except for scientific purposes, held in detention pending the determination of their health status. These procedures are also used to confirm identifications before large-scale regulatory actions are taken, or to confirm the report of a new occurrence of a pathogen.

The principal problem facing quarantine services is a lack of test materials that must be developed with purified, or partially purified preparations for pathogens of QS. Usually the scientists of one country do not have access to the pathogens located in distant countries unless special safeguards (e.g. importation of purified virus) can be established, or the scientists work in other countries. Frequently, scientists obtain preparations from cooperating scientists in other countries (usually by trading materials of different agents) or from commercial firms. Even so, since test materials are not available for many pathogens of QS, the new procedures currently have limited application for exclusion activities.

International donor agencies, professional society committees, quarantine services, agribusiness interests, crop- or pathogen-oriented working groups g. see ref. (4) should be encouraged to increase their participation in the development of test materials for pathogens of QS. Such outside support is needed because while scientists are willing to exchange small amounts of these materials, they are reluctant to furnish larger amounts for routine purposes. Once scientists have developed a new methodology and developed new test materials, they often move on to other areas of research. Commercial firms should take up the production of test materials for a wider spectrum of pathogens than are now available. Most commercial kits are geared to certification programs to develop planting stock that is tested against common domestic pathogens, many of which are ubiquitous and not of QS to other countries. The funds made available by international donor agencies for this purpose should be increased to cover additional equipment and staffing costs at centers of excellence. Such agencies have a deep commitment to assist farmers and increase food production, but the funds are most often placed in more direct channels such as dealing with the host/pathogen interaction at the farmer's level. Funding the development of test materials for exclusion purposes is an indirect method of assisting farmers by lightening their pest and pathogen control burdens.

EDUCATING THE PUBLIC ABOUT EXOTIC PEST AND PATHOGEN RISKS

Although quarantine services through out the world are aware of the importance of educating the public about exotic pest and pathogen risk, many agencies do not have enough funds to implement or expand existing public information programs, or program managers dealing with the short-term do not assign a high enough priority to these long-term programs. The goal of these programs should be to reduce the incidence of the illegal importation of plant materials, particularly those of intermediate- to- high-risk, by the traveling, scientific, commercial, and general publics. Such programs are not expected to eliminate all illegal or inadvertent importations, but rather to significantly reduce their incidence to an acceptable level. Any reduction in exotic inoculum reduces the chances of the synchronization of the three factors, i.e. sufficient inoculum, susceptible host, and favorable environment (Figure 1), as discussed under biological concepts.

An example of the incidence of illegal traffic in agricultural products may be found in the results of a recent cooperative APHIS and California Department of Agriculture study known as the "airport blitz." During one week in May, 1990, Federal, State, and County inspectors conducted a 100% inspection of passenger and flight crew (the latter usually exempted) baggage on foreign flights arriving in Los Angeles from designated areas (10). For 153 targeted flights, baggage inspection, which involved 16,997 passengers, revealed 677 interceptions of prohibited fruit and vegetables, and 140 prohibited animal products (2828 pounds). Data for 367 nontargeted flights were also presented. The conclusion drawn was that "The results validate the current expenditure of resources by the USDA for inspection of passenger baggage at foreign arrival airport terminals. However, the results also demonstrate that the pathway is not adequately closed and that more resources need to be expended to close this important pathway and build a stronger deterrent to the illegal activity which is now common."

The author proposes that more resources be expended in public education and information programs to reduce smuggling and inadvertent importation. The consensus of exclusion or quarantine officers is that a determined smuggler can find a way to bring in small quantities of plant material. However, high-risk and prohibited plant material can also be brought in by uninformed persons who might not be inclined to do so if they were aware of the regulations and agricultural risks. In the United States, such uninformed persons are given an opportunity to turn in plant material before handing in the customs form that includes a declaration as to whether the traveler is carrying agricultural materials.

Travelers would be more willing to comply if they were more aware of: (*a*) the reason for regulatory actions; (*b*) the risk to agriculture and the environment should an exotic pest or pathogen become established; and (*c*) the economic impact of such damage on travelers as consumers should agricultural raw materials be adversely affected.

Government quarantine service officers are well aware of what needs to be done in these public relations activities but are often constrained by limited funding. Program managers in allocating scarce resources to areas of highest known or perceived risk usually have funds only for short-term public relations activities in the form of press releases in response to some emergency situation (e.g. appearance of exotic fruit flies or bacteria that infect citrus). Either funds are not available for an all-out campaign or the cooperation of managers of radio and television stations, airports, and airlines is often difficult to enlist.

Notwithstanding, much can be done by government or agricultural university or experiment station scientists who publish articles in trade, garden, or plant hobbyists magazines, or speak to special interest groups. While some coverage is already given to exotic pest risks by some scientists, an active campaign is needed to reduce the tendency to smuggle plant material. For example, two occurrences of chrysanthemum white rust (caused by *Puccina horiana* Henn., a fungus of quarantine significance to the United States) have been associated with the importations of elite chrysanthemum plants or cuttings by hobbyist groups. One of these establishments was eradicated while the other is under an eradication program.

SUMMARY

Exclusion as a component of regulatory control, also known as plant quarantine or plant protection and quarantine, is a major disease control strategy. The goal of exclusion is to prevent or delay the entry of pest and pathogens along man-made pathways, with emphasis on those organisms that do not have a natural means of entry. Quarantine services determine the entry status of imported agricultural articles based on perceived or known risk. A biological position is that if the risk (i.e. chances of entry) is perceived to be high, the entry status should be conservative: conversely, if the risk is perceived to be low, the entry status should be liberal. The biological, geographical, economic, political, social, legal, and administrative basis of quarantine and their interactions are discussed.

Since regulatory actions sometimes adversely affect the scientific, commercial, and general publics, plant quarantine has a negative image. This image is reinforced when affected persons do not understand the biological basis of regulatory actions and consider them to be another form of bureaucratic red tape. These and other factors that contribute to the negative image are discussed and suggestions made to convert the negative image to a more positive one to enlist public support and cooperation. To make other exclusion practices more meaningful in preventing or delaying the entry of pathogens, educational programs should be increased to promote voluntary compliance by reducing smuggling and the incidence of travelers importing plant materials in ignorance of the regulations and risk involved.

Areas of research that could improve the efficacy of exclusion are discussed, including the need for quantitative risk assessment methodology, threshold determinations, information storage and retrieval, eradicant treatments, and the application of advanced pathogen detection and identification methodology to exclusion activities.

ACKNOWLEDGMENT

The author greatfully acknowledges the computerized information retrieval services furnished by Eileen Welch, and reviews by B. Glenn Lee, Joseph Foster, John Lightfield, and Frank Cooper, Plant Protection and Quarantine, Animal and Plant Health Inspection Service, USDA, Hyattsville, Maryland and Washington, DC.

Literature Cited

1. Agrios, G. N. 1980. Escape from disease. In *Plant Disease: An Advanced Treatise*. ed. J. G. Horsfall, E. B. Cowling, 5:18–37. NY: Academic. 534 pp.
2. Baker, C. R. B., Bailey, A. G. 1979. Assessing the threat to British crops from alien diseases and pests. In *Plant Health,* ed. D. E. Ebbels, J. E. King. pp. 43–54. Oxford: Blackwell. 322 pp.
3. Berg, G. H. 1977. Post-entry and intermediate quarantine stations. In *Plant Health and Quarantine in the International Transfer of Genetic Resources,* ed. W. B. Hewitt, L. Chiarappa, pp. 315–25. Cleveland: CRC. 347 pp.
4. Chiarappa, L. 1979. International collaboration for the study and control of plant diseases. *Rev. Plant Pathol.* 58: 391–98
5. Chock, A. K. 1979. International Plant Protection Convention. See Ref. 2, pp. 1–12
6. Chock, A. K. 1983. International cooperation on controlling exotic pests. In *Exotic Plant Pests and North American Agriculture,* ed. C. L. Wilson, C. L. Graham, pp. 480–98. NY:Academic. 522 pp.
7. Counc. Agric. Sci. Technol. 1987. Pests of plant and animals: their introduction

and spread. *Rep. No. 112*. Ames, Iowa: CAST, 40 pp.
8. Crooks, E., Havel, K., Shanon, M., Snyder, G., Wallenmaier, T. 1981. Stopping pest introductions. See Ref. 6, pp. 240–59
9. Curtis, B. C. 1986. Plant quarantine systems: the impact on international research. *Diversity*. Nov. 9:37–39
10. Div. Plant Ind. CDFA. 1990. Pest exclusion. *Calif. Plant Pest Dis. Rep.* 9:63–66
11. FAO. 1951. International Plant Protection Convention, FAO, Rome. 34 pp.
12. Foster, J. A. 1982. Plant quarantine problems in preventing the entry into the United States of vector-borne pathogens. In *Pathogens, Vectors, and Plant Diseases,* ed. K. F. Harris, K. Maramorosch, pp. 151–85. NY: Academic. 310 pp.
13. Foster, J. A. 1988. Regulatory action to exclude pests during the international exchange of plant germplasm. *Hort. Sci.* 23:60–66
14. Foster, J. A. 1991. Exclusion of plant pests by inspections, certifications, and quarantines. In *CRC Handbook of Pest Management in Agriculture*. 1:311–38. 2nd ed.
15. Gram, E. 1960. Quarantines. *In Plant Pathology,* ed. J. G. Horsfall, A. E. Dimond, 3:313–56. NY: Academic
16. Holdeman, Q., ed. 1986. *Plant Pests of Quarantine Significance to Importing Countries and States*. Sacramento: Calif. Dept. Food Agric. 152 pp. 5th ed.
17. Johnston, A. 1979. Information requirements for effective plant quarantine. See Ref. 2, pp. 55–61
18. Kahn, R. P. 1977. Plant quarantine: Principles, methodology, and suggested approaches. See Ref. 3, pp. 290–307
19. Kahn, R. P. 1979. A concept of pest risk analysis. *EPPO Bull.* 9:119–30
20. Kahn, R. P. 1982. The host as a vector: exclusions as a control. See Ref. 12, pp. 123–49
21. Kahn, R. P. 1985. Technologies to maintain biological diversity: assessment of plant quarantine practices. *Off. Technol. Assess. Food Renew. Res. Rep.* Washington, DC: US Congr. 77 pp.
22. Kahn, R. P. 1989. *Plant Protection and Quarantine,* Vol. 1. *Biological Concepts*. Boca Raton, FL: CRC. 226 pp.
23. Kahn, R. P., ed. 1989. *Plant Protection and Quarantine,* Vol. 2, *Selected Pests and Pathogens of Quarantine Significance*. Boca Raton, FL: CRC. 265 pp.
24. Kahn, R. P., Monroe, R. L. 1970. Virus infection in plant introductions collected as vegetative propagations: I. Wild vs. cultivated *Solanum* species. *FAO Plant Prot. Bull.* 18:97–101
25. Kahn, R. P., Monroe, R. L., Hewitt, W. B., Goheen, A. C., Wallace, A. C., et al. 1967. Incidence of virus detection in vegetatively propagated plant introductions under quarantine in the United States 1957–1967. *Plant Dis. Reptr.* 51:715–19
26. Kahn, R. P., Sowell, G. Jr. 1970. Virus infection in plant introductions collected as vegetative propagations: I. Wild vs. cultivated Arachis species. *FAO Plant Prot. Bull.* 18:142–44
27. Kahn, R. P., Waterworth, H. E., Gillaspie, A. G., Foster, J. A., Goheen, A. C., et al. 1979. Detection of viruses and virus-like agents in vegetatively propagated plants under quarantine in the United States, 1968–1978. *Plant Dis. Rptr.* 63:775–79
28. Karpati, J. F. 1983. Plant quarantine on a global basis. *Seed Sci. Technol.* 11: 1145–57
29. Kim, K. C. 1983. How to detect and combat exotic pests. See Ref. 6, pp. 262–319
30. Lee, E. B. 1989. Plant inspection stations: Concepts, facilities, and staffing. In *Plant Protection and Quarantine,* ed. R. P. Kahn, 3:195–207. Boca Raton, FL: CRC. 215 pp
31. Mathys, G. 1975. Thoughts on quarantine problems. *EPPO Bull.* 5:55–64
32. Mathys, G. 1977. Society supported disease management activities. In *Plant Disease,* ed. J. G. Horsfall, E. B. Cowling, 1:363–80. NY: Academic. 426 pp.
33. Mathys, G. 1977. European and Mediterranean Plant Protection Protection Organization. *FAO Plant Prot. Bull.* 24:152–56
34. Mathys, G., Baker, E. A. 1980. An appraisal of the effectiveness of quarantines. *Annu. Rev. Phytopathol.* 18:85–101
35. McCubbin, W. A. 1950. Plant pathology in relation to Federal domestic plant quarantines. *Plant Dis. Suppl.* 191:67–91
36. McCubbin, W. A. 1954. The plant quarantine problem. *Annu. Cryptogram. Phytopathol.* Vol. 11. 255 pp.
37. McGregor, R. C. 1973. *The Emigrant Pests. A Report to the Administrator.* APHIS, USDA, Hyattsville, Md. 167 pp.
38. McGregor, R. C. 1978. People-placed pathogens: The Emigrant pests. See Ref. 1, 2:383–95. NY: Academic
39. Morschel, J. R. 1971. *Introduction to*

Plant Quarantine. Canberra: Austr. Gov. Publ. Serv. 71 pp.

40. Morschel, J. R. 1973. *An Outline of Plant Quarantine*. Canberra: Aust. Dept. Health, Aust. Gov. Publ. Serv. 36 pp.
41. Morschel, J. R. 1979. Controlling the movement of plant diseases into, out of, and within Australia. See Ref. 2, pp. 35–42
42. Morschel, J. R. 1983. Quarantine procedures for seed imported into Australia. *Seed Sci. Technol.* 11:1231–37
43. Mulders, J. M. 1979. The International Plant Protection Convention—25 years old. *FAO Plan Prot. Bull.* 25:149–51
44. Nichols, C. W. 1984. *Principles of Plant Quarantine*. Presented at Ann. Meet. Natl. Plant Board, 58th, Cherry Hill, NJ
45. Parliman, B. J., White, G. A. 1985. The plant introduction and quarantine system of the United States. *Plant Breed. Rev.* 3:361–434
46. Plant Health Div. 1978–83. *Intercepted Plant Pests*. Ottawa: Agriculture Canada
47. Plant Prod. Prot. Div. 1984. *List of National Plant Quarantine Services,* AGPP:Misc/38, FAO, Rome. 113 pp.
48. Plant Prod. Prot. Div. 1989. Update on the International Plant Protection Convention (IPPC). *FAO Plant Prot. Bull:* 37:137–38
49. Plant Prot. Quarantine. 1987–1990. *Export Certification Manual*. Washington, DC: USDA
50. Plant Prot. Quarantine. 1980–84. *Lists of Intercepted Pests* APHIS 82–8 to 82–11, Washington, DC:USDA
51. Plant Protection and Quarantine, and Cooperating State Departments of Agriculture. 1982–1988. *Action Plans*. Washington, DC: USDA
52. Rainbow, A. F. 1977. Plant quarantine philosophy in New Zealand. *Comb. Proc. Ann. Meet. Int. Plant Prop. Soc.* 27:335–39
53. Rohwer, G. G. 1979. Plant quarantine philosophy of the United States. See Ref. 2, pp. 33–34
54. Royer, M. H. 1989. Global pest information systems—Can we make them work? See Ref. 30, pp. 37–58
55. Rosenberg, D. Y. 1989. The interaction of State and Federal quarantines. See Ref. 30, pp. 60–74
56. Santacroce, N. G. 1978. An evaluation of the USDA preshipment clearance program for propagative material. *EPPO Bull.* 8:67–72
57. Schoulties, C. L., Seymour, C. P., Miller, J. W. 1983. Where are the exotic disease threats? See Ref. 6, pp. 139–81
58. Smith, I. M. 1979. EPPO: the work of a regional plant protection organization, with particular reference to phytosanitary regulations. See Ref. 2, pp. 13–22
59. Sutherst, R. W., Maywald, G. F. 1985. Computerized system for matching climates in ecology. *Agric. Ecosyst. Env.* 13:281–99
60. Todd, J. M., Howell, P. J. 1979. The scientific basis of plant quarantine in the United Kingdom and Western Europe. See Ref. 2, pp. 79–86

Annu. Rev. Phytopathol. 1991. 29:247–78

MOLECULAR AND GENETIC ANALYSIS OF TOXIN PRODUCTION BY PATHOVARS OF *PSEUDOMONAS SYRINGAE*

Dennis C. Gross

Department of Plant Pathology, Washington State University, Pullman, Washington 99164

KEY WORDS: antibiotics, phytotoxins, plant signal molecules, plasmids, biosynthesis, gene cluster

---

INTRODUCTION

Plant pathogenic bacteria are equipped with an arsenal of virulence factors that first condition the host for colonization and then escalate the intensity of disease. Nonhost-specific toxins are generally acknowledged to be an element of virulence for many bacteria, and this is especially evident among the pathovars of *Pseudomonas syringae*. But considerable debate has focused on whether toxin production has a primary role in virulence because phytotoxic symptoms occur long after lesion formation (20, 75). Only recently with the advent of molecular genetic analysis of site-specific mutants of phytobacteria has the extent to which toxin production contributes to virulence been tested rigorously. Studies of nontoxigenic (Tox^-) mutants strongly implicate toxins in phytopathogenesis, but the functional roles of these genes associated with toxin production are unclear. Consequently, genetic research is beginning to probe the secrets of toxin biosynthesis and the regulatory network that links toxin production to expression of plant pathogenicity.

The nonhost-specific toxins produced by pathovars of *P. syringae* are a

0066-4286/91/0901-0247$02:00

family of structurally diverse compounds, usually peptide in nature, that in some cases display wide-spectrum antibiotic activity (73). The fact that most if not all pathovars of *P. syringae* have the genetic machinery to produce one or more phytotoxins suggests that there is strong selective pressure for toxigenic strains in the plant environment. Within the past few years chemical studies have revealed the structures of toxins produced by the more prevalent pathovars (10, 75). This is of particular significance because the structure of a toxin yields clues about the biosynthetic process and the mechanism of action. Because the toxins are likely to be synthesized by multistep pathways, the genetic organization will likely prove to be complex. Nevertheless, progress has been made in identifying gene clusters associated with toxin production (54, 65, 84, 92), and recent evidence (65, 80) suggests that some toxin genes are activated in response to specific plant signals.

Several superb reviews focus on the genetics (20, 90, 121), chemical properties (73, 75), and physiological roles (29, 40, 75) of bacterial phytotoxins. Although the relevance of toxin production to virulence was discussed with insight by Mitchell (75), recent genetic discoveries warrant the further examination of the implications of toxin production to plant pathogenicity. This review examines the genetic basis of toxin production by pathovars of *P. syringae* with the aim of linking toxigenicity to emerging principles inherent to plant-microbe interactions.

STRUCTURAL FEATURES OF TOXINS SUGGESTIVE OF THE MECHANISM OF BIOSYNTHESIS

Five structurally distinct classes of toxins that cause either chlorotic or necrotic symptoms in infected plant tissues are produced by pathovars of *P. syringae*. The chlorosis-inducing toxins are the most common among pathovars of *P. syringae* and are classified broadly as either tabtoxin, phaseolotoxin, coronatine, or tagetitoxin. In contrast, the necrosis-inducing toxins are restricted to pathovar *syringae* and encompass a family of structurally related lipopeptides called the syringomycins. Pathovars of *P. syringae* that produce a particular kind of toxin appear to constitute distinct taxonomic clusters that share a high degree of genomic DNA relatedness (26, 46). The overall genetic differences coupled with the apparent dichotomy in toxin biosynthetic pathways suggest that toxigenesis has evolved independently among the various toxigenic classes of pseudomonads. Accordingly, genes required for coronatine and phaseolotoxin biosynthesis are conserved among strains of pathovars known to produce the toxins, and the clones can be used as highly specific probes in disease diagnosis (23, 101).

One is amazed by the vast structural variation among the toxins from pathovars of *P. syringae* (Figure 1). Peptidolactone (syringomycins), polyketide (coronatine), monocyclic β-lactam (tabtoxin), sulpho-diamino-phosphinyl peptide (phaseolotoxin), and hemithioketal (tagetitoxin) structures are observed. Although little is known about their biosynthetic pathways, studies of antibiotics from *Streptomyces* and *Bacillus* that either structurally resemble one of the pseudomonad toxins or exhibit common chemical features can serve as a guide to the steps involved in assembly and the underlying genetic organization (56, 57, 68). In all cases, a nonribosomal pathway composed of a series of enzymatic steps involved in either peptide chain derivation or modification reactions are required for toxin assembly.

Syringomycins

The syringomycins are produced exclusively by strains of *P. s. syringae*, the most widespread pathovar in nature since it is recovered as a resident epiphyte from innumerable plant species and exhibits little restriction in host range as a pathogen (17). Structural analysis of syringomycins (10, 34, 49, 50, 102) reveal them to be cyclic lipodepsinonapeptides (Figure 1) that are members of the polypeptin class of antibiotics produced by a few strains of *Bacillus circulans* and *Erwinia herbicola* (7, 108). The syringomycins differ from the structurally related polymyxins produced by *Bacillus* spp. by virtue of 9, rather than 10, amino acid residues joined into a cyclic configuration by an ester linkage rather than an amide linkage (108).

The structure of syringomycin was resolved by chemical analysis of toxin preparations from strains of *P. s. syringae* isolated from such disparate hosts as pome fruit trees and grasses (10, 34, 102). The nonapeptide moiety of syringomycin contains two moles each of serine and 2,4-diaminobutyric acid (a constituent characteristic of polypeptin and polymyxin antibiotics (56, 108)), and one mole each of arginine, phenylalanine, and a trio of uncommon amino acids, dehydrothreonine, β-hydroxyaspartic acid, and 4-chlorothreonine. The N-terminal serine residue is linked by a lactone ring to the terminal carboxyl group of 4-chlorothreonine; serine also is acylated with either a 10-, 12- or 14-member 3-hydroxy fatty acid. Differences in fatty acid chain length have no apparent effect on the biological activity of syringomycin; physiological rather than genetic factors are presumed to be responsible for variations in chain length (118).

Syringostatin and syringotoxin are amino acid analogs of syringomycin from lilac and citrus strains of *P. s. syringae*, respectively (10, 33, 49, 50). Syringostatin has homoserine, ornithine, and threonine substituted for the arginine, phenylalanine, and one mole of serine contained in syringomycin (Figure 1). Syringotoxin differs from syringostatin solely by the substitution of glycine for one of the residues of 2,4-diaminobutyric acid (10). Syringotoxin also contains D-isomers of 2,4-diaminobutyric acid and homoserine,

CO—CH$_2$—CH(OH)—(CH$_2$)$_{10}$—CH$_3$

Ser—Ser—Dab—Dab—Arg—Phe—Dhb—(3-OH)Asp—(4-Cl)Thr

Syringomycin

CO—CH$_2$—CH(OH)—(CH$_2$)$_{10}$—CH$_3$

Ser—Dab—Dab—Hse—Orn—Thr—Dhb—(3-OH)Asp—(4-Cl)Thr

Syringostatin

CO—CH$_2$—CH(OH)—(CH$_2$)$_{10}$—CH$_3$

Ser—Dab—Gly—Hse—Orn—Thr—Dhb—(3-OH)Asp—(4-Cl)Thr

Syringotoxin

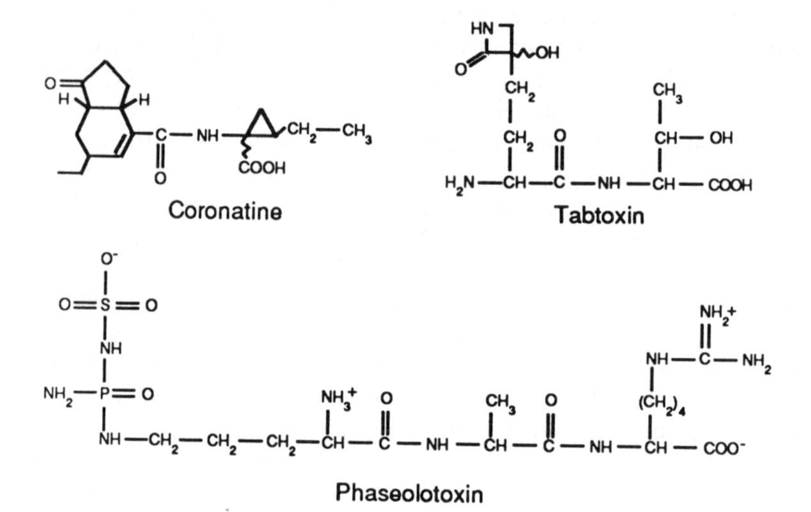

H$_3$C—C(=O)—O—CH—CH(OH)—C(OH)(COOH)—S

H$_3^+$N—CH—CH(O—HPO$_3^-$)—C(OH)(COOH)—CH$_2$

Tagetitoxin

Figure 1 Toxin structures. Abbreviations: Arg, arginine; (3-OH)Asp, 3-hydroxyaspartic acid; (4-Cl)Thr, 4-chlorothreonine; Dab, 2,4-diaminobutyric acid; Dhb, 2,3-dehydro-2-aminobutyric acid; Gly, glycine; Hse, homoserine; Orn, ornithine; Phe, phenylalanine; Ser, serine; Thr, threonine.

suggesting that D-amino acids are found in all analogs of syringomycin and contribute to toxin activity and stability (56). The overall structural similarities of syringomycin, syringostatin, and syringotoxin strongly suggest that they are biosynthesized by a singular mechanism, hence they are generically referred to as syringomycins.

Biosynthesis of the syringomycins likely occurs via a thioltemplate multienzyme mechanism, a model widely used to describe the synthesis of peptide antibiotics such as the polymyxins (55–57). According to the model, peptide synthetases, consisting of large multifunctional proteins between 100 and 500 kd, serve as templates for the sequential addition of amino acids. The associated enzymatic reactions include activation and stabilization of the carboxyl group of amino acids as thiolesters. A cofactor, 4'-phosphopantetheine, is covalently linked to the multienzyme and assists in the sequential polymerization and transfer of intermediates. In the synthesis of polymyxin, the N-terminal amino acid, 2,4-diaminobutyric acid, is activated and bound to a 300-kd activating enzyme, and then acylated by acyltransferase and octanoyl coenzyme A before peptide formation and cyclization (59).

Experimental evidence supportive of such a biosynthetic mechanism for the syringomycins is not yet available, but cells grown at conditions permissive to toxigenesis express up to five large proteins, ranging from about 130–470 kd in size (84, 124). Both protein formation and toxin production require iron at concentrations greater than the threshold needed to maintain cellular growth, which closely corresponds with the iron regulatory effects on antibiotic biosynthesis in many bacteria (120). Furthermore, Tox^- transposon mutants of *P. s. syringae* commonly are deficient either in one or more of the large proteins or form truncated proteins. Several steps would be required for the formation of a lipodepsinonapeptide such as syringomycin. Based on the polymyxin biosynthetic model (56, 59), the first step likely would be the activation and binding of the N-terminal serine to one of the syringomycin synthetases. The amino group of serine would then be acylated to form a conjugate with 3-hydroxydodecanoic acid prior to elongation of the peptide chain. Reactions associated with activation of the carboxyl group of substrate amino acids by adenylation, stabilization of peptide intermediates as thiolesters, and phosphopantetheine-mediated transfer of intermediates between multienzymes would likely resemble the general biosynthetic mechanism of peptide antibiotics as detailed in the review of Kleinkauf & von Döhren (56, 57). Several modification reactions would also be associated with synthetase activity including the epimerization of some L-amino acids to the D-form (56), the formation of dehydroaminobutyrate via β-elimination from threonine (56), chlorination of threonine by a chloroperoxidase (66), and cyclization to form a lactone ring by linking the hydroxyl group of serine with the carboxyl group of 4-chlorothreonine.

The lipopeptide structure of syringomycin permits it to insert within the lipid membrane layers. Accordingly, several effects of syringomycin on ion transport across the plasmalemma of plant and fungal cells have been identified (15, 16, 86, 128), including hyperpolarization, rapid efflux of K^+, and stimulation of a proton pump ATPase. The various membrane effects are thought to be caused by the activation of a protein kinase, resulting in the phosphorylation of several proteins including the proton pump ATPase (15). Minute doses of syringomycin close stomata of leaves by activating a K^+ export system similar to the way abscisic acid regulates stomatal closure (86). Thus, syringomycin does not appear to cause a general breakdown of membrane integrity as suggested by early studies (8), but rather the rapid and prolonged ion transport effects would be lethal to affected cells. Similar membrane responses occur during the hypersensitive reaction of tobacco induced by *P. s. syringae* (9), but the effects are not attributed to syringomycin because Tox^- strains also induce hypersensitivity (123, 124). Atkinson & Baker (5, 6) also noted the occurrence of the K^+/H^+ exchange response during disease development in bean. A fundamental difference is that the initial exchange rate is lower or delayed in diseased tissues, thus permitting the systematic release of cellular metabolites into the intercellular spaces to support a crescendo of bacterial growth. Considering these and other observations (16, 128), the apparent benefit of syringomycin production during phytopathogenesis is that the cellular release of K^+ would raise the relatively acidic pH of intercellular fluids to near neutrality, an environment more conducive to bacterial multiplication. Furthermore, interference with active transport would lead to the accumulation of sucrose, amino acids, and inorganic ions within the intercellular spaces.

Coronatine

Coronatine was first associated with chlorosis of Italian ryegrass caused by *P. s. atropurpurea,* (48), and has since been implicated in diseases caused by *P. s. tomato, P. s. glycinea, P. s. morsprunorum,* and *P. s. maculicola* (23, 75). The phytotoxin is composed of an α-amino acid, coronamic acid, linked by an amide bond to a polyketide moiety, coronafacic acid (Figure 1). Mitchell (74) and Mitchell & Frey (77) analyzed the carboxylic acid fraction from cultures of *P. s. atropurpurea* and identified minor quantities of naturally occurring L-amino acid analogs of coronatine that suggested a possible biosynthetic pathway for coronamic acid. Isoleucine resembles coronamic acid, and its addition to cultures yielded *N*-coronafacoyl-L-isoleucine that was then converted to coronatine upon formation of a cyclopropyl ring. Coronafacic acid was coupled nonspecifically with L-amino acids having aliphatic constituents (e.g. L-alanine, L-α-aminobutyric acid, L-norvaline), but amino acids with longer side chains (e.g. L-isoleucine) were incorporated at a greater efficiency. Except for *N*-coronafacoyl-L-isoleucine, none of the *N*-coronafacoyl-

L-amino acids were cyclized to form the 2-ethylcyclopropane ring (77). Cyclization has a strong effect on biological activity since *N*-coronafacoyl-L-isoleucine is less active than coronatine (32, 105). The enzymatic steps involved in cyclization are unknown.

Coronafacic acid is synthesized from a branched polyketide composed of five molecules of acetate (91). A novel feature in the synthesis of one of the polyketide chains is that pyruvate rather than acetate serves as the starter unit. The biosynthetic mechanism for polyketide antibiotics from *Streptomyces* spp., such as tetracenomycin C (14) and granaticin (103), may provide key information concerning the enzymology of coronafacic acid biosynthesis. Polyketide biosynthesis is similar to fatty acid biosynthesis except for the frequent lack of reduction of β-ketoacyl intermediates in the polyketide pathway, and the lack of restriction to using acetate as starter and a malonate unit for every extender (14, 47). There are several steps involved in polyketide assembly, including condensation, reduction, dehydration, and release of acylthioester intermediates. In polyketide-producing bacteria a dissociable multienzyme complex functionally related to the type II fatty acid synthase of *Escherichia coli* is most frequently encountered (14, 103).

The structural resemblance of coronamic acid to the ethylene precursor 1-aminocyclopropane-1-carboxylic acid suggested that toxin-induced chlorosis resulted from interference with the regulation of ethylene production (75). Not surprisingly, coronatine at concentrations as low as 5×10^{-7} M stimulated ethylene production in bean leaf discs (32). Stimulation of the methionine pathway prior to conversion of 1-aminocyclopropane-1-carboxylic acid to ethylene, and not release of endogenous ethylene bound to tissue or the conversion of coronatine or coronamic acid to ethylene, was responsible for the induction. Both chlorosis and ethylene production were induced by coronafacoylvaline, a naturally occurring analog of coronatine, but neither coronamic acid nor coronafacic acid were active. These observations suggested that the 2-ethylcyclopropyl ring is not necessary for activity, but rather that the amide bond coupling coronamic acid to coronafacic acid is essential. It is uncertain whether ethylene stimulation is directly involved in chlorosis or whether the response is an indirect consequence of coronatine toxicity. Because coronatine also is reported (98) to cause auxin-like hypertrophic growth in plant tissues, stimulation of ethylene may have resulted from enhanced effects of auxin on ethylene biosynthesis. This is unlikely to be resolved until the specific biochemical mechanism by which coronatine causes phytotoxicity is defined.

Tabtoxin

Tabtoxin is a monocyclic β-lactam produced by at least three pathovars, namely *P. s. tabaci, P. s. coronafaciens,* and *P. s. garcae* (75). The dipeptide toxin contains the unusual amino acid tabtoxinine-β-lactam linked by a peptide bond to either threonine or serine to form, respectively, tabtoxin (Figure 1) and its analog (2-serine)tabtoxin. The chlorosis-inducing activity occurs only after hydrolysis of the peptide bond to liberate tabtoxinine-β-lactam, presumably by exposure to peptidases of either bacterial or plant origin (29, 30).

The biosynthetic precursors for tabtoxin, identified by incorporation of ^{13}C-labelled compounds, consist of L-threonine and L-aspartate for the side chain and pyruvic acid and the methyl group of L-methionine for the β-lactam moiety (97, 117). The proposed biosynthetic model for the tabtoxinine-β-lactam backbone resembles that of lysine, whereby there is a condensation of pyruvate and aspartate semialdehyde. Tabtoxinine-β-lactam synthesis proceeds along the lysine pathway until an unknown intermediate is reached; branching off in the pathway occurs prior to the formation of 2,6-diaminopimelic acid (97). The steps involved in formation of the β-lactam ring are unknown for tabtoxin, and comparisons to possible biosynthetic mechanisms of β-lactam antibiotics in actinomycetes offer few clues as to how the β-lactam constituent of tabtoxin might be derived. For example, although the lysine pathway is involved in the biosynthesis of cephamycin (88), the precursor amino acid α-aminoadipic acid is a catabolic derivative of lysine.

Tabtoxinine-β-lactam irreversibly inhibits glutamine synthetase, which leads to the accumulation of toxic concentrations of ammonia to cause chlorosis. The assortment of harmful effects attributed to ammonia include disruption of the thylakoid membrane of the chloroplast and uncoupling of photophosphorylation (29, 75, 115). Activity of tabtoxinine-β-lactam is lost upon the opening of the β-lactam ring following exposure to β-lactamase to yield tabtoxinine (58).

Phaseolotoxin

Production of phaseolotoxin is restricted to *P. s. phaseolicola*, cause of halo blight of a wide range of legumes. The tripeptide structure of phaseolotoxin was elucidated by Mitchell (72), with minor revision by Moore et al (82), as N^{δ}-(N'-sulpho-diaminophosphinyl)ornithylalanylhomoarginine (Figure 1). In addition, lesser amounts (about 5–10%) of a 2-serine analog of phaseolotoxin are consistently produced by strains of *P. s. phaseolicola* (75).

Much is known about the chemistry and mode of action of phaseolotoxin and its contribution to phytopathogenicity, but little is known about its biosynthesis. Because the N^{δ}-(N'-sulpho-diaminophosphinyl)-L-ornithine portion, called octicidin (82), is biologically active and present in all amino acid analogs of phaseolotoxin, a critical step in biosynthesis is the formation of the N^{δ}-(N'-sulpho-diaminophosphinyl) moiety and its linkage to the δ-amino of ornithine. Unfortunately, the biosynthetic precursors of octicidin have not been identified, and there is no precedent for the formation of such a chemical species within the broad spectrum of antibiotics produced by microorganisms. The homoarginine and ornithine moieties of phaseolotoxin are synthesized by a transamidination reaction from arginine and lysine (67). An amidinotransferase with a M_r of about 200,000 and with high substrate affinities for arginine and lysine is restricted to phaseolotoxin-producing strains of *P. s. phaseolicola*. The enzyme from *P. s. phaseolicola* biochemically resembles the amidinotransferase isolated from *Streptomyces griseus*, which is a key enzyme in streptomycin biosynthesis (68).

The octicidin portion of phaseolotoxin binds to ornithine carbamoyltransferase to competitively inhibit binding of the structurally related carbamoyl phosphate, thus blocking conversion of ornithine to citrulline, a precursor of arginine (75). The ensuing chlorosis is due to reduced chlorophyll synthesis, an indirect consequence of a depleted arginine pool in developing leaves (114). The chlorotic effects induced by phaseolotoxin can be readily reversed simply by the application of arginine to affected tissues. Intact phaseolotoxin is less biologically active than octicidin (75) and, accordingly, octicidin is the primary toxic entity in diseased tissues (76). Peptidases of plant origin sequentially cleave homoarginine and alanine from phaseolotoxin to release the more potent octicidin.

Tagetitoxin

Tagetitoxin is the most recent pseudomonad toxin described. It is produced only by *P. s. tagetis*, a pathogen of several members of the Compositae family including marigolds and sunflowers (75). Mitchell & Hart (78) proposed a novel eight-membered hemithioketal ring structure for tagetitoxin that bares no obvious resemblance to the peptide-containing pseudomonad toxins (Figure 1). Biosynthetic precursors have not been identified for tagetitoxin, and it is premature to speculate about its biosynthesis.

The biochemical mechanism by which tagetitoxin causes the conspicuous apical chlorosis in susceptible plants is unknown. Leaves treated with tagetitoxin do not accumulate either the large or small subunits of ribulose 1,5-bisphosphate carboxylase, presumably due to an inhibition of gene transcription resulting from the marked depletion in plastid ribosomes and their component rRNAs (64). These results together with the assortment of ultrastructural and physiological changes that occur in plastids indicate that the toxin may have a general role in the repression of chloroplast genes. Consequently, only developing leaf tissues are affected and the phytotoxic response is reversed by removal of the toxin.

CLONING OF GENES INVOLVED IN TOXIN PRODUCTION

Genes required for the production of syringomycin (syringotoxin), coronatine, tabtoxin, and phaseolotoxin have been cloned and are now being functionally analyzed (12, 52, 54, 81, 83, 92, 124). The standard approach has been to isolate Tox$^-$ mutants by transposon mutagenesis and identify the relevant gene from a genomic library of the toxigenic strain. It is common for antibiotic resistance genes, responsible for encoding a mechanism of self protection from toxic activity, to be physically linked to biosynthetic genes (68). Consequently, biosynthetic genes can be cloned along with genes required for toxin resistance, which is an easily selectable phenotype. Alternatively, "reverse genetics" can be used whereby a DNA probe is synthesized based on an amino acid sequence from the purified protein product of the toxin gene and used to screen a genomic library for homologous sequences (70).

Mutational Cloning of Toxin Genes

Nontoxigenic mutants are obtained at a seemingly high frequency by random transposon mutagenesis of various pathovars of *P. syringae*. For example, Tn*5* mutagenesis of *P. s. syringae* strains that produced either syringomycin or syringotoxin yielded Tox$^-$ phenotypes at a frequency of about 0.3% (83, 123). Similar high rates of nontoxigenicity among Tn*5* mutants were reported after mutagenesis of strains that produced either coronatine (1.3%), tabtoxin (0.3%) or phaseolotoxin (0.1%) (54, 81, 92). Although a large proportion of the Tox$^-$ mutants arise due to mutations in toxin-biosynthetic genes, mutations in genes affecting precursor formation, toxin transport, or the regulation of genes for toxin biosynthesis also could lead to a Tox$^-$ phenotype. Amino acid auxotrophy is one example of a mutation that reduces the intracellular pool of precursors necessary for toxigenesis. In *P. s. syringae*, an assortment of auxotrophies for *ser, arg, cys, met,* and *trp* were either Tox$^-$ or low producers of syringotoxin (83).

Hypoproducers and hyperproducers are two other classes of mutants that, respectively, produce lower and higher quantities of toxin. Hypoproducers are recovered after transposon mutagenesis at relatively high frequencies. They can arise due to one of several possibilities, including auxotrophy, reduced cellular growth, impaired toxin transport, and inactivation of genes that regulate formation of toxin precursors or synthetases (20). For example, 2.8% of the Tn*5* mutants of a syringomycin-producing strain were hypoproducers (123). The syringomycin hypoproducers were largely indistinguishable from the wild-type strain in virulence, which further suggests that a large proportion of the genes affected have a peripheral role in toxin production. A similar conclusion can be drawn from studies by Peet et al (92) of mutants of *P. s. phaseolicola* that proved, after modifications of the bioassay method, to produce low amounts of residual phaseolotoxin.

Thus far, only syringomycin hyperproducers have been reported (123). Approximately 1.2% of all Tn*5* mutants of *P. s. syringae* in bioassays with the syringomycin-sensitive *Geotrichum candidum* produced zones of inhibition on agar media that were nearly double in size. Although hyperproducing strains might be expected to exhibit greater virulence, in fact, they were either indistinguishable from the wild type or markedly reduced in virulence. Therefore, one should not presume that hyperproducers are of trivial interest in elucidating the toxigenic mechanism. Mutation of a negative regulatory gene controlling syringomycin synthesis would give elevated toxin levels if a functional synthetase system is in place and adequate biosynthetic precursors are available. Repressors of antibiotic biosynthetic genes occur, for example, in *Streptomyces griseus* and apparently fine-tune expression of the biosynthetic gene cluster for streptomycin (68). Some hyperproducers of syringomycin exhibit less than 10% of the virulence of the wild-type strain, suggesting that the negative regulator likewise controls other virulence genes (N. B. Quigley & D. C. Gross, unpublished data). Recently, Tang et al (113) showed that a negative regulatory gene in *Xanthomonas campestris* pv. *campestris* coordinately controlled production of several factors associated with virulence, including production of extracellular polysaccharide (EPS) and a variety of extracellular enzymes. Although relative levels of extracellular enzymes and EPS were increased threefold in some cases, the mutant caused less disease in turnip leaves. The specific mechanism by which negative regulatory genes modulate expression of virulence in both *Pseudomonas* and *Xanthomonas* may prove to be a key element in the global regulatory network controlling pathogenicity.

Identification of the above three mutant classes—Tox$^-$, hypoproducers, and hyperproducers—requires a sensitive assay for detecting altered levels of toxin production. Microbiological assays are convenient and widely used to rapidly screen large numbers of mutant colonies for changes in toxin production. The exceptions are for coronatine and tagetitoxin that do not exhibit antimicrobial activity and must therefore be painstakingly screened for ability to elicit symptoms in plant tissues. Phaseolotoxin production can be detected in picogram quantities by growth inhibition of *E. coli* strain K-12, levels substantially lower than those required in assays for in vitro inhibition of OCTase and in leaf bioassays (110). One possible limitation in using *E. coli* for bioassay is that octicidin, which lacks the tripeptide backbone, exhibits only 0.01% of the activity of phaseolotoxin (79). This is because the oligopeptide permease of *E. coli* responsible for uptake of phaseolotoxin does not transport octicidin (75). Accordingly, Tox$^-$ phenotypes would be scored for colonies able to produce octicidin, the primary and more toxic form found in diseased tissues (76).

Microbiological assays for syringomycin production present a different problem. Although highly sensitive, *G. candidum* is less sensitive to the toxin than leaf tissues (41), and therefore weakly toxigenic mutants may be scored erroneously as Tox$^-$. Unfortunately, no biochemical assay is available for detecting minute quantities of syringomycin or precursors of the toxin. In genetic studies of coronatine production, organic acid extracts from cultures can be assayed by reverse-phase high pressure liquid chromatography for the presence of coronatine or its nonphytotoxic biosynthetic intermediate, coronafacic acid (11, 81). Although this method is too laborious for extensive screening of random transposon mutants of wild-type strains for Tox$^-$ phenotypes, it does clarify whether such phenotypes identified in bioassay fail to produce one or more of the coronafacoyl compounds. In addition, either organic acid culture extracts or the mutant pseudomonad strains themselves can be screened for ability to induce hypertrophy on potato slices, a presumptive test for coronatine production reportedly more sensitive than bioassays for chlorosis of leaves (36).

Mutational cloning may be confounded by the spontaneous loss of genes required for toxigenesis. Chromosomal regions required for tabtoxin production appear to be very unstable. Independent studies by Turner & Taha (116) and Kinscherf et al (54) estimated that the spontaneous Tox$^-$ mutants arise at a frequency of about 10^{-3} in culture. A similar problem occurs in mutational cloning with coronatine-producing strains. Because the polyketide biosynthetic genes are located on large (90 to 105 kb) plasmids in nearly all strains (13), Tox$^-$ strains can simply arise due to plasmid loss or its partial deletion. Bender et al (11) analyzed 38 chlorosis-defective mutants of *P. s. tomato* strain PT23.2; only five mutants contained Tn*5* in plasmid pPT23A required for coronatine biosynthesis, 29 were lacking pPT23A altogether, and four contained deletion derivatives of pPT23A.

Cloning of Biosynthesis Gene Clusters

Genes encoding enzymes responsible for antibiotic biosynthesis are usually organized in clusters on the bacterial chromosome or on plasmids. Clusters of biosynthesis genes have been reported for several classes of antibiotics produced by Streptomyces and *Bacillus* spp., including β-lactams such as cephamycin C (68, 107), polyketides such as granaticin (103), and peptides such as gramicidin S (60). Genes required for the synthesis of syringomycin (syringotoxin), coronatine, tabtoxin, and phaseolotoxin likewise appear to be organized in one or more clusters (54, 65, 80, 84, 92) that range from ~7 kb for a cluster of *syr* genes involved in syringomycin production (N. B. Quigley, Y.-Y. Mo & D. C. Gross, unpublished) to ~32 kb for a cluster of genes associated with formation of large proteins hypothesized to function as syringotoxin synthetases (84).

Genes conferring resistance to the antibiotic generally are linked with antibiotic gene clusters (22), suggesting that their expression is highly integrated to prevent autotoxicity and facilitate a quick response by the bacterium to a fluctuating environment. Genes associated with phaseolotoxin production are linked within a ~22-kb DNA clone to a gene, *argK*, encoding an OCTase that confers immunity to the toxin in *P. s. phaseolicola* (85, 93). Phaseolotoxin immunity is also expressed in *E. coli* HB101 upon introduction of cosmid clones containing the phaseolotoxin gene cluster, a situation that permits rapid screening of cosmid libraries of genomic DNA. Based on the amino acid sequence, the OCTase resembles the catabolic OCTase of *Pseudomonas aeruginosa* and is distinct from the toxin-sensitive OCTase also found in *P. s. phaseolicola* that is encoded by a gene not linked to those associated with phaseolotoxin production (71). Correspondingly, tabtoxin resistance is physically linked with a tabtoxin biosynthetic cluster spanning ~25 kb (54). Although the cosmid-cloned tabtoxin cluster neither expressed tabtoxin production nor resistance in *E. coli*, its transfer to strain Cit7 of *P. s. syringae* or an "angulata" strain of *P. s. tabaci* resulted in expression of both properties. Anzai et al (4) characterized a tabtoxin-resistance gene (*ttr*) within a 700-bp fragment that encoded an acetyltransferase, which presumably acts by acetylating the amino group of the tabtoxinine β-lactam moiety.[1] There also is circumstantial evidence that a gene(s) for β-lactamase, an enzyme that hydrolyses the monocyclic β-lactam ring to yield tabtoxinine, is part of the tabtoxin cluster because Tox^- strains lack β-lactamase activity (58).

Although uncharacterized, autoresistance likely occurs for strains of *P. s. syringae* that produce syringomycin or one of its amino acid analogs, based on its resistance to the final concentration of toxin that accumulates in culture media (42).

GENETIC FEATURES OF TOXIN PRODUCTION

Large Protein Complexes Associated with the Biosynthesis of the Syringomycins

Production of the syringomycins by *P. s. syringae* is associated with the formation of large protein complexes ranging up to 470 kd (84, 124). Although direct enzymatic evidence is lacking, genetic analysis of Tox^- mutants suggests that the proteins may function as synthetases fundamentally similar to the large polyenzymatic protein complexes responsible for the biosynthesis of the classical peptide antibiotics. This is perhaps best ex-

[1]However, it was recently shown that the *ttr* gene does not contain sequences homologous to the tabtoxin biosynthetic cluster (54).

emplified by the work of Morgan & Chatterjee (84), who showed that two large proteins of ~470 and ~435 kd, respectively, and called ST1 and ST2, were formed only when strain B457 was grown at conditions permissive to syringotoxin production. Subsequent analysis of 13 Tox$^-$ Tn*5* mutants of B457 revealed alterations in the presence of either ST2 or both ST1 and ST2. Truncated forms of ST1 or ST2 were detected in several instances, presumably due to translational stops within Tn*5*. It appears that the structural genes for ST1 and ST2 are located on a DNA segment spanning ~32 kb, and that the translational start site for ST2 is located near the C-terminal end of ST1. The region was subjected to Tn*3*HoHo1 mutagenesis to generate *lacZ* fusions for subsequent use in complementation analysis with the various Tn*5* mutants (19). Surprisingly, four possible transciptional units were identified; two (T1 and T2) were associated with ST1 and another two (T3 and T4) with ST2. This suggests that both ST1 and ST2 are multimeric proteins since more than one gene product is required for their formation.

Large proteins are likewise associated with syringomycin production, although they differ in number (five) and size (ranging from ~130 to ~470 kd) compared to those associated with syringotoxin production (124). Protein analysis of most Tox$^-$ mutants obtained by Tn*5* mutagenesis of the syringomycin-producing strain B301D consistently shows protein alterations, such as the occurrence of truncated forms of the ~470-kd protein (Y-Y. Mo & D. C. Gross, unpublished data) or deficiencies in one or more protein species (124). Three syringomycin genes, called *syrB*, *syrC*, and *syrD*, are clustered within a ~7-kb DNA segment (N. B. Quigley, Y.-Y. Mo & D. C. Gross, unpublished data). Transposon insertions in either *syrB* or *syrC* cause a deficiency in two proteins of ~350 and ~130 kd (80), called SR4 and SR5, respectively. The *syrB* and *syrC* genes are about 3.1 and 1.8 kb in size, respectively, with *syrC* located immediately downstream of *syrB*. Furthermore, *syrB* exerts polar effects on the expression of *syrC* similar to the polar effects associated with expression of the ST1 and ST2 proteins in syringotoxin-producing strains. Because neither *syrB* nor *syrC* are physically large enough to directly encode either SR4 or SR5, it remains to be determined whether the SyrB and SyrC proteins might somehow be covalently linked as a multimeric protein. Alternatively, the *syrB* and *syrC* genes might regulate expression of the SR4 and SR5 proteins.

A *syrD* mutant is deficient in all five proteins associated with syringomycin production (N. B. Quigley, Y-Y. Mo & D. C. Gross, unpublished data). The *syrD* gene is ~1.2 kb in size and is divergently transcribed relative to *syrB*. Evidence suggests that *syrD* is a regulatory gene since it is responsible for about a threefold increase in expression of a fusion between *syrB* and a promoterless *lacZ* gene. One can therefore speculate that the *syrD* gene product coordinately regulates expression of *syr* genes that encode the other three large proteins associated with syringomycin production.

Considering the structural similarities shared by syringomycin and syringotoxin, it is unclear why the large proteins postulated to have synthetase activity would differ noticeably in size and number in these two systems. Indeed, DNA sequences associated with the formation of either ST1 or ST2 found in a syringotoxin-producing strain share homology with sequences found in syringomycin-producing strains (M. M. Morgan & A. K. Chatterjee, personal communication); moreover, a *syrB*::*lacZ* fusion can be introduced into the genome of a syringotoxin-producing strain by marker exchange confirming that *syrB* sequences are highly conserved (124). Nevertheless, there are indications that significant differences exist in the genetic organization responsible for the biosynthesis of the two toxins as evidenced by syringotoxin production at wild-type levels by a mutant obtained by marker exchange of a *syrB*::*lacZ* fusion into the chromosome. It also appears that *syr* genes are located in at least two gene clusters not amenable to cloning as a single biosynthetic unit.

It is easy to envision a kinetic advantage for cells to synthesize the syringomycins on multienzymatic protein complexes because intermediates would be directly transferred between enzymes and not released into the cytoplasm (68). It is less obvious why the synthetase genes themselves appear to be organized in tight clusters on the chromosome. An attractive explanation is that the clustering of genes encoding related functions into operons permits coordinate regulation of expression of the whole pathway.

An operative model for the gene organization of the syringomycin synthetases may be derived by comparisons to the granaticin synthetase gene cluster of *S. violaceoruber* (103). Short nucleotide overlaps occur between the 3' ends of open reading-frame (ORF)1 and ORF5 with the 5' ends of ORF2 and ORF6, respectively. Such gene arrangements facilitate translational coupling, a phenomenon observed in many bacterial operons to ensure equimolar production of two gene products (126). Because each ORF encodes a protein component of the granaticin synthetase complex, translational coupling promotes an ordered assembly into an oligomeric protein (103). The possible occurrence of an overlap between the 3' end of *syrB* and the 5' end of *syrC* is suspected based on mapping of transposon mutations and tests for functional complementation of *syrB* and *syrC* mutants with subclones (80; Y-Y. Mo & D. C. Gross, unpublished data). Moreover, physical evidence for a linkage between the *syrB* and *syrC* gene products was observed in Western blots when a translational *syrC*::*lacZ* fusion was analyzed. A ~300 kd fusionprotein was observed that was surmised to contain the 116-kd *B*-galactosidase protein fused to SyrB and the N-terminal portion of SyrC. Thus, it appears that the *syrC* gene can be transcribed either by readthrough of the *syrB* gene or from its own promoter independent of *syrB* expression. Nucleotide sequence analysis of possible gene overlaps should help resolve intergenic relationships between structural genes encoding syringomycin synthetases.

Plasmids and Coronatine Biosynthesis

The spontaneous loss of toxigenicity long has been observed to occur at relatively high frequencies when some strains of *Pseudomonas* are subcultured and maintained in vitro (31, 41, 116). This suggests that a mechanism other than simple point mutation is responsible. Early studies reported a correlation between the occurrence of plasmids with production of either syringomycin (37) or phaseolotoxin (35), but more rigorous genetic analysis failed to validate plasmid involvement (24, 51, 95). Plasmids also were excluded as genetic determinants of tabtoxin production based on the retention of a toxigenic phenotype after loss of all plasmid DNA sequences (89). Nevertheless, cryptic plasmid species are diverse and widespread among strains of several *P. syringae* pathovars (21). The frequent integration and imprecise excision of plasmids into the chromosome of *P. s. phaseolicola*, due to recombination at homologous repetitive sequences (112), suggests that plasmids confer plasticity to strains. Up to ~130 kb of chromosomal DNA sequences were found by Szabo & Mills (111) in an excision derivative of pMC7105 from *P. s. phaseolicola*. Thus, one cannot rule out the possibility that one or more genes required for production of phaseolotoxin, tabtoxin, or the syringomycins reside on plasmids in some strains.

An association of plasmid DNA with coronatine production was described initially by Sato et al (100) for *P. s. atropurpurea*; this was later confirmed by conjugal transfer of the pCOR1 plasmid in planta to Tox$^-$ mutants and retrieval of a toxigenic (Tox$^+$) phenotype (99). More extensive surveys (13) showed that the coronatine biosynthetic genes (*cor* genes) are clustered on large plasmids of ~100 kb in most strains of pathovars *tomato, glycinea, atropurpurea,* and *morsprunorum*. Three unique *cor*::Tn*5* clones contained within the 30-kb *cor* cluster from plasmid pPT23A of *P. s. tomato* were used to demonstrate conserved homology in coronatine plasmids of the various pathovars. Subsequent marker exchange of *cor*::Tn*5* sequences yielded mutants that failed to produce coronafacic acid, coronafacoylvaline, and coronatine. In addition, coronatine production was conferred upon transfer of pPT23A in vitro to a strain of *P. s. syringae* (11). Although it is unknown whether the transconjugant strain also produced syringomycin, the ability to produce two phytotoxins forseeably could increase virulence. An argument against such transfer occurring in nature is that coronatine production is strictly limited to specific pathovars that do not produce either syringomycin, phaseolotoxin, or tabtoxin. Cuppels et al (23) used a 5.3-kb sequence from the *cor* gene cluster as a DNA hybridization probe and showed exclusive hybridization to all 244 field-isolates of *P. s. tomato* and representative strains of pathovars *glycinea, atropurpurea,* and *maculicola*; several strains of *P. s. syringae* showed no homology to the probe along with strains from other coronatine-negative pathovars.

Plasmids frequently carry genes that encode traits that give greater flexibility to the bacterium in responding to changes in the environment (21). Accordingly, there would be strong selective pressure for horizontal transfer of a coronatine-encoding plasmid, since coronatine production promotes lesion expansion and a corresponding multiplication of the bacterium (12). This simplistic theory does not answer why *P. s. syringae*, which spatially occupies the same epiphytic niche as coronatine-producing pathovars (23, 26), does not obtain the coronatine plasmid in nature. Moreover, the *cor* gene cluster is chromosomally determined in strain DC3000 of *P. s. tomato* (81) and in many strains of *P. s. maculicola* (D. A. Cuppels, personal communication). Evidence (26) that there is less than 50% DNA-DNA homology between strains of *P. s. tomato* and *P. s. syringae* further suggests that the two pathovars are evolving independently with little or no genetic exchange.

Six complementation groups spanning over 30 kb of DNA were observed by Ma et al (65) for the *cor* cluster from strain DC3000. Not surprisingly, the *cor* gene products are restricted to the cytoplasm since none of the *cor*::TnphoA fusions throughout the 30-kb cluster produced membrane proteins. The genes required for formation of coronafacic acid are associated with the *cor* cluster as judged by absence of the precursor in organic acid extracts from Tox^- mutants. Because coronafacic acid is a polyketide, the synthesis of which presumably resembles that of fatty acids in bacteria (47, 68), a multifunctional polyketide synthase(s) can be hypothesized to be encoded by the *cor* genes. Nucleotide sequence analysis of polyketide synthase genes in *Streptomyces* has uncovered high amino acid homologies to type II fatty acid synthases, and has yielded clues as to possible gene function (14, 47, 103). Such analyses may also help resolve the biosynthetic functions of several *cor* genes.

Spontaneous Deletion of the Tabtoxin Gene Cluster

A curious phenomenon associated with tabtoxin-producing strains is the spontaneous deletion, at a frequency of $\sim 10^{-3}$, of a ~25-kb region containing perhaps all of the tabtoxin biosynthetic gene cluster together with genes for resistance to tabtoxinine-β-lactam (121). Using genes cloned from the tabtoxin gene cluster of bean wildfire strain BR2, Kinscherf et al (54) found conserved sequences in all tabtoxin-producing strains including representatives of pathovars *tabaci, coronafaciens,* and *garcae*. Comparable sequences were absent from all spontaneous Tox^- mutants, as well as from "angulata" and "striafaciens" strains thought to be naturally occurring Tox^- members of *P. s. tabaci* and *P. s. coronafaciens*, respectively, that have retained pathogenicity. Cosmid pRTBL823 converted Tox^- strains, including "angulata" and "striafaciens" strains, to Tox^+. This demonstrates that the entire tabtoxin biosynthetic gene cluster is located on one contiguous DNA fragment of about 25 kb. The pRTBL328 clone will prove invaluable in delineating the physical and functional organization of tabtoxin genes of which little is known. It can also be used to resolve the mechanism responsible for deletion, including the possibility that excision is facilitated by recombination between repetitive sequences flanking the tabtoxin gene cluster.

What might be the biological significance for the spontaneous deletion of the tabtoxin gene cluster at such a high frequency? One can argue that such spontaneous deletions commonly occur in nature as exemplified by the occurrence of "angulata" and "striafaciens" strains. Perhaps the tabtoxin gene cluster is contained within a mobile genetic element responsible for horizontal dissemination of toxigenicity to related plant-associated pseudomonads. Evidence that the cluster is excised precisely, with the deleted sequence varying less than 200 bp in length, is consistent with this possibility (121). Excision and integration of chromosomal genes into plasmids is well documented in *P. s. phaseolicola* (112). Such a mechanism of gene movement may be widespread among the pathovars of *P. syringae*, especially for genes associated with virulence such as the tabtoxin gene cluster.

The Phaseolotoxin Gene Cluster

Genes required for biosynthesis of phaseolotoxin appear to be clustered on a ~22-kb DNA sequence in *P. s. phaseolicola*, but little is known about the organization and functions of these genes (93). Because the six Tox^- mutants obtained by Peet et al (92) subsequently proved to produce residual toxin, it could be that the mutants arose due to Tn*5* insertions either in regulatory genes or genes that only indirectly affect toxin production. Nevertheless, a 2.6-kb fragment that functionally complemented two of the six mutants is conserved among strains of *P. s. phaseolicola*, showing potential utility as a DNA hybridization probe for detection and identification of strains that produce phaseolotoxin (101). Subsequent studies (93) also showed linkage to the cloned region of the OCTase gene (i.e. *argK* (85)) encoding phaseolotoxin immunity, which indicates close proximity to genes encoding key enzymes involved in biosynthesis. Another gene likely to occur in the phaseolotoxin gene cluster is that for amidinotransferase, an enzyme responsible for the formation of homoarginine and ornithine (67). Consequently, the OCTase gene encoding a toxin-resistant enzyme also has an important role in phaseolotoxin biosynthesis since it supplies critical levels of arginine which serves as a substrate for amidinotransferase. Naturally occurring strains of *P. s. phaseolicola* reported to lack both phaseolotoxin production and immunity (31, 92) may have arisen from spontaneous deletion of all or part of the phaseolotoxin gene cluster.

Kamdar et al (52) identified a class of unusual Tox^- mutants of *P. s. phaseolicola* produced by either UV or chemical (ethylmethane sulfonate

[EMS]) mutagenesis. Surprisingly, cosmid clones, such as pDC938, functionally complemented all 80 of the UV and EMS mutants tested. The region responsible for complementation was mapped to 350–400 bp of DNA; insertion of the Ω interposon into the BamHI site within this region disrupted the ability to restore mutants to a Tox$^+$ phenotype. Curiously, marker exchange of the Ω-mutated DNA into the wild-type strain failed to yield a Tox$^-$ phenotype. Kamdar et al (52) proposed that the cloned sequence may contain a DNA binding site for a repressor protein that regulates toxin production.

THE CONTRIBUTION OF TOXIGENICITY TO VIRULENCE

Only minute quantities of the pseudomonad toxins are necessary to cause symptoms distinctive to the corresponding disease (40, 75). Thus, toxins are presumed to have significant roles in disease development leading to enhanced bacterial growth and persistence in the plant environment. There has been much conjecture, however, as to whether the toxins simply act as virulence factors to increase disease intensity or exhibit a function necessary for plant pathogenicity. Genetic analysis of Tox$^-$ mutants indicates that toxin production is a virulence factor in the plant pseudomonads. It now appears that the common occurrence (54, 116, 123, 124) of Tox$^-$ mutants that also are nonpathogenic is due to mutation or spontaneous deletion of genes that regulate various processes associated with pathogenicity in addition to toxigenesis.

The contribution of syringomycin production to the virulence of *P. s. syringae* is substantial based on pathogenicity tests of Tox$^-$ transposon mutants in immature cherry fruits (123). Strains with mutations in either the *syrB* or *syrC* genes formed small necrotic lesions that were only about 60% as large as those produced by the parent restored by marker exchange of the intact gene into the chromosome to a Tox$^+$ phenotype (80; Y-Y. Mo & D. C. Gross, unpublished data). Moreover, a *syrB*::*lacZ* transcriptional fusion was highly expressed in cherry tissues with appreciable activity occurring within 24 hr of inoculation (80). This indicates that the *syr* genes are transcriptionally activated shortly after bacteria penetrate host tissues. However, syringomycin production did not stimulate bacterial growth in planta, based on the fact that the populations of either *syrB* and *syrC* mutants were not significantly different from those of the parental strain. This underscores the involvement of other bacterial products in pathogenesis that are necessary for promotion of bacterial growth and lesion formation. Syringomycin apparently accentuates the disease process by killing a larger number of host cells during pathogenesis, as reflected by larger lesion size.

Coronatine production likewise is a virulence factor in *P. s. tomato* and is not essential for pathogenicity (12). Both the *cor*::Tn*5* mutant PT23.20 and the Tox$^-$ mutant PT23.21, which lacks the entire *cor*-containing pPT23A plasmid, caused necrotic lesions in tomato leaves that were only about one third the size of those produced by the wild-type strain. In contrast to syringomycin, however, coronatine production resulted in significantly higher bacterial populations in mature lesions.

Pathogenicity also is retained by mutants of *P. s. phaseolicola* that fail to produce phaseolotoxin, although the genes affected have not been characterized (92). Current evidence suggests that phaseolotoxin production does not contribute to either lesion size or ability of the bacterium to grow in plant tissues.

In contrast to the other pseudomonad toxins, tabtoxin production by *P. s. tabaci* was concluded by Turner & Taha (116) to be a pathogenicity determinant based on analysis of Tox$^-$ mutants of strain ATCC11528 that failed to cause necrotic lesions in tobacco leaves, although this was not confirmed in a subsequent study (54). Kinscherf et al (54) also noted loss of pathogenicity for Tox$^-$ mutants of the bean wildfire strain BR2, but Tox$^-$ transposon or deletion mutants of strains ATCC11528 and Pc27 of pathovars *tabaci* and *coronafaciens,* respectively, retained lesion formation. Nevertheless, all Tox$^-$ mutants were indistinguishable from parental strains in ability to grow *in planta* (54; D. K. Willis, personal communication). The nonpathogenicity by Tox$^-$ derivatives of strain BR2 is puzzling since recombinational exchange of the *tblA9::*Tn*5* mutation from BR2 into *P. s. tabaci* ATCC11528 yeilds a Tox$^-$ homogenote that forms lesions on tobacco (D. K. Willis, personal communication). This demonstrates that pathogenicity is not directly linked to the *tblA9* locus necessary for tabtoxin production. One can thus speculate that the nonpathogenic phenotype resulted from inactivation of a regulatory gene that controls expression of genes associated with both pathogenesis and tabtoxin production in strain BR2 or that the *tblA9::*Tn*5* mutation disrupted an operon necessary for expression of pathogenicity. Search for an alternative explanation seems justified because Tox$^-$ forms of *P. syringae* that are pathogenic commonly occur in nature, as exemplified by the "angulata" and "striafaciens" strains (121). In addition, tabtoxin production was conferred on the nonpathogenic Cit7 strain of *P. syringae* upon acquisition of cosmid pRTBL823 that contains all of the genes essential for tabtoxin biosynthesis (54); however, strain Cit7(pRTBL823) remained nonpathogenic to bean and tobacco. Finally, because tabtoxinine-β-lactam is strictly limited in its mode of action to inhibit glutamine synthetase (29, 75, 115), it seems inconceivable that this solely would be necessary for expression of pathogenicity.

Caution must be exercised in using Tox$^-$ mutants for critical tests of toxin involvement in plant pathogenesis. Unless the specific function of the mutated gene is restricted to toxin biosynthesis, one cannot discount that all or part of the observed effects on virulence resulted from other causes. Examples include regulatory genes that exhibit pleiotropic effects on virulence, or genes involved in the synthesis of a toxin precursor essential for other biosynthetic processes. In syringomycin studies, for example, several Tox$^-$ transposon mutants (containing unique Tn*5* insertions) were both nonpathogenic and severely restricted in ability to grow in plant tissues (123). One of the genes, called *syrA* (124), subsequently cloned, was also necessary for arginine prototropy (N. B. Quigley & D. C. Gross, unpublished data). Although the Tox$^-$/nonpathogenic phenotype of the *syrA* mutant cannot be attributed simply to arginine deficiency, results from pathogenicity tests of the *syrA* mutant could lead erroneously to the conclusion that syringomycin is a pathogenicity determinant necessary for growth of the bacterium in plant tissues. One must also consider the occurrence of secondary mutations, such as IS*50* insertions (2), or the possibility that the mutant is Tox$^-$ in vitro but Tox$^+$ *in planta*. A potentially serious problem in using transposons is that their insertion in an operon can have a polar effect on expression of all the downstream genes of the operon, thus preventing expression of several genes at once.

Finally, the question "What does the bacterium gain from toxin production?" requires further scrutiny. Strong evidence that toxin production leads to higher populations exists only for strains that produce coronatine (12). Nevertheless, because of the acute physiological effects of syringomycin, phaseolotoxin, and tabtoxin in plant tissues, it is difficult to consider them to be mere virulence factors. The complex genetic basis of toxin production simply would not have evolved and been maintained throughout the ages if toxins were not of critical importance to long-term survival in the plant habitat. Because toxin production may exert only subtle effects on bacterial growth, it can be argued that the methodology used to evaluate Tox$^-$ mutants in pathogenicity tests is inadequate. In support of this contention, the effects of syringomycin on ion transport in plant membranes would be expected to make the intercellular fluids more conducive to bacterial multiplication (5, 6, 15, 16). Another consideration is that syringomycin, phaseolotoxin, and tabtoxin display antimicrobial activity (54, 110, 123). Could it be that the bacterium obtains a selective advantage by antagonizing microorganisms that compete for the same niche?

ENVIRONMENTAL CONTROL OF GENE EXPRESSION

Nutritional Factors

Bacteria employ sophisticated mechanisms for sensing nutrient concentrations that control their behavior in the environment. The regulatory influence of nutrient balance on toxin synthesis was discussed by Chatterjee & Vidaver (20); consequently, this review is limited to mechanisms by which specific nutrients might regulate expression of toxin genes.

Inorganic phosphate commonly down-regulates genes encoding antibiotic synthetases in bacteria (62, 68). Syringomycin production is repressed by phosphate concentrations of 1 mM and higher (39), and the regulatory mechanism resembles the model proposed for phosphate-regulated enzymes by acting at the transcriptional level (62). Examination of the promoters of antibiotic biosynthesis genes regulated by phosphate has revealed phosphate control sequences that are strikingly homologous to the *pho* boxes observed for genes in many bacteria including *P. aeruginosa*. The mechanism by which phosphate selectively blocks transcription at phosphate-regulated promoters is uncertain, but DNA-binding proteins that recognize the *pho* box may affect the specificity of the interaction between RNA polymerase and DNA.

Syringomycin production is positively controlled by iron concentration (39), and expression of the large proteins associated with toxin biosynthesis reflects this iron dependency (84, 124). It would seem that metal ions regulate the complex process of antibiotic biosynthesis at the level of gene transcription, possibly by interaction with a key *trans*-acting regulatory protein. Iron is known to serve as a cofactor that activates the Fur protein of enteric bacteria, which functions as a repressor of transcription of several iron-regulated proteins (25, 44). The mode by which metal-binding proteins function as either repressors or activators of transcription in bacteria is reviewed by Hennecke (44). An amino acid domain, which appears to be conserved in iron-binding proteins (43, 109), contains cysteine residues that seem to be involved in metal binding. It is possible that a similar iron-binding motif may occur in the amino acid sequence of a transcriptional activator of syringomycin biosynthesis.

The regulatory effects of carbon and nitrogen sources on toxin biosynthesis in the plant pseudomonads are poorly defined. This is unfortunate because carbon catabolite regulation is exerted at the gene transcription level for several antibiotics (68), and nitrogen-metabolite control of antibiotic biosynthesis probably occurs at the levels of translation and posttranslation as well as transcription (1). Although there is a serious void in understanding nutritional regulation of toxin production in the pseudomonads, one might anticipate significant progress as the structural, functional, or regulatory features of toxin genes are defined.

Activation by Plant Signals

Bacteria respond to a variety of environmental stimuli by coinducing or corepressing operons within a regulatory network called a stimulon (87). There is growing evidence that virulence genes in bacteria respond to environmental stimuli (for reviews see 28, 69), and apparently plant pseudomonads are no exception (61, 65, 80). Because virulence determinants are not constitutively expressed in most bacteria, activation of virulence genes upon perception of a specific chemical or physical stimulus imparts order and balance to pathogenesis that will optimize the bacterium's chances for long-term survival. Such regulatory control of virulence is observed for *A. tumefaciens,* a bacterium that infects plants only after induction of its virulence (*vir*) genes by specific phenolic compounds associated with wounded plant tissues (127). Although the molecular events responsible for transmitting the plant signal to the transcriptional apparatus of the *vir* genes are not completely known, the crown gall induction system serves as a model for other plant-microbe interactions. Gene expression resulting from plant-microbe interactions has been recently reviewed (94). Consequently, this discussion is limited to evidence that toxigenesis in *Pseudomonas* is regulated by environmental stimuli.

A *syrB::lacZ* fusion in strain B3AR132 of *P. s. syringae* (80) showed that a gene required for syringomycin production was transcriptionally activated in response to plant signals. Of over 30 phenolics tested, only certain phenolic glucosides, including arbutin, salicin, and esculin activated the *syrB::lacZ* fusion as measured as β-galactosidase activity (Figure 2) (Y-Y. Mo & D. C. Gross, unpublished data). Inducers of the *vir* genes in *A. tumefaciens,* such as acetosyringone (127), and flavonoid inducers of nodulation (*nod*) genes in *Rhizobium* spp., such as apigenin and naringenin (125), did not activate the *syrB::lacZ* fusion. Consequently, the *syr* genes of *P. s. syringae* respond to a different chemical type of phenolic signal. Phenolic glucosides derived from a single benzene ring (e.g. arbutin, salicin and phenyl-β-D-glucoside) are about three times more active than esculin, which contains a coumarin ring (Figure 2). Furthermore, the β-glucosidic linkage is necessary for activity since aglucone derivatives, such as the hydroquinone moiety of arbutin, lack activity. Of relevance to pathogenesis by *P. s. syringae* is that host tissues can contain high concentrations of aromatic-*O*-glucosides, e.g. arbutin at ~8,000 ppm in pear leaves (*Pyrus communis*) (106), well in excess of the amount needed for maximum expression of *syrB*. Furthermore, crude extracts of cherry leaves (*Prunus avium)* contain sufficient concentrations of signal molecules per gram (fresh weight) to yield over 17,000 units of *syrB*-inducing activity (Y-Y. Mo & D. C. Gross, unpublished data). However, the production of β-glucosidases by nearly all strains of *P. s. syringae* (45, 96) superficially appears inconsistent with the signal transduction process because cleavage of the β-glucosidic linkage destroys activity. The relationship of β-glucosidase production to toxigenicity and disease development warrants genetic exploration.

A few mono- and disaccharides amplify the *syrB*-inducing activities of phenolic glucoside signal molecules found in plant tissues (Y-Y. Mo & D. C. Gross, unpublished data). Sucrose and D-fructose are the most active, causing about a fivefold enhancement in activity when 10 μM of arbutin is present. More significantly, ~250 units of syringomycin were produced by strain B3A-R cultured in a defined medium supplemented with both arbutin and

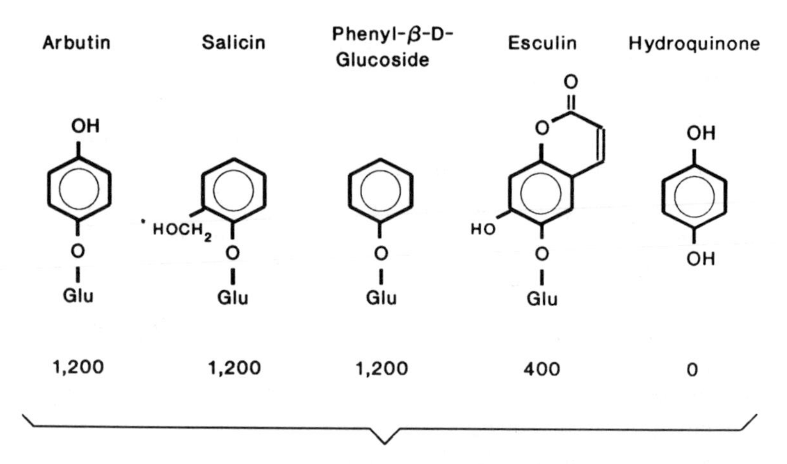

Figure 2 Structures of phenolic β-glucosides that exhibit signal activity in *P. s. syringae* to induce genes required for syringomycin production. Approximate β-galactosidase units of activity are shown after induction of a *syrB::lacZ* fusion in strain B3AR132 (80). Hydroquinone is an aglucone derivative of arbutin. Abbreviation: glu, glucose.

fructose, but no production occurred in the medium without fructose, even when arbutin was added. This confirms the importance of two classes of plant metabolites, i.e. sugars and phenolic glucosides, for expression of toxigenicity in at least some strains of *P. s. syringae*. Such observations correspond to the effects of sugars on induction of *vir* genes of *A. tumefaciens* when limited concentrations of acetosyringone are present (3, 18, 104). However, the spectra of sugars enhancing the phenolic signal activity differ for the two bacteria. Unlike *vir* genes, the *syrB* gene is not activated by D-galactose, D-xylose, and L-arabinose. It is unclear why only certain sugars amplify induction of *syrB*, but it may be related in part to cellular uptake. For example, the ChvE protein of *A. tumefaciens* is a periplasmic glucose- and galactose-binding protein that is necessary for sugar-mediated induction of *vir* genes (3, 18). Because *P. s. syringae* obtains nutrients in the intercellular spaces of leaf and shoot tissues, the supply of sugars with inducing activity should not be limiting. Indeed, fructose and sucrose are abundant nonstructural carbohydrates in cherry, each normally composing 0.5 to 3% of the dry weight of shoots (53).

Coronatine biosynthesis by *P. s. tomato* also appears to be plant inducible. Using Tn*3*-Spice to make transcriptional fusions of *inaZ* to *cor* genes, Ma et al (65) observed that a fusion within the CorII region of the coronatine gene cluster expressed about 370 times more ice nucleation activity *in planta* than in vitro. The plant signal compounds have not been characterized. Possibly the *cor* gene is transcriptionally activated in response to both phenolic and sugar signals.

Obviously a complex genetic network occurs in the pseudomonads that is responsible for the perception and transduction of signals to the transcriptional apparatus of genes involved in virulence. Genetic analysis of the regulatory mechanism responsible for signal transduction offers exciting opportunities for uncovering key elements of plant-bacteria interactions. Could it be that the host specificity observed for many toxigenic strains results from differences in the spectrum of plant signals to which a strain can respond? What other sets of genes besides toxin genes are part of the stimulon? The sequential expression of regulatory and structural genes within a stimulon undoubtedly promotes a dynamic response by the bacterium to transient changes in the plant habitat without undue expenditure of metabolic energy.

PERSPECTIVES

A common assumption in plant pathology is that the host specificity exhibited by most strains of *P. syringae* does not involve toxigenesis since all pseudomonad phytotoxins exhibit activity that is indiscriminate of plant species. Recent observations that specific plant metabolites activate expression of toxin genes support the alternative premise that toxigenicity is linked to host specificity at the regulatory level. Host specificity is dictated in part by the recognition and transduction of appropriate plant signals by strains of *Agrobacterium* and *Rhizobium* (63, 127), and similar constraints may apply to the plant pseudomonads. If a plant should lack sufficient signal activity the pseudomonad may be diminished in virulence due to its failure to produce toxin, or because other virulence genes coordinately controlled by the same regulatory mechanism are not expressed. The relevance to phytopathogenesis will become clear as we learn more about the properties of plant signals and the mechanism by which bacteria receive and transmit the message to effect a particular response.

Studies of gene regulation promise to reveal the hierarchy of gene expression to disclose the sequence by which the plant pseudomonads express virulence during pathogenesis. The upper echelons of gene control are beginning to be analyzed and are exemplified by work on genes required for pathogenicity such as *hrpS* of *P. s. phaseolicola* (38) and *lemA* of *P. s. syringae* (122). Soon, these and other genetic systems will be linked to the various facets of virulence including toxigenesis. Such developments will have tremendous impact on understanding the fundamental processes that permit the pseudomonads to exploit the plant habitat so successfully.

Now that the major toxin gene clusters have been identified and cloned, efforts must be directed at the more difficult task of resolving gene function. Elucidation of the nucleotide sequence offers much hope for identifying possible gene functions since antibiotic synthetase genes frequently share conserved sequences from disparate antibiotic-producing bacteria (60, 70, 103, 107). In addition, much can be learned about differential expression of genes by analyzing the promoter regions of toxin genes. It is common for the promoters of genes involved in antibiotic biosynthesis to mediate expression only under specific environmental conditions (68). Deretic et al (27) reviewed the common denominators of promoter control in *Pseudomonas* and described several functional groups, some of which could forseeably control toxin genes. Because almost nothing is known about the promoter regions of the toxin genes in *P. syringae*, one cannot yet speculate about global mechanisms of gene expression. But the differential recognition of promoters for antibiotic genes in other bacteria by alternative forms of RNA polymerase suggests that a similar mechanism might be responsible for the selective expression of pseudomonad toxin genes in a relatively stringent environment.

Genetic analysis of Tox^- mutants has yielded convincing evidence that toxigenicity is crucial to plant pathogenicity. Vining (119) recently noted that because of the complexity displayed by such biosynthetic and regulatory processes, it was incomprehensible to suggest that such pathways would not confer selective value to the bacterium. Toxin production can be lost upon mutation of one of many possible genes, and this is certainly true for the pseudomonad toxins. If this was of no consequence to the bacterium, then the associated genetic baggage would be shed. We also know that gene sequences in toxin production are conserved among clusters of *P. syringae* strains. Thus, it is not surprising that gene sequences required for coronatine or phaseolotoxin production are among the best probes used in disease diagnosis and for differentiating pathovars of *P. syringae* (23, 101). Finally, transgenic tobacco plants harboring a presumptive tabtoxin resistance gene, *ttr*, from *P. s. tabaci* proved to be resistant to the wildfire disease (4). Such novel approaches to disease resistance portend exciting avenues for disease control resulting from studies of toxigenicity.

ACKNOWLEDGMENTS

I wish to thank those investigators who provided unpublished information. The helpful discussions and suggestions of Drs. N. B. Quigley and D. M. Weller are gratefully acknowledged.

Literature Cited

1. Aharonowitz, Y. 1980. Nitrogen metabolite regulation of antibiotic biosynthesis. *Annu. Rev. Microbiol.* 34: 209–33
2. Anderson, D. M., Mills, D. 1985. The use of transposon mutagenesis in the isolation of nutritional and virulence mutants in two pathovars of *Pseudomonas syringae*. *Phytopathology* 75:104–8
3. Ankenbauer, R. G., Nester, E. W. 1990. Sugar-mediated induction of *Agrobacterium tumefaciens* virulence genes: Structural specificity and activities of monosaccharides. *J. Bacteriol.* 172:6442–46
4. Anzai, H., Yoneyama, K., Yamaguchi, I. 1989. Transgenic tobaccoresistant to a bacterial disease by the detoxification of a pathogenic toxin. *Mol. Gen. Genet.* 219:492–94
5. Atkinson, M. M., Baker, C. J. 1987. Alteration of plasmalemma sucrose

transport in *Phaseolus vulgaris* by *Pseudomonas syringae* pv. *syringae* and its association with K^+/H^+ exchange. *Phytopathology* 77:1573–78

6. Atkinson, M. M., Baker, C. J. 1987. Association of host plasma membrane K^+/H^+ exchange with multiplication of *Pseudomonas syringae* pv. *syringae* in *Phaseolus vulgaris*. *Phytopathology* 77:1273–79
7. Aydin, M., Lucht, N., König, W. A., Lupp, R., Jung, G., et al. 1985. Structure elucidation of the peptide antibiotics herbicolin A and B. *Liebigs Ann. Chem.* 2285–2300
8. Backman, P. A., DeVay, J. E. 1971. Studies on the mode of action and biogenesis of the phytotoxin syringomycin. *Physiol. Plant Pathol.* 1:215–34
9. Baker, C. J., Atkinson, M. M., Collmer, A. 1987. Concurrent loss in Tn5 mutants of *Pseudomonas syringae* pv. *syringae* of the ability to induce the hypersensitive response and host plasma membrane K^+/H^+ exchange in tobacco. *Phytopathology* 77:1268–72
10. Ballio, A., Bossa, F., Collina, A., Gallo, M., Iacobellis, N. S., et al. 1990. Structure of syringotoxin, a bioactive metabolite of *Pseudomonas syringae* pv. *syringae*. *FEBS Lett.* 269:377–80
11. Bender, C. L., Malvick, D. K., Mitchell, R. E. 1989. Plasmid-mediated production of the phytotoxin coronatine in *Pseudomonas syringae* pv. *tomato*. *J. Bacteriol.* 171:807–12
12. Bender, C. L., Stone, H. E., Sims, J. J., Cooksey, D. A. 1987. Reduced pathogen fitness of *Pseudomonas syringae* pv. *tomato* Tn5 mutants defective in coronatine production. *Physiol. Mol. Plant Pathol.* 30:273–83
13. Bender, C. L., Young, S. A., Mitchell, R. E. 1991. Conservation of plasmid DNA sequences in coronatine-producing pathovars of *Pseudomonas syringae*. *Appl. Environ. Microbiol.* 57:993–99
14. Bibb, M. J., Biró, S., Motamedi, H., Collins, J. F., Hutchinson, C. R. 1989. Analysis of the nucleotide sequence of the *Streptomyces glaucescens tcmI* genes provides key information about the enzymology of polyketide antibiotic biosynthesis. *EMBO J.* 8:2727–36
15. Bidwai, A. P., Takemoto, J. Y. 1987. Bacterial phytotoxin, syringomycin, induces a protein kinase-mediated phosphorylation of red beet plasma membrane polypeptides. *Proc. Natl. Acad. Sci. USA* 84:6755–59
16. Bidwai, A. P., Zhang, L., Bachmann, R. C., Takemoto, J. Y. 1987. Mechanism of action of *Pseudomonas syringae* phytotoxin, syringomycin. *Plant Physiol.* 83:39–43
17. Bradbury, J. F. 1986. *Guide to Plant Pathogenic Bacteria*, pp. 175–77. Farnham Royal, England: CAB Int. Mycol. Inst.
18. Cangelosi, G. A., Ankenbauer, R. G., Nester, E. W. 1990. Sugars induce the *Agrobacterium* virulence genes through a periplasmic binding protein and a transmembrane signal protein. *Proc. Natl. Acad. Sci. USA* 87:6708–12
19. Chatterjee, A. K., Somlyai, G., Nordeen, R. O. 1990. Molecular cloning and expression of syringotoxin (*syt*) genes of *Pseudomonas syringae* pv. *syringae*. In *Pseudomonas: Biotransformations, Pathogenesis, and Evolving Biotechnology*, ed. S. Silver, A. M. Chakrabarty, B. Iglewski, S. Kaplan, 6:58–63. Washington: Am. Soc. Microbiol. 423 pp.
20. Chatterjee, A. K., Vidaver, A. K. 1986. Genetics of pathogenicity factors: Application to phytopathogenic bacteria. In *Advances in Plant Pathology*, ed. D. S. Ingram, P. H. Williams, Vol. 4. London: Academic. 224 pp.
21. Coplin, D. L. 1989. Plasmids and their role in the evolution of plant pathogenic bacteria. *Annu. Rev. Phytopathol.* 27: 187–212
22. Cundliffe, E. 1989. How antibiotic-producing organisms avoid suicide. *Annu. Rev. Microbiol.* 43:207–33
23. Cuppels, D. A., Moore, R. A., Morris, V. L. 1990. Construction and use of a nonradioactive DNA hybridization probe for detection of *Pseudomonas syringae* pv. *tomato* on tomato plants. *Appl. Environ. Microbiol.* 56:1743–49
24. Currier, T. C., Morgan, M. K. 1983. Plasmids of *Pseudomonas syringae:* No evidence of a role in toxin production or pathogenicity. *Can. J. Microbiol.* 29: 84–89
25. de Lorenzo, V., Wee, S., Herrero, M., Neilands, J. B. 1987. Operator sequences of the aerobactin operon of plasmid ColV-K30 binding the ferric uptake regulation (*fur*) repressor. *J. Bacteriol.* 169:2624–30
26. Denny, T. P., Gilmour, M. N., Selander, R. K. 1988. Genetic diversity and relationships of two pathovars of *Pseudomonas syringae*. *J. Gen. Microbiol.* 134:1949–60
27. Deretic, V., Konyecsni, W. M., Mohr, C. D., Martin, D. W., Hibler, N. S. 1989. Common denominators of promoter control in *Pseudomonas* and other bacteria. *Bio/Technology* 7:1249–54

28. DiRita, V. J., Mekalanos, J. J. 1989. Genetic regulation of bacterial virulence. *Annu. Rev. Genet.* 23:455–82
29. Durbin, R. D., Langston-Unkefer, P. J. 1988. The mechanism for self-protection against bacterial phytotoxins. *Annu. Rev. Phytopathol.* 26:313–29
30. Durbin, R. D., Uchytil, T. F. 1984. The role of intercellular fluid and bacterial isolate on the in vivo production of tabtoxin and tabtoxinine-β-lactam. *Physiol. Plant Pathol.* 24:25–31
31. Ferguson, A. R., Johnston, J. S., Mitchell, R. E. 1980. Resistance of *Pseudomonas syringae* pv. *phaseolicola* to its own toxin, phaseolotoxin. *FEMS Microbiol. Lett.* 7:123–25
32. Ferguson, I. B., Mitchell, R. E. 1985. Stimulation of ethylene production in bean leaf discs by the pseudomonad phytotoxin coronatine. *Plant Physiol.* 77:969–73
33. Fukuchi, N., Isogai, A., Nakayama, J., Suzuki, A. 1990. Structure of syringotoxin B, a phytotoxin produced by citrus isolates of *Pseudomonas syringae* pv. *syringae*. *Agric. Biol. Chem.* 54:3377–79
34. Fukuchi, N., Isogai, A., Yamashita, S., Suyama, K., Takemoto, J. Y., et al. 1990. Structure of phytotoxin syringomycin produced by a sugar cane isolate of *Pseudomonas syringae* pv. *syringae*. *Tetrahedron Lett.* 31:1589–92
35. Gantotti, B. V., Patil, S. S., Mandel, M. 1979. Apparent involvement of a plasmid in phaseotoxin production by *Pseudomonas phaseolicola*. *Appl. Environ. Microbiol.* 37:511–16
36. Gnanamanickam, S. S., Starratt, A. N., Ward, E. W. B. 1982. Coronatine production in vitro and in vivo and its relation to symptom development in bacterial blight of soybean. *Can. J. Bot.* 60:645–50
37. Gonzalez, C. F., Vidaver, A. K. 1979. Syringomycin production and holcus spot disease of maize: Plasmid-associated properties in *Pseudomonas syringae*. *Curr. Microbiol.* 2:75–80
38. Grimm, C., Panopoulos, N. J. 1989. The predicted protein product of a pathogenicity locus from *Pseudomonas syringae* pv. *phaseolicola* is homologous to a highly conserved domain of several procaryotic regulatory proteins. *J. Bacteriol.* 171:5031–38
39. Gross, D. C. 1985. Regulation of syringomycin synthesis in *Pseudomonas syringae* pv. *syringae* and defined conditions for its production. *J. Appl. Bacteriol.* 58:167–74
40. Gross, D. C., Cody, Y. S. 1985. Mechanisms of plant pathogenesis by *Pseudomonas* species. *Can. J. Microbiol.* 31:403–10
41. Gross, D. C., DeVay, J. E. 1977. Population dynamics and pathogenesis of *Pseudomonas syringae* in maize and cowpea in relation to the in vitro production of syringomycin. *Phytopathology* 67:475–83
42. Gross, D. C., DeVay, J. E. 1977. Production and purification of syringomycin, a phytotoxin produced by *Pseudomonas syringae*. *Physiol. Plant Pathol.* 11:13–28
43. Henderson, N., Austin, S., Dixon, R. A. 1989. Role of metal ions in negative regulation of nitrogen fixation by the *nifL* gene product from *Klebsiella pneumoniae*. *Mol. Gen. Genet.* 216: 484–91
44. Hennecke, H. 1990. Regulation of bacterial gene expression by metal-protein complexes. *Mol. Microbiol.* 4:1621–28
45. Hildebrand, D. C., Schroth, M. N. 1964. β-Glucosidase activity in phytopathogenic bacteria. *Appl. Microbiol.* 12:487–91
46. Hildebrand, D. C., Schroth, M. N., Huisman, O. C. 1982. The DNA homology matrix and non-random variation concepts as the basis for the taxonomic treatment of plant pathogenic and other bacteria. *Annu. Rev. Phytopathol.* 20:235–56
47. Hopwood, D. A., Sherman, D. H. 1990. Molecular genetics of polyketides and its comparison to fatty acid biosynthesis. *Annu. Rev. Genet.* 24:37–66
48. Ichihara, A., Shiraishi, K., Sato, H., Sakamura, S., Nishiyama, K., et al. 1977. The structure of coronatine. *J. Am. Chem. Soc.* 99:636–37
49. Isogai, A., Fukuchi, N., Yamashita, S., Suyama, K., Suzuki, A. 1989. Syringostatins, novel phytotoxins produced by *Pseudomonas syringae* pv. *syringae*. *Agric. Biol. Chem.* 53:3117–19
50. Isogai, A., Fukuchi, N., Yamashita, S., Suyama, K., Suzuki, A. 1990. Structures of syringostatins A and B, novel phytotoxins produced by *Pseudomonas syringae* pv. *syringae* isolated from lilac blights. *Tetrahedron Lett.* 31:695–98
51. Jamieson, A. F., Bieleski, R. L., Mitchell, R. E. 1981. Plasmids and phaseolotoxin production in *Pseudomonas syringae* pv. *phaseolicola*. *J. Gen. Microbiol.* 122:161–65
52. Kamdar, H. V., Rowley, K. B., Clements, D., Patil, S. S. 1991. *Pseu-*

domonas syringe pv. *phaseolicola* genomic clones harboring heterologous DNA sequences suppress the same phaseolotoxin-deficient mutants. *J. Bacteriol.* 173:1073–79

53. Keller, J. D., Loescher, W. H. 1989. Nonstructural carbohydrate partitioning in perennial parts of sweet cherry. *J. Am. Soc. Hortic. Sci.* 114:969–75
54. Kinscherf, T. G., Coleman, R. H., Barta, T. M., Willis, D. K. 1991. Cloning and expression of the tabtoxin biosynthetic region from *Pseudomonas syringae*. *J. Bacteriol.* 173:In press
55. Kleinkauf, H., von Döhren, H. 1983. Peptides. In *Biochemistry and Genetic Regulation of Commercially Important Antibiotics,* ed. L. C. Vining, 5:95–145. Reading, MA: Addison-Wesley. 370 pp.
56. Kleinkauf, H., von Döhren, H. 1987. Biosynthesis of peptide antibiotics. *Annu. Rev. Microbiol.* 41:259–89
57. Kleinkauf, H., von Döhren, H. 1990. Nonribosomal biosynthesis of peptide antibiotics. *Eur. J. Biochem.* 192:1–15
58. Knight, T. J., Durbin, R. D., Langston-Unkefer, P. J. 1987. Self-protection of *Pseudomonas syringae* pv. *"tabaci"* from its toxin, tabtoxinine-β-lactam. *J. Bacteriol.* 169:1954–59
59. Komura, S., Kurahashi, K. 1980. Biosynthesis of polymyxin E. III. Total synthesis of polymyxin E by a cell-free enzyme system. *Biochem. Biophys. Res. Commun.* 95:1145–51
60. Krätzschmar, J., Krause, M., Marahiel, M. A. 1989. Gramicidin S biosynthesis operon containing the structural genes *grsA* and *grsB* has an open reading frame encoding a protein homologous to fatty acid thioesterases. *J. Bacteriol.* 171: 5422–29
61. Lindgren, P. B., Frederick, R., Govindarajan, A. G., Panopoulos, N. J., Staskawicz, B. J., et al. 1989. An ice nucleation reporter system: Identification of inducible pathogenicity genes in *Pseudomonas syringae* pv. *phaseolicola*. *EMBO J.* 8:1291–301
62. Liras, P., Asturias, J. A., Martin, J. F. 1990. Phosphate control sequences involved in transcriptional regulation of antibiotic biosynthesis. *Trends Biotechnol.* 8:184–89
63. Long, S. R. 1989. *Rhizobium* genetics. *Annu. Rev. Genet.* 23:483–506
64. Lukens, J. H., Mathews, D. E., Durbin, R. D. 1987. Effect of tagetitoxin on the levels of ribulose 1,5-bisphosphate carboxylase, ribosomes, and RNA in plastids of wheat leaves. *Plant Physiol.* 84:808–13
65. Ma, S-W., Morris, V. L., Cuppels, D. A. 1991. Characterization of a DNA region required for production of the phytotoxin coronatine by *Pseudomonas syringae* pv. *tomato*. *Mol. Plant-Microbe Interact.* 4:69–74
66. Malik, V. 1983. Chloramphenicol. See Ref. 55, 12:293–309
67. Märkisch, U., Reuter, G. 1990. Biosynthesis of homoarginine and ornithine as precursors of the phytoeffector phaseolotoxin by the amidinotransfer from arginine to lysine catalyzed by an amidinotransferase in *Pseudomonas syringae* pv. *phaseolicola*. *J. Basic Microbiol.* 30:425–33
68. Martín, J. F., Liras, P. 1989. Organization and expression of genes involved in the biosynthesis of antibiotics and other secondary metabolites. *Annu. Rev. Microbiol.* 43:173–206
69. Miller, J. F., Mekalanos, J. J., Falkow, S. 1989. Coordinate regulation and sensory transduction in the control of bacterial virulence. *Science* 243:916–22
70. Miller, J. R., Ingolia, T. D. 1989. Cloning and characterization of beta-lactam biosynthetic genes. *Mol. Microbiol.* 3:689–95
71. Mindrinos, M. N., Rahme, L. G., Frederick, R. D., Hatziloukas, E., Grimm, C., et al. 1990. Structure, function, regulation, and evolution of genes involved in pathogenicity, the hypersensitive response, and phaseolotoxin immunity in the bean halo blight pathogen. See Ref. 19, pp. 74–81
72. Mitchell, R. E. 1976. Isolation and structure of a chlorosis-inducing toxin of *Pseudomonas phaseolicola*. *Phytochemistry* 15:1941–47
73. Mitchell, R. E. 1981. Structure: Bacterial. In *Toxins in Plant Disease,* ed. R. D. Durbin, pp. 259–93. New York: Academic. 515 pp.
74. Mitchell, R. E. 1984. A naturally-occurring structural analogue of the phytotoxin coronatine. *Phytochemistry* 23: 791–93
75. Mitchell, R. E. 1984. The relevance of non-host-specific toxins in the expression of virulence by pathogens. *Annu. Rev. Phytopathol.* 22:215–45
76. Mitchell, R. E., Bieleski, R. L. 1977. Involvement of phaseolotoxin in halo blight of beans. Transport and conversion to functional toxin. *Plant Physiol.* 60:723–29
77. Mitchell, R. E., Frey, E. J. 1986. Production of *N*-coronafacoyl-L-amino acid analogues of coronatine by *Pseudomonas syringae* pv. *atropurpurea* in liquid cultures supplemented with

L-amino acids. *J. Gen. Microbiol.* 132: 1503–7
78. Mitchell, R. E., Hart, P. A. 1983. The structure of tagetitoxin, a phytotoxin of *Pseudomonas syringae* pv. *tagetis*. *Phytochemistry* 22:1425–28
79. Mitchell, R. E., Johnston, J. S., Ferguson, A. R. 1981. Phaseolotoxin and other phosphosulphamyl compounds: Biological effects. *Physiol. Plant Pathol.* 19:227–35
80. Mo, Y-Y., Gross, D. C. 1991. Expression in vitro and during plant pathogenesis of the *syrB* gene required for syringomycin production by *Pseudomonas syringae* pv. *syringae*. *Mol. Plant-Microbe Interact.* 4:28–36
81. Moore, R. A., Starratt, A. N., Ma, S-W., Morris, V. L., Cuppels, D. A. 1989. Identification of a chromosomal region required for biosynthesis of the phytotoxin coronatine by *Pseudomonas syringae* pv. *tomato*. *Can. J. Microbiol.* 35:910–17
82. Moore, R. E., Niemczura, W. P., Kwok, O. C. H., Patil, S. S. 1984. Inhibitors of ornithine carbamoyltransferase from *Pseudomonas syringae* pv. *phaseolicola*. *Tetrahedron Lett.* 25: 3931–34
83. Morgan, M. K., Chatterjee, A. K. 1985. Isolation and characterization of Tn*5* insertion mutants of *Pseudomonas syringae* pv. *syringae* altered in the production of the peptide phytotoxin syringotoxin. *J. Bacteriol.* 164:14–18
84. Morgan, M. K., Chatterjee, A. K. 1988. Genetic organization and regulation of proteins associated with production of syringotoxin by *Pseudomonas syringae* pv. *syringae*. *J. Bacteriol.* 170:5689–97
85. Mosqueda, G., Van den Broeck, G., Saucedo, O., Bailey, A. M., Alvarez-Morales, A., Herrera-Estrella, L. 1990. Isolation and characterization of the gene from *Pseudomonas syringae* pv. *phaseolicola* encoding the phaseolotoxin-insensitive ornithine carbamoyltransferase. *Mol. Gen. Genet.* 222:461–66
86. Mott, K. A., Takemoto, J. Y. 1989. Syringomycin, a bacterial phytotoxin, closes stomata. *Plant Physiol.* 90:1435–39
87. Neidhardt, F. C., Ingraham, J. L., Schaechter, M. 1990. Regulation of gene expression: Multigene systems and global regulation. *In Physiology of the Bacterial Cell: A Molecular Approach,* 13:351–88. Sunderland, MA: Sinauer. 506 pp.
88. Nüesch, J., Heim, J., Treichler, H-J. 1987. The biosynthesis of sulfur-containing β-lactam antibiotics. *Annu. Rev. Microbiol.* 41:51–75
89. Obukowicz, M., Shaw, P. D. 1985. Construction of Tn*3*-containing plasmids from plant-pathogenic pseudomonads and an examination of their biological properties. *Appl. Environ. Microbiol.* 49:468–73
90. Panopoulos, N. J., Peet, R. C. 1985. The molecular genetics of plant pathogenic bacteria and their plasmids. *Annu. Rev. Microbiol.* 23:381–419
91. Parry, R. J., Mafoti, R. 1986. Biosynthesis of coronatine, a novel polyketide. *J. Am. Chem. Soc.* 108:4681–82
92. Peet, R. C., Lindgren, P. B., Willis, D. K., Panopoulos, N. J. 1986. Identification and cloning of genes involved in phaseolotoxin production by *Pseudomonas syringae* pv. "*phaseolicola*". *J. Bacteriol.* 166:1096–105
93. Peet, R. C., Panopoulos, N. J. 1987. Ornithine cabamoyltransferase genes and phaseolotoxin immunity in *Pseudomonas syringae* pv. *phaseolicola*. *EMBO J.* 6:3585–91
94. Peters, N. K., Verma, D. P. S. 1990. Phenolic compounds as regulators of gene expression in plant-microbe interactions. *Mol. Plant-Microbe Interact.* 3:4–8
95. Quigley, N. B., Lane, D., Bergquist, P. L. 1985. Genes for phaseolotoxin synthesis are located on the chromosome of *Pseudomonas syringae* pv. *phaseolicola*. *Curr. Microbiol.* 12:295–300
96. Roos, I. M. M., Hattingh, M. J. 1987. Pathogenicity and numerical analysis of phenotypic features of *Pseudomonas syringae* strains isolated from deciduous fruit trees. *Phytopathology* 77:900–8
97. Roth, P., Hädener, A., Tamm, C. 1990. Further studies on the biosynthesis of tabtoxin (wildfire toxin): Incorporation of [2,3-$^{13}C_2$]pyruvate into the β-lactam moiety. *Helv. Chim. Acta* 73:476–82
98. Sakai, R. 1980. Comparison of physiological activities between coronatine and indole-3-acetic acid to some plant tissues. *Ann. Phytopathol. Soc. Jpn.* 46:499–503
99. Sato, M. 1988. *In planta* transfer of the gene(s) for virulence between isolates of *Pseudomonas syringae* pv. *atropurpurea*. *Ann. Phytopathol. Soc. Jpn.* 54:20–24
100. Sato, M., Nishiyama, K., Shirata, A. 1983. Involvement of plasmid DNA in the productivity of coronatine by *Pseudomonas syringae* pv. *atropurpurea*.

Ann. Phytopathol. Soc. Jpn. 49:522–28

101. Schaad, N. W., Azad, H., Peet, R. C., Panopoulos, N. J. 1989. Identification of *Pseudomonas syringae* pv. *phaseolicola* by a DNA hybridization probe. *Phytopathology* 79:903–7
102. Segre, A., Bachmann, R. C., Ballio, A., Bossa, F., Grgurina, I., et al. 1989. The structure of syringomycins A_1, E and G. *FEBS Lett.* 255:27–31
103. Sherman, D. H., Malpartida, F., Bibb, M. J., Kieser, H. M., Bibb, M. J., et al. 1989. Structure and deduced function of the granaticin-producing polyketide synthase gene cluster of *Streptomyces violaceoruber* Tü22. *EMBO J.* 8:2717–25
104. Shimoda, N., Toyoda-Yamamoto, A., Nagamine, J., Usami, S., Katayama, M., et al. 1990. Control of expression of *Agrobacterium vir* genes by synergistic actions of phenolic signal molecules and monosaccharides. *Proc. Natl. Acad. Sci. USA* 87:6684–88
105. Shiraishi, K., Konoma, K., Sato, H., Ichihara, A., Sakamura, S., et al. 1979. The structure-activity relationships in coronatine analogs and amino compounds derived from (+)-coronafacic acid. *Agric. Biol. Chem.* 43:1753–57
106. Smale, B. C., Keil, H. L. 1966. A biochemical study of the intervarietal resistance of *Pyrus communis* to fire blight. *Phytochemistry* 5:1113–20
107. Smith, D. J., Burnham, M. K. R., Bull, J. H., Hodgson, J. E., Ward, J. M., et al. 1990. β-Lactam antibiotic biosynthetic genes have been conserved in clusters in prokaryotes and eukaryotes. *EMBO J.* 9:741–47
108. Sogn, J. A. 1976. Structure of the peptide antibiotic polypeptin. *J. Med. Chem.* 19:1228–31
109. Spiro, S., Roberts, R. E., Guest, J. R. 1989. FNR-dependent repression of the *ndh* gene of *Escherichia coli* and metal ion requirement of FNR-regulated gene expression. *Mol. Microbiol.* 3: 601–8
110. Staskawicz, B. J., Panopoulos, N. J. 1979. A rapid and sensitive microbiological assay for phaseolotoxin. *Phytopathology* 69:663–66
111. Szabo, L. J., Mills, D. 1984. Characterization of eight excision plasmids of *Pseudomonas syringae* pv. *phaseolicola*. *Mol. Gen. Genet.* 195:90–95
112. Szabo, L. J., Mills, D. 1984. Integration and excision of pMC7105 in *Pseudomonas syringae* pv. *phaseolicola*: Involvement of repetitive sequences. *J. Bacteriol.* 157:821–27
113. Tang, J. L., Gough, C. L., Daniels, M. J. 1990. Cloning of genes involved in negative regulation of production of extracellular enzymes and polysaccharide of *Xanthomonas campestris* pathovar *campestris*. *Mol. Gen. Genet.* 222:157–60
114. Turner, J. G. 1986. Effect of phaseolotoxin on the synthesis of arginine and protein. *Plant Physiol.* 80:760–65
115. Turner, J. G., Debbage, J. M. 1982. Tabtoxin-induced symptoms are associated with accumulation of ammonia formed during photorespiration. *Physiol. Plant Pathol.* 20:223–33
116. Turner, J. G., Taha, R. R. 1984. Contribution of tabtoxin to the pathogenicity of *Pseudomonas syringae* pv. *tabaci*. *Physiol. Plant Pathol.* 25:55–69
117. Unkefer, C. J., London, R. E., Durbin, R. D., Uchytil, T. F., Langston-Unkefer, P. J. 1987. The biosynthesis of tabtoxinine-β-lactam: Use of specifically ^{13}C-labeled glucose and ^{13}C NMR spectroscopy to identify its biosynthetic precursors. *J. Biol. Chem.* 262:4994–99
118. Vater J. 1989. Lipopeptides, an interesting class of microbial secondary metabolites. In *Biologically Active Molecules: Identification, Characterization and Synthesis*, ed. U. P. Schlunegger, pp. 27–38.Berlin: Springer-Verlag. 252 pp.
119. Vining, L. C. 1990. Functions of secondary metabolites. *Annu. Rev. Microbiol.* 44:395–427
120. Weinberg, E. D. 1977. Mineral element control of microbial secondary metabolism. In *Microorganisms and Minerals*, ed. E. D. Weinberg, 7:289–316. New York: Marcel Dekker. 492 pp.
121. Willis, D. K., Barta, T. M., Kinscherf, T. G. 1991. Genetics of toxin production and resistance in phytopathogenic bacteria. *Experentia*. In press
122. Willis, D. K., Hrabak, E. M., Rich, J. J., Barta, T. M., Lindow, S . E., et al. 1990. Isolation and characterization of a *Pseudomonas syringae* pv. *syringae* mutant deficient in lesion formation in bean. *Mol. Plant-Microbe Interact.* 3:149–56
123. Xu, G-W., Gross, D. C. 1988. Evaluation of the role of syringomycin in plant pathogenesis by using Tn*5* mutants of *Pseudomonas syringae* pv. *syringae* defective in syringomycin production. *Appl. Environ. Microbiol.* 54:1345–53
124. Xu, G-W., Gross, D. C. 1988. Physical and functional analyses of the *syrA* and *syrB* genes involved in syringomycin

production by *Pseudomonas syringae* pv. *syringae*. *J. Bacteriol.* 170:5680–88

125. Zaat, S. A. J., van Brussel, A. A. N., Tak, T., Pees, E., Lugtenberg, B. J. J. 1987. Flavonoids induce *Rhizobium leguminosarum* to produce *nodDABC* gene-related factors that cause thick, short roots and root hair responses on common vetch. *J. Bacteriol.* 169:3388–91
126. Zalkin, H., Ebbole, D. J. 1988. Organization and regulation of genes encoding biosynthetic enzymes in *Bacillus subtilis*. *J. Biol. Chem.* 263:1595–98
127. Zambryski, P. 1988. Basic processes underlying *Agrobacterium*-mediated DNA transfer to plant cells. *Annu. Rev. Genet.* 22:1–30
128. Zhang, L., Takemoto, J. Y. 1987. Effects of *Pseudomonas syringae* phytotoxin, syringomycin, on plasma membrane functions of *Rhodotorula pilimanae*. *Phytopathology* 77:297–303

Annu. Rev. Phytopathol. 1991. 29:279–303

BREEDING FOR DISEASE RESISTANCE IN PEANUT (*ARACHIS HYPOGAEA* L.)

J. C. Wynne and M. K. Beute

Departments of Crop Science and Plant Pathology, North Carolina State University, Raleigh, North Carolina 27695

S. N. Nigam

Groundnut Improvement Program, International Crops Research Institute for the Semi-Arid Tropics, Patancheru, A.P. 500324, India

KEY WORDS: groundnut, genetics of peanut, oilseed crop, pathogen variability, food legume

---

INTRODUCTION

The peanut (*Arachis hypogaea* L.) is an important oilseed crop and food legume grown on approximately 20 million ha in warm tropical or subtropical areas throughout the world. Diseases of the peanut reduce yield and quality and increase the cost of production wherever the crop is grown. Because of the economic impact of diseases, much effort has been given to developing both chemical and nonchemical disease management strategies. The use of crop protection chemicals in areas where the crop has high value such as the United States, not only adds to production costs but is becoming controversial because of environmental and food safety concerns. In many developing countries where chemicals are not readily available, diseases cause significant losses in spite of management strategies designed to manage or control the disease. Thus, one major objective of peanut breeding programs throughout the world is to develop disease-resistant or tolerant cultivars that can be used in managing or controlling peanut diseases.

0066-4286/91/0901-0279$02:00

Diseases of Peanut

During the early years of cultivation in the US, the peanut was regarded as relatively free from diseases (45). However, diseases of the peanut now occur throughout the growing season and into the postharvest period, and attack all parts of the plant. Diseases of the peanut are caused by fungi, bacteria, viruses, nematodes, and a mycoplasm (109). Only a few diseases are of worldwide importance. Most diseases and pathogens are of local or sporadic importance (4). Of the fungal diseases, early leafspot (CA) caused by *Cercospora arachidicola,* late leafspot (CP) caused by *Cercosporidium personatum,* and rust caused by *Puccinia arachidis* have widespread occurrence and are of great economic importance (24). The aflatoxin problem associated with the fungi *Aspergillus flavus* and *A. parasiticus* is also of great economic importance. Two virus diseases, rosette and tomato spotted wilt, can cause severe losses even though the peanut mottle virus is more widespread. Although several nematodes attack peanut, the two root knot nematodes, *Meloidogyne arenaria* and *M. hapla,* are the most important.

Breeding for Resistance

Progress in breeding for disease resistance in the peanut has lagged behind that for many major crops for several reasons including: (*a*) the late initiation of breeding programs for the crop; (*b*) the regional importance of the crop in the US; (*c*) the relatively few scientists assigned to peanuts; and (*d*) the lack of financial resources for peanut research (148). Most of the literature on peanut diseases prior to 1970 was published by American scientists (89). Because of the high value of the crop in the US and the availability of chemicals for disease control, breeding for disease resistance was not given a high priority until the late 1970s. There was also a prevalent idea that variability for resistance among cultivated peanuts was lacking (56). However, during the past 15 years, numerous germplasm accessions among cultivated peanuts have been identified as sources of disease resistance. Breeding for disease resistance has received high priority by scientists at the International Crops Research Institute for the Semi-Arid Tropics (ICRISAT), which adopted peanut as a mandate crop in 1976 (48) and received additional emphasis in the US when the United States Agency for International Development (USAID) funded the Peanut Collaborative Research Support Program (CRSP) in 1982. Breeders have tended to focus on the major diseases of peanut, although breeding for locally important diseases such as rosette virus in Africa or Cylindrocladium black rot (CBR) in North Carolina and Virginia has also received considerable attention.

HOST VARIABILITY

One of the first and most important phases of breeding for disease resistance is to identify a source of resistance. In peanut, as in other crops, it is most desirable to find resistance in closely related materials such as local or foreign cultivars or landraces. The less related the source of resistance to the germplasm being improved, the more difficult it is to transfer resistance without also transferring undesirable genes or gene complexes. Much of the variability found in the cultigen, including genes for disease resistance, arose in South America where the peanut is native (56). Six centers of genetic diversity (gene centers) for the cultivated peanut are recognized in South America: (*a*) Peru, (*b*) northeastern Brazil, (*c*) the Guarani region (Paraguay and southeast Brazil), (*d*) Rondonia and northwest Mato Grosso (Brazil), and (*e*) the eastern foothills of the Andes in Bolivia. The cultivated peanut has been classified into two subspecies, *A. hypogaea hypogaea* Krap. et Rig. and *A. hypogaea fastigiata* Waldron. These two subspecies have been subdivided into two botanical varieties, with three of the four varieties being grown commercially in the US (56). In addition to the cultivated peanut, considerable variation for disease resistance exists among the species of *Arachis* (68). All species of *Arachis* are native to South America and provide potential sources of disease resistance for use in genetic improvement. The genus probably originated in central Brazil and was first domesticated in northern Argentina and eastern Bolivia. Peanut was thought to have been taken from Brazil to Africa, India, and the Far East by the Portuguese; from the west coast of South America to the Western Pacific, Indonesia, and China by the Spaniards early in the 16th century. By the mid-16th century, peanuts were grown in North America and were distributed worldwide (24). Africa is an important secondary center of genetic variation despite evidence that introductions to Africa were from a single center in South America (60).

Since the initial introduction of peanut to the US, additional germplasm has been collected through expeditions sponsored by the US Department of Agriculture, with the cooperation of state experiment stations and several foreign countries. Approximately 6000 accessions have been assembled by the Southern Regional Plant Introduction Station at Experiment, Georgia. Nearly 12,000 collections are being maintained by ICRISAT at Hyderabad, India. Collections of cultivated and wild species of *Arachis* (maintained at ICRISAT, in Argentina and Brazil, and at several state experiment stations in the US) provide a major genetic resource for resistance to diseases, insect pests, and other desirable traits (148).

The first attempt to use these genetic resources to develop a disease-resistant cultivar was made in 1927 in East Java (Indonesia) by Dutch scientists who developed Schwarz 21, a cultivar resistant to *Pseudomonas solanacearum* (23). The cultivar NC 2, released in 1953, was selected for partial resistance to *Sclerotium rolfsii* (30). Characterization of components of partial resistance of NC 2 indicated that reduced disease was related to plant phenology; however, NC 2 also suppressed inoculum production (117).

In spite of these early successes in exploiting host-plant variability, breeding for disease resistance did not receive high priority until the late 1970s. The increasing cost of production, the failure of chemical control methods to work effectively against diseases such as CBR caused by *Cylindrocladium crotalariae,* the availability of new accessions of cultivated peanuts, and the adoption of peanut as a mandate crop by ICRISAT in 1976 resulted in an increased priority for breeding for disease resistance.

With increased emphasis in screening peanut for disease resistance, numerous genotypes with resistance to several pathogens have now been identified.

Soil-Borne Pathogens

Screening for resistance to the CBR disease, identified in North Carolina and Virginia in 1970, was initiated in greenhouse tests in 1973 (149). Over 1200 genotypes were evaluated in preliminary screening. Although field screening was not extensive because of the necessity of measuring near-homozygous lines for percent visibly infected plants in replicated trials (53) (i.e. less than 150 genotypes), a few virginia types and several spanish types were confirmed as resistant. Subsequent heritability studies indicated that resistance was quantitative, with only additive genetic effects important in the inheritance of resistance (55, 58).

NC 3033, a line resistant to CBR and also found to be resistant to *Sclerotium rolfsii* (15), was used as a parent to develop breeding lines resistant to both CBR and southern stem rot (115). Genotypes resulting from crosses with commercial cultivars expressed partial resistance to southern stem rot in the field and greenhouse; combining phenological suppression (disease escape) and metabolic resistance was considered possible (117). However, because of limited resources, it was decided to emphasize breeding for CBR resistance and only evaluate genotypes that were resistant to CBR for resistance to southern stem rot.

Exploitation of PI 341885, Toalson, and TxAG-3 (a selection from PI 365553), which are resistant to southern stem rot and pythium pod rot caused by *Pythium myriotylum,* has been pursued in the Texas peanut breeding program (122). Breeding lines using these sources of resistance as parents varied in reaction to the two diseases, with some lines showing considerable resistance to both pathogens. Apparently the two mechanisms of resistance differ and are independent. However, progeny was not as resistant as the resistant parents, and quality was reduced (122).

Resistance has also been found for sclerotinia blight caused by *Sclerotinia minor* (29). The cultivar Va 81B was selected for resistance to Sclerotinia blight in Virginia. Resistance in Va 81B is primarily phenological. Additional sources of resistance to Sclerotinia including Chico, germplasm from Texas (TX 498731, TX 798736, TX 804475) and Virginia (TRC 02056-1), and seven plant introductions from China (PIs 467829, 476831, 476834, 476835, 476842, 467843, 467844) are presently being evaluated as resistant sources in Virginia and North Carolina.

In the past, aflatoxin was considered predominantly a postharvest problem and as such, received little attention in crop improvement programs. In 1973, two germplasm lines, PI 337409 and PI 337394F, were found to be resistant to seed invasion and colonization by *A. flavus* (97). The screening was done on rehydrated sound mature seeds inoculated with conidia of *A. flavus* in an environment favorable to fungus development. Resistance to invasion and colonization by *A. flavus,* located in the seed coat, was suggested to be an effective means of preventing aflatoxin contamination. Differences have also been reported for the ability of peanut seeds to support the production of aflatoxins (11, 99, 111, 138). Several sources of resistance have been reported from Senegal (150), US (12, 84, 97), India (46, 90), China (137), and the Philippines (110). Sources resistant to preharvest field infection have also been reported from India (91), Senegal (144, 150), and the US (85). Three resistant genotypes—PI 337409, PI 337394F, and J 11—have been evaluated in more than one country. J 11 was found to be resistant to infection in the US and India. PI 337409 showed resistance in Senegal and India, but was susceptible in the US (85).

Percent colonization of seeds of F_1 and F_2 plants of reciprocal crosses between PI 337409 (resistant) and PI 331326 (susceptible) indicated a high broad-sense heritability (96). From the preliminary studies at ICRISAT on combining ability, using line x tester and diallel mating designs, Vasudeva Rao et al (142) reported UF 71513, Ah 7223, PI 337409, and PI 337394F as good combiners for seed coat resistance. Recent studies on heritability of all three types of resistances indicated that there was no correlation among mechanisms, suggesting that the three mechanisms are controlled by different genes (140).

Foliar Diseases

Considerable effort has been given to identifying sources of resistance to rust and early and late leafspot because of their worldwide importance. Sources of resistance to rust were identified more than two decades ago (22). In 1977, Hammons (59) reported resistant sources that consisted of three lines: (*a*) Tarapoto (PIs 259747, 341879, 350680, 381622, 405132), (*b*) Israeli line 136 (PIs 298115 and 315608) and (*c*) DHT 200 (PI 314817). Tarapoto and DHT 200 both originated from Peru.

Resistance of PI 298115 to rust was controlled by two recessive genes (85). Resistance was also shown to be controlled by duplicate recessive genes in additional resistant sources, although some sources of resistance could not be explained by a two-gene system. Fourteen progeny from the cross of PI 298115 and an unknown pollen source were released as rust-resistant germplasm, FESR 1-14. ICRISAT has screened over 12,000 lines of *A. hypogaea* in the field using infector rows to develop disease pressure. One hundred fifty-three lines, including the 14 rust-resistant germplasm lines (FESR 1-14), have been identified as resistant (130). Subsequent studies indicated that many of the resistant lines also possess "rate-reducing" components, i.e. "slow rusting" epidemiological mechanisms. These genotypes had increased incubation period, decreased infection frequency, and reduced pustule size, spore production, and spore germinability (132, 133). Additive genetic effects and additive types of epistasis have been found for crosses of rust resistant by susceptible crosses (121).

Many wild *Arachis* species and their interspecific derivatives with the cultivated types have also been screened for resistance to rust under both field and laboratory conditions (135). Many species were found to be immune to rust. These include *A. batizocoi* (PI 298639, PI 338312), *A. duranensis* (PI 219823), *A. cardenasii* (PI 262141), *A. chacoense* (PI 276235), *A. pusilla* (PI 338449), *A. villosa* (PI 210554), and *A. correntina* (PI 331194), among others (131). Most of the interspecific derivatives showed a high degree of resistance to rust. They had small and slightly depressed uredosori that did not rupture to release the comparatively few urediospores produced. In some diploid wild *Arachis* species, rust resistance is partially dominant (121).

In recent years, germplasm for resistance to leafspots has been intensively screened in different parts of the world. Several sources of resistance to both CA and CP leafspots have been reported (40, 41, 52, 64, 92, 124). Screening for CP resistance has been most extensive at ICRISAT where genotypes screened for rust have also been screened for CP resistance. A total of 83 lines of *A. hypogaea* have been identified with some resistance and/or tolerance to CP while using infector rows for inoculation. Twenty-nine of these lines are also resistant to rust.

Among the many wild *Arachis* species screened for CP resistance, *A. chacoense* (PI 276325), *A. cardenasii* (PI 262141), and *A. stenosperma* (PI 338280) in section *Arachis* (cross-compatible with the cultivated *A. hypogaea*) showed either an immune or a highly resistant reaction to the pathogen. Highly resistant species in other sections included *A. repens, A. appressipila, A. paraguariensis, A. villosulicarpa, A. hagenbeckii,* and *A. glabrata* (134).

Only limited screening for CA resistance has been possible at ICRISAT Center. Germplasm already planted in the field is evaluated for CA whenever the disease incidence is high, which occurs every few years. Germplasm lines NC 3033, PI 270806, PI 259747, and PI 350680, which possess epidemiological components of rate-reducing resistance in the US (64, 124), did not maintain their resistance in India using infector row inoculation techniques (70). These lines were also found susceptible to the disease in Malawi using similar inoculation techniques (105). In 1987, screening on a limited scale was started in Pantnagar, India and, unexpectedly, disease at the ICRISAT Center was severe. This provided an opportunity to evaluate 3000 genotypes planted in the field. Several genotypes showed moderate levels of field resistance to early leafspot at Pantnagar and the ICRISAT Center. These include ICG 2711 (NC 5), ICG 6709 (NC Ac 16163), ICG 7291 (PI 262128), ICG 7406 (PI 262121), ICG 7630 and ICG 7892 (PI 393527-B), and ICG 9990 (145).

More than 1000 selected germplasm lines of cultivated types have been screened for CA resistance in Malawi using infector row inoculation. All failed to show any appreciable level of resistance to the disease. A "bulk" testing of germplasm was followed to include a large number of germplasm lines (S. N. Nigam, personal communication). One hundred and ten bulks, each with 500 seeds, were formed by compositing five seeds per germplasm line. This method allowed representation of 11,000 germplasm lines in the screening. Only two bulks had a few plants that warranted further testing. Component lines of these two bulks were planted separately and scored for the disease. Only three germplasm lines—ICG 50, ICG 84 and ICG 11282—have been retained for further testing (71). Other germplasm lines of *A. hypogaea* that retain a higher proportion of foliage in spite of heavy disease pressure are ICGM 189, ICGM 197, ICGM 281, ICGM 284, ICGM 285, ICGM 286, ICGM 292,ICGM 300, ICGM 473, ICGM 500, and ICGM 525. Thirty-five lines reported to have resistance to CA at the ICRISAT Center did not maintain their disease reaction in Malawi (71).

Many wild *Arachis* species and interspecific derivatives have also been screened for CA resistance in Malawi. Only *Arachis* species 30003 has consistently shown a high level of disease resistance using infector row inoculation. Other species, *A. chacoense* and *A.* sp. 30085, which showed high promise in the first year of screening, were subsequently rated as susceptible. *Arachis stenosperma,* which was rated as resistant in the US, was highly susceptible in Malawi (72). Several interspecific derivatives retained more foliage than the susceptible control.

Eight lines of *A. hypogaea* with moderate to high levels of resistance to the major foliage diseases—rust, CP, and CA—have been identified at the ICRISAT Center. These sources (with multiple foliar resistances) should be used to produce new cultivars with resistance to the major foliar diseases. The rust and CP resistances of these genotypes are generally stable over a wide range of geographic locations. Only for NC Ac 17090 and PI 298115 has variation in rust scores been observed across locations.

Due to the observation that a high level of resistance to CA was not available for breeding, emphasis on developing CA resistance in the US has focused on incorporation of multiple components of rate-reducing resistance in commercial cultivars. Concerns about variability in virulence characteristics of both CA and CP were considered to be minimized by this strategy. Several sources of resistance to CA were identified in *A. hypogaea* (PI 109839, PI 270806, GP-NC 343, NC 3033, etc). Two diploid wild species, *A. chacoense* and *A. stenosperma,* were rated highly resistant (40). Studies on inheritance of resistance components for CA and CP were initiated (86). Substantial additive genetic effects have been found for both CA and CP among early generation progenies (9).

In studies conducted primarily in North Carolina, both additive and additive-by-additive epistasis have been found to be significant for progenies in late generation (54). Dominant genetic variance for CP in early generations was significant for several components of resistance (75). Estimates of narrow-sense heritabilities have ranged from low to high for components of resistance, with estimates varying considerably depending upon the component and cross. Estimates of broad-sense heritabilities have been higher indicating that nonadditive effects are also important.

Progenies in F_2 generation from crosses between two resistant and three susceptible cultivars were screened for components of resistance to CP in detached leaf tests at ICRISAT (104). A five-locus polygenic system assuming resistance to be completely recessive was proposed to explain the frequency of resistant plants in the F_2 generation. Nonadditive gene action was reported to be extremely important, but its nature could not be elucidated due to the absence of F_1 generation.

Except for rosette virus, evaluation of peanut for virus resistance has only recently occurred. Resistance to rosette virus was discovered in local land races in Burkina Faso in the 1950s (33). Resistance was reported to be due to the production of an antivirus substance by resistant plants (34). Resistant lines are not immune and individual plants can become infected with the disease under heavy inoculum pressure. This resistance apparently operates against both chlorotic (34) and green (61) rosette. Recent studies (98) have shown that resistance is directed against both the virus and its satellite RNA. Wild *Arachis* species are now being screened for rosette resistance by a southern African regional program of ICRISAT. Of seven species tested, two, *A.* sp. 30003 and *A.* sp. 30017, remained symptom-free throughout the season. Plant samples of these species were assayed and found free from both the virus and the assistor virus. The apparent immunity of *A.* sp. 30003 is of great interest, particularly as this species is also highly resistant to early leafspot (19). Resistance to rosette in cultivated types is controlled by two independent recessive genes (33, 106). However, Misari et al (95) reported that it might not be simply inherited.

More than 7000 germplasm lines have been screened at ICRISAT for field resistance to tomato spotted wilt virus. Many lines have been identified that have shown consistently low disease incidence in the field. Some of these lines are C102, C121, C136, NC Ac 343, NC Ac 1741, NC Ac 2232, NC Ac 2242, NC Ac 17888, and Gujrat Narrow Leaf Mutant (70). Germplasm and breeding lines with resistance to thrips and low disease incidence in the field were also tested under laboratory conditions for resistance to tomato spotted wilt virus. Only two breeding lines, ICGV 86029 and ICGV 86031, showed tolerance to the virus (72). Forty-two wild species have been tested in the glasshouse by mechanical and thrips (*F. schultzei)* inoculation. Only *A. chacoense* remained free from virus infection in these tests, but became infected with the virus under graft inoculation (128). *Arachis chacoense* and three other species—*A. pusilla* (12922), *A. correntina* (9530), and *A. cardenasii* (10017)—when infected by mechanical and thrips inoculation, show no infection under field conditions.

Over 2500 germplasm lines belonging to the cultivated species *A. hypogaea* have been screened in the fields at ICRISAT for resistance to the peanut mottle virus (PMV). No line showed resistance to the virus. However, many germplasm lines had much lower yield loss than control cultivars. Two germplasm lines, NC Ac 2240 and NC Ac 2243, have shown insignificant yield loss due to disease over the years (70). A few breeding lines have also shown tolerance to the disease. Fifty wild *Arachis* species accessions have also been screened for virus resistance under glasshouse conditions using mechanical leaf rub and air brush inoculations. Of these, only two species, *A. chacoense* (10602) and *A. pusilla* (12911), remained free from infection even after repeated graft inoculations (128).

Seeds of PMV-infected plants of several germplasm lines were screened in the laboratory for virus presence using ELISA techniques. Two rust-resistant germplasm lines, EC 76446(292) and NC Ac 17133(RF), have failed to show any seed transmission in repeated tests over years on more than 13,000 seeds (72). Many breeding lines involving these rust-resistant parents in their parentage have also shown no seed transmission. An inheritance study on nonseed transmission is in progress at ICRISAT. Lines with low yield loss and no seed transmission have been crossed and advanced generation lines are in field tests for measuring yield loss due to the disease. Promising lines from these tests will be studied for nonseed transmission in the laboratory.

Over 7000 germplasm lines of the cultivated peanut *A. hypogaea* have been screened by ICRISAT scientists for resistance to the peanut clump virus in farmers' fields in the Punjab and Andhra Pradesh, India. None of these lines showed resistance to the virus. A few lines showed tolerance to the disease, i.e. they did not suffer severely in growth and yield. Of 38 wild *Arachis* species and their 200 interspecific derivatives tested, only *Arachis* species 30036 did not become infected in the field.

PATHOGEN VARIABILITY

Breeding for disease resistance is generally considered more difficult than breeding for morphological or agronomic traits since efficacy of disease resistance is not static and is influenced simultaneously by host characteristics (magnitude of effectiveness, stability over environments, etc), and genetic variability of the pathogen (isolate aggressiveness, genetic selection for increased virulence, etc). The concept of "physiologic races," identified by pathogenic response on differential host genotypes, has assisted in understanding pathogenic variability in relation to breeding for major gene disease resistance. Major gene resistance has not been found, however, for most peanut pathogens. Effectiveness of minor gene resistance in peanut (oligogenic or polygenic; additive effects) is strongly influenced by exposure to inoculum density (16, 83) and conduciveness of physical environments that influence host-defensive responses and/or pathogen activity (17, 25, 119). In addition to pathogen aggressiveness, other "fitness" characteristics (geographic adaptation, inoculum survival, etc) are of concern to the breeder (13, 114, 123, 129). Variability of peanut pathogens generally is poorly understood and much additional research is needed to optimize breeding efforts. A brief summary of current knowledge of pathogenic variability for peanut pathogens follows.

Foliar Diseases

It is generally assumed that peanut rust (*P. arachidis)* has inherent capabilities for development of physiologic races when confronted by genotypes possessing major gene resistance. However, rust resistance presently used by peanut breeders appears to include factors for "slow rusting" (polygenic, minor genes) and no authenticated report of physiologic specialization is known (69, 129). Resistance to both early (CA) and late (CP) leafspot pathogens may also be based on additive genetic effects (9). Nevertheless, host specificity has been suggested for both CA and CP (47, 93). Inoculation techniques in testing, i.e. spreader rows, may mask levels of partial resistance, however. Indeed, putative resistance for both leafspot pathogens failed when genotypes were evaluated in diverse geographic sites (129, 131). Gibbons suggested that both host specificity factors and pathogen adaptation to local growing conditions could occur (47). Although evidence for host specificity has not been conclusively demonstrated, pathogen adaptation to local environments has been reported (123). Additional information on differential sensitivity of partially resistant genotypes to temperature effects indicated that "location effects" on variability in severity or incidence of disease may be the result of alteration in host resistance metabolism rather than pathogen specificity (116). The possibility of environmental adaptation of local pathogen populations (penetration rate, survival of germinating spores, etc) should not, however, be discounted entirely (3, 13).

Although only minimal information is available on resistance to peanut virus diseases, extremely high variability obviously occurs in most plant and animal virus pathosystems. Peanut viruses are similar to many other viruses in this respect; several viruses of peanut are reported to have at least two to five "strains" based on symptomology and serology (35, 107, 109). Rapid development of virulent strains, specific to major gene resistance, is common in many crop species (43), and occurrence of new virulent strains of peanut viruses can be expected in the future.

Soil-Borne Pathogens

Peanut roots, pods, and stems are parasitized by a wide array of soil-borne fungi and nematodes. Resistance to soil-borne fungal pathogens is, to date, attributed to polygenic, additive gene effects. Physiologic specialization is generally not considered to develop under these circumstances (44, 141). However, increased virulence has been reported for *C. crotalariae* and other soil-borne pathogens in the presence of resistant genotypes (57). Effectiveness of partial resistance to soil-borne pathogens is influenced by inoculum density as well as conduciveness of environment (119). Aggressiveness characteristics of pathogen populations may, possibly, be of equal importance (31, 57, 122). Interactions between inoculum densities (ID) and pathogen aggressiveness characteristics often determine the effectiveness of resistance (16, 17, 25, 83). Knowledge of ID, range of aggressiveness for each major pathogen pest, as well as the potential for physiologic race development, are essential for long-term breeding efforts and control strategies (25, 30).

Peanut is host to several destructive ectoparasitic and endoparasitic nematode pests (109). At present no commercially available peanut cultivar in the US has any significant level of resistance to nematodes (102, 109). In other crops, however, development of resistant cultivars has quickly led to the recognition of new races or bio-types of the nematodes that place new cultivars in a pathologically vulnerable position (39). As Triantaphyllou (136) indicated, variability exists within each nematode species in both parasitism and aggressiveness, i.e. pathogenesis (136). Despite recurring problems in the development of hostspecific virulent races with most crops, breeding crop plants for resistance to the most destructive pathogenic nematodes has been an effective and economical method of minimizing crop loss (136). Both hypersensitive necrosis and reproductive resistance to *Meloidogyne* sp. has recently been reported in complex hybrids (tetraploid) of *A. hypogaea,* which should be useful in peanut breeding while also preventing race development (66, 101).

Bacterial wilt is found wherever peanuts are grown but generally is considered to be a disease of minor importance except in certain areas of Asia and Africa (109). Strains of *P. solanacearum* have been differentiated by both host range and biochemical tests (65). Five biovars and three pathogenic races of the bacterium are recognized (80) with Race 1 having a comprehensive host range, including several plants in the Leguminosae (23, 80). However, strains of Race 1 can be cross-inoculated to tobacco and may differ in pathogenicity to both crops in some instances. Peanut represents the first crop in which resistance has been successfully employed against *P. solanacearium.* The high level of resistance in Schwarz 21 has held up for over 60 years, and the bacterium has not developed new virulent strains (21a, 23, 109). Two additional resistant cultivars have recently been released in China (21a). However, because of the incomplete (partial) nature of resistance, environmental conditions influence expression and effectiveness of resistance (21a, 32).

NATURE OF RESISTANCE

Although the biochemical nature of resistance to peanut pathogens has been postulated for several diseases, the physiological mechanisms for resistance are not well understood. It can be assumed, however, that metabolic resistance mechanisms in peanut are similar to those reported for other hosts (37, 38, 42, 49, 62). Current strategies in disease resistance breeding use polygenic, additive gene effects to provide varying levels of "partial," i.e. incomplete, resistance (49). Performance of partially resistant cultivars in the field is based both on physiological reduction in disease severity or incidence, and "escape" mechanisms (29, 36, 114, 117, 119). Plant anatomy, e.g. canopy density or branching habit, as well as root growth dynamics, may be important in avoidance or compensation for disease. With monocyclic root or stem pathogens, inhibition of initial infection and/or restriction of lesion development are considered primary epidemiological mechanisms of resistance (62, 63). With polycyclic leafspot pathogens, several components of "rate-reducing" resistance may function singly or in combination (26, 44, 77, 103).

Foliar Diseases

The principal phytoalexin of peanut leaves infected with any of four leafspot pathogens was identified as medicarpin (126). However, eleven other antifungal compounds were also isolated from infected foliage (127). Although the role of phytoalexins in resistance to fungal pathogens is often debated, these antifungal compounds are assumed to inhibit pathogen ingress and/or reproduction in peanut tissues. Several anatomical and morphological characteristics of peanut tissue have also been associated with resistance to leafspot diseases. Resistance to both CA and CP was reported to be associated with formation of pectic substances and a thickening of cell walls (2). "Directed" growth of germ tubes toward stomata in susceptible cultivars has also been reported, whereas no directed growth was detected in resistant genotypes and less on moderately susceptible genotypes (2). Size of stomatal apertures has been correlated with resistance of field-grown *Arachis* sp.; however, Hassan et al and Cook did not find stomatal size to be a mechanism of resistance in later studies (30, 64). Ketring & Melouk (82) demonstrated that two peanut cultivars inoculated with *C. arachidicola* produced ethylene and had enhanced leaflet abscission, but an immune wild species produced only background levels of ethylene and retained its leaves (82). Recognition of epidemiological "components" of rate-reducing resistance in leafspot diseases of peanut has provided a major strategy for current breeding efforts (7, 26, 92). The "infection frequency" component of resistance has been used for both rust and leafspot diseases of peanut. Epidemics of peanut rust are apparently inhibited by a reduction in effectiveness of inoculum infecting resistant leaflets; thus, reproductive cycling of the pathogen is reduced even though only minor differences are detected in latent period (LP) or sporulation characteristics (7). Reduction in LP, lesion size, and duration of sporulation may also contribute, however, to inhibition of disease progress when infection occurs early in growing seasons.

A poor correlation is often reported between field performance of genotypes of *A. hypogaea* and number of CA and CP lesions developing in greenhouse tests when inoculated with a known range of inoculum densities (77, 103, 131). Resistance in other *Arachis* sp., however, does involve inefficiency of inoculum to induce lesions (40, 134). Reduction in size of lesions, and a corresponding reduction in number of conidia produced per lesion, is also a major resistance component for both CA and CP (7, 26, 77). Length of LP and percent of lesions that sporulate have both been reported to be important components of resistance in several studies (7, 92). Johnson et al concluded that the effects of several resistance components for CA were additive, and the higher magnitude of effectiveness of certain components could compensate for lesser effectiveness of others, resulting in a similar reduction in disease progress in field tests when interplot interference was minimal.

Biochemical and physiological mechanisms that determine host range and nonhost immunity to virus diseases are poorly understood (42), especially for peanut. However, most plants are nonhosts for most viruses and many viruses are limited to a few species of plants as hosts. The limited host range of a particular virus might be the result of its specialized adaptation to several aspects of the biology of a plant that could be transferred by breeding or genetic transformation technology (14, 108). Field resistance to peanut virus diseases may result from characteristics of the plant that inhibit vector contact or virus transmission (42, 81, 112). Resistance to helper (assistor) virus, or vector specificity for either the primary or a helper virus, may also be possible mechanisms of field resistance to peanut viruses (10, 42, 43). Additional mechanisms of resistance to virus diseases include restricted virus replication or accumulation in host tissue, virus localizing mechanisms, and the hypersensitive reaction (10, 108). A direct-interference phenomenon, which inhibits virus infection, i.e. "cross protection," can be induced by closely related viruses (108). Virus coat-protein mediated resistance has been demonstrated for several plant pathogens but mechanism(s) providing protection are not totally understood (14). It has recently been demonstrated that transgenic plants that express coat-protein genes of one virus can interfere with disease development of other nonrelated viruses (6).

Soil-Borne Pathogens

The great diversity of pathogens attacking below-ground portions of roots, pods and pegs, makes it difficult to generalize about common mechanisms of resistance. Fungal pathogens—i.e. *Pythium* sp., *C. crotalariae, Sclerotium rolfsii, Fusarium* sp., *Aspergillus* sp., etc—each have unique characteristics for virulence. Similarly, peanut tissues have evolved physical, metabolic, and anatomical adaptations for escape, or resistance, to attack by these microbes. Resistance to parasitic nematodes is similar in some mechanisms of resistance to fungal pathogens but quite different in other aspects.

Constitutive anatomical traits in mature peanut shells have been associated with resistance to *P. myriotylum* and *Rhizoctonia solani* pod-rotting diseases in Texas (50). Induction of periderm formation in tap and fibrous roots has been associated with resistance to *C. crotalariae* (62, 63). In this disease, however, it was suggested that containment of the pathogen by periderm tissue occurred subsequent to partial inhibition of the fungus by unknown metabolic factors (62). As indicated for foliar pathogens of peanut, a number of antimicrobial compounds can be produced by roots and stems of *Arachis* sp. when challenged by microbe infection (87, 126, 127). Many workers have associated the seed coat resistance to *Aspergillus* sp. (*A. flavus/A. parasiticus*) with the presence of different chemicals—5,7-dimethoxyisoflavone (139), tannin (79, 87, 113), and total soluble amino compounds and arabinose (5). However, Jambunathan et al (73) did not find significant correlation between seed colonization and polyphenol content in the seed coat. Necrotrophic pathogens, i.e. *S. rolfsii* and *Sclerotinia minor*, use a combination of oxalic acid and enzymatic degradation to parasitize limbs and stems. Mechanisms of field resistance for peanut and similar crops include escape due to canopy morphology, physical barriers to toxic compounds (waxy layers on stems, thick-walled cortical cells, cork cambium activity), and phytoalexin induction (1, 21, 29, 31). Physiological resistance to Sclerotinia blight of peanut appeared to be at least partially controlled by a cytoplasmic factor (29).

Extremely high levels of resistance or immunity to bacterial wilt have been reported for three peanut introductions (32, 109). Resistance in Schwarz 21 and two new Chinese cultivars is effective in field tests but can be overcome with high inoculum densities under conducive environmental conditions (21a, 32, 74). Although physiological mechanisms of bacterial wilt resistance in peanut have not been described, general mechanisms for resistance to bacterial wilt in other crops probably also function in peanut. It has been demonstrated in tobacco that postinfection host-responses limit multiplication (and subsequent distribution within host tissues) of virulent populations as well as incompatible races of *P. solanacearum* (23). Chemical alterations in host tissues during pathogenesis involved production or inhibition of several phytoalexins and related enzymatic pathways (23). Host respiration increased as water uptake and transport decreased. Mechanisms of resistance to plant parasitic nematodes of peanut are postulated to include both preinfection and postinfection phenomena (37, 94, 102, 143). Root leachates from healthy plants are reported to either enhance or suppress egg hatch, chemotaxis, and physiological behavior of some nematodes (67, 78). These mechanisms should be effective against both ectoparasitic and endoparasitic nematodes.

Several additional mechanisms of resistances are postulated to occur after penetration and feeding have been initiated by various nematode species (51, 67). A positive correlation between concentration of phenolics in plants and resistance to root-knot nematodes has been reported (120). Nutritional status of plants is thought to be important in resistance to nematodes in two ways: first, absence of certain nutrients may influence nematodes to move out of infected roots; and second, host nutrition may affect reproduction of ectoparasites or alter the sex ratio of root-knot nematodes within infected roots (67). Although production of phytoalexins has been investigated primarily in relation to fungal diseases, phytolexins also have been reported to be important in bacterial, viral, and nematode diseases (67, 78). Accumulation of phytoalexins in resistant plants has been found in root-knot infected soybean and cotton cultivars (67). Hypersensitive necrosis of peanut cells may localize invading pathogens and prevent further development of disease (101, 102). An incompatible host reaction to endoparasitic nematodes (hypersensitivity) is considered to be a defense mechanism by many pathologists (37). However, growth and development of juvenile nematodes in infected roots, as well as gall formation, can be inhibited without necrosis of host cells (101, 102).

Recognition by plant tissues of invading pathogens stimulates phytoalexin production, as well as hypersensitive necrosis. Breeders should be aware that dependence on hypersensitivity alone to provide resistance may result in the rapid development of new races of plant parasitic nematodes. Although it is desirable to provide high levels of protection in peanut to serious nematode pests, it may be more useful to incorporate levels of reproductive resistance in genotypes, whenever possible, to provide durable cultivar performance.

BREEDING STRATEGY

Although breeders are aware of the hazards of using sources of resistance with major genes, peanut breeders, like all breeders, use resistant sources that become available. However, only a few major genes for disease resistance have been identified in the peanut. Resistance to rosette virus, controlled by two independent recessive genes, is relatively easy to transfer to agronomically desirable types. Genotypes with resistance can be developed by backcrossing or by use of a breeding method that results in homozygous genotypes after hybridization. Some sources of rust resistance are also controlled by two recessive genes. These sources can be used in a manner similar to that for rosette resistance.

Most sources of resistance to soil-borne fungi in peanut show low levels of resistance or tolerance. Such partial resistance is presumably governed by polygenes and is assumed to be similar to horizontal resistance (44). There are practical difficulties in incorporating this type of resistance into germplasm with desired agronomic traits such as large fruit, high yield, a high oleic/linoleic ratio, and with superior organoleptic characteristics. Most resistant sources among cultivated genotypes originated from native landraces and generally have low yields and undesirable fruit and seed sizes, especially for the sophisticated market of the US. In the process of selecting plants for better agronomic traits in crosses involving these resistant sources, levels of resistance are often diluted. The persistent association of poor quality with southern stem rot and pythium pod rot resistance has concerned those breeding for resistance in peanut (122). A similar situation has been observed in breeding for CBR resistance. Resistant parents produce low yields and small irregular shaped fruit (149). Selection for larger fruit and higher yields reduces resistance (147). The highest yielding lines with large fruit (resulting from crosses of resistant and agronomically desirable types) are generally the least resistant of the progeny. The strategy has been to breed for a low level of host-pathogen coexistence that is stable, environmentally balanced, and economically useful; however, selections have been compromises between resistance and yield. Successful use of such cultivars requires excellent management skills that simultaneously reduce disease severity and inoculum reproduction (18, 28, 119).

Resistance to late and early leafspots has most often been considered quantitative with a high heritability (9, 76); however, resistance to late leafspot has been reported to be governed by five loci (104). Regardless of inheritance, levels of resistance to the leafspots appear to be higher than for soil-borne fungi and are easier to manipulate genetically.

When multiple disease resistance is needed, it is difficult to accumulate enough polygenes to provide good levels of resistance to all diseases if the genes governing resistance are inherited independently. Attempts to incorporate polygenes for resistance to two diseases may result in the loss of resistance to one disease as selection occurs for the second disease. Exceptions to this will occur if the same genes confer resistance to more than one disease, as may be the case in peanuts. During the development of resistance to *Pythium* pod rot, variation in reaction to *S. rolfsii* was also found (122). Similar observations were made in screening for CBR resistance. Several genotypes resistant to CBR were also resistant to southern stem rot (115, 117).

There has also been interest in selecting simultaneously for both early and late leafspot resistance. Early leafspot is the primary leafspot in North Carolina, but evidence suggests that the development of cultivars resistant to CA may increase the incidence of CP. Two strategies are being used to attempt to combine resistance to both leafspots in a single genotype. One strategy is to select for CP resistance among germplasm already selected for CA resistance. NC 5 and GP-NC 343, originally identified as being resistant to CA, and progenies from crosses involving these parents that were first selected for CA, were also found to be resistant to CP (86, 146). The level of partial resistance to both pathogens was only moderate but may be sufficient to manage the disease if cultural methods to reduce inoculum density are used. A second approach is to combine individual sources of resistance to CA and CP into a single genotype. Genes for resistance to CA and CP are inherited independently and can be incorporated into a single genotype (8, 9, 86). Germplasm with resistance to both rust and CP is being used at ICRISAT in an attempt to develop high-yielding cultivars with resistance to both diseases (71). Since only one of the resistance is polygenic, this approach should be successful (71).

Principal component analysis, tree diagrams, and biplots have been used to assess the potential for selecting components of resistance to rust and both leafspots simultaneously (7). Selection of genotypes resistant to the three diseases can be made based on different levels of partial components of resistance, depending on the goals of the breeder.

PROGRESS IN BREEDING FOR DISEASE RESISTANCE

Considerable progress in breeding for disease-resistant peanuts has been made since the release of the bacterial wilt-resistant Schwarz 21 in 1927 and the release of NC 2, a cultivar with resistance to southern stem rot in 1953.

Since 1956, breeders have developed and released rosette-resistant cultivars RG1, KH 149A, KH 241D, 69-101, RMP 12, and RMP 91 in West or southern Africa (20, 118). Rosette resistance also has been successfully transferred by backcrossing to a released cultivar, 28-206(R) (88). The transfer of rosette resistance (which is governed by major genes) to newer cultivars, continues in several West African and southern African locations (100).

Although Va 81B (resistant to Sclerotinia blight) and NC 8C and NC 10C (resistant to CBR) have been released in recent years in the US, progress in breeding for resistance to the soil-borne fungi has been difficult and slow. All of these cultivars have partial resistance with field performance dependent on inoculum density. These cultivars also have compromised one or more agronomic traits that make the cultivars less competitive in absence of the disease. Considerable cooperative breeding and pathology research is needed if soil-borne pathogens of peanut are to be managed using the available sources of partial resistance.

Progress in breeding for resistance to early and late leafspots and rust has been accelerated in the past decade. Southern Runner, a high-yielding CP-resistant cultivar, was released for use in the US. Greater progress has been made at ICRISAT, where cultivar development can be targeted to a less sophisticated market. Breeding at ICRISAT has concentrated on developing high-yielding cultivars with resistance or tolerance to both rust and CP. From early generation material supplied by ICRISAT to cooperators in India, resistant cultivars such as Girnar-1, DOR 8-10, and ALR 1 have been developed. High-yielding lines, ICG(FDRS)4 and ICG(FDSR)10, resistant to rust and moderately resistant to CP, are being considered for release in India. Use of the diploid species, *A. cardenasii,* has resulted in several breeding lines with levels of CP resistance exceeding that found in cultivated peanuts. One line, 259-2, has excellent resistance to CP and has also shown resistance to CA (71).

Breeding for resistance to CA has not led to the release of a cultivar in the US to date; however, adequate levels of resistance were found among progenies from crosses of GP-NC 343 by NC 5. Inbred lines from this cross have been used as resistant parents and selection is now being practiced in several crosses. Selection also is being practiced in crosses with four other resistant lines—PI 109839, PI 270806, PI 269685, and Kanyoma.

Several laboratories have initiated breeding programs for aflatoxin resistance, tomato spotted wilt virus resistance, nematode resistance, and other locally important diseases. Despite the intensified effort on screening for sources of resistance and the transfer of resistance genes to agronomically desirable genotypes, few cultivars with good agronomic traits and high levels of disease resistance have been developed. Perhaps the research effort has not been of suitable duration; however, the lack of qualitative sources of resistance for most diseases may partly be responsible for the slow progress in developing disease-resistant cultivars in peanut (4).

FUTURE EXPECTATIONS

Much progress in breeding for disease resistance in peanut can be expected during the next decade. Many sources of disease resistance among cultivated germplasm have been identified and are being used in breeding programs. Advanced breeding lines with disease resistance are currently being evaluated at numerous locations in the US and around the world; breeding for disease resistance has become a priority in most peanut breeding programs.

Considerable attention is also being devoted to using the wild species of *Arachis* for disease resistance. The high levels of resistance or immunity to early and late leafspots, rust, nematodes, peanut stunt virus, and tomato spotted wilt virus must be transferred to cultivated peanuts (125). Differences in ploidy level, and incompatibility of species outside section *Arachis* with the cultivated peanut, make it difficult to use the wild species. Stable 40-chromosome hybrid derivatives have been obtained for only a few interspecific hybrids. Because of the high levels of leafspot and rust resistance for these derivatives, the wild species will eventually contribute substantially to disease resistance in peanut.

Finally, several researchers are developing methodologies to use molecular techniques for improvement of the peanut. Perfection of a transformation and regeneration protocol for peanut, which is being evaluated at present, will allow researchers to incorporate genes from sources outside the genus (27). The first successes will probably involve cross-protection against virus diseases.

Literature Cited

1. Abawi, G. S., Provvidenti, R., Crosier, D. C., Hunter, J. E. 1978. Inheritance to white mold disease in *Phaseolus coccineus*. *J. Hered.* 69:200–2
2. Abdou, Y. A-M., Gregory, W. C., Cooper, W. E. 1974. Sources and nature of resistance to *Cercospora arachidicola* and *Cercosporidium personatum* in *Arachis* spp. *Peanut Sci.* 1:6–11
3. Alderman, S. C., Beute, M. K. 1987. Influence of temperature, lesion water potential, and cyclic wet-dry periods on sporulation of *Cercospora arachidicola* on peanut. *Phytopathology* 77:960–63
4. Allen, D. J. 1983. The pathology of tropical food legumes. In *Groundnut Diseases*, pp. 98–111. New York: Wiley
5. Amaya, F-J., Young, C. T., Norden, A. J., Mixon, A. C. 1980. Chemical screening for *Aspergillus flavus* resistance in peanut. *Oléagineux* 35:255–57
6. Anderson, E. J., Stark, D. M., Nelson, R. W., Powell, P. A., Tumer, N. E., Beachy, R. N. 1989. Transgenic plants

that express coat protein genes of tobacco mosaic virus or alfalfa mosaic virus interfere with disease development of some nonrelated viruses. *Phytopathology* 79:1284–90

7. Anderson, W. F., Beute, M. K., Wynne, J. C., Wongkaew, S. 1990. Statistical procedures for assessment of resistance in a multiple foliar disease complex of peanut. *Phytopathology* 80: 1451–59
8. Anderson, W. F., Wynne, J. C., Green, C. C. 1986. Potential for incorporation of early and late leafspot resistance in peanut. *Plant Breed.* 97:163–70
9. Anderson, W. F., Wynne, J. C., Green, C. C., Beute, M. K. 1986. Combining ability and heritability of resistance to early and late leafspot of peanut. *Peanut Sci.* 13:10–14
10. Ansa, O. A., Misari, S. M., Kuhn, C. W., Demski, J. W., Breyel, E., Casper, R. 1984. The current status of research on groundnut rosette virus disease in Nigeria. In *Proc. ICRISAT Reg. Groundnut Meet. W. Afr., 1st,* Patancheru, A. P. India: ICRISAT
11. Aujla, S. S., Chohan, J. S., Mehan, V. K. 1978. The screening of peanut varieties for the accumulation of aflatoxin and their relative reaction to the toxigenic isolate of *Aspergillus flavus* Link ex Fries. *J. Res. Punjab Agric. Univ.* 15: 400–3
12. Bartz, Z. A., Norden, A. J., LaPrade, J. C., Demuynk, T. J. 1978. Seed tolerance in peanut (*Arachis hypogaea* L.) to members of the *Aspergillus flavus* group of fungi. *Peanut Sci.* 5:53–56
13. Bashi, E., Rosen, J. 1974. Adaptation of four pathogens to semiarid habitats as conditioned by penetration rate and germinating spore survival. *Phytopathology* 64:1035–39
14. Beachy, R. N., Loesch-Fries, S., Turner, N. 1990. Coat protein-mediated resistance against virus infection. *Annu. Rev. Phytopathol.* 28:451–74
15. Beute, M. K., Wynne, J. C., Emery, D. A. 1976. Registration of NC 3033 peanut germplasm. *Crop Sci.* 16:887
16. Black, M. C., Beute, M. K. 1984. Relationship among inoculum density, microsclerotium size, and inoculum efficiency of *Cylindrocladium crotalariae* causing root rot on peanut. *Phytopathology* 74:1128–32
17. Black, M. C., Beute, M. K. 1985. Soil components that affect severity of Cylindrocladium black rot of peanut. *Plant Dis.* 69:36–39
18. Black, M. C., Pataky, J. K., Beute, M. K., Wynne, J. C. 1984. Management tactics that complement host resistance for control of Cylindrocladium blackrot of peanut. *Peanut Sci.* 11:70–73
19. Bock, K. R. 1989. ICRISAT Regional Groundnut Pathology Program: A review of research progress during 1985–87 with special reference to groundnut streak necrosis disease. In *Proc. Reg. Groundnut Workshop S. Afr., 13–18 March 1988, Lilongwe, Malawi, 3rd,* pp. 13–20. Patancheru, India: ICRISAT
20. Bockelee-Morvan, A. 1983. Les différentes variétés d'arachide: Répartition-géographique et climatique, disponibilité. *Oléagineux* 38:73–76. In French, summary in English
21. Boland, G. J., Hall, R. 1987. Evaluating soybean cultivars for resistance to *Sclerotinia sclerotiorum* under field conditions. *Plant Dis.* 71:934–36

21a. Boshou, L., Yuying, W., Xingming, X., Guiying, T., Yujun, T., Darong, S. 1990. Genetic and breeding aspects of resistance to bacterial wilt in groundnut. In *ACIAR Proc. No. 31,* pp. 39–47. Kingaroy, Aust.: Burnett Printing

22. Bromfield, K. R., Cevario, S. J. 1970. Greenhouse screening of peanut (*Arachis hypogaea*) for resistance to peanut rust (*Puccinia arachidis*). *Plant Dis. Rep.* 54:381–83
23. Buddenhagen, I. W., Kelman, A. 1964. Biological and physiological aspects of bacterial wilt caused by *Pseudomonas solanacearum*. *Annu. Rev. Phytopathol.* 2:203–30
24. Bunting, A. H., Gibbons, R. W., Wynne, J. C. 1985. Groundnut (*Arachis hypogaea* L.). In *Grain Legume Crops,* ed. R. J. Summerfield, E. H. Roberts, pp. 747–800. London: Collins
25. Carlin, D. E., Leiner, R. H. 1990. Effects of temperature on virulence of *Rhizoctonia solani* and other Rhizoctonia on potato. *Phytopathology* 80:930–34
26. Chiteka, Z. A., Gorbet, D. W., Knauft, D. A., Shokes, F. M., Kucharek, T. A. 1988. Components of resistance to late leafspot in peanut. II. Correlations among components and their significance in breeding for resistance. *Peanut Sci.* 15:76–81
27. Clemente, T. E., Weissinger, A. K., Moyer, J. M., Beute, M. K. 1990. Development of a transformation system for peanut. *Phytopathology* 80:1022
28. Cline, W. O., Beute, M. K. 1986. Effect of metam sodium, peanut genotype and inoculum density on incidence of Cylindrocladium black rot. *Peanut Sci.* 13:41–45
29. Coffelt, T. A., Porter, D. M. 1982.

Screening peanuts for resistance to Sclerotinia blight. *Plant Dis.* 66:385–87
30. Cook, M. 1981. Susceptibility of peanut leaves to *Cercosporidium personatum*. *Phytopathology* 71:787–91
31. Cooper, W. E. 1961. Strains of, resistance to, and antagonists of *Sclerotium-rolfsii*. *Phytopathology* 51:113–16
32. Darong, S., Chuenrung, C., Yuring, W. 1981. Resistance evaluation of bacterial wilt (*Pseudomonas solanacearum* E. F. Sm.) of peanut (*Arachis hypogaea* L.) in the Peoples' Republic of China. *Proc. Am. Peanut Res. Educ. Soc.* 13:21–28
33. de Berchoux, C. 1958. Etude sur la résistance de l'arachide en Haute Volta. Premiers résultats. *Oléagineux* 13:237–39
34. de Berchoux, C. 1960. La rosette de l'arachide en Haute Volta. Comportement des lignes résistantes. *Oléagineux* 15:237–39
35. Demski, J. W., Reddy, D. V. R., Wongkaew, S., Kaneya-Iwaki, M., Saleh, N., Xu, Z. Y. 1988. Naming of peanut stripe virus. *Phytopathology* 78:631–32
36. Dow, R. L., Powell, N. L., Porter, D. M. 1988. Effects of modification of the plant canopy environment on Sclerotinia blight of peanut. *Peanut Sci.* 15:1–5
37. Dropkin, V. H., Helgeson, J. P., Upper, C. D. 1969. The hypersensitivity reaction of tomatoes resistant to *Meloidogyne incognita:* Reversal by cytokinins. *J. Nematol.* 1:55–61
38. Ebel, J. 1986. Phytoalexin synthesis: The biochemical analysis of the induction process. *Annu. Rev. Phytopathol.* 24:235–64
39. Fassuliotis, G. 1987. Genetic basis of plant resistance to nematodes. In *Vistas on Nematology*, ed. J. A. Veech, D. W. Dickson, pp. 364–71. DeLeon Springs, FL: Painter Printing Co. 509 pp.
40. Foster, D. J., Stalker, H. T., Wynne, J. C., Beute, M. K. 1981. Resistance of *Arachis hypogaea* L. and wild relatives to *Cercospora arachidicola* Hori. *Oléagineux* 36:139–43
41. Foster, D. J., Wynne, J. C., Beute, M. K. 1980. Evaluation of detached leaf culture for screening peanuts for leafspot resistance. *Peanut Sci.* 7:98–100
42. Fraser, R. S. S. 1985. *Mechanisms of Resistance to Plant Diseases*. Boston, MA: Nijhoff/Junk. 462 pp.
43. Fraser, R. S. S. 1987. *Biochemistry of Virus-Infected Plants*. New York: Wiley. 259 pp.
44. Fry, W. E. 1982. *Principles of Plant Disease Management*. New York: Academic. 378 pp.
45. Garren, K. H., Wilson, C. 1951. Plant diseases. In *The Peanut, the Unpredictable Legume*, pp. 262–324. Washington, DC: Natl. Fertilizer Assoc.
46. Ghewande, M. P., Nagaraj, G., Reddy, P. S. 1989. Aflatoxin research at the Indian National Research Center for Groundnut. In *Aflatoxin Contamination of Groundnut: Proc. Int. Workshop, 6–9 Oct. 1987*, pp. 237–43. Patancheru, India: ICRISAT
47. Gibbons, R. W. 1966. Mycosphaerella leafspots of groundnuts. *FAO Plant Prot. Bull.* 15:25–30
48. Gibbons, R. W. 1980. The ICRISAT Groundnut Program. In *Proc. Int. Workshop Groundnuts, 13–17 Oct.*, pp. 12–16. Patancheru, India: ICRISAT
49. Giebel, J. 1982. Mechanism of resistance to plant nematodes. *Annu. Rev. Phytopathol.* 20:257–79
50. Godoy, R., Smith, O. D., Taber, R. A., Pettit, R. E. 1985. Anatomical traits associated with pod rot resistance in peanut. *Peanut Sci.* 12:77–82
51. Gommers, F. J., Bakker, J., Smits, L. 1980. Effects of singlet oxygen generated by the nematocidal compound α-terthienyl from Tagetes on the nematode *Aphelenchus avenae*. *Nematologica* 26: 369–75
52. Gorbet, D. W., Shokes, F. M., Jackson, L. J. 1982. Control of peanut leafspot with a combination of resistance and fungicide treatment. *Peanut Sci.* 9:87–90
53. Green, C. C., Beute, M. K., Wynne, J. C. 1983. A comparison of methods of evaluating resistance to *Cylindrocladium crotalariae* in peanut field tests. *Peanut Sci.* 10:66–69
54. Green, C. C., Wynne, J. C. 1987. Genetic variability and heritability for resistance to early leafspot in four crosses of Virginia-type peanut. *Crop Sci.* 27:18–21
55. Green, C. C., Wynne, J. C., Beute, M. K. 1983. Genetic variability and heritability estimates based on the F2 generation from crosses of large-seeded virginia-type peanuts with lines resistant to Cylindrocladium black rot. *Peanut Sci.* 10:47–51
56. Gregory, W. C., Gregory, M. P., Krapovickas, A., Smith, B. W., Yarbrough, J. A. 1974. Structure and genetic resources of peanuts. In *Peanuts—Culture and Uses*. Stillwater, OK: Am. Peanut Res. Educ. Assoc. Inc.
57. Hadley, B. A., Beute, M. K., Leonard, K. J. 1979. Variability of *Cylindrocladium crotalariae* response to resistant host plant selection pressure in peanut. *Phytopathology* 69:1112–14

58. Hadley, B. A., Beute, M. K., Wynne, J. C. 1979. Heritability of Cylindrocladium black rot resistance in peanut. *Peanut Sci.* 6:51–54
59. Hammons, R. O. 1977. Groundnut rust in the United States and the Caribbean. *PANS* 23:300–4
60. Hammons, R. O. 1982. Origin and early history of the peanut. In *Peanut Science and Technology*, ed. H. E. Pattee, C. T. Young, pp. 1–20. Yoakum, TX: Am. Peanut Res. Educ. Soc. Inc.
61. Harkness, C. 1977. The breeding and selection of groundnut varieties for resistance to rosette virus disease in Nigeria. *Inst. Agric. Res. Rep.* Ahmadu Bello Univ., Zaria, Nigeria
62. Harris, N. E., Beute, M. K. 1982. Histological responses of peanut germplasm resistant and susceptible to *Cylindrocladium crotalariae* in relationship to inoculum density. *Phytopathology* 72: 1250–56
63. Harris, N. E., Beute, M. K. 1982. *Cylindrocladium crotalariae*-induced periderm formation in tap root and fibrous roots of *Arachis hypogaea*. *Peanut Sci.* 9:82–86
64. Hassan, H. N., Beute, M. K. 1977. Evaluation of resistance to cercospora leaf spot in peanut germplasm potentially useful in a breeding program. *Peanut Sci.* 4:78–83
65. Hayward, A. C. 1991. Biology and epidemiology of bacterial wilt caused by *Pseudomonas solanacearum*. *Annu. Rev. Phytopathol.* 29:65–87
66. Holbrook, C. C., Noe, J. P. 1990. Resistance to *Meloidogyne arenaria* in *Arachis* spp. and its implication on development of resistant peanut cultivars. *Peanut Sci.* 17:35–38
67. Huang, J. C. 1985. Mechanisms of resistance to root-knot nematodes. In *An Advanced Treatise on Meloidogyne*, Vol. 1, ed. J. N. Sasser, C. C. Carter. Raleigh, NC: NC State Univ. Graphics. 422 pp.
68. International Board for Plant Genetic Resources (IBPGR). 1990. International crop network series. 2. Report of a workshop on the genetic resources of wild Arachis species including preliminary descriptors for Arachis(IBPGR/ICRISAT). Rome
69. International Crops Research Institute for the Semi-Arid Tropics. (ICRISAT) 1983. Rust disease of groundnut. *Info. Bull. No. 13,* Patancheru, India.15 pp.
70. International Crops Research Institute for the Semi-Arid Tropics. (ICRISAT) 1984.*Annual report 1983*. Patancheru, India: ICRISAT. 186 pp.
71. International Crops Research Institute for the Semi-Arid Tropics. (ICRISAT) 1986. *Annual report 1985*. Patancheru, India: ICRISAT. 250 pp.
72. International Crops Research Institute for the Semi-Arid Tropics. (ICRISAT) 1989. *Annual report 1988. pp. 119, 140*. Patancheru, India: ICRISAT
73. Jambunathan, R., Mehan, V. K., Gurtu, S. 1989. Aflatoxin contamination of groundnut. See Ref. 46, pp. 357–64
74. Jenkins, S. F., Hammons, R. O., Dukes, P. D. 1966. Disease reaction and symptom expression of seventeen peanut cultivars to bacterial wilt. *Plant Dis. Report.* 50:520–23
75. Jogloy, S. 1988. *Estimates of gene action on leafspot resistance and agronomic traits in peanut(* Arachis hypogaea *L.*). PhD thesis. NC State Univ., Raleigh. 121 pp.
76. Jogloy, S., Wynne, J. C., Beute, M. K. 1987. Inheritance of late leafspot resistance and agronomic traits in peanut. *Peanut Sci.* 14:86–90
77. Johnson, C. S., Beute, M. K., Ricker, M. D. 1986. Relationship between components of resistance and disease progress of early leaf spot on Virginia-type peanut. *Phytopathology* 76:495–99
78. Kaplan, D. T., Davis, E. L. 1987. Mechanisms of plant incompatibility with nematodes. See Ref. 39, pp. 267–76
79. Karchesy, J. J., Hemingway, R. W. 1986. Condensed tannins (48B– > 8; 2B– > O– > 7)-linked procyanidins in *Arachis hypogaea* L. *J. Agric. Food Chem.* 34:966–70
80. Kelman, A., Person, L. H. 1961. Strains of *Pseudomonas solanacearum* differing in pathogenicity to tobacco and peanut. *Phytopathology* 51:158–61
81. Kennedy, G. G. 1976. Host plant resistance and the spread of plant viruses. *Environ. Entomol.* 5:827–32
82. Ketring, D., Melouk, H. 1982. Ethylene production in leaflet abscission of three peanut genotypes infected with *Cercospora arachidicola*. *Plant Physiol.* 69: 789–92
83. Kinsbursky, R. S., Weinhold, A. R. 1988. Influence of soil inoculum density disease incidence relationships of *Rhizoctonia solani*. *Phytopathology* 78: 127–30
84. Kisyombe, C. T., Beute, M. K., Payne, G. A. 1985. Field evaluation of peanut genotypes for resistance to infection by *Aspergillus parasiticus*. *Peanut Sci.* 12:12–17
85. Knauft, D. A. 1987. Inheritance of rust resistance in groundnut. In *Groundnut*

Rust Disease: Proc. Discuss. Group Meet., 24–28 Sept. 1984, pp. 183–87. Patancheru, India: ICRISAT

86. Kornegay, J. L., Beute, M. K., Wynne, J. C. 1980. Inheritance of resistance to *Cercospora arachidicola* and *Cercosporidium personatum* in six virginia-type peanut lines. *Peanut Sci.* 7:4–9
87. Lansden, J. A. 1982. Aflatoxin inhibition and fungistatis by peanut tannins. *Peanut Sci.* 9:17–20
88. Mauboussin, J. C., Laurent, P., Delafond, G. 1970. Les variétés d'arachides recommandées au Sénégal et leur emploi. *Cah. Agric. Prat. Pays Chauds* 2:63–89
89. McDonald, D., Raheja, A. K. 1980. Pests, diseases, resistance, and crop protection in groundnuts. In *Advances in Legume Science,* ed. R. J. Summerfield, A. H. Bunting, pp. 501–14. London: HMSO
90. Mehan, V. K., McDonald, D., Ramakrishna, N. 1986. Varietal resistance in peanut to aflatoxin production. *Peanut Sci.* 13:7–10
91. Mehan, V. K., McDonald, D., Ramakrishna, N., Williams, J. H. 1986. Effects of genotype and date of harvest on infection of peanut seed by *Aspergillus flavus* and subsequent contamination with aflatoxin. *Peanut Sci.* 13:46–50
92. Melouk, H. A., Banks, D. J., Fanous, M. A. 1984. Assessment of resistance to *Cercospora arachidicola* in peanut genotypes in field plots. *Plant Dis.* 68:395–97
93. Miller, L. I. 1953. *Studies of the parasitism of* Cercospora arachidicola *Hori* and *Cercosporidium personatum* (Berk. & Curt.) Ell. and Ev. PhD thesis. Univ. Minn., St. Paul. 120 pp.
94. Minton, N. A., Hammons, R. O. 1975. Evaluation of peanut for resistance to the peanut root-knot nematode, *Meloidogyne arenaria*. *Plant Dis. Report.* 59:944–45
95. Misari, S. M., Ansa, O. A., Demski, J. W., Kuhn, C. W., Casper, O. F. R., Breyel, E. 1988. Groundnut rosette: Epidemiology and management in Nigeria. In *Coordinated Research on Groundnut Rosette Virus Disease: Summary Proc. Consult. Group Meet. Discuss. Collab. Res. Groundnut Rosette Virus Disease,* 8–10 March, Lilongwe, Malawi, pp. 15–16. Patancheru, India: ICRISAT
96. Mixon, A. C. 1979. Developing groundnut lines with resistance to seed colonization by toxin producing strains of *Aspergillus* species. *PANS* 25:394–400
97. Mixon, A. C., Rogers, K. M. 1973. Peanut accessions resistant to seed infection by *Aspergillus flavus*. *Agron. J.* 65:560–62
98. Murant, A. F., Rajeshwari, R., Robinson, D. J., Raschke, J. H. 1988. A satellite RNA of groundnut rosette virus that is largely responsible for symptoms of groundnut rosette disease. *J. Gen. Virol.* 69:1479–86
99. Nagrajan, V., Bhat, R. V. 1973. Aflatoxin production in peanut varieties by *Aspergillus flavus* Link and *Aspergillus parasiticus* Speare. *Appl. Microbiol.* 25:319–21
100. Ndunguru, B. J., Greenberg, D. C., Subrahmanyam, P. 1989. Groundnut improvement program at the ICRISAT Sahelian Center: Research problems, priorities, and strategies. See Ref. 19, pp. 105–13
101. Nelson, S. C., Simpson, C. E., Starr, J. L. 1989. Resistance to *Meloidogyne arenaria* in *Arachis* sp. germplasm. *J. Nematol.* 21:654–60
102. Nelson, S. C., Starr, J. L., Simpson, C. E. 1990. Expression of resistance to *Meloidogyne arenaria* in *Arachis batizocoi* and *A. cardenasii*. *J. Nematol.* 22: 423–25
103. Nevill, D. J. 1981. Components of resistance to *Cercospora arachidicola* and *Cercosporidium personatum* in groundnuts. *Annu. Appl. Biol.* 99:77–86
104. Nevill, D. J. 1982. Inheritance of resistance to Cercosporidium personatum in groundnuts: A genetic model and its implications for selection. *Oléagineux* 37:355–62
105. Nigam, S. N. 1987. A review of the present status of the genetic resources of the ICRISAT Regional Groundnut Improvement Program, of the Southern African Cooperative Regional Yield Trials, and of rosette virus resistance breeding. In *Proc. Reg. Groundnut Workshop S. Afr., 2nd, 10–14 Feb. 1986,* Harare, Zimbabwe, pp. 15–30. Patancheru, A. P., India: ICRISAT
106. Nigam, S. N., Bock, K. R. 1990. Inheritance of resistance to groundnut rosette virus in groundnut (*Arachis hypogaea*). *Ann. Appl. Biol.* 117:553–60
107. Paguio, O. R., Kuhn, C. 1974. Survey for peanut mottle virus in peanut in Georgia. *Plant Dis. Report.* 58:107–10
108. Ponz, F., Bruening, G. 1986. Mechanisms of resistance to plant viruses. *Annu. Rev. Phytopathol.* 24:355–81
109. Porter, D. M., Smith, D. H., Rodriguez-Kabana, R. 1982. Peanut diseases. In *Peanut Science and Technology,* ed. H. E. Pattee, C. T. Young, pp. 326–

410. Yoakum, TX: Am. Peanut Res. Educ. Soc.
110. Pua, A. R., Medalla, E. C. 1986. *Screening for resistance to* Aspergillus flavus *invasion in peanut*. Presented at *Anniv. Ann. Conv. Pest Control Counc. Philippines, 17th. 8–10 May,* Iloila City, Philippines
111. Rao, K. S., Tulpule, P. G. 1967. Varietal differences of groundnut in the production of aflatoxin. *Nature* 214:738–39
112. Rossell, H. W. 1977. Preliminary investigations on the identity and ecology of legume virus diseases in northern Nigeria. *Trop. Grain Legume Bull.* 8: 41–46
113. Sanders, T. H., Mixon, A. C. 1978. Effect of peanut tannins on percent seed colonization and in vitro growth by *Aspergillus parasiticus*. *Mycopathologia* 66:169–73
114. Shew, B. B., Beute, M. K. 1984. Effects of crop management on the epidemiology of southern stem rot of peanut. *Phytopathology* 74:530–35
115. Shew, B. B., Beute, M. K., Bailey, J. E. 1985. Potential for improved control of southern stem rot of peanut with resistance and fungicides. *Peanut Sci.* 12:4–7
116. Shew, B. B., Beute, M. K., Wynne, J. C. 1988. Effects of temperature and relative humidity on expression of resistance to *Cercosporidium personatum* in peanut. *Phytopathology* 78:493–98
117. Shew, B. B., Wynne, J. C., Beute, M. K. 1987. Field, microplot, and greenhouse evaluations of resistance to *Sclerotium rolfsii* in peanut. *Plant Dis.* 71: 188–91
118. Sibale, P. K., Kisyombe, C. T. 1980. Groundnut production, utilization, research problems, and further research needs in Malawi. See Ref. 89, pp. 249–53
119. Sidebottom, J. R., Beute, M. K. 1989. Control of Cylindrocladium black rot of peanut with cultural practices that modify soil temperature. *Plant Dis.* 73:672–76
120. Singh, B., Choudury, B. 1973. The chemical characteristics of tomato cultivars resistant to root-knot nematodes (*Meloidogyne* spp.). *Nematologica* 19: 443–48
121. Singh, A. K., Subrahmanyam, P., Moss, J. P. 1984. The dominant nature of resistance to *Puccinia arachidis* in certain wild *Arachis* species. *Oléagineux* 39:535–38
122. Smith, O. D., Boswell, T. E., Gricher, W. J., Simpson, C. E. 1989. Reaction of select peanut (*Arachis hypogaea* L.) lines to southern stem rot and Pythium pod rot under varied disease pressure. *Peanut Sci.* 16:9–13
123. Sommartya, T., Beute, M. K. 1986. Temperature effects on germination and comparative morphology of conidia for Thai and USA isolates of *Cercosporidium personatum*. *Peanut Sci.* 13:67–70
124. Sowell, G., Smith, D. H., Hammons, R. O. 1976. Resistance of peanut plant introductions to *Cercospora arachidicola*. *Plant Dis. Report.* 60:494–98
125. Stalker, H. T., Moss, J. P. 1987. Speciation, cytogenetics, and utilization of *Arachis* species. *Adv. Agron.* 41:1–40
126. Strange, R. N., Edwards, C., Richards, S. 1988. Phytoalexins of groundnuts and their role in resistance to fungal parasites. *Phytoparasitica* 16:202–3
127. Subba Rao, P. V., Geiger, J. P., Einhorn, J., Rao, B., Malosse, C., et al. 1988. Host defense mechanisms against groundnut rust. *Int. Arachis Newsl.* 4:16–18
128. Subrahmanyam, P., Ghanekar, A. M., Nolt, B. L., Reddy, D. V. R., McDonald, D. 1985. Resistance to groundnut diseases in wild Arachis species. In *Proc. Int. Workshop Cytogenet. Arachis,* 31 Oct.–Nov. 1983, pp. 49–55. Patancheru, India: ICRISAT
129. Subrahmanyam, P., Hammons, R. O., Nigam, S. N., McDonald, D., Gibbons, R. W., et al. 1983. International cooperative screening for resistance of peanut to rust and late leaf spot. *Plant Dis.* 67:1108–11
130. Subrahmanyam, P., McDonald, D. 1983. Rust disease of groundnut. International Crops Research Institute for the Semi-Arid Tropics (ICRISAT) *Info. Bull.* No. 13. Patancheru, India
131. Subrahmanyam, P., McDonald, D., Gibbons, R. W., Nigam, S. N., Nevill, D. J. 1982. Resistance to rust and late leafspot diseases in some genotypes of *Arachis hypogaea*. *Peanut Sci.* 9:6–10
132. Subrahmanyam, P., McDonald, D., Gibbons, R. W., Subba Rao, P. V. 1983. Components of resistance to *Puccinia arachidis* in peanuts. *Phytopathology* 73:253–56
133. Subrahmanyam, P., McDonald, D., Subba Rao, P. V. 1983. Influence of host genotype on uredospore production and germinability in *Puccinia arachidis*. *Phytopathology* 73:726–29
134. Subrahmanyam, P., Moss, J. P., McDonald, D., Subba Rao, P. V., Rao, V. R. 1985. Resistance to leafspot caused by *Cercosporidium personatum* in wild *Arachis* species. *Plant Dis.* 69:951–54

135. Subrahmanyam, P., Moss, J. P., Rao, V. R. 1983. Resistance to peanut rust in wild *Arachis* species. *Plant Dis.* 67: 209–12
136. Triantaphyllou, A. C. 1987. Genetics of nematode parasitism on plants. See Ref. 39, pp. 354–63
137. Tsai, A. H., Yeh, C. C. 1985. Studies on aflatoxin contamination and screening for disease resistance in groundnuts. *J. Agric. Res. China* 34:79–86
138. Tulpule, P. G., Bhat, R. V., Nagraj, V. 1977. Variations in aflatoxin in production due to fungal isolates and crop genotypes and their scope in prevention of aflatoxin production. *Arch. Inst. Pasteur Tunis* 54:487–93
139. Turner, R. B., Lindsey, D. L., Davis, D. D., Bishop, R. D. 1975. Isolation and identification of 5,7-dimethoxyisoflavone, an inhibitor of *Aspergillus flavus* from peanut. *Mycopathologia* 57:39–40
140. Utomo, S. D. 1990. *Inheritance of resistance to* Aspergillus parasiticus *Speare in peanut (*Arachis hypogaea *L.)*. MS thesis, NC State Univ., Raleigh. 56 pp.
141. Van der Plank, J. E. 1968. *Disease Resistance in Plants*. New York: Academic
142. Vasudeva Rao, M. J., Nigam, S. N., Mehan, V. K., McDonald, D. 1989. *Aspergillus flavus* resistance breeding in groundnut: Progress made at ICRISAT Center. See Ref. 46, pp. 345–55
143. Veech, J. A. 1982. Phytoalexins and their role in resistance of plants to nematodes. *J. Nematol.* 14:2–9
144. Waliyar, F., Bockelee-Morvan, A. 1989. Resistance of groundnut varieties to *Aspergillus flavus* in Senegal. See Ref. 46, pp. 305–10
145. Waliyar, F., McDonald, D., Nigam, S. N., Subba Rao, P. V. 1989. Resistance to early leafspot of groundnut. See Ref. 19, pp. 49–54
146. Walls, S. B., Wynne, J. C., Beute, M. K. 1985. Resistance to late leafspot peanut of progenies selected for resistance to early leafspot. *Peanut Sci.* 12:17–22
147. Wynne, J. C., Beute, M. K., Bailey, J. E., Mozingo, R. W. 1990. Registration of 'NC 10C' peanut. *Crop Sci.* 30:In press
148. Wynne, J. C., Halward, T. 1989. Cytogenetics and genetics of *Arachis*. In *Critical Reviews in Plant Science,* ed. B. V. Conger, pp. 189–220. Boca Raton, FL: CRC Press
149. Wynne, J. C., Rowe, R. C., Beute, M. K. 1975. Resistance of peanut genotypes to *Cylindrocladium crotalariae*. *Peanut Sci.* 2:54–56
150. Zambettakis, C., Waliyar, F., Bockelee-Morvan, A., de Pins, O. 1981. Results of four years of research on resistance of groundnut varieties to *Aspergillus flavus*. *Oléagineux* 36:377–85

Annu. Rev. Phytopathol. 1991. 29:305–23

THE PHYTOPATHOLOGICAL SIGNIFICANCE OF MYCELIAL INDIVIDUALISM

A. D. M. Rayner

School of Biological Sciences, University of Bath, Claverton Down, Bath BA2 7AY, England

KEY WORDS: incompatibility in fungi, intraspecific antagonism, developmental variation, genetic variation, life history strategies

---

INTRODUCTION

Phytopathologists have long been aware of the considerable intraspecific variation that can exist in natural populations of phytopathogenic mycelial fungi. However, the significance of this variation has largely been perceived in terms of fungal-plant rather than interfungal relationships. Correspondingly, questions about variation in the pathogenicity and virulence of different fungal lineages have received more attention than questions about the way in which these lineages have the potential to interact antagonistically or synergistically with one another (55). The purpose of this article is to explore the many ways in which mycelial interactions may, nonetheless, be relevant to our understanding of the etiology and control of plant disease.

An important step in understanding mycelial interactions is to have a clear perspective of the units of selection in natural fungal populations. Historically, this issue has been clouded both by the indeterminate nature of fungal mycelia, which have the potential to continue extension indefinitely, and by the concept that hyphal anastomosis could allow extensive cooperation between colonies of diverse genetic origin (22, 57, 58). The resultant view that members of the same species could pool their genetic resources in a versatile

0066-4286/91/0901-0305$02.00

manner, and hence, in effect, be subject to "group selection" (23) gained widespread acceptance. However, at about the time that the group-selection mechanisms in general evolutionary biology was being fully exposed as illogical (33), an alternative concept of the "individualistic mycelium" was proposed for fungi (68, 70, 71, 81). This concept emphasized the fundamentally territorial nature of fungal mycelia, or, perhaps more appropriately, of the nuclear and mitochondrial genomic populations that inhabit the protoplasmic domains within mycelia (61).

This territoriality is epitomized by the occurrence of often overtly antagonistic nonself rejection or somatic (= vegetative) incompatibility reactions. In higher fungi, capable of somatic hyphal fusion, these reactions entail activation of cell death or "senescence" pathways and commonly follow encounters between genetically different mycelia, helping to maintain their physiological, ecological, and genetic integrity. In many basidiomycetes, and some ascomycetes, the rejection responses are opposed to "access" mechanisms that allow invasion of nonself genes, for example, via nuclear migration (60). The resultant dynamic interplay or "tension" between rejection and access may sometimes result in evolutionarily creative instabilities (61); and in basidiomycetes without specialized sex organs, such tension is central to the process of sexual outcrossing. In interbreeding populations of these latter fungi, somatic incompatibility is typically expressed between secondary mycelia (stable heterokaryons containing populations of two types of nuclei with compatible mating-type alleles), but is usually overridden between sexually compatible homokaryons (68). By contrast, somatic incompatibility is often expressed directly between homokaryons in ascomycetes, which can nonetheless outcross by means of specialized gametangia that circumvent the somatic rejection response.

Somatic nonself access mechanisms in ascomycetes with such sex organs do not therefore determine mating compatibility as such, as they do in somatogamous basidiomycetes. They can, however, lead to mycelial or intrahyphal invasion and takeover by nonself genomes, and may also increase the fecundity of compatible matings by enhancing the range over which sexual exchange can occur (17, 48, 75, 76). Moreover, it is important to recognize that the operation of rejection following hyphal fusion need not absolutely prevent somatic genetic interactions at the boundaries between mycelia (60, 62). For this reason we should be wary of the different interpretations of fungal population structure that could arise if demarcation zones between colonies and the absolute inability of two colonies to form a heterokaryon ("heterokaryon incompatibility") are interchanged indiscriminately as criteria of somatic incompatibility (60, 62). Heterokaryon incompatibility may sometimes reflect inability to anastomose, rather than expression of a postfusion rejection response (31).

Recognition of genetic individualism in fungal mycelia allows an approach to fungal population biology that follows basic principles similar to those used in studying the population biology of plants and animals. An essential part of this approach involves the characterization of patterns of distribution of genetic variation among individual population members, or, to use a term derived from plant population biology (47), individual "genets". This term is valuable because it helps to convey the idea of genetic individualism in organisms that develop indeterminately and are capable of asexual reproduction, and avoids the ambiguity often implicit in the use of the term "clone" (20). Accordingly, for fungi an individual genet encompasses all descendants (ramets), whether propagated asexually or by hyphal proliferation, from an identical genetic source. Naturally occurring genets are probably predominantly homokaryotic in ascomycetes, whereas in outcrossing basidiomycete populations they are heterokaryotic secondary mycelia.

An important corollary of fungal individualism is that the ability to produce developmental variation, ultimately involving differential gene expression within genets, rather than collectivistic pooling between genets, is necessary to provide versatility. Versatility, i.e. the ability to perform different functions in different spatiotemporal locations, is an important attribute of many phytopathogenic fungi, especially those with necrotrophic or saprotrophic nutritional phases, as they face contrasting selection pressures or demands at different stages in their life cycles.

What follows, then, is an attempt to analyze the significance of interactions between genets, and developmental variation within genets, in relation to the infectivity, survival, and evolution of phytopathogenic fungi. It will become evident that many of the concepts draw on experience with wood and bark-inhabiting fungi, where their relevance is arguably clearest. However, it is hoped that the arguments may stimulate interest in how well these concepts apply to pathogens of nonperennating plants or plant parts.

INDIVIDUALISM AND THE DEVELOPMENT OF INOCULUM POTENTIAL

A long-standing phytopathological concept, arguably more applicable to necrotrophic than biotrophic infections, is that of the invasive force or inoculum potential of phytopathogenic fungi (42). This inoculum potential is set against the defensive, or, as I often prefer to think of it, the reparative potential of the host, with the often delicately counterpoised balance between these two potentials determining the outcome of infection.

Implicit in the idea of inoculum potential is the need for a successful pathogen to have or to establish a sufficient inoculum base from which to gain access to host tissues and create a favorable microenvironment for itself therein. This requirement may often entail positive interactions among fungal entities, either directly with one another or through their joint action on the host, and hence depends critically on modes of arrival and establishment of the pathogen.

Going one step further back, it is also important to note that, in correspondence with their predominantly indeterminate patterns of development, many phytopathogenic fungi are to varying degrees capable of an independent saprotrophic existence. Inoculum may therefore arise both from infected plants and from other, nonliving sources, where competition with resident, purely saprotrophic organisms has an important effect on fitness, and may be manipulable in biological control systems. Moreover, the dividing line between pathogenic and nonpathogenic behavior is by no means always precise, and small changes in conditions may often be sufficient to cause a switch from one category of behavior to another. For these reasons, the interrelations between genets colonizing nonliving resources may be equally important to interrelations on or in living hosts in our overall understanding of many pathogenic fungi and their competitors.

Modes of Arrival—Spores or Mycelium?

Because individual host plants are discontinuously distributed, phytopathogenic fungi require means of arrival at them, generally either by means of spores and other propagules or by means of migratory mycelium. These distinct modes of arrival generate contrasting patterns of distribution and interaction of individual genets (29, 63, 73).

Where arrival can only be achieved by means of spores, the physiological domains of individual genets will inevitably be constrained within the boundaries of individual host plants or nonliving resource units (food bases). Such genets may be described as "resource-unit-restricted" (29, 73).

There is clearly no scope for cooperative interactions through mycelial connections between resource-unit-restricted genets in different plants. Moreover, the scope for cooperation, through hyphal anastomosis and other means, between separate spores or infections from separate spores on the same plant may depend in part on their relatedness.

Asexual reproduction, nonoutcrossing breeding strategies, and a predominance of local over long-range spore inputs will enhance the probability that adjacent infections on the same plant or in the same disease patch or nonliving resource unit will be genetically identical or related to one another. All these factors may therefore be significant in the enhancement of inoculum potential through cooperative interactions. By the same token, outcrossing breeding strategies and long-range input of spores will enhance the probability of antagonistic mycelial interactions. Some evidence suggests that such antagonistic interactions may reduce expression of disease caused by certain phytopathogenic fungi (49, 53), and that both the rate of decay and production of fruit bodies may be inhibited by such interactions in wood-inhabiting fungi (28). However, the effect of such interactions between mycelia in reducing disease expression may be modified by indirect interactions arising, for example, through the influence of one infection in enhancing susceptibility or resistance of plant tissue to another infection. Studies on the influence on disease expression of antagonistic and synergistic interactions between species of microorganisms have a long history (32). The influence of genetic interrelationships on such interactions within fungal species, and their bearing on infection processes now seems ripe for further investigation.

Fungal genets capable of mycelial extension beyond the physical boundaries of individual plants or resource units (hence "non-unit-restricted") have the potential to form enormous genetic territories, measurable in hectares and centuries-old in the case of some wood-inhabiting forms such as *Armillaria* spp. and *Phellinus weirii* (e.g. 35, 79). Small genets establishing alongside such large ones are liable to become engulfed, thus reducing overall population diversity.

Arrival by means of mycelium obviously confers greater inoculum potential than arrival by spores (73), and is often associated with an "ectotrophic" infection habit, as exhibited by *Gaeumannomyces graminis* and *Heterobasidion annosum* (42), in which production of superficial non-assimilative mycelium precedes penetration of host tissues by trophic mycelium (80). There can be import of resources from food bases located elsewhere, as well as joint action and buffering against adverse microenvironmental factors among hyphal ramets of the same genet at the arriving mycelial front. The migratory mycelium may also be organized into specialized rhizomorphic organs or mycelial cords through which resources can be channelled during growth through soil or litter. Alternatively, specialized bridging mycelium may form at root or stem contacts; this mycelium may be strengthened and protected from adverse biotic and abiotic influences by the action of phenoloxidases causing melanization, and formation of hydrophobic, sclerotized tissues—as in *Hymenochaete corrugata* (5).

Mycelial outgrowth patterns between food bases can be interpreted ecologically in terms of foraging strategies, parallel to those, for example, of army ant raids or stoloniferous plants (36, 37, 39). An important consideration in understanding these patterns may be their cost-effectiveness in locating appropriate hosts or food bases with minumum expenditure of biomass. Such cost-effectiveness depends on the probability of locating new bases within the vicinity of present bases, which in turn is likely to reflect both saprotrophic and parasitic potential. Depending on this probability, the commitment of biomass into foraging mycelium may be varied from densely branched systems spreading over a short range, to sparsely branched systems distributed over a long range. Short-range foraging systems tend to exhibit a high degree of lateral communication, and large-scale reallocation of resources may follow successful colonization of a new food base. By contrast, long-range foraging units emerging from a food base may act more independently from one another and perhaps have a greater sensitivity to environmental stimuli, allowing directed growth towards suitable food sources. Variations in apical dominance between truly rhizomorphic structures and marginally effuse mycelial cord systems may be interpretable in these terms (69).

Whereas mycelial interactions between unit-restricted genets can only occur within or upon food bases, those between non-unit-restricted genets can also occur between foraging mycelial systems (38). The spatial pattern of foraging systems may therefore be critical in determining the complexity of boundaries between non-unit-restricted genets; generally, the less branched the foraging system, the more complex the boundary may be expected to be. Such effects may also be worth consideration in terms of the use of non-unit-restricted saprotrophs for biological control of non-unit-restricted pathogens, e.g. the use of mycelial-cord-forming decomposer basidiomycetes against *Armillaria* spp. (39).

Modes of Establishment: Blitzkrieg or Trojan Horse?

Establishment of fungi in plant tissues may or may not give rise directly to overt symptoms. Where establishment does not induce symptoms, at least initially, it can be regarded as cryptic, latent, or endophytic. Current use of the term "endophytism" tends to subsume latency (26; O. Petrini, personal communication), although some prefer to reserve the latter term for infections that eventually give rise to symptoms.

The distribution patterns and interactions of genets resulting from overt and endophytic modes of establishment are liable to be distinctive. In cases of overt infection, and also during colonization of nonliving-resource units, arrival by genetically different spores is more likely to result in the presence of competitive neighboring genets than is arrival by mycelium.

In analyzing the effects of the resulting competition on the distribution patterns of genets in resource units it is useful to distinguish between two distinct aspects or phases of competition, which can be described as primary resource capture and combat (29). Primary resource capture involves the processes of gaining initial access to and influence over resource supplies and is dependent on such factors as rates of arrival and growth as well as host responses to infection. For example, in a study of competition between physiologic races of *Puccinia hordei* on barley, it was found that at high inoculum density, the race with most rapid germination, and hence most likely to succeed in competition for stomatal entry, predominated. By contrast, at low inoculum density, the race with the higher spore viability, the greatest pustule size, and the higher spore production per pustule predominated (41).

A general consequence to be expected from primary capture competition following multiple arrivals on a surface of a readily colonizable resource unit is that the number of genets present will be progressively reduced as access to the interior of the structure is gained. This effect can be particularly evident in bulky units such as the trunks and branches of trees (1, 28). An interesting outcome is that since the surface area of an object increases in proportion to volume when the linear scale against which it is measured is reduced, self-similar patterns of population structure may be expected in units of vastly different sizes. Methodologically, this means that the sampling patterns used to analyze population structures need to be adjusted to be appropriate to the scale of the object being studied.

As primary resource capture progresses, so combative interactions based on directly antagonistic mechanisms at the boundaries between neighboring genets become increasingly important in determining distribution patterns. Where these interactions result in mutual exclusivity or deadlock, the boundaries may remain stable indefinitely. Somatic incompatibility reactions commonly result in such stable boundaries, both in culture and in nature. However, in some fungi, and under some circumstances, the boundaries may become unstable, allowing invasion of one genet by another (17, 18, 75, 76) and leading to changing distribution patterns. Interspecific combat leading to replacement and deadlock may also increasingly influence distribution patterns as time and the transition to purely saprotrophic nutrition within plant residues proceeds.

Endophytic establishment results in a more or less extended period during which the host remains healthy until a change in microenvironmental conditions within the tissue heralds a switch to a mode of mycelial development that is actively parasitic and/or destructive of plant cells. The change in conditions may be brought about by the activity of the fungus itself, the onset of plant senescence, some form of stress to the host, or a combination of all three factors. It is often very difficult to disentangle the contributions of each of these factors and consequently to discern the degree to which the fungus should be regarded as truly parasitic, lowering the evolutionary fitness of its host. Correspondingly, the boundaries between endophytism and various types of biotrophic and necrotrophic infections, including vascular wilts, may be ill-defined (12). Equally, it is important to appreciate the competitive benefits in primary resource capture following host stress or death that are afforded by endophytic establishment (29).

A large number of wood- and bark-inhabiting fungi appear to become established endophytically in healthy trees, with water shortage, variously caused, in the host or host organs, probably being the major factor directly or indirectly eliciting a change to active mycelial development (9, 12, 14, 26, 27, 63, 64).

Where fungi are established endophytically in bark, a high probability of multiple infections arising from independent spore arrivals would be expected. The resulting population on any one tree or branch, assuming that the spores are genetically different but not subject to strong directional selection, would contain many, spatially localized genets. An example of this type of population structure is provided by an outcrossing species of *Phomopsis*, which inhabits the bark of elm, *Ulmus* (21).

Many fungi causing localized cankers also presumably establish by means of independent spore arrivals at the bark surface (e.g. 2). However, fungi associated with another pattern of bark death, which extends for several or even tens of meters along the length of branches or trunks and either occupies the entire circumference or is confined to elongated strips, may have a very different origin. Where death is confined to strips of bark, the latter may be termed "strip cankers", and characteristically follow the axial orientation of the tree's conductive system. Observations suggest that strip cankers originating in high order branches may extend onto lower order (finer) branches, and indeed lead to complete die-back of the latter, but not vice versa (65).

Current evidence suggests that individual fungal genets associated with strip cankers and equivalent die-back phenomena tend to be axially very extensive in the underlying wood, indeed sometimes as long as the cankers themselves. Examples include a range of wood-decaying basidiomycetes, including *Stereum gausapatum, Vuilleminia comedens, Peniophora quercina, Phellinus ferreus* and *Phlebia rufa* in attached branches of *Quercus* (13, 15, 16), *Piptoporus betulinus* in trunks and branches of *Betula* (3), and *Stereum rugosum* (72) and *Hymenochaete corrugata* (5) in coppiced poles of *Corylus avellana*. Ascomycetous examples include several members of the Xylariaceae, for example, *Daldinia concentrica* and *Hypoxylon rubiginosum* in *Fraxinus* (11, 74) and *Biscogniauxia nummularia* and *Hypoxylon "purpureum"* in *Fagus* (76, 77), as well as Diatrypaceae (65; J. K. Mukiu & S. Hendry, personal communication) and the deuteromycete, *Cryptostroma corticale* (10).

The axially extensive distribution of these fungal genets, which appear often to become active within a single growing season, strongly suggests that they are systemically preestablished in the wood of healthy trees, either as propagules or sparse mycelium proliferated from rare entry sites. In some sense, their behavior could therefore be considered to resemble that of vascular wilt fungi, except in that they only become active under conditions of stress affecting the tree.

Another pattern of endophytic establishment appears to occur in certain strongly host-selective species of *Hypoxylon*, notably *H. fragiforme* in *Fagus* (26, 27, 74; A. Inman, personal communication) and *H. fuscum* in *Corylus avellana*. Here numerous individual genets, each occupying a relatively small (< 2 cm^3) pocket of decay, become active as the moisture content of the sapwood is reduced below a critical threshold. The small size of the perithecial stromata produced externally by these fungi may be an indicator of the small domains occupied by each individual genet. Experimental studies have shown that these genets can become established both from bark and from foci deep within the wood (27), whence the competitive elimination during primary capture when numerous genets invade a resource from an external surface (see above) do not occur. How access is gained to such internal locations is not clear. I. H. Chapela (personal communication) has shown that ascospores of *H. fragiforme* specifically recognize, become attached to, and then penetrate bark of *Fagus*. Subsequently, the fungus lives within single medullary ray cells before growing out, when the host is stressed, initially as microhyphae and then as macrohyphae invading vessel lumina. However, the sequence from bark penetration to the endophytic phase has not been documented. Evidence also indicates that the ascospores of *H. fragiforme* germinate preferentially on leaves of *Fagus*, leading to endophytic establishment therein (O. Petrini, personal communication). It seems conceivable then that dormant infections established in leaf traces could become progressively buried as secondary thickening continues, until a change in microenvironment allows active mycelial development. This would parallel the behavior suggested for the heartrot fungus, *Echinodontium tinctorium*, which establishes from infections in buried branch stubs (40).

INDIVIDUALISM AND THE LIFE HISTORY STRATEGIES OF PHYTOPATHOGENIC FUNGI

In trying to develop a predictive understanding of the way in which natural selection mechanisms determine the distinctive population structures arising from varying modes of arrival and establishment of phytopathogenic fungi, it is clearly important to appreciate how these modes are related to individual life styles or niches. This problem raises in turn the question of how best to categorize the various types of behavior to make appropriate comparisons. One approach is through identification of life history or ecological strategies evolved in response to *r*- and *K-selection* (8). The particular value of this approach rather than, for example, focusing on the particular modes of pathogenesis or nutrition that it subsumes (50, 51), lies in the ability to draw on concepts and principles developed from general ecological and evolutionary theory.

R-, C- and S-Selection in Phytopathogenic Fungi

According to the general premise of strategy theory, *r*-selection favors organisms with a short life span and rapid commitment to reproduction, whereas *K*-selection favors a long life span and slow or intermittent commitment to reproduction. Within the *r–K* continuum it is useful, following Grime (45), but with some modifications, to identify three opposing forms of selection, which result in three primary strategies. R-selection, due to recurrent ecological disturbance, i.e. any more or less sudden environmental event that renders new domains available for exploitation, leads to ruderal strategies. S-selection, due to stress, i.e. any form of continuously imposed environmental extreme that tends to restrict biomass production by most organisms in question, leads to stress-tolerant or stress-adapted strategies. C-selection, due to a high potential incidence of competitors under conditions relatively free from stress, leads to combative strategies (29).

In examining how ecological strategy concepts may be applied to phytopathogenic fungi, note that strategies are defined in terms of responses to different types of selection operating within a particular context. Strategies are therefore relative, not absolute, and should not be thought of as a way of classifying species, or indeed indeterminate individual genets, in any taxonomic sense.

With respect to phytopathogenic fungi, R-, C-, and S-selection are largely determined by the nature of the host plant or organ in terms of its longevity and susceptibility and responsiveness to infection, which in turn are conditioned by the life history strategy of the plant itself. To some degree, all long-lived, mature, undamaged, nonsenescent plant tissues can be thought of as extreme environments, imposing S-selection, which greatly reduces the range of fungi able to colonize them, sometimes to the extent of determining the individual genotypes of successful colonists (30). On the other hand, the production of juvenile, damaged, and ephemeral tissues imposes R-selection, favoring organisms capable of early arrival and rapid primary resource capture. Senescence and cell death may either impose R-selection or, in cases where it is associated with the production of secondary metabolites, as in heartwood formation and hypersensitive responses, S-selection. C-selection occurs on or in long-lasting plant tissues or remains where, for whatever reason, the intensity of S-selection is reduced.

Whereas the action of S-selection on phytopathogenic fungi may be expected to result in dependency and unit-restriction, R-selection and C-selection are liable to be associated with enhanced independent saprotrophic capabilities and non-unit-restriction. Moreover, while R-selection will favor clonal modes of reproduction, both C- and S-selection may favor production of variable offspring, but with the likelihood of direct interactions between the latter being reduced under S-selection.

In these terms, it becomes possible to envisage how the balance between R-, S-, and C-selection critically influences the potential for mycelial interactions following varying modes of arrival and establishment in plant tissues. Note also that this balance is liable to change significantly over the life span of individual fungal genets not subject to purely S-selection, eliciting fundamental changes in behavior such as those associated with hemibiotrophy (52) and the activation of endophytic infections. In these latter cases, establishment under highly S-selective conditions allows a considerable advantage in primary resource capture as these are superseded by R- and C-selective conditions following damage, stress, or senescence affecting host tissues.

Routine and Episodic Selection

Although S-, C-, and R-selection all involve some degree of local environmental unpredictability, they all have sufficently recurrent effects on a temporal scale to be encompassed within what Brasier (19) has termed the "routine" selection characteristic of stable ecosytems. Such temporally recurrent selection pressures can be contrasted with irregular or nonrecurrent "episodic" selection that occurs when populations are, due to internal or external mechanisms, subject to a radical departure from previous conditions.

In essence, episodic selection provides evolutionary opportunities that enable derivatives from pre-existing populations to emerge into a new niche (62). Human intervention in natural ecosystems, through the propagation of crop monocultures, and intercontinental introductions, have been important sources of such opportunities for phytopathogenic fungi, most particularly within the twentieth century.

Episodic selection characteristically opens up a much wider niche or window of opportunity for individual genets than is available in populations subject to routine C- and S-selection pressures. The fitness of offspring from a genet able to enter this niche may be undermined by recombinatorial mechanisms. Clonal reproduction will therefore, at least initially, be favored prior to the reimposition of routine selection pressures as the niche stabilizes and/or becomes internally diversified (19, 62).

The Maintenance, Regulation, and Role of Genetic Variation in Pathogen Populations

Having introduced the probable relationship of genetic variation to life history strategies and mycelial interactions in phytopathogenic fungi, it becomes important to appreciate how such variation is sustained and regulated, and how it influences the fitness of individual genets.

On current evidence, the principal mechanism sustaining substantial heterozygosity in fungal populations involves sexual outcrossing and meiotic recombination. The diverse mechanisms of somatic or parasexual recombination that have been detected seem unlikely to contribute significantly to natural variation (25), at least not under routine selection conditions.

A key feature of many fungi is their ability to reproduce both asexually, via conidia and sporangiospores, and by sexual outcrossing. Moreover, there is increasing evidence for an ability, within a taxonomic species, to interconvert between outcrossing and nonoutcrossing breeding strategies (e.g. 60). The relative contribution of outcrossing, nonoutcrossing and asexual reproductive pathways to succeeding generations, in effect, provides a sensitive control over patterns of population variation.

A principal factor promoting variable population structures is microhabitat diversity (54, 56), such as that found in strongly S-selected pathogens infecting a variable host population. In such situations, individual host plants may be strongly selective for particular pathogen genotypes. The problem facing a fungal genet inhabiting an individual plant may then be presented as follows. Do its genes stand a better chance of becoming established in a neighboring plant if it reproduces clonally or by outcrossing. If it reproduces clonally the danger of arriving on a resistant host has to be carefully balanced against the benefits of potentially increased inoculum potential and of avoiding any of the genetic or energetic cost of sex.

An important consideration here is that adjacent plants are more likely to be related to one another than distant plants. One expectation would therefore be for clonal dispersal mechanisms to be favored over a short range, and outcrossing mechanisms to be favored over long range. Consequently, small-scale diversity may be disproportionately lower than large scale diversity within a population. Several investigators are currently looking into this possibility (e.g. 2; S. L. Anagnostakis, personal communication). It is a relatively small step from the situation just described to one where outcrossing mechanisms are dispensed with altogether, particularly under the influence of clonal agriculture. This leads to the prediction that while pathogens infecting crops may tend towards nonoutcrossing and asexual reproduction, their relatives inhabiting wild host-plant populations may tend towards outcrossing. Some indications for such effects are provided by *Septoria nodorum* infecting Graminae (25) and *Crinipellis perniciosa* infecting cocoa and tropical lianes (44).

In situations where S-selection is insufficiently strong to prevent different genets from coming into proximity, the benefits of outcrossing may be further called into question. One widely held view is that the resulting variation and incompatibility mechanisms help to reduce the transfer of somatic infections in long-lived organisms (in R-selected organisms infections are not such a major threat to individual fitness). Examples where somatic incompatibility limits the spread of infectious agents in fungi, and hence incidentally limits the use of such agents as means of biological control, have been found in

Aspergillus amstelodami (24), *Ophiostoma ulmi* (17, 18), and *Cryphonectria parasitica* (7). Attractive though this argument may be, however, it is important to recall that while outcrossing may lower the risk of somatic infections, at the same time it enhances the possibility of sexually transmitted infectious agents (6; cf 46).

Although a reduction in S-selection may in one sense be thought to reduce microhabitat diversity, and hence increase the possibility of intraspecific conflict, it also enhances the probability of interspecific competition. Such competition introduces another form of uncertainty, in terms of where, when, and what sort of competitors are likely to be encountered, and hence another context in which genetic variation and outcrossing may be beneficial. However, as with all other arguments presented so far, this possibility focuses on the implications of outcrossing in terms of its major perceived consequence, i.e. genetic variation, at the population level of organization. It may also be appropriate to consider the implications of outcrossing in terms of genomic rather than organismal interrelationships.

Genomic Symbiosis

Sexual outcrossing is commonly thought of as an interaction, to their mutual benefit, between two different nuclear genomes. This benefit is perceived both in terms of the production of variable offspring, just discussed, and in terms of heterosis. The latter occurs where complementation occurs between two different haploid genomes, or genomic populations, which coexist either within the same nuclear membrane or, as in heterokaryotic cells, within the same plasma membrane. Such positive genomic interactions are, however, irrelevant for the many outcrossing organisms, including many fungi, with predominantly haploid life cycles, and so can most readily be envisaged as a secondary outcome of coevolutionary positive feedback and development of dependency (6).

Another perspective comes from considering the relationships between associated genomes following similar principles to those used in interpreting interorganism symbiosis, using the latter term in the original sense of De Bary (34) as any form of living together, regardless of consequences. Underlying this interpretation is the idea that any biological entity, if access can be gained to it, represents a potentially exploitable resource or habitat for another entity. For the genomes inhabiting one mycelial genet, another fungal genet represents potentially exploitable domain. Entry into that domain opens up possibilities for a wide spectrum of stable and unstable outcomes of interactions with resident nuclear and mitochondrial populations.

Basidiomycetes, in which somatic hyphal fusion is a prerequisite for sexual outcrossing, have recently been used as model systems for examining these varying outcomes of interactions between genomic populations within the same protoplasm. The results have been reviewed elsewhere (60, 61) and so will only be summarized here.

At one end of the spectrum are the stable compatible matings between homokaryons belonging to the same breeding population; at the other end are the mutual rejection responses characteristic of interactions between heterokaryotic secondary mycelia, and between homokaryons that are not competent to mate. In between these extremes are responses between homokaryons from geographically or ecologically isolated populations, or populations with different breeding strategies, leading to genomic takeovers, degeneration, and complex phenotypic emergence. These unstable outcomes appear to depend strongly on cytoplasmic background. A current idea is that nuclear-mitochondrial relationships are instrumental in determining the stability or instability of genomic associations. A further proposal is that unstable genomic associations may be "creative" under conditions of episodic selection (62).

As already implied, the ability to interconvert between outcrossing and nonoutcrossing breeding strategies may be a vital ingredient in responses to episodic selection. Genomic interactions between representatives of these different breeding strategies may therefore have a considerable evolutionary significance. Such interactions between representatives from outcrossing and nonoutcrossing populations have been studied in the basidiomycete, *Stereum hirsutum* (6). It was found that while nonoutcrossing homokaryons were resistant to invasion by nonself nuclear DNA, commensurate with expression of a rejection response, outcrossing homokaryons were sometimes susceptible to invasion by nonoutcrossers. Genomic takeover, degeneration associated with a change in resident mitochondrial DNA, and production of variable basidiospore progeny were all found in different examples of outcrossing–nonoutcrossing interactions. These results demonstrate the negative consequences of allowing nonself access and resultant intrinsic vulnerability of outcrossing populations. However, since viable, variable progeny were also sometimes generated, they also indicate one route by which variation can arise in emergent nonoutcrossing populations. This is of interest because nonoutcrossing fungal populations that are nonetheless variable, being subdivided into few or many clonal subpopulations, have been detected in several species, including the phytopathogens *Crinipellis perniciosa* (44), *Stereum sanguinolentum* (4, 72), and *Rosellinia desmazieresii* (78).

The Control of Developmental Variation and its Relation to Life Cycle Predictability

Although genetic variability may confer a degree of versatility at the population level of organization, at the level of the individual genet, versatility may largely be expected to be based on epigenetic mechanisms that regulate the expression rather than the type of genetic instructions. Although it is sometimes envisaged that versatility could arise through mutation and extensive heterokaryosis, it is difficult to imagine how such mechanisms could either be relied upon or controlled. It is of course true that basidiomycetes commonly form stable heterokaryons as part of their life cycle, but there is little evidence that these can contain more than two genomic types. Chaos theory (43) may provide an interesting explanation for this in that, while two attractor systems can have simple solutions, three attractor systems and above have complex, locally unpredictable solutions.

The complex life cycles of many phytopathogenic fungi as they respond to changing selection pressures inside and outside their host necessitate sometimes drastic alterations in phenotype. In terms of patterns of mycelial morphogenesis these can include transitions between determinate and indeterminate development; slow, dense and fast, effuse branching patterns; assimilative and nonassimilative mycelium; diffuse and aggregated hyphal systems, and juvenile and senescent phases (65–67). The importance of such transitions between mycelial developmental modes during the life cycle of a particular pathogen, *Heterobasidion annosum,* have been discussed by Stenlid & Rayner (80). In this case, while the different selection pressures acting on this only partly S-selected fungus during its life cycle are predictable, the sequence in which these pressures occur is unpredictable. By contrast, in uredospores of *Puccinia graminis* normally germinating on wheat, the requisite infection structures are produced in a highly prescribed sequence even in the absence of host cues (82). In the latter case, an endogenously controlled program that fully anticipates each stage in development is apposite. However, under more unpredictable circumstances, the use of external environmental cues may become more important in controlling development.

The mechanisms controlling switching between distinctive mycelial modes are not understood. However, an analysis of the relationships between nonself recognition phenomena and developmental regulation in fungi allows an hypothesis to be advanced (69a). This is that in indeterminate developmental systems, the mitochondrion provides a very sensitive, manipulable executive because of its interfaces with both calcium- and cAMP-based second messenger systems and active transport systems. Moreover, disfunction of Krebs cycle-associated primary metabolism represents a simple way of switching cells into secondary metabolism, activating, or derepressing, for example, the shikimic acid and acetate pathways, and phenoloxidase enzyme systems. The resultant production of highly hydrophobic compounds and free radicals would then have important effects on cell behavior, including cell death responses and the control of turgor potential gradients between sources and sinks within mycelial networks. Such a feedback mechanism, in addition to explaining the close relationship between developmental regulation and mycelial interaction responses (67), may also underlie the apparently close parallels between the genetical and physiological mechanisms involved in interfungal and host-pathogen interactions (59). An understanding of what these parallels imply at the level of cellular and molecular feedback mechanisms may yet allow studies of mycelial individualism to be of significance to phytopathology (or vice versa) in an unexpected way.

CONCLUSIONS

This review has focused on the problems facing an individual, plant-infecting, fungal mycelial genet in terms of the maintenance of evolutionary fitness, both of itself, and of its progeny. The extent to which competition with other genets of the same or different species features as a major determinant of its life style depends largely on the degree to which it is adapted to S-, R-, and C-selection in stable ecosystems, or to episodic selection in unstable ecosystems. Developmental versatility is an important component of fitness where changes in microenvironmental conditions attend colonization and exploitation of host and nonliving resources. The degree to which such versatility is internally or externally regulated may depend on the predictability of microenvironmental change. Knowledge of the relationship between developmental regulation and self-nonself recognition systems may contribute to better understanding of the evolutionary biology of phytopathogens in terms of interfungal relationships. It may also help in the understanding of inter- and intracellular interactions of general significance in eukaryote biology.

ACKNOWLEDGMENTS

My thanks to BP Venture Research Unit for supporting some of the work referred to in this review. Thanks also to Drs A. M. Ainsworth, L. Boddy, J. M. Clarkson, R. C. Cooke, and M. Mogie for commenting on a first draft.

Literature Cited

1. Adams, D. H., Roth, L. F. 1969. Intraspecific competition among genotypes of *Fomes cajanderi* decaying young growth Douglas-fir. *Forest Sci.* 15:327–31
2. Adams, G., Hammar, S., Proffer, T. 1990. Vegetative compatibility in *Leucostoma persoonii*. *Phytopathology* 80:287–91
3. Adams, T. J. H. 1982. Piptoporus betulinus, *some aspects of population biology*. PhD thesis. Univ. Exeter, UK
4. Ainsworth, A. M. 1987. Occurrence and interactions of outcrossing and non-outcrossing populations in *Stereum, Phanerochaete* and *Coniophora*. In *Evolutionary Biology of the Fungi,* ed. A. D. M. Rayner, C. M. Brasier, D. Moore, pp. 285–99. Cambridge: Cambridge Univ. Press
5. Ainsworth, A. M., Rayner, A. D. M. 1990. Aerial mycelial transfer by *Hymenochaete corrugata* between stems of hazel and other trees. *Mycol. Res.* 94:263–88
6. Ainsworth, A. M., Rayner, A. D. M., Broxholme, S. J., Beeching, J. R. 1990. Occurrence of unilateral genetic transfer

and genomic replacement between strains of *Stereum hirsutum* from non-outcrossing and outcrossing populations. *New Phytol.* 115:119–28
7. Anagnostakis, S. L. 1987. Chestnut blight: the classical problem of an introduced pathogen. *Mycologia* 79:23–37
8. Andrews, J. H., Rouse, D. I. 1982. Plant pathogens and the theory of *r*- and *K*-selection. *Am. Nat.* 120:283–96
9. Bevercombe, G. P. 1980. *Diseases affecting sycamore bark*. PhD thesis. Univ. Exeter, UK
10. Bevercombe, G. P., Rayner, A. D. M. 1982. Population structure of *Cryptostroma corticale,* the cau¬al fungus of sooty bark disease of sycamore. *Plant Pathol.* 33:211–17
11. Boddy, L., Gibbon, O. M., Grundy, M. A. 1985. Ecology of *Daldinia concentrica:* effect of abiotic variables on mycelial extension and interspecific interactions. *Trans. Br. Mycol. Soc.* 85: 201–11
12. Boddy, L., Griffith, G. S. 1989. Role of endophytes and latent invasion in the development of decay communities in sapwood of angiospermous trees. *Sydowia Ann. Mycol.* 41:41–73
13. Boddy, L., Rayner, A. D. M. 1982. Population structure, intermycelial interactions and infection biology of *Stereum gausapatum*. *Trans. Br. Mycol. Soc.* 78:337–51
14. Boddy, L., Rayner, A. D. M. 1983. Origins of decay in living deciduous trees: the role of moisture content and re-appraisal of the expanded concept of tree decay. *New Phytol.* 94:623–41
15. Boddy, L., Rayner, A. D. M. 1983. Mycelial interactions, morphogenesis and ecology of *Phlebia radiata* and *Phlebia rufa* from oak. *Trans. Br. Mycol. Soc.* 80:437–48
16. Boddy, L., Rayner, A. D. M. 1983. Ecological roles of basidiomycetes forming decay communities in attached oak branches. *New Phytol.* 93:177–88
17. Brasier, C. M. 1984. Inter-mycelial recognition systems in *Ceratocystis ulmi:* their physiological properties and ecological importance. In *The Ecology and Physiology of the Fungal Mycelium,* ed. D. H. Jennings, pp. 451–97. Cambridge: Cambridge Univ. Press
18. Brasier, C. M. 1986. The population biology of Dutch elm disease: its principal features and some implications for other host-pathogen systems. *Adv. Plant Pathol.* 5:53–118
19. Brasier, C. M. 1987. The dynamics of fungal speciation. See Ref. 4, pp. 231–60
20. Brasier, C. M., Rayner, A. D. M. 1987. Whither terminology below the species level in fungi? See Ref. 4, pp. 379–88
21. Brayford, D. 1990. Vegetative incompatibility in *Phomopsis* from elm. *Mycol. Res.* 94:745–52
22. Burnett, J. H. 1976. *Fundamentals of Mycology,* 2nd ed. London: Arnold
23. Carlile, M. J. 1987. Genetic exchange and gene flow: their promotion and prevention. See Ref. 4, pp. 203–14
24. Caten, C. E. 1972. Vegetative incompatibility and cytoplasmic infection in fungi. *J. Gen. Microbiol.* 72:221–29
25. Caten, C. E. 1987. The genetic integration of fungal life styles. See Ref. 4, pp. 215–29
26. Chapela, I. H. 1989. Fungi in healthy stems and branches of American beech and aspen: a comparative study. *New Phytol.* 113:65–75
27. Chapela, I. H., Boddy, L. 1988. Fungal colonization of attached beech branches. II. Spatial and temporal organization of communities arising from latent invaders in bark and functional sapwood under different moisture regimes. *New Phytol.* 110:47–57
28. Coates, D., Rayner, A. D. M. 1985. Fungal population and community development in cut beech logs. I. Establishment via the aerial cut surface. *New Phytol.* 101:153–71
29. Cooke, R. C., Rayner, A. D. M. 1984. *Ecology of Saprotrophic Fungi*. London/New York: Longman
30. Cooke, R. C., Whipps, J. M. 1987. Saprotrophy, stress and symbiosis. See Ref. 4, pp. 137–48
31. Correll, J. C., Klittisch, C. J. R., Leslie, J. F. 1989. Heterokaryon self-incompatibility in *Gibberella fujikuroi (Fusarium moniliforme)*. *Mycol. Res.* 93:21–27
32. D'Aeth, H. R. X. 1939. A survey of interaction between fungi. *Biol. Rev.* 14:105–31
33. Dawkins, R. 1976. *The Selfish Gene*. Oxford: Oxford Univ. Press
34. De Bary, A. 1887. *Comparative Morphology and Biology of the Fungi, Mycetozoa and Bacteria*. Oxford: Clarendon Press
35. Dickman, A., Cook, S. 1989. Fire and fungus in a mountain hemlock forest. *Can. J. Bot.* 67:2005–16
36. Dowson, C. G., Rayner, A. D. M., Boddy, L. 1986. Outgrowth patterns of mycelial cord-forming basidiomycetes from and between woody resource units in soil. *J. Gen. Microbiol.* 132:203–11
37. Dowson, C. G., Rayner, A. D. M., Boddy, L. 1988. Foraging patterns of

Phallus impudicus, Phanerochaete laevis and *Steccherinum fimbriatum* between discontinuous resource units in soil. *FEMS Microbiol. Ecol.* 53:291–98
38. Dowson, C. G., Rayner, A. D. M., Boddy, L. 1988. The form and outcome of mycelial interactions involving cord-forming decomposer basidiomycetes in homogeneous and heterogeneous environments. *New Phytol.* 109:423–32
39. Dowson, C. G., Springham, P., Rayner, A. D. M., Boddy, L. 1989. Resource relationships of foraging mycelial systems of *Phanerochaete velutina* and *Hypholoma fasciculare* in soil. *New Phytol.* 111:501–9
40. Etheridge, D. E., Craig, H. M. 1976. Factors influencing infection and initiation of decay by the Indian paint fungus *(Echinodontium tinctorium)* in western hemlock. *Can. J. For. Res.* 6:299–318
41. Falahati-Rastegar, M., Manners, J. G., Smartt, J. 1983. Factors determining results of competition between physiologic races of *Puccinia hordei*. *Trans. Br. Mycol. Soc.* 81:233–39
42. Garrett, S. D. 1970. *Pathogenic Root-Infecting Fungi*. Cambridge: Cambridge Univ. Press
43. Gleick, J. 1988. *Chaos*. London: Heinemann
44. Griffith, G. W. 1989. *Population biology of the cocoa pathogen,* Crinipellis perniciosa. PhD thesis, Univ. Wales, Aberystwyth
45. Grime, J. P. 1979. *Plant Strategies and Vegetation Processes*. Chichester/New York: Wiley
46. Hamilton, W. D. 1980. Sex versus non-sex versus parasite. *Oikos* 35:282–90
47. Harper, J. L. 1977. *Population Biology of Plants*. London: Academic
48. Kolmer, J. A., Leonard, K. J. 1986. Loci independent of mating type that condition secondary sexual incompatibility in the fungus *Cochliobolus heterostrophus*. *Heredity* 57:85–91
49. Lebeau, J. B. 1975. Antagonism between isolates of a snow mould pathogen. *Phytopathology* 65:877–80
50. Lewis, D. H. 1973. Concepts in fungal nutrition and the origin of biotrophy. *Biol. Rev.* 48:261–78
51. Lewis, D. H. 1974. Micro-organisms and plants: the evolution of parasitism and mutualism. *Symp. Soc. Gen. Microbiol.* 24:367–92
52. Luttrell, E. S. 1974. Parasitism of fungi on vascular plants. *Mycologia* 66:1–15
53. Matsumoto, N., Tajimi, A. 1983. Intra- and intertaxon competition among dikaryons of *Typhula incarnata* and *T. ishikariensis* biotypes A, B, and C. *Trans. Mycol. Soc. Jpn* 24:459–65
54. Maynard-Smith, J. 1978. *The Evolution of Sex*. Cambridge: Cambridge Univ. Press
55. McDonald, B. A., McDermott, J. M., Goodwin, S. B., Allard, R. W. 1989. The population biology of host-pathogen interactions. *Annu. Rev. Phytopathol.* 27:77–94
56. Michod, R. E., Levin, B. R., eds. 1988. *The Evolution of Sex*. Sunderland, MA: Sinauer
57. Raper, J. R. 1966. *Genetics of Sexuality in Higher Fungi*. New York: Ronald Press
58. Raper, J. R., Flexer, A. S. 1970. The road to diploidy with emphasis on a detour. *Symp. Soc. Gen. Microbiol.* 20: 401–32
59. Rayner, A. D. M. 1986. Mycelial interactions—genetic aspects. In *Natural Antimicrobial Systems,* ed. G. W. Gould, M. E. Rhodes-Roberts, A. K. Charnley, R. M. Cooper, R. G. Board, pp. 277–96. Bath: Bath Univ. Press
60. Rayner, A. D. M. 1990. Natural genetic transfer systems in higher fungi. *Trans. Mycol. Soc. Jpn* 31:75–87
61. Rayner, A. D. M. 1991. The challenge of the individualistic mycelium. *Mycologia* 83:48–71
62. Rayner, A. D. M. 1991. Monitoring genetic interactions between fungi in terrestrial habitats. In *Genetic Interactions Between Microorganisms in the Microenvironment,* ed. E. M. Wellington, J. D. Van Elsas. Manchester: Manchester Univ. Press. In press
63. Rayner, A. D. M., Boddy, L. 1986. Population structure and the infection biology of wood-decay fungi in living trees. *Adv. Plant Pathol.* 5:119–60
64. Rayner, A. D. M., Boddy, L. 1988. Fungal communities in the decay of wood. *Adv. Microbial Ecol.* 10:115–66
65. Rayner, A. D. M., Boddy, L. 1988. *Fungal Decomposition of Wood: Its Biology and Ecology*. Chichester/New York: Wiley. 587 pp.
66. Rayner, A. D. M., Boddy, L., Dowson, C. G. 1987. Genetic interactions and developmental versatility during establishment of decomposer basidiomycetes in wood and tree litter. *Symp. Soc. Gen. Microbiol.* 41:83–123
67. Rayner, A. D. M., Coates, D. 1987. Regulation of mycelial organisation and responses. See Ref. 4, pp. 115–36
68. Rayner, A. D. M., Coates, D., Ainsworth, A. M., Adams, T. J. H., Williams, E. N. D., Todd, N. K. 1984. The

biological consequences of the individualistic mycelium. See Ref. 17, pp. 509–40
69. Rayner, A. D. M., Powell, K. A., Thompson, W., Jennings, D. H. 1985. Morphogenesis of vegetative organs. In *Developmental Biology of Higher Fungi*, ed. D. Moore, L. A. Casselton, D. A. Wood, J. C. Frankland, pp. 249–79. Cambridge: Cambridge Univ. Press
69a. Rayner, A. D. M., Ross, I. K. 1991. Sexual politics in the cell. *New Sci.* 129:30–33
70. Rayner, A. D. M., Todd, N. K. 1977. Intraspecific antagonism in natural populations of wood-decaying basidiomycetes. *J. Gen. Microbiol.* 103:85–90
71. Rayner, A. D. M., Todd, N. K. 1979. Population and community structure and dynamics of fungi in decaying wood. *Adv. Bot. Res.* 7:333–420
72. Rayner, A. D. M., Turton, M. N. 1982. Mycelial interactions and population structure in the genus *Stereum: S. rugosum, S. sanguinolentum* and *S. rameale*. *Trans. Br. Mycol. Soc.* 78:483–93
73. Rayner, A. D. M., Watling, R., Frankland, J. C. 1985. Resource relations—an overview. See Ref. 69, pp. 1–40
74. Sharland, P. R. 1987. *Mycelial biology of Xylariaceous fungi*. PhD thesis, Univ. Bath, UK
75. Sharland, P. R., Rayner, A. D. M. 1986. Mycelial interactions in *Daldinia concentrica*. *Trans. Br. Mycol. Soc.* 86:643–50
76. Sharland, P. R., Rayner, A. D. M. 1989. Mycelial interactions in outcrossing populations of *Hypoxylon*. *Mycol. Res.* 93:187–98
77. Sharland, P. R., Rayner, A. D. M. 1989. Mycelial ontogeny and interactions in non-outcrossing populations of *Hypoxylon*. *Mycol. Res.* 93: 273–81
78. Sharland, P. R., Rayner, A. D. M., Ofong, A. U., Barrett, D. K. 1988. Population structure of *Rosellinia desmazieresii* causing ring-dying of *Salix repens*. *Trans. Br. Mycol. Soc.* 90:654–56
79. Shaw, C. G. III, Roth, L. F. 1976. Persistence and distribution of a clone of *Armillaria mellea* in a ponderosa pine forest. *Phytopathology* 66:1210–13
80. Stenlid, J., Rayner, A. D. M. 1989. Environmental and endogenous controls of developmental pathways: variation and its significance in the forest pathogen, *Heterobasidion annosum*. *New Phytol.* 113:245–58
81. Todd, N. K., Rayner, A. D. M. 1980. Fungal individualism. *Sci. Prog.* 66: 331–54
82. Wanner, R., Förster, H., Mendgen, K., Staples, R. C. 1985. Synthesis of differentiation-specific proteins in germlings of the wheat stem rust fungus after heat shock. *Exp. Mycol.* 9:279–83

Annu. Rev. Phytopathol. 1991. 29:325–42

MAINTAINING GENETIC DIVERSITY IN BREEDING FOR RESISTANCE IN FOREST TREES

Gene Namkoong

USDA Forest Service, Southeastern Forest Experiment Station, Genetics Department, North Carolina State University, Raleigh, North Carolina 27695-7614

KEY WORDS: genetic, conservation, diversity, resistance breeding

---

INTRODUCTION

As with other crop species (24), resistance to diseases and insects is a prime target in breeding operations with trees (7, 38, 48, 97, 133). Reliable and repeatable resistance-testing techniques have been developed (see e.g. 2, 6, 65, 89–91, 141). High correlations between nursery tests and field trials have been achieved (118), and disease resistance has been increased markedly in the first generation of selection (151, 160). Most tree breeders try to capture an easily identifiable kind of resistance in breeding populations and hope that this resistance will endure and increase in subsequent generations. This approach tends to identify single-gene effects. Such genes must either be present in the evolving breeding populations, or they must be incorporated by backcrossing or otherwise transgenically transferred into these populations. The approach was widely adopted despite the fact that no specific genes for resistance had been identified (58), and no genes for resistance had been transgenically transferred in any plant species (24).

Management of single-gene effects remains the central objective of resistance breeding. The hope is often expressed that even if individual tree resistance can be broken, other genes for resistance can be found and that selection for pathogen virulence may be sufficiently weak for breeding populations to continue to be useful (116, 150). However, neither have such hopes been critically examined nor has the evolution of any forest-tree/ pathogen system been investigated. The relative frequencies of virulence types could shift in pathogen populations, reestablishing high rates of disease incidence or even epidemic states. Breeding for general or specific resistances in single, large breeding populations might never achieve more than a few years respite in recurring rounds of a virulence/resistance dance.

New resistance might not always be found in available breeding populations, and if it is not, it seems doubtful that the tree breeder can backcross for 6 to 10 generations, as breeders of some agronomic crops can. Intercrossing improved with unimproved populations would be costly because it would dilute the gain in all the other economically useful traits of the breeding population. These costs would increase as generations of breeding advance. New biotechnological techniques might allow for continued and multiple infusions of genes for resistance if they can be identified and transferred. Regardless of which breeding or transgenic introduction techniques are used, it is not clear that genes can be selected that would create an enduring pathosystem with an acceptably low amount of disease loss. In tree breeding, short-term resistance is only marginally useful. Tree breeders need resistances that can endure into future generations and they need carefully thought out methods for managing these genetic resources. Several types of resistances and several breeding methods may be needed to maintain the genetic variation required for long-term progress.

It would be convenient if the gene-conservation problem could be separated from the breeding problem. However, there is a circularity in the maintenance of genetic diversity in breeding for resistance in forest trees. Since tree breeders are still in the initial stages of domesticating relatively wild plant species, our breeding methods and the structure of our breeding populations are still developing. At this time, our choice of breeding method can affect the kinds of resistances that we can effectively use, and the kinds and nature of resistances we wish to use can affect the sampling and testing that we can efficiently use. Our sampling and prebreeding programs, such as provenance testing and selection, affect the kind of breeding that we can effectively do. Hence, how we plan to conserve genetic diversity affects how we use that variation, and vice versa. The existence of such feedback implies a need for joint analysis of the breeding system, the nature of resistance, and the conservation program. Analysis of any one or two, taken in isolation, can be misleading, and merely adopting an ad hoc set of seed-storage or breeding programs or relying on large numbers of random collections can be wasteful. Therefore, more directed and coordinated gene conservation may be necessary to find the kinds of resistances we seek, to capture them in well-designed samples, and to develop them in efficient gene-management systems.

In discussing these issues, I start by examining the nature of heritable resistances and what is known of their distribution. I then consider the types of breeding that can be used for resistance, and how tree breeding is related to gene-conservation programs. Finally, I discuss the size and design of gene conservation systems and attempt to integrate conservation with resistance breeding. I conclude that broader conservation programs should be coupled with active breeding programs to manage the pathosystems.

DISTRIBUTION AND TYPES OF RESISTANCE

Heterogeneity in the distribution of resistance genes among related species is well known and has served as a basis for species selection and hybridization (see e.g. 36, 47, 113). It also could be used in transgenic single-gene-transfer programs (66). Conservation of resistance is an objective of long-term breeding programs (16), and sampling species from their centers of origin is often recommended for increasing the probability of finding resistant alleles (77). Among the white pines (29, 72), for example, the Eurasian species generally display moderate to higher resistance to blister rust than the American species (143). Also, among the several native pines in the southeastern United States, species differences in resistance to fusiform rust are strong (61, 115, 132, 140). When species are moved to new environments, however, resistances can change (44, 59, 65). Hence, simple virulence/resistance relationships are not common, even at the species level. Differences in resistance among species may be due to alleles unique to species at particular loci, or to the collective effect of species differences in growth, phenology, metabolism, etc. At single loci, it is feasible to consider single trait transfers (66), and, if particular hybrids are directly useful, species-level differences can be used. If they are available, intraspecific variations are easier to use than interspecific variations that require hybridizing and selecting in advanced intermating or backcross generations in which genotypes segregate different trait combinations. If resistance is caused by many developmental events, selection for resistance will change many traits, and the interspecies mix of desired traits must be specified.

Heterogeneous distributions of resistances, which are common among subspecific variants, have been used to develop multiline varieties in crop breeding (13). However, while populations often differ in resistance levels and types, we know few generalizations about clustering of resistance alleles and sampling for them. Indeed, the concept that resistances may be found in the centers of a species' range is flatly contradicted by the relatively low frequency of resistance to fusiform rust in the center of loblolly pine's range (78). It seems that complexities of population structure (22), local population dynamics (14), and environmental stress (27) can all affect the natural distribution of resistance. Patterns of resistance may exist, but the causes and geographic distribution patterns are not obvious at the outset. Only after testing do patterns of resistance appear in loblolly pine (115, 117), in Douglas-fir (88), or in radiata pine (109). Interestingly, in lodgepole pine, some strong geographic patterns exist (50, 148), but the patterns seem to shift for different diseases and over time (85, 154, 157). For some species, such as ponderosa pine, provenances differ widely in resistance, but without discernible geographic trends (50), and families within provenance often vary widely in resistance (85). In fact, some species show both patterned and unpatterned variation in resistance to different diseases (154). Lacking known correlations of diseases with geographic or environmental variables, the use of monoterpenes has been suggested since their use may both serve as an identifiable marker and may be causally related to resistance to some diseases (32, 96, 129, 130). However, strong provenance differences in resistance and in monoterpenes such as β-phellandrine are confounded with other provenance differences, and the causal relationship of the marker and resistance may not be strong (130). Similarly, isozyme markers are potentially useful, but non-equilibrium associations of markers with resistance alleles are needed (18, 156).

Since it is difficult if not impossible to distinguish single- from multiple-locus differences among populations of a species, variation among individuals within populations can be useful, especially if it is stronger than variation between populations. For example, Powers (116) has consistently found heritable variation useful for selection at all levels of organization in loblolly pine. High intrapopulational variation in resistance also exists in Norway spruce (26), in cottonwood (60), and in loblolly pine (8). It is also at the individual level of variation that single-locus effects can be discerned. When heritable variations in resistance are found among individuals within populations and segregate into classes, more exact selections for specific kinds of resistance may be possible. It may be possible to find single-locus qualitatively effective genes at the individual level of biotic organization. In two cases, the frequency of such possible resistance alleles has not been high enough to be useful (42, 66). While distributional patterns for alleles directly related to resistances are not known, they may be located in advanced segregating generations originating from interpopulation hybrids.

The scale of distributional heterogeneity seems to run the gamut from individual tree to population and species differences. The types of resistance that have been observed in forest trees also seem to extend from single locus (67) to more purely quantitative (17), and from resistance to a highly specific pathogen source (i.e. vertical) to general resistances that appear to be effective across a set of test pathogen sources (41, 159). Some evidence of direct effect exists for a gene product and pathogen growth, but exact tests are lacking (32, 130, 137). Tissue hypersensitivity seems to be frequently implicated in resistance to *Cronartium* attacks, but several distinct virulence and resistance types for the same disease often exist, making gene-for-gene models difficult to analyze (66). The unique interaction of a single gene that segregates for resistance to a single source of pathogen virulence from a single tree (68) is rare. It is not likely to be observed in current testing programs (66), despite improvements in analytical techniques (56, 57). However, at the population level, observations of differences among provenances and families generally support the existence of some degree of interaction between host sources of resistance and pathogen sources of virulence (74, 89-91, 116). Thus, no cases have yet been demonstrated where uniform resistance to all pathogen sources exist for any family or population, but the definition of pathogen populations and their variability is seldom as clear as for the host plant. When specific mutants or races have been studied, as in *Melampsora* infections on *Populus* species (119–121, 138), specific variations in reactions to the mutants were observed and one poplar family was seen to have reaction types that segregated at different ratios. Thus, different loci and numbers of loci determine reaction to mutant types. Gibson et al (41) conjectured that a wide range of resistance alleles with specific to general effects might exist. That possibility seems entirely feasible because resistance may often be a secondary effect of alleles selected for other characteristics that more directly affect fitness.

Similarly, other evolutionary forces also act on pathogen populations and their evolutionary dynamics. Examples include fitness on alternate hosts (30) and the adaptability of hosts and pathogens to environmental stresses (27). Since resistances may exist at multiple sites (107, 131, 136), virulences may also, and pathogenicity and virulence may simultaneously involve many different mechanisms. It is not surprising, therefore, that even if gene-for-gene effects are present, inconsistencies in the interaction of host and pathogen abound (10). In many cases the dynamics of virulence and resistance may be secondarily induced effects of adaptations by both host and pathogen to other environmental factors (94, 112). Resistance may be more a matter of general tree health and population density dynamics than any specific mechanism (126, 134), and selection for specific resistances may not work (73). Nevertheless, family level resistances can be stable (142), even "defeated" resistances can be useful (107), and the cumulative effects of an array of specific and general resistances might provide broad protection against multiple pathogens (41).

In addition to physiological mechanisms of resistance and virulence, and of host tissue interactions with fungal hyphae and their chemistry, population-level dynamics among the host, pathogen, and other species can also affect epidemiology in a surprisingly short time. Over generations of naturally or artificially regenerated stands, changes in gene frequency and population density of both host and pathogen will affect the evolution and stability of the natural system, and can affect future coexistence or exclusion of any of the elements of the system. While it is clear that the structure of the host population can affect the distribution of virulence alleles (22, 75), and that the structure of pathogen variability can affect host evolution, specific analyses have been conducted only rarely. It usually is not clear which of several resistance/virulence states can exist either sequentially or simultaneously in any natural system (22). However, it can be expected that a mixture of specific and general resistances and tolerance and escape mechanisms exist simultaneously (135). Within controlled breeding systems, both specific and general resistances can be used in a system of stabilizing selection (21). In both naturally and artificially produced stands, a mosaic of different resistances can exist in patches such that pathogen density and migration rates are reduced and the development of epidemics is inhibited (14).

It is not generally known for naturally or artificially regenerated populations whether any genes or individual or population processes generate stasis. Pathosystems may be in continuous unstable arms races in which at least local genes and populations become extinct. Alternatively, we may be witnessing long periods of stasis and stability with stochastic variations of little long-term significance. Some models suggest that stable coexistence of virulence and resistance genes is possible (76). Even if single gene-for-gene systems are unstable, multilocus systems may have been selected into natural pyramids of resistance alleles that permit long periods of quasi-stasis before the possible evolution of pathogens virulent to many resistances simultaneously. Alternatively, such pyramiding may change the cost of multiple virulence sufficiently to evolve a stable host multilocus vs pathogen multilocus system. It is also possible that the dynamics of otherwise virulent pathogen populations could be moderated by the density and patchy distribution of their hosts (41, 73, 95, 114) and by their temporal escape mechanisms (132, 157). In the absence of uniformly strong gene-for-gene systems, many possible states of coexistence can evolve simultaneously, and subdivided populations can maintain different states of stasis and rates of change.

Multispecies dynamics may also strongly influence both host resistance and pathogen virulence. The susceptibility of more than one host species can have affected the evolution of virulence in *Cronartium* (61, 117). In addition, changes in populations of alternate hosts generate different complexities in a three-species pathosystem (83, 92, 147,148).

Several types of multiple-species systems involving pathogens exist in forest trees (31, 146), and these could generate complex dynamics for all of the species involved (10, 52, 53, 86, 122, 139, 145). Furthermore, in resistance systems that are reactions to other organisms (64) and systems with multiple types of immunity (81), different coexistence dynamics can be generated. The interactions of species at different "trophic" levels can also force dynamics that would otherwise be difficult to anticipate (87). In fact, some insect and disease interactions may be important for both insect and disease to succeed (37). Also, in some forest trees, pathogen resistance may be based primarily on reaction to a secondary rather than a primary pathogen (5, 123, 137).

With such complexities of attack and defense in tree species that are not yet fully domesticated, there is no need to limit breeding to only one level and type of defense, or to conserve just one type of gene or response function. Given the wide distribution of alleles among species and populations, there also is no need to limit breeding to one source population or to conserve just one population. A breeding strategy may limit use of resistance to one type and one source of resistance genes. Clearly, however, broad choices exist, and how we breed for resistance will determine how we conserve genes.

RESISTANCE AND BREEDING

Even though resistance is a conglomerate phenomenon with a variety of genes affecting specific and general resistance mechanisms (41), host selection can be directed to stabilize both host and pathogen populations (21). There seems little question that breeding can change host phenotypes in ways that affect the pathogen. It also seems clear that both special and general resistances can respond to selection for qualitatively or quantitatively acting genes (17), and that multiple resistance and multiple trait selection can be effective (39, 144). Hence, many breeding options would appear to be available and various types of resistance may be developed by tree breeders.

In the absence of uniformly effective, single-locus resistances, selection for resistance in the host favors virulence alleles in pathogens and even mild selection pressure for virulence should favor ultimate victory of the virulent pathogens (68). While it may be useful in the short run, accumulating single loci into pyramids of resistance can be an unstable strategy, especially if the frequency of resistance is low (42, 93). Nevertheless, selection for single loci or single forms of resistance may not lead immediately to heightened virulence in pathogen populations (35). If host selection is for a form of resistance for which the energy costs of virulence are unavoidably high (75, 107) or if the individual plant defenses and population deployment require less than maximum virulence by the pathogen (19, 80, 111), development of supervirulence may never be an optional evolutionary strategy for the pathogen.

While all types of individual tree resistance can be useful to breeders, it is desirable to determine the kinds of resistance/virulence systems that can be induced or become self-regulating. High precision is required to test for these phenomena. While single alleles have not yet been transferred (24), techniques exist for detecting loci that can directly affect resistance (e.g. 158), and for using tissue culture in new and unique regeneration systems (151). Such detail has not yet been developed for any forest tree, but families have been detected that segregate single-locus resistances (69). Isolated cells and tissues are not sufficient measures of resistance (23), but in vitro testing of cultured loblolly pine tissues (2) has uncovered multiple resistance mechanisms, even at the tissue level of defense. The net effect of the multiple resistances observed in vitro is clearly correlated with field test results (33). Other tests, such as tissue water stress reactions, can also be useful (28) if what is measured is directly related to the resistance reaction. Phenological mechanisms of escape might also be detected easily. They could be simply inherited and could present adaptational demands on the pathogen that would be difficult to overcome (e.g. see 55). It is also possible that directly associated and detectable differences in monoterpenes exist and can be used at least for indirect selection (9, 32, 129, 130).

For these purposes, and for breeding in future recurrent selection systems, populations must be large enough to avoid inbreeding depression (17). In a single breeding population, a few tens of parents can be sufficient if selection for resistance and other traits does not reduce the effective population size (104). However, in most intensive breeding programs, an objective is to maintain genetic variation in all important traits. For this purpose, an effective population size of several hundred is required (102).

These two objectives can be simultaneously satisfied in any of several breeding systems. A hierarchy of populations as in agricultural crop breeding (63) is often discussed (99). In this approach, a large base population is maintained and nonintensively selected. More intensively selected subsets are periodically segregated for immediate and rapid selection. However, few programs of this sort have actually been executed (17). The direction of germplasm flow can also be bidirectional, with parents from the selected and reduced populations contributing gametes to the large base population (20). In this system, the effective population sizes of all of the populations must be monitored. An alternative structure includes multiple populations that are originated, maintained, and bred for different objectives (106). In all of these populations, inbreeding depression is avoided by maintaining effective population sizes above some minimum (e.g. 20 to 50). With 10 or more such populations, the total genetic variance is at least as large as in a single population. For selection of a new trait or for production, the populations may be recomposed with a few or many parents. This system avoids some of the operational problems that tree breeders face in attempting to incorporate desirable genes from unproductive individuals or populations into advanced populations by backcrossing. All populations are advanced to some degree by contemporaneous selective breeding.

To ensure that more alleles are included in the set of breeding populations,

Table 1 Minimum number of genotypes required for an average loss of one allele at any of the loci with one rare allele per locus

No. of loci (s)	Frequency of rare allele at each locus			
	.05	.01	.005	.001
10	45	227	455	2278
50	77	389	780	3906
100	90	458	919	4601

larger effective population sizes are required (43, 62). This requirement, however, is not qualitatively different from that for maintaining genetic variances.

Effective population sizes of several thousand are needed, and these may be maintained in single or multiple populations. Programs of similar size and form have been recommended for climatic and geographic adaptations in Sweden (124, 125) and in the United States (128). Separate breeding populations are maintained for each breeding zone. In Japan, zones have been recommended for plantings of *Cryptomeria* that have been subdivided for economic and disease resistance priorities as well as for climatic adaptability (108). Populational subdivisions could as easily be created for different resistance mechanisms or allelic sets (13) and used as multiple lines that are used in varietal mixtures to enhance population-level resistance (152). If such efforts are targeted to high-hazard areas, they could be especially effective (34, 46, 116). Multiple resistances and other traits of economic importance can be obtained from index selection (39) or from multiple-index selection (100) with independent culling levels (144).

The use of multiple lines (14, 84) makes sense if different sets of resistance alleles and mechanisms are expected to reduce the capacity of any pathotype to generate epidemics. This expectation is partly based on the effect that mixtures create by reducing susceptible host density or by dispersing susceptible patches (81, 82, 110) and partly on the effects that multiple adaptability requirements place on pathogen evolution (10, 15, 70, 135). Increasing the number and variety of resistant host biotypes decreases the chances of the pathogen being catastrophic on any one and increases the likelihood that stabilizing selection will develop in the pathogen.

The analogous use of a large number of clones or genotypes as protection against pathogen attacks does not carry the same logic, however. For any pathotype, if the frequency of resistance is not changed, merely incorporating large numbers of genotypes in plantings only ensures that the frequency of susceptible individuals will be close to the average breeding population frequency for that pathotype. If resistance is rare, then high susceptibility is ensured. If resistance is frequent, adding more genotypes will do little to ensure further resistance (54, 79, 127). Since this is true for all potential pathotypes, the probability of a catastrophic epidemic due to any one pathotype is high. Though some resistant hosts will survive and the natural evolution of resistance can proceed, breeders will be faced with a continual sequence of pathogen outbreaks and the need to continually generate new resistances.

To escape this form of arms race, multiple forms of resistance may be found and developed in different populations. They can then be incorporated as different selection objectives and deployed in the field as multiple population mixtures—as individuals, in blocks, or in temporally varying sequences. For forest trees as for many agricultural crops, testing the available germplasm is critical (11). Because long generations in forest trees inhibit backcrossing to incorporate resistances, multiple resistances must be developed contemporaneously. The problem for gene conservationists is therefore not only to find any existing genetic variation that may require generations of recombination to use, but also to find or develop resistance in a form that is readily available to tree breeders. Therefore, detecting heritable variations in resistance mechanisms at all levels of organization is critical for breeding and conservation because multiple mechanisms can exist for one pathogen, and certainly do exist for the several pathogens that can attack any one population. At the gene-locus level, many potential alleles and several loci can affect different resistances to one pathogen (24). At the tissue level, many forms of reaction can be detected (33). At the whole-tree level, many heritable resistance mechanisms have been detected (97). At the population and community-composition levels, heritable variations in response to stand densities and the frequencies of other species can affect pathogen dynamics (14, 82). Therefore, multiple levels of resistance can be detected, and several screening programs have already successfully selected an array of resistance types in forest trees (see e.g. 3, 98). However, their optimal incorporation into breeding programs for developing long-term resistance has rarely been examined critically (17, 103).

CONSERVATION AND RESISTANCE BREEDING

In forest tree breeding, gene conservation must include more than just prescribing minimum breeding-population and collection sizes. To be useful, genetic diversity must be made readily available to breeders. Resistance may be expressed at the cell, tissue, whole-tree, and/or stand levels, and it involves different host/pathogen interaction dynamics at each level. The capture techniques and measures of diversity, therefore, should include all these different levels. Since resistance involves dynamic systems and the evolution of pathogens, hosts, and other biotic factors, it cannot be captured in a fixed state. Since breeding involves the evolution of one or several populations, the use of genetic diversity for resistance breeding must also evolve. Hence, both the use and conservation of resistance must be dynamic processes. Neither can be optimally maintained in a fixed state.

It is impossible to completely capture even the existing variation. There are virtually an infinite number of possible mutants at each gene locus and among the 10,000 to 50,000 loci that may exist in a conifer, any combination of allelic variants can exist as a unique combination. Any finite population must therefore miss some alleles at some loci. If we sample that population for gene-conservation purposes, losses will inevitably occur. The only objective that sampling can achieve is to reduce the probability of loss below some acceptable maximum level. The greatest probabilities of random loss are with rare alleles. In a sample of one tree from a random mating population, the probability of loss of an allele at frequency q is $(1\text{-}q)^2$. In a random sampling of n trees from that population, the Pr (loss) = $(1\text{-}q)^{2n}$. Clearly, higher sample sizes are needed if q is low. In a subdivided or inbred population, the frequency of homozygotes and the probability of loss are higher, even if the average allele frequency is unchanged. Furthermore, if there are several alleles at low frequency at the resistance loci such as for some mildew resistance genes in barley (24), the probability of loss of at least one of them is high, and saving all of them requires still larger sample sizes (43). Thus, large sample sizes are necessary to reduce the probability of loss to an average loss of only one allele. In Table 2, a few examples are listed that show how minimum sample size increases as the allele frequency decreases and the number of loci increases. A few thousand samples are needed to save most alleles in most populations, even if allele frequency distributions differ among loci (11).

For alleles that are broadly distributed, reasonably finite sample sizes can capture at least one representative in a random collection. But if allele distribution is patchy, and some alleles are present only in one population

Table 2 Minimum number of genotypes required for an average loss of one allele at any of the loci with four rare alleles per locus

No. of loci (s)	Frequency of rare alleles at each locus			
	.05	.01	.005	.001
10	72	367	735	3681
50	104	528	1057	5295
100	117	597	1196	5988

due to their recent origin or limited migration (25), then the way the random sampling is structured can affect the probability of capture. For example, if the allele frequency is q only in a portion p of the populations, then there is a probability $(1\text{-}p)$ that a random sampling of populations will miss the allele entirely. Thus, using N random population samples, and using n trees per population, the probability of loss is $[(1\text{-}p)+p(1\text{-}q)^{2n}]^{N}$. Dispersing samples among more populations is clearly advantageous if the present natural populations and alleles are dispersed and gene flow between populations is limited (1). If any knowledge exists about the heterogeneity of allele distributions, then targeted sampling among populations is desirable. For resistance genes, allele distributions might be discerned from direct tests or inferred from pathogen distribution and behavior. Near the center of a species' distribution, larger populations may be expected. One might also expect maximum numbers of alleles and advanced host/pathogen coevolution there (77). However, this rule may not be generally useful (78, 154). Marginal populations that could be selected for different adaptabilities can have higher frequencies of unique and possibly useful alleles. Distribution of alleles depends strongly on whether they are nearly neutral and hence are distributed only by mating patterns, or whether they are selected for similar genotypic composition among areas or for disparate patterns of adaptability (18, 105). For alleles that are uniformly distributed, the dispersal of sampling sites does not diminish the probability of their capture, but for alleles that have heterogeneous distribution, dispersal is highly desirable.

If the present distribution of alleles is captured best by dispersed sampling, the pattern of populations in which they are maintained can subsequently affect the distribution of genetic variation within and among populations. By either in situ or ex situ techniques, the structure of variation can be either reduced to homogeneity or be increased by selection and restriction on gene flow. Unlike agricultural crop species that are bred in short generations and intensively controlled environments, trees are not amenable to the collection of germplasm into a single, undifferentiated population. Also, since trees are still largely reproduced naturally, their evolution with pathogens of native origin is likely to continually change the distribution of resistance alleles and mechanisms. Under such conditions, in situ conservation is desirable for continuing evolution of the present host/pathogen system, and is especially useful for independently evolving populations. For some forest tree species in which breeding is not yet a well-established practice, in situ conservation may be the only practical means of ensuring the continued existence of reproductively viable populations (101). In conservation programs within in situ stands, it is possible to exert selection pressure on the populations by selectively thinning some stands. The stands must be selected initially to contain a broad array of potentially diverse evolutionary states.

For species that are threatened in a significant part of their native range, ex situ collections and stands may be necessary. If they are included in breeding programs, populations must often be maintained in reproductively capable but reproductively isolated populations. The distinction between in situ and ex situ stands may well be blurred in future generations because both may be managed for multiple objectives, including diverse resistances. The stands should not be permitted to develop without considering how they could be optimally differentiated. In contrast to agronomic crop species, tree-seed collections will play a less central role in conservation programs due to the time requirements for regeneration. Instead, seeds will serve as backup reserves. The need for long-duration tests is likely to make the economies of combined testing and conservation sufficiently attractive that the distinction between breeding and conservation will be blurred. Thus, gene conservation can involve more than simply maintaining genetic variation. It also can involve developing more useful genetic variation in more readily available populations. To develop multiple resistance mechanisms and to use multiple levels of resistance, breeders require more than only maintenance of genetic variation. They require an active combination of directed conservation and breeding.

Literature Cited

1. Allard, R. W. 1970. Population structure and sampling methods. In *Genetic Resources in Plants: Their Exploration and Conservation*, ed. O. H. Frankel, E. Bennett, IBP Handbook 11:97–107. Oxford: Oxford Univ. Press
2. Amerson, H. V., Frampton, L. J. Jr., Mott, R. L. 1985. In vitro methods for the study of fusiform rust in association with loblolly pine. See Ref. 3, pp. 103–5
3. Barrows-Broaddus, J., Powers, H. R., eds. 1985. *Proc. Rusts Hard Pines Work. Party Conf., Athens, GA. 1984*. Athens: GA Cent. Contin. Ed. 331 pp.
4. Baumgartner, D. M., Krebill, R. G., Arnott, J. T., Weetman, G. F., eds. 1985. *Lodgepole Pine, The Species and Its Management. Symp. Proc., Spokane, WA*
5. Berryman, A. A. 1972. Resistance of conifers to invasion by bark beetle-fungus associations. *BioScience* 22: 598–602
6. Bey, C. F. 1982. Progeny testing the southern pines. In *Breeding Insect and Disease Resistant Forest Trees, Proc. Servicewide Genet. Workshop*, pp. 300–2. Eugene, OR. USDA For. Serv.
7. Bingham, R. T., Hoff, R. J., McDonald, G. I. 1971. Disease resistance in forest trees. *Annu. Rev. Phytopathol.* 11:433–52
8. Blair, R. L. 1970. *Quantitative inheritance of resistance to fusiform rust in loblolly pine*. PhD thesis. NC State Univ., Raleigh
9. Bridges, J. R. 1987. Effects of terpenoid compounds on growth of symbiotic fungi associated with the southern pine beetle. *Ecol. Epidemiol.* 77:83–85
10. Brittain, E. G. 1982. Stability in a multicomponent host-pathogen model. See Ref. 49, pp. 304–11
11. Brown, A. H. D. 1989. The case for core collections. See Ref. 12, pp. 136–56
12. Brown, A. H. D., Frankel, O. H., Marshall, D. R., Williams, J. T., eds. 1989. *The Use of Plant Genetic Resources*. Cambridge: Cambridge Univ. Press. 382 pp.
13. Browning, J. A., Frey, K. J. 1969. Multiline cultivars as a means of disease control. *Annu. Rev. Phytopathol.* 7:355–82
14. Burdon, J. J. 1987. *Diseases and Plant Population Biology*, ed. R. S. K.

Barnes, H. J. B. Birks, E. F. Connor, J. L. Harper, R. T. Paine. Cambridge: Cambridge Univ. Press. 208 pp.
15. Burdon, J. J., Jarosz, A. M. 1989. Wild relatives as sources of disease resistance. See Ref. 12, pp. 280–96
16. Callaham, R. Z. 1972. Exchanging and conserving tree breeding materials. See Ref. 51, pp. 281–82
17. Carson, S. D., Carson, M. J. 1989. Breeding for resistance in forest trees: A quantitative genetic approach. *Annu. Rev. Phytopathol.* 27:373–95
18. Caten, C. E. 1987. The concept of race in plant pathology. See Ref. 153, pp. 21–37
19. Christiansen, E., Ericsson, A. 1986. Starch reserves in *Picea abies* in relation to defence reaction against a bark beetle transmitted blue-stain fungus, *Ceratocystis polonica*. *Can. J. For. Res.* 16:78–83
20. Cotterill, P., Dean, C., Cameron, J., Brindbergs, M. 1989. Nucleus breeding: A new strategy for rapid improvement under clonal forestry. In *Breeding Tropical Trees: Population structure and genetic improvement strategies in clonal and seedling forestry. Proc. Conf. IUFRO Work. Part. Pattaya, Thailand,* 1988. pp. 39–51. 503 pp.
21. Crill, P. 1977. An assessment of stabilizing selection in crop variety development. *Annu. Rev. Phytopathol.* 15:185–202
22. Crute, I. R. 1987. The geographical distribution and frequency of virulence determinants in *Bremia lactucae*. See Ref. 153, pp. 193–212
23. Daub, M. E. 1986. Tissue culture and the selection of resistance to pathogens. *Annu. Rev. Phytopathol.* 24:159–86
24. Day, P. R., Barrett, J. A., Wolfe, M. S. 1983. The evolution of host-parasite interaction. See Ref. 71, pp. 419–30
25. Devlin, B., Ellstrand, N. C. 1990. The development and application of a refined method for estimating gene flow from angiosperm paternity analysis. *Evolution* 44:248–59
26. Dimitri, L. 1982. Some host/parasite relationships between Norway spruce (*Picea abies*) and *Fomes annosus*. See Ref. 49, pp. 260–67
27. Dinoor, A., Eshed, N. 1987. The analysis of host and pathogen populations in natural ecosystems. See Ref. 153, pp. 75–88
28. Driver, C. H., Morse, D. R., Edmonds, R. L. 1982. Detection of disease resistance in Douglas-fir seedlings and variation in pathogenicity in *Phellinus weirii* by monitoring water stress after inoculation. See Ref. 49, pp. 275–82
29. Duffield, J. W. 1972. Limitations and advantages of conifers vs. agronomic crops as resistance breeding materials. See Ref. 51, pp. 19–25
30. Dwinell, L. D. 1985. Oak hosts affect the pathogenic variability of fusiform rust fungus. See Ref. 3, pp. 251–58.
31. Dwinell, L. D., Barrows-Broaddus, J. 1985. Infection of fusiform rust galls on slash and loblolly pines by the pitch canker fungus *Fusarium moniliforme* var. *subglutinans*. See Ref. 3, pp. 239–48
32. Flodin, K. 1979. Effects of monoterpenes on *Fomes annosus* (Fr.) Cooke and its phenol oxidase activity. *Eur. J. For. Pathol.* 9:1–6
33. Frampton, L. J. Jr., Amerson, H. V., Moyer, J. W. 1985. Development of in vitro techniques for screening loblolly pine for fusiform rust resistance. See Ref. 3, pp. 125–39
34. Froelich, R. C. 1982. Rating sites for fusiform rust hazard and trees for rust resistance. See Ref. 98, pp. 381–93
35. Gale, J. S. 1987. Factors delaying the spread of a virulent mutant of a fungal pathogen: some suggestions from population genetics. See Ref. 153, pp. 55–62
36. Gallo, L. A., Stephan, B. R., Krusche, D. 1985. Genetic variation of *Melampsora* leaf rust resistance in progenies of crossings between and within *Populus tremula* and *P. tremuloides* clones. *Silvae. Genet.* 34:208–14
37. Gara, R. I., Littke, W. R., Agee, J. K., Geiszler, D. R., Stuart, J. D., Driver, C. H. 1985. Influence of fires, fungi and mountain pine beetles on development of a lodgepole pine forest in south-central Oregon. See Ref. 4, pp. 153–62
38. Gerhold, H. D. 1966. In quest of insect-resistant forest trees. See Ref. 40, pp. 305–18
39. Gerhold, H. D. 1972. Multiple trait selection in white pine breeding systems: Blister rust resistance, weevil resistance, timber yield. See Ref. 51, pp. 591–97
40. Gerhold, H. D., Schreiner, E. J., McDermott, R. E., Winieski, J. A., eds. 1966. *Breeding Pest-Resistant Trees. Proc. NATO NSF Adv. Study Inst. Genet. Improv. Dis. Insect Resist. For. Trees.* 1964. University Park, PA/New York: Pergamon
41. Gibson, I. A. S., Burley, J., Speight, M. R. 1982. The adoption of agricultural practices for the development of heritable resistance to pests and

pathogens in forest crops. Ref. 49, pp. 9–21
42. Goddard, R. E. 1981. The University of Florida cooperative forest genetics research program. See Ref. 45, pp. 31–42
43. Gregorius, H-R. 1980. The probability of losing an allele when diploid genotypes are sampled. *Biometrics* 36:643–52
44. Gremmen, J. 1972. Relative blister rust resistance of *Pinus strobus* in some parts of Europe. See Ref. 51, pp. 241–49
45. Guries, R. P., Kang, H., eds. 1981. *Research Needs in Tree Breeding. Proc. N. Am. Quant. For. Gen. Group Workshop, 15th*. 1981. Coeur d'Alene, Idaho. NC State Univ., Raleigh: Sch. For. Res. 136 pp.
46. Hadfield, J. S. 1982. Rating forest stands for disease and insect damage potential in the Pacific Northwest. See Ref. 98, pp. 378–80
47. Heimburger, C. 1972. Relative blister rust resistance of native and introduced white pines tested in eastern North America. See Ref. 51, pp. 257–69
48. Heybroek, H. M. 1982. Monoculture versus mixture: interactions between susceptible and resistant trees in a mixed stand. See Ref. 49, pp. 326–41
49. Heybroek, H. M., Stephan, B. R., von Weissenberg, K., eds. 1982. *Resistance to Diseases and Pests in Forest Trees. Proc. Int. Workshop Genet. Host-Parasite Interact. For.*, 1980. Wageningen. Wageningen: Cent. Agric. Publ. Doc. 503 pp.
50. Hoff, R. J. 1985. Resistance of Lodgepole and Ponderosa pine to western gall rust. See Ref. 3, pp. 307–19
51. Hoff, R. J., McDonald, G. I., Program coordinators. 1972. *Biology of Rust Resistance in Forest Trees. Proc. NATO-IUFRO Adv. Study Inst.*, 1969. Moscow, Idaho. Publ. 1221. Washington: USDA For. Serv. 681 pp.
52. Hsu, S. B., Hubbell, S. P., Waltman, P. 1978. A contribution to the theory of competing predators. *Ecol. Monogr.* 48: 337–49
53. Hsu, S. B., Hubbell, S. P., Waltman, P. 1978. Competing predators. *SIAM. J. Appl. Math.* 35:617–25
54. Huhn, M. 1986. Theoretical studies on the necessary number of components in mixtures. *Theor. Appl. Genet.* 72:211–18
55. Hunt, R. A., Ying, C. C., Ashbee, D. 1987. Variation in damage among *Pinus contorta* provenances caused by the needle cast fungus *Lophodermella concolor*. *Can. J. For. Res.* 17:594–97
56. Jenns, A. E., Leonard, K. J. 1985. Reliability of statistical analyses for estimating relative specificity in quantitative resistance in a model host-pathogen system. *Theor. Appl. Genet.* 69:503–15
57. Jenns, A. E., Leonard, K. J., Moll, R. H. 1982. Stability analyses for estimating relative durability of quantitative resistance. *Theor. Appl. Genet.* 63:183–92
58. Johnson, E. C. 1982. Is resistance real? See Ref. 98, pp. 7–38
59. Johnson, R. 1987. Selected examples of relationships between pathogenicity in cereal rusts and resistance in their hosts. See Ref. 153, pp. 181–92
60. Jokela, J. J. 1966. Incidence and heritability of *Melampsora* rust in *Populus deltoides*. See Ref. 40, pp. 111–17
61. Kais, A. G., Snow, G. A. 1972. Host response of pines to various isolates of *Cronartium quercuum* and *Cronartium fusiforme*. See Ref. 51, pp. 495–503
62. Kang, H., Namkoong, G. 1988. Inbreeding effective population size under some artificial selection schemes. *Theor. Appl. Genet.* 75:333–39
63. Kannenberg, L. W. 1984. Utilization of genetic diversity in crop breeding. See Ref. 155, pp. 93–109
64. Karban, R., Adamchak, R., Schnathorst, W. C. 1987. Induced resistance and interspecific competition between spider mites and a vascular wilt fungus. *Science* 235:678–80
65. Kinloch, B. B. 1972. Genetic variation in resistance to *Cronartium* and *Peridermium* rusts in hard pines. See Ref. 51, pp. 445–63
66. Kinloch, B. B. 1982. Mechanisms and inheritance of rust resistance in conifers. See Ref. 49, pp. 119–29
67. Kinloch, B. B. 1982. Resistance to white pine blister rust in sugar pine: Research at the Pacific Southwest Forest and Range Experiment Station. See Ref. 98, pp. 186–91
68. Kinloch, B. B., Dupper, G. E. 1987. Restricted distribution of a virulent race of the white pine blister rust pathogen in the western United States. *Can. J. For. Res.* 17:448–51
69. Kinloch, B. B., Littlefield, J. L. 1977. White pine blister rust: Hypersensitive resistance in sugar pine. *Can. J. Bot.* 55:1148–55
70. Kiyosawa, S. 1982. Genetics and epidemiological modeling of breakdown of plant disease resistance. *Annu. Rev. Phytopathol.* 20:93–117
71. Kosuge, T., Meredith, C. P., Hollaender, A., Wilson, C. M., eds. 1983. *Genetic Engineering of Plants, An Agri-*

cultural Perspective. *Basic Life Sci*. Vol. 26. New York: Plenum. 499 pp.
72. Kriebel, H. B. 1972. White pines and North and Central America: *Pinus strobus* and introduced Asian and European species. See Ref. 51, pp. 201–14
73. Krusche, D. 1982. Selection in host-parasite systems. See Ref. 49, pp. 312–17
74. Kuhlman, E. G., Matthews, F. R. 1985. Pine host and rust source affect sporulation of *Cronartium quercuum* f. sp. *fusiforme*. See Ref. 3, pp. 259–70
75. Leonard, K. J. 1987. The host population as a selective factor. See Ref. 153, pp. 163–79
76. Leonard, K. J., Czochor, R. 1980. Theory of genetic interactions among populations of plants and their pathogens. *Annu. Rev. Phytopathol*. 18:237–58
77. Leppik, E. E. 1970. Gene centers of plants as sources of disease resistance. *Annu. Rev. Phytopathol*. 10:323–44
78. Lewis, R. A. 1982. Maintaining genetic variation in resistant populations—How much variation is necessary? See Ref. 98, pp. 347–54
79. Libby, W. J. 1982. What is a safe number of clones per plantation? See Ref. 49, pp. 342–60
80. Loehle, C., Namkoong, G. 1987. Constraints on tree breeding: growth tradeoffs, growth strategies, and defensive investment. *For. Sci*. 33:1089–97
81. Longini, I. M. Jr. 1988. A mathematical model for predicting the geographic spread of new infectious agents. *Math. Biosci*. 90:367–83
82. Lorimer, N. 1982. Pest outbreaks as a function of variability in pests and plants. See Ref. 49, pp. 287–94
83. Mac Key, J. 1981. Alternative strategies in fungal race-specific parasitism. *Theor. Appl. Genet*. 59:381–90
84. Marshall, D. R., Burdon, J. J. 1981. Multiline varieties and disease control. III. Combined use of overlapping and disjoint gene sets. *Aust. J. Biol. Sci*. 34:81–95
85. Martinsson, O. 1980. Stem rusts in Lodgepole pine provenance trials. *Silvae Genet*. 29:23–26
86. Matessi, C., Gatto, M. 1984. Does K-selection imply prudent predation? *Theor. Popul. Biol*. 25:347–63
87. Matsuda, H., Kawasaki, K., Shigesada, N., Teramoto, E., Ricciardi, L. M. 1986. Switching effect on the stability of the prey-predator system with three trophic levels. *J. Theor. Biol*. 122:251–62
88. McDermott, J. M., Robinson, R. A. 1989. Provenance variation for disease resistance in *Pseudotsuga menziesii* to the Swiss needle-cast pathogen, *Phaeocryptopus gaeumarnnii*. *Can. J. For. Res*. 19:244–46
89. McDonald, G. I. 1982. Genetics of *Cronartium ribicola*. See Ref. 98, pp. 259–71
90. McDonald, G. I. 1982. Use of blister rust resistance. See Ref. 98, pp. 465–76
91. McDonald, G. I. 1982. Resistance-hazard alignment: A blister rust management philosophy. See Ref. 98, pp. 355–77
92. McDonald, G. I., Andrews, D. S. 1981. Genetic interaction of *Cronartium ribicola* and *Ribes hudsonianum* var. *petiolare*. *For. Sci*. 27:758–63
93. McDonald, B. A., McDermott, J. M., Goodwin, S. B., Allard, R. W. 1989. The population biology of host-pathogen interactions. *Annu. Rev. Phytopathol*. 27:77–94
94. McNabb, H. S. Jr., Hall, R. B., Ostry, M. E. 1982. Biological and physical modifications of the environment and the resulting effect upon the host-parasite interactions in short-rotation tree crops. See Ref. 49, pp. 60–71
95. McNair, J. N. 1986. The effects of refuges on predator-prey interactions: A reconsideration. *Theor. Popul. Biol*. 29:38–63
96. Michelozzi, M., Squillace, A. E., White, T. L. 1990. Monoterpene composition and fusiform rust resistance in slash pine. *For. Sci*. 36:470–75
97. Miller, D. 1982. Description of rust resistance mechanisms. See Ref. 98, pp. 243–48
98. Miller, R. G., ed. 1982. *Breeding Insect and Disease Resistant Forest Trees, Proc. Servicewide Genet. Workshop*. Eugene, Ore: USDA For. Serv. 498 pp.
99. Namkoong, G. 1974. Breeding for future generations. pp. 29-30. In *Proc. IUFRO Joint Meet. Work. Part. Popul. Ecol. Genet., Breed. Theor. Progeny Testing*. Stockholm: R. Coll. For. 497 pp.
100. Namkoong, G. 1976. A multiple index selection strategy. *Silvae Genet*. 25: 199–201
101. Namkoong, G. 1978. Choosing strategies for the future. *Unasylva* 31:38–41
102. Namkoong, G. 1984. Strategies for gene conservation in forest tree breeding. See Ref. 155, pp. 79–90
103. Namkoong, G. 1988. Sampling for germplasm collections. *Hort. Sci*. 23: 79–81

104. Namkoong, G. 1989. Population genetics and the dynamics of conservation. In *Biotic Diversity and Germplasm Preservation, Global Imperatives, Beltsville Symp. Agric.*, ed. L. Knutson, A. K. Stoner, 13:161–81. 1988. Dordrecht: Kluwer Acad. Publ. 530 pp.
105. Namkoong, G., Kang, H. 1990. Quantitative genetics of forest trees. In *Plant Breeding Rev.* 8:139–88. Portland, OR: Timber Press. 388 pp.
106. Namkoong, G., Kang, H. C., Brouard, J. S. 1988. *Tree Breeding: Principles and Strategies.* New York: Springer-Verlag. 179 pp.
107. Nelson, R. R. 1982. On genes for disease resistance in plants. See Ref. 49, pp. 84–93
108. Ohba, K. 1984. Genetics and breeding strategy of *Cryptomeria*. In *Genetics: New Frontiers, Proc. Int. Congr. Genet., 15th*, ed. V. L. Chopra, B. C. Joshi, R. P. Sharma, H. C. Bansal, 1983. 4:361–71. New Delhi: Oxford & IBH Publ. 398 pp.
109. Old, K. M., Libby, W. J., Russell, J. H., Eldridge, K. G. 1986. Genetic variability in susceptibility of *Pinus radiata* to western gall rust. *Silvae Genet.* 35:145–49
110. Oestergaard, H. 1983. Predicting development of epidemics on cultivar mixtures. *Phytopathology* 73:166–72
111. Parker, M. A. 1988. Genetic uniformity and disease resistance in a clonal plant. *Am. Nat.* 132:538–49
112. Parry, W. H. 1982. The role of the environment in host-insect interactions. See Ref. 49, pp. 22–31
113. Patton, R. F., Riker, A. J. 1966. Lessons from nursery and field testing of eastern white pine selections and progenies for resistance to blister rust. See Ref. 40, pp. 403–14
114. Post, W. M., DeAngelis, D. L., Travis, C. C. 1983. Endemic disease in environments with spatially heterogeneous host populations. *Math. Biosci.* 63:289–302
115. Powers, H. R. Jr. 1972. Testing for pathogenic variability within *Cronartium quercuum* and *Cronartium fusiforme*. See Ref. 51, pp. 505–11
116. Powers, H. R. Jr. 1982. Pathogenic variation within *Cronartium quercuum* f. sp. *fusiforme*. See Ref. 98, pp. 249–58
117. Powers, H. R. Jr. 1985. Response of sixteen loblolly pine families to four isolates of *Cronartium quercuum* f. sp. *fusiforme*. See Ref. 3, pp. 89–96
118. Powers, H. R. Jr., Kraus, J. F. 1982. Seedling seed orchards for the production of fusiform rust resistant slash and loblolly pine. See Ref. 98, pp. 325–35
119. Prakash, C. S., Heather, W. A. 1985. Reaction of cultivars of *Populus* spp. to radiation induced virulent mutants of *Melampsora medusae*. *Euphytica* 34:309–15
120. Prakash, C. S., Heather, W. A. 1986. Inheritance of resistance to races of *Melampsora medusae* in *Populus deltoides*. *Silvae Genet.* 35:74–77
121. Prakash, C. S., Heather, W. A. 1986. Relationship between increased virulence and the aggressiveness traits of *Melampsora medusae*. *Phytopathology* 76:266–69
122. Price, P. W., Westoby, M., Rice, B. 1988. Parasite-mediated competition: some predictions and tests. *Am. Nat.* 131:544–55
123. Raffa, K. F., Berryman, A. A. 1982. Physiological differences between Lodgepole pines resistant and susceptible to the mountain pine beetle and associated microorganisms. *Environ. Entomol.* 11:486–92
124. Raymond, C. A., Lindgren, D. 1990. Genetic flexibility: a model for determining the range of suitable environments for a seed source. *Silvae Genet* 39:112–21
125. Raymond, C. A., Namkoong, G. 1990. Optimizing breeding zones: Genetic flexibility or maximum value. *Silvae Genet* 39:110–12
126. Ridé, M. 1982. Environmental factors affecting host-bacteria interactions in bacterial diseases of forest trees and shrubs. See Ref. 49, pp. 48–59
127. Roberds, J. H., Namkoong, G., Skrøppa, T. 1990. Genetic analysis of risk in clonal populations. *Theor. Appl. Genet.* 79:841–48
128. Roberds, J. H., Hyun, J. O., Namkoong, G., Rink, G. 1990. Analysis of height growth response functions for white ash provenances. *Silvae Genet* 39:121–29
129. Rockwood, D. L. 1973. Monoterpene-fusiform rust relationships in loblolly pine. *Phytopathology* 63:551–53
130. Rockwood, D. L. 1974. Cortical monoterpene and fusiform rust resistance relationships in slash pine. *Phytopathology* 64:976–79
131. Roth, L. F. 1966. Foliar habit of ponderosa pine as a heritable basis for resistance to dwarf mistletoe. See Ref. 40, pp. 221–28
132. Schmidtling, R. C. 1985. Coevolution of host/pathogen/alternate host systems

in fusiform rust of loblolly and slash pine. See Ref. 3, pp. 13–19
133. Schreiner, E. J. 1972. Mass production of improved forest tree planting stock through synthetic varieties. See Ref. 51, pp. 571–90
134. Schutt, P. 1982. Trees, fungi and environment. See Ref. 49, pp. 39–47
135. Segal, A., Manisterski, J., Browning, J. A., Fischbeck, G., Wahl, I. 1982. Balance in indigenous plant populations. See Ref. 49, pp. 361–70
136. Shigo, A. L. 1984. Compartmentalization: A conceptual framework for understanding how trees grow and defend themselves. *Annu. Rev. Phytopathol.* 22:189–214
137. Shrimpton, D. M., Whitney, H. S. 1968. Inhibition of growth of blue stain fungi by wood extractives. *Can. J. Bot.* 46:757–61
138. Singh, S. J., Heather, W. A. 1982. Temperature sensitivity of qualitative race-cultivar interactions in *Melampsora medusae thum.* and *Populus* species. *Eur. J. For. Pathol.* 12:123–27
139. Smith, H. L. 1982. The interaction of steady state and hope bifurcations in a two-predator-one-prey competition model. *SIAM. J. Appl. Math.* 42:27–43
140. Snow, G. A. 1985. Coevolution of *Cronartium quercuum* with hard pines in eastern North America. See Ref. 3, pp. 1–12
141. Snow, G. A., Kais, A. G. 1972. Technique for inoculating pine seedlings with *Cronartium fusiforme*. See Ref. 51, pp. 325–26
142. Snow, G. A., Nance, W. L., Snyder, E. B. 1982. Relative virulence of *Cronartium quercuum* f. sp. *fusiforme* on loblolly pine from Livingston parish. See Ref. 49, pp. 243–50
143. Soegaard, B. F. 1972. Relative blister rust resistance of native and introduced white pine species in Europe. See Ref. 49, pp. 233–39
144. Steenackers, V. 1972. The state of knowledge in breeding rust resistant poplars. See Ref. 49, pp. 419–30
145. Taylor, A. D. 1988. Parasitoid competition and the dynamics of host-parasitoid models. *Am. Nat.* 132:417–36
146. Theisen, P. A. 1982. Some genetic considerations in the maintenance and enhancement of balance in host-pest systems. See Ref. 98, pp. 442–57
147. van Buijtenen, J. P. 1982. A population genetic model of the alternate host system of *Cronartium quercuum* f.sp. *fusiforme*. *For. Sci.* 28:745–52
148. van der Kamp, B. J. 1988. Susceptibility of Lodgepole pine provenances to geographically separate western gall rust spore sources. *Can. J. For. Res.* 18:1203–5
149. van der Kamp, B. J., Hawksworth, F. G. 1985. Damage and control of the major diseases of Lodgepole pine. See Ref. 4, pp. 125–31
150. Weir R. J. 1981. North Carolina State University-industry cooperative tree improvement program. See Ref. 45, pp. 57–70
151. Wenzel, G. 1985. Strategies in unconventional breeding for disease resistance. *Annu. Rev. Phytopathol.* 23:149–72
152. Wolfe, M. S. 1985. The current status and prospects of multiline cultivars and variety mixtures for disease resistance. *Annu. Rev. Phytopathol.* 23:251–73
153. Wolfe, M. S., Caten, C. E., eds. 1987. *Populations of Plant Pathogens. Their Dynamics and Genetics*. London: Blackwell Sci. 280 pp.
154. Yanchuk, A. D., Yeh, F. C., Dancik, B. P. 1988. Variation of stem rust resistance in a Lodgepole pine provenance-family plantation. *For. Sci.* 34:1067–75
155. Yeatman, C. W., Kafton, D., Wilkes, G., eds. 1984. *Plant Genet. Resour. A Conservative Imperative. Proc. AAAS Symp. Toronto, Ontario, 1981*. Boulder, CO: Westview Press. 164 pp.
156. Yeh, F. C. 1986. Recent advances in the application of biochemical methods to tree improvement. In *New Ways in Forest Genet. Proc. Meet. Can. Tree Improv. Assoc. 20th*, ed. F. Caron, A. G. Corriveau, T. J. B. Boyle, 2:29–39. Ottawa: Can. For. Serv. 200 pp.
157. Ying, C. C., Hunt, R. S. 1987. Stability of resistance among *Pinus contorta* provenances to *Lophodermella concolor* needle cast. *Can. J. For. Res.* 17:1596–601
158. Yoder, O. C. 1983. Use of pathogen-produced toxins in genetic engineering of plants and pathogens. See Ref. 71, 335–53
159. Zadoks, J. C. 1972. Reflections on disease resistance in annual crops. See Ref. 51, pp. 43–62
160. Zobel, B. J. 1972. The world's need for pest-resistant forest trees. See Ref. 49, pp. 1–8

Annu. Rev. Phytopathol. 1991. 29:343–60

DEVELOPMENT, IMPLEMENTATION, AND ADOPTION OF EXPERT SYSTEMS IN PLANT PATHOLOGY

J. W. Travis

Department of Plant Pathology, The Pennsylvania State University, University Park, Pennsylvania 16802

R. X. Latin

Department of Botany and Plant Pathology, Purdue University, Lilly Hall of Life Sciences, West Lafayette, Indiana 47907

KEY WORDS: decision support systems, knowledge-based systems, computer software implementation, artificial intelligence, integrated pest management

---

BEGINNINGS OF EXPERT SYSTEMS IN PLANT PATHOLOGY

Computerized decision support is not new to plant pathology. The weather-based computerized disease forecasters initially developed in the 1970s were the precursors to expert systems. Blitecast (25) and the apple scab predictive system (19) are examples of forecasters that are currently used to help farmers make decisions about the management of potato late blight and apple scab, respectively. They are similar to expert systems in that the rationale behind their development and application is to aid in the implementation of economically and environmentally sound control practices for a particular disease.

Expert systems were formally introduced into plant pathology in the U.S. at

0066-4286/91/0901-0343$02:00

the annual meeting of the American Phytopathological Society in Cincinnati, Ohio in 1987. At the discussion session on the "Development and Practical Use of Expert Systems for Pest Management" participants reviewed concepts in artificial intelligence applications and demonstrated their expert systems and the hardware and software required for operation. Interest in this new technology was very high. However, few operational expert systems have been developed in plant pathology since that time. There may be various reasons for the slow development of expert systems in plant pathology (21), at least some of which may involve a basic lack of understanding of expert systems and the potential benefit to plant disease management and the motivation of farmers and farm advisors to use this technology. This review is intended to provide a discussion of expert systems as they apply to plant pathology by briefly describing what expert systems are, how they should be constructed, and how they can be moved into production agriculture. Moreover, we review recent progress in their development and application. General references are available on the subject of expert systems (7, 13, 21, 30, 43, 44).

Expert systems are computer programs that emulate the logic and problem-solving proficiency of a human expert. They are most successful in addressing specific problems in well-defined areas in which experts can articulate the required knowledge. Expert systems often are applied as management tools that can interpret complex information and help the user make informed decisions. Expert systems, like human experts, can employ heuristics, a problem-solving technique that uses experience-based rules-of-thumb to achieve appropriate solutions. Similarly, this intuitive problem-solving approach is not subject to the constraints of conventional computer programming, but instead expands the usefulness of the computer. Expert systems can use data in almost any format and have the added advantage of suggesting a solution to the problem even though the information about the problem may not be complete.

COMPONENTS OF EXPERT SYSTEMS

All expert systems are composed of several basic components: a user interface, database, knowledge base, and an inference mechanism. Their relationship to each other is diagrammed in Figure 1. Moreover, expert system development usually proceeds through several phases, including problem selection, knowledge acquisition, knowledge representation, programming, testing, and evaluation.

Knowledge in the Expert System

The knowledge the expert uses to solve a problem must be so represented that it can be used to code into the computer and then be available for decision

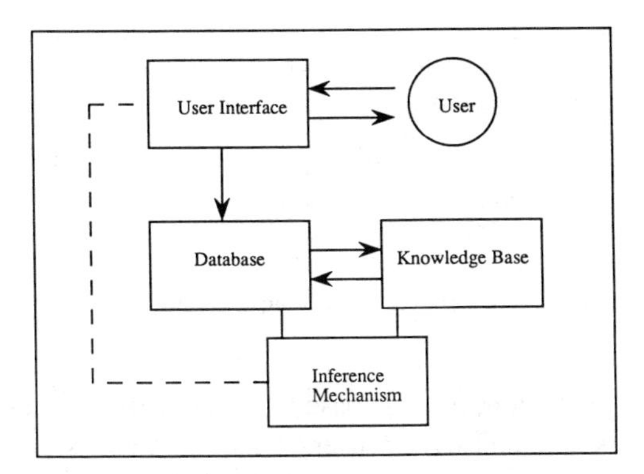

Figure 1 Diagram of expert system architecture.

making by the expert system. There are various formal methods for representing knowledge (3, 9, 11, 16) and usually the characteristics of a particular problem will determine the appropriate representation techniques employed. Knowledge bases can be represented by production rules, consisting of a condition or premise followed by an action or conclusion (IF condition, THEN action). Production rules permit the relationships that make up the knowledge base to be broken down into manageable units. For example, consider a simple rule for recommending an appropriate control action for muskmelon powdery mildew (Figure 2). In this rule, data are structured into parameter/value combinations. The parameters are, "Disorder," and "Weeks-before-Harvest," and "Previous-Fungicide-Applications." These parameters are factors that an expert would consider to determine a specific control recommendation. The value for each parameter is determined by relevant facts in the data base. The farmer provided the facts during a session with the computer. In this case, "Disorder" = powdery mildew, "Weeks-before-Harvest" =0 (harvesting of fruit has not yet begun), and "Previous Fungicides" = 0 (no fungicides effective against powdery mildew had been applied). Once the conditional parts of the rule are satisfied, then the rule is executed and the conclusion (a situation-specific action advisory) is drawn. If the facts are changed, the conditional part would change causing a different rule to be executed and possibly a different recommendation (conclusion). This rule (Figure 2) has three conditions, but there is no limit to the number of conditions for any given rule. Having a knowledge base that consists of hundreds or thousands of rules can cause a problem with management and organization of the rules. Organizing rules and visualizing their interconnectedness can be accomplished through dependency networks (Figure 3).This method of knowledge representation is demonstrated later in the chapter under knowledge acquisition and representation.

IF the DISORDER is POWDERY MILDEW, and
WEEKS-BEFORE-HARVEST > 0, and
PREVIOUS-FUNGICIDE-APPLICATIONS =
BAYLETON, BENLATE, OR TOPSIN,
THEN recommend the application of BAYLETON 50 DF at
4 oz.acre within the next 24 hours.

Figure 2 An example of a rule for recommending a control action for powdery mildew of muskmelon.

During the consultation, the rule base is searched for conditions that can be satisfied by facts supplied by the user. This operation is performed by the inference engine. Once all conditions (i.e. the IF parts) of a rule are matched, the rule is executed and the appropriate conclusion is drawn. Based upon the conclusions drawn and the facts obtained during consultation, the inference mechanism determines which questions will be asked and in what order. Various inferencing methods are available to perform the tasks of searching, matching, and execution (13, 15, 43).

A characteristic of expert systems distinct from conventional programs is their ability to use incomplete or incorrect data. Given only a partial data set, an expert is likely to have less than absolute certainty in his or her conclusion. The degree of certainty can be quantified in relative terms and included in the knowledge base. The certainty values are assigned by the expert during the knowledge acquisition phase of developing the system. By incorporating rules in the knowledge base with different certainty values, the system will be able to offer solutions to problems without a complete set of data. The capacity to deal with uncertainty is available in development software.

Explanation of the Decision Making Process

One attractive feature of expert systems is the program's ability to review a consultation and provide the user with an explanation for how its conclusion was derived. The explanation function is essentially a record of the reasoning process used by the expert to resolve the problem. It provides for a better understanding of how the conclusion was reached and instills in the user a greater confidence in the conclusion and the expert system. The accumulation of facts to be presented when an explanation is asked for is usually part of the development shell or software.

Making the Expert System Easy to Use

Whether or not an expert system achieves success may be determined by the nature of its user interface. This is the part of the expert system that requires interaction with the user. Even the most powerful expert system will not be

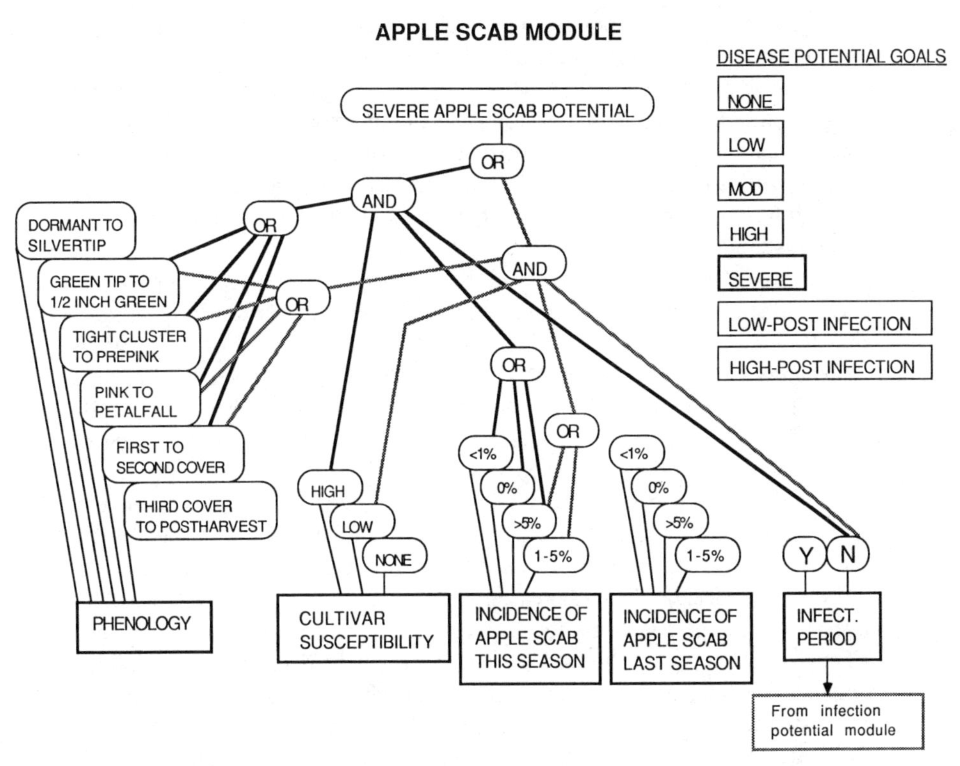

Figure 3 Dependency network of severe apple scab potential.

applied if it requires too much effort on the part of the user (5, 14, 24). For this reason, it is important to make the computer as easy for the user to operate as possible. Almost all modern development programs offer the capacity to interact with the system through both text and graphics. Several of the expert systems reviewed later in this article (22, 23, 37, 42) rely heavily on mouse operation, buttons, menus, and graphics to facilitate use of the expert system.

HARDWARE AND SOFTWARE NEEDED TO DEVELOP AN EXPERT SYSTEM

Many of the early concerns of plant pathologists involved in the development of expert systems centered on the type of hardware and software to use. Within the past five years, improvements in software and hardware have occurred so rapidly that the issue is no longer important. Expert systems can be developed in almost any computer language available, however object-oriented or symbolic languages such as LISP and PROLOG are the traditional artificial intelligence languages. Recently introduced Object C and PASCAL languages have become popular because they lack some of the operational constraints of the AI languages, e.g. they are faster, consume less memory and are more compatible with PC-based operations. An advantage to using a programming language for development of an expert system is that the system's components can be tailored to perform in an optimal fashion for the designed task of the program. A distinct disadvantage, however, is that the programming language approach requires a great deal of time to build an expert system from scratch.

Most expert systems are developed using an expert system building tool or "shell" rather than a programming language. Shells are written in programming languages such as LISP and Object C and include all of the generic components of an expert system. The user interface, inference mechanism, and method of knowledge representation are preprogrammed into the shell. Developers who use a shell to create an expert system need only to acquire and structure the domain knowledge.

At one time, the development of artificial intelligence applications was restricted to mainframes or mini-computers. Although large complex expert systems may still require such costly machines for their development, current hardware advances and the increased sophistication of expert system shells have allowed expert systems to be developed on personal computers.

The requirements of the end-user will ultimately determine the kind of computer selected for delivery or application of the expert system. Therefore, most successfully deployed systems will be delivered on personal computers. Many targeted users either already have the hardware or can readily purchase personal computers at relatively low cost.

DEVELOPING EXPERT SYSTEMS IN PLANT PATHOLOGY

The power of an expert system is derived from the knowledge of the expert. The lack of computer skills should not inhibit anyone from using this tool in implementing their disease management program. There are published procedures that one can follow to develop expert systems (13), and that can be adapted to individual style of thinking, resources, and problem area. Listed below are several examples of deviations by the developers of the Penn State Apple Orchard Consultant (PSAOC) (42) from accepted developmental procedures that nevertheless produced a successful expert system for advising Pennsylvania apple growers.

Selecting the Problem

In this section we discuss the development of expert systems in three parts: (*a*) selecting a problem; (*b*) knowledge acquisition and representation; and (*c*) evaluation and adoption. We concentrate on the aspects of developing expert systems that relate to plant pathology and spend little time on discussing expert system software tools, which are described elsewhere (3, 15, 44). In addition, the Penn State Apple Orchard Consultant (42) is used as an example to further illustrate each of the developmental phases. Other expert systems in plant pathology are reviewed later in this article.

The most critical step in developing an expert system is identifying a suitable problem. Criteria to determine the feasibility of developing expert systems and selecting the problem have been discussed previously (13, 38, 39, 44). Expert systems are best suited to problems that require experience, knowledge, judgment, and complex interactions to arrive at a solution (44). One of the first tests to determine if a subject area is suitable for an expert system is whether the solution of the problem requires the knowledge and expertise of a human expert. For many agricultural situations a reasonably knowledgeable person with the correct information such as a production guide, or an available data base management system, can determine the solution without the help of an expert or an expert system. The abilities inherent in the expert system software may not be necessary to solve problems such as determining days-to-harvest limitations for a pesticide, or calculating infection periods, which may be better accomplished by using a production manual, or a conventional computer program such as a spreadsheet or database manager. In addition, in crop areas where little is known about certain diseases, the power of an expert system is really not necessary.

Expert systems are expensive, requiring resources, expertise, and time to build. Costs include software, hardware, and personnel. It should be determined if the proposed expert system can be justified in potential savings or other benefits. If current loss due to disease or reduced yield or quality is great enough, then the successful implementation of an expert system may offset cost of development. Sometimes other benefits are realized that more than justify costs: these include reduced pesticide inputs, enhanced IPM training, effective delivery of IPM strategies not possible by other delivery methods, and identification of gaps in knowledge to direct research programs. Expert systems are also valuable as a constant source of expertise for decision makers in agriculture, available at all times and to all growers.

Knowledge Acquisition and Representation

After the problem has been selected, the knowledge acquisition phase of expert system development is begun. The knowledge that the expert uses to solve the problem must now be displayed logically for computer coding. There are many discussions in the literature on the extraction of knowledge from domain experts (3, 16, 18, 29). Most expert system developers suggest that a knowledge engineer (someone who is trained in the extraction of information from experts) extract the knowledge and design the expert system. In some cases, a knowledge engineer is needed to "tease apart" highly compiled expert knowledge because the expert is too familiar with the area to break it into its logical components. Software is available to help in self-knowledge engineering (32). A different method of knowledge acquisition was undertaken to develop the PSAOC. The domain experts were first trained in knowledge acquisition and representation, then when they had represented their knowledge in dependency networks (diagrams of the rules), the networks were given to programmers who coded it into the computer. The domain experts in this case were both the source of knowledge and the designers of the expert system. The most obvious advantage here was the saving in the cost of a knowledge engineer for the extraction and representation of knowledge. However, using the domain experts as the knowledge engineers presents an even more important advantage since only they know the extent of their knowledge, how they think about problems and what variables are important in the decision process. Rather than the knowledge engineer painfully extracting the information from experts, the knowledge flows from expert as they articulate from internal examination how decisions are made. The domain expert becomes an active participant in the knowledge acquisition and representation process rather than a passive participant who is acted upon. The domain expert becomes more involved in the development of the expert system, facing the challenge of representing on paper and then programming into the computer, the methods and processes by which problems are solved. In addition, domain experts can better communicate their logic to other domain experts—an important consideration where different domains overlap and there must be agreement among domains. Also, extension specialists who serve as the domain experts are more familiar with how the end users prefer to think about and solve problems. Their experience with the user clientele helps them to format the expert system in a manner that county agents or growers will best understand and use.

Throughout the development of PSAOC, knowledge was structured in the form of dependency networks (Figure 3). Dependency networks are decision trees that use Boolean logical approaches (AND/OR diagrams) and are valuable because they provide a "hardcopy" reservoir of system logic and a means of communication among domain experts and programmers. For example, the apple scab module in PSAOC essentially determines the disease potential of the orchard since the last pesticide application (Figure 3). The goals none, low, moderate, high, and severe (i.e. module end points) rate the disease potential since the last spray if no new infections have occurred. The low and high postinfection goals represent the disease potential when there was the possibility of infection since the last application. For apple scab this determination is based on five factors: the current phenology of the orchard; the cultivar susceptibility; the incidence of scab during the current and previous seasons; and the potential of infection since the last fungicide application. The current phenology is provided by the user according to a standard classification scheme illustrated within the system. The cultivar susceptibility to apple scab is assigned by the system. Incidence of scab this season and last season are supplied by the grower within the orchard description. The potential for infection since the last fungicide application is determined by a subroutine that involves a separate dependency network for infection potential.

The dependency networks are the primary mechanism of communication between the domain experts and the programmers. This record is most useful in the logic updating process done by the expert as new knowledge is acquired. The PSAOC pest management modules alone represent 135 dependency networks, each of which may involve more than one situation. For example, the "severe" disease potential goal is described in a network as two situations (Figure 3) shown as lines drawn from the "and" statements. One situation that describes the "severe" goal is shown by the darker lines. In this case, the severe disease potential for apple scab can occur from green tip to second cover phenological stages if the cultivar susceptibility is high and the incidence of apple scab this season is greater than 0% and there has been no infection potential since the last fungicide application. The second situation is illustrated with the lighter lines. Here, if a less susceptible cultivar is involved, then a higher incidence of disease must occur for the system to conclude that scab potential is severe. The network describing the severe goal is relatively simple, involving only two distinct situations. Some networks have many individual situations to be displayed.

The knowledge diagrams are translated into expert system code (9). Prototype expert systems can then be reviewed by developers and users.

Evaluation and Adoption

The prototype expert system must be tested and evaluated to determine internal consistency in the logic and to confirm that it was built according to planned specifications (40). This first step of system evaluation has been termed verification and is conducted by the developers of the system. The next step, which should involve people other than the developers, is to validate the system or determine that it behaves like a human expert (28). The final step is evaluation by potential users in realistic situations. For PSAOC a pilot test group consisting of 26 Pennsylvania apple growers was used for evaluation (33). The number of times a grower used the PSAOC expert system and the duration of each session were used as indicators of the degree of adoption of the expert system. These measures of system use taken together indicate one aspect of adoption—use of the innovation. While number of times the PSAOC system is accessed shows how frequently the system is being used, the actual amount of time spent using the system may be a more significant indicator of adoption. Some growers reported using the system primarily as a quick validation of their own knowledge regarding a decision. These growers would report a relatively high number of accesses and a low number of hours used. Conversely, the growers reporting high numbers of hours using the system are presumably more fully engaging the logic of the system in their decision-making process. Change in a grower's production practices is a second aspect of adoption. According to this study, 65.2% of the growers indicated some change in production practices as a result of using the expert system. Included within this study was a feedback loop to allow incorporation of grower comments and suggestions into the expert system. This technique, in addition to improving operation and interface, gave the growers a feeling of ownership of the expert system. Their efforts were clearly reflected in the improved operation of new versions of the expert system.

Based on this study, adoption of expert systems will be dependent upon (*a*) grower access to a microcomputer capable of running the system; and (*b*) proficiency in computer operation. Although the software was designed for use by people with little or no computer experience, growers with the least computer proficiency also had the lowest rate of system use.

These attendant prerequisites are termed the technology's "access conditions" (1, 10). Other access conditions include the availability of help for problems, an infrastructure to update software, and a structured educational program to teach computer and expert system skills. Computer technology in the absence of access conditions will be much less likely to be adopted.

To summarize the procedures for developing an expert system:

Select a problem that is not too complex and is important to the clientele;
Secure the cooperation of the most credible expert(s) available;
Represent the expert's knowledge in a form translatable into computer code;
Test the prototype expert system for consistency in logic and accuracy in building specifications;
Review the expert system by experts and revise as necessary;
Test the expert system with the intended users and make revisions;
Release the expert system and supply support for the users;
Regularly revise and update the expert system.

EXPERT SYSTEMS IN PLANT PATHOLOGY

The following review of plant pathology expert systems is only partially complete and does not include several expert systems that were developed in Europe (2, 4, 8, 12, 17, 41). The expert systems developed in Europe are primarily diagnostic in nature and have features similar to those described later in this chapter. The expert systems reviewed below contain the major features common to plant pathology expert systems at the present time.

Several are diagnostic expert systems whose conclusions are often weighted by certainty factors that express the degree of confidence that the system has in its conclusion. This feature enables the system to draw conclusions with incomplete data. For example, if all essential symptoms of a disease were expressed and recorded by the user during a consultation, then the maximum certainty factor would be expected (i.e. the system is absolutely certain of its conclusion). However, if a symptom is not expressed or not observed by the user, then the system would be less certain of its conclusion and would indicate the decrease in confidence with a lower than maximum certainty value. Users may also use certainty values while recording their observations. For example, if the user is not absolutely certain that a symptom is present, a less than maximum certainty value may be assigned to the response. The software then uses an algorithm inherent to the shell of the expert system to combine the various certainty values into a single meaningful measure of confidence in the conclusion drawn by the system. Developers can establish threshold values for certainty, below which the conclusion is not reported to the user. Several algorithms can be used to represent uncertainty (6).

PLANT/ds

The first expert system in plant pathology was developed by Michalski et al (27) in 1983 to diagnose 17 soybean diseases in Illinois. The Plant/ds expert system offers advice about soybean diseases on the basis of symptoms communicated by the user. Descriptors are used to characterize symptoms such as condition of leaves and leaf spots. The user answers multiple choice questions through key selection to provide information. The order of questions is not fixed but varies according to the user's responses. At any point in the session the next question selected is the one that will affect the most rules still under consideration. In a typical session the system asks about half of the 41 possible questions. The system was developed on a mainframe computer using PASCAL computer language and uses uncertainty to express the degree to which a rule matches the description of the diseased plant. The uncertainty associated with each response serves as an approximate indicator of the confidence in the advice provided by the expert system.

PLANT/ds has reached one of the most advanced stages of development of expert systems in plant pathology (Table 1). The prototype system was field tested and is commercially available.

POMME

POMME was developed in Virginia to help manage diseases and insects on apples (34, 35). POMME was one of the first expert systems to incorporate the decision making process of the expert to advise producers in making

Table 1 Comparison of expert systems in plan pathology

System name	Date publ.	Reference	Crop	Type system	Hardware[a]	Software[b]	Stage of Development[c]				
							Proto type	Valid.	Implem. study	Comm.	Regular revis.
PLANT/ds	1983	27	Soybean	Diagnositc	DOS	PASCALL	+	+		+	
POMME	1987	34	Apple	Disease management	Main-frame	PROLOG	+				
GrapES	1987	37	Grape	Disease management	MAC	Rule-Master	+	+			
White Pine Blisterust	1988	36	White Pine	Disease management	DOS	Insight2	+	+			
COUNSELLOR	1988	20	Wheat	Disease management	DOS	Savior	+	+			
Apple Pest and Disease Diagnosis	1988	22	Apple	Diagnostic	MAC	PROLOG	+	+			
CALEX/ Peaches	1989	31	Peach & Nectarine	Diagnostic	DOS	CALEX	+	+			
Muskmelon Disorder Management System	1990	26	Muskmelon	Diagnostic	DOS	Personal Consultant Plus	+	+		+	
Penn State Apple Orchard Consultant	1991	42	Apple	Disease management	MAC, DOS	PennShell	+	+	+	+	+

[a]Type of computer the expert system operates on, MAC = Macintosh, DOS = IBM or compatible
[b]Language or shell used to develop the expert system
[c]Stages of development are prototype, validation of the expert system, study of implementation process, commercially available, and regular revision

disease management decisions. POMME advises growers on disease potential, insect problems, drought control, winter injury, and spray recommendations. An infection model for apple scab was incorporated into the knowledge base that allows the system to make recommendations based on the actual occurrence of infection periods. This system also provides pesticide recommendations and application timing suggestions. The expert system was validated by extension specialists.

POMME was not developed past the prototype and early evaluation stage. POMME could not be implemented because it is only available on a mainframe computer on which it was developed for use with PROLOG computer language. Despite its limitations, POMME demonstrated that expert systems could incorporate extensive knowledge (e.g. infection period requirements) into its knowledge base and thus enable disease management expert systems to be built.

GrapES

GrapES (37) was developed about the same time as POMME and included many similarities. GrapES is a pest management expert system that includes both the determination of disease potential and fungicide recommendations and timings of applications. Insect problems are also considered, and insecticides are incorporated in the recommendation, along with fungicides. The procedures used to develop GrapES were early versions of the techniques used to develop the PSAOC, which have already been discussed.

White Pine Blister Rust

This is a limited prototype expert system developed in Idaho to predict the hazard of white pine blister rust to white pines planted on a particular site (36). The expert system uses site factors that influence the suitability of the habitat for *Ribes* germination and growth to predict blister rust potential. Once the hazard rating is determined, the system provides a recommendation on the suitability of planting white pine on the site and the recommended level of rust resistance of the planting stock. The users are intended to be foresters responsible for regenerating white pine sites. There was no validation or field testing of the system. The author of the expert system has some reservations about the representation and completeness of the knowledge base. This expert system was developed on a DOS computer using the expert system shell Insight 2+.

COUNSELLOR

COUNSELLOR was developed in England by ICI Agrochemicals (Haslemere, Surrey) to manage insects and diseases on wheat (20). It was developed on a DOS computer using the expert system shell Savoir. This shell is unique in that it uses evidence nets to predict risk of disease and to optimize treatment recommendations. The risk of each disease is determined separately and combined to estimate yield loss. The control recommendation includes efficacy rating for the recommended pesticides and a cost effectiveness rating. COUNSELLOR displays a cost benefit analysis for the treatment recommendation. COUNSELLOR was tested in field trials by measuring yield and economic benefits and found to be effective and beneficial.

Apple Pest and Disease Diagnosis

This expert system was developed in New Zealand to diagnose disorders caused by insects, diseases, and nutrient imbalances (22, 23). It was written using PROLOG and is available on a Macintosh computer. The system operates by interaction of the grower with the computer concerning a particular apple problem. From grower-supplied information, a list of possible problems is established and the likelihood of each is estimated. The grower is asked to agree or alter the initial set of potential problems listed by the computer. The computer then examines the remaining possibilities more completely with the user and generates a final list of potential causes. No treatment recommendation is given.

The user interface was very important to the authors and is well developed. The system makes full use of the graphics capabilities of the Macintosh computer to help the user in understanding and answering questions. Menus, windows, and buttons are used extensively throughout the system. This system is also unique in that the authors have tried to simplify questions. For example, the screen for inputting soil and leaf test information is designed to resemble the form that an orchardist might receive from the testing laboratory.

CALEX/Peaches

CALEX/Peaches diagnoses 120 disorders of peaches in California, including insects, diseases, and cultural problems (31). The user begins a session by identifying an area on the plant where the problem occurs. The expert system uses "certainty factor" to arrive at conclusions. At the end of a session the expert system displays all conclusions reached with corresponding levels of certainty.

The system was validated first by extension specialists and farm advisors. A follow-up evaluation was conducted by having two groups of graduate students use the system to determine how many correct answers they would obtain. One group comprised students of plant protection and the other of plant pathology. The plant protection students identified 99% of the problems correctly while the plant pathology students identified 96% correctly. The latter students made incorrect responses on some problems dealing with insects.

Muskmelon Disorder Management System (MDMS)

MDMS serves as a computerized consultant to help the user in identifying 17 disorders of muskmelons (26). The disorders include 11 infectious diseases, 3 nutrient imbalances, and 3 noninfectious problems. The expert system was built on a DOS computer using the expert system shell Personal Consultant Plus. Certainty has been built into the knowledge base to help the expert system reach a conclusion when there is incomplete data or unknown information. The certainty values were assigned by the domain expert as the knowledge base was being built. The expert system can diagnose more than one problem at a time and list all possibilities when the session is complete. MSDS was validated by comparing the responses obtained by technicians, graduate students, and county agents to the domain expert's responses to the same muskmelon disorders. The nonexperts arrived at 71% correct responses with 100% certainty, and 13% of the responses were correct diagnoses with less than 100% certainty.

As previously discussed, the success of an expert system depends on the degree of adoption by users. Adoption of the expert system is dependant not only on the correctness of the knowledge base and user friendliness of the system but also on how well the user understands its use. MSMD includes a very complete user's guide, in addition to 28 color photographs to aid in decision making.

Penn State Apple Orchard Consultant (PSAOC)

The development and structure of PSAOC (42) has already been described above. PSAOC manages 8 diseases and 17 insects in an integrated fashion. Weather information is used within the expert system to determine the potential for disease development, calculate the effect of rainfall on chemical residues, and screen chemicals for potential phytotoxicity. The expert system considers biological, cultural, and chemical control options when arriving at a pest control recommendation. A chemical management expert system also operates within the PSAOC to select chemicals, adjust rates, and set application intervals. It also checks recommended chemicals for compatibility and days-to-harvest limitations. PSAOC is capable of arriving at unique management recommendations for literally thousands of pest control scenarios. PSAOC was developed on a Macintosh computer and employs a frame-based expert system tool, Pennshell ©, which was written in the C programming language. It was found in the early development of expert systems at Pennsylvania State University that many of the commercially available expert system shells did not address all of the needs of agricultural expert systems. One important area of weakness in many commercial expert system shells is the lack of ability to adequately handle weather information for use in disease prediction models and other management decisions. Although parts of PSAOC were built directly from the C language, Pennshell © is designed as a "tool box" of often-used functions so that little direct coding is necessary. PSAOC was made commercially available to apple producers in Pennsylvania in 1990. It is available for use on DOS or Macintosh computers and is revised and expanded annually. An effort is currently underway to expand its applicability to the New England states.

THE FUTURE OF EXPERT SYSTEMS IN PLANT PATHOLOGY

Expert systems are most appropriately used with high value crops (horticultural crops) whose management options include the application of fungicides and insecticides. Growers of these crops are experiencing increasing difficulty in making the complex disease and insect management decisions that are required almost daily throughout the season. Expert systems will also become an essential resource to individuals who influence farmers' decisions (county agents, ag-chemical distributors, industry representatives). They serve to decentralize diagnostic knowledge and distribute it where it is needed most. Expert systems for disease management will have the best chance for success if the experts (plant pathologists) play an integral role in their design and development. As methods for knowledge representation become less technical, the development of expert systems by the experts will become more common. A major challenge in developing successful systems will involve filling gaps in our knowledge about disease management. We will need to expand our research in epidemiology to better understand the behavior of populations of plant pathogens and to include fungicide resistance management.

To be successful expert systems must deal with important problems, must be accurate and reliable, and easy to use. In addition, the success of an expert system may depend on the "access conditions" that effect adoption. Access conditions include the availability of help for user's problems, an infrastructure to provide software updates and a structured educational program to teach computer and expert system skills.

An increasingly important objective of disease management expert systems will be to use very site-specific information to improve pesticide application and reduce pesticide use. Given the benefits of rational disease management practices, expert systems development and implementation probably will continue as long as plant pathologists are serious in their concerns about environmentally sound crop production and management.

ACKNOWLEDGMENTS

This article represents concepts, ideas, and methods developed through the cooperation of colleagues working in artificial intelligence at The Pennsylvania State University and Purdue University. We especially thank Edwin Rajotte, Richard Bankert, Robert Crassweller, and Michael Foster for their contributions to this article.

Literature Cited

1. Audirac, I., Beaulieu, L. J. 1986. Microcomputers in agriculture: A proposed model to study their diffusion/adoption. *Rur. Soc.* 51(1):60–77
2. Badia, J. 1988. MalHerb: identification des mauvaises herbes des cultures. *Tech. Rep. Stn. Biom. Intell. Artif.* INRA, Castanet-Tolosan, 31 pp.
3. Barrett, J. R., Jones, D. D., eds. 1989. *Knowledge Engineering in Agriculture.* ASAE Monogr. No. 8, ASAE: St. Joseph, MI. 214 pp.
4. Blancard, D., Bonnet, A., Colena, A. 1985. TOM, un systeme expert en maladies des tomates. *Rev. Hortic.* 216:7–13
5. Berry, D. C., Broadbent, D. E. 1986. Expert systems and the man-machine interface. *Expert Syst.* 3:228–31
6. Buchanan, B. G., Shortliffe, E. H., eds. 1984. *Rule-based Expert Systems, the MYCIN Experiments of the Stanford Heuristic Programming Project.* Reading, MA: Addison-Wesley. 748 pp.
7. Buchanan, B. G., Smith, R. G. 1988. Fundamentals of expert systems. *Annu. Rev. Comp. Sci.* 3:23–58
8. Cervo, R. 1988. Diagnose von Krankheits- und Schadlingsbefall bei Oliven. In *Wissensbasierte Systeme in der Landwirtschaft—Auf dem Weg zum Anwender. 2. Int. DLG- Computerkongr., Frankfurt, Bad Soden* 19–22. Juni. pp. 506–14
9. Cullen, J., Bryman, A. 1988. The knowledge acquisition bottleneck—Time for reassessment? *Expert Syst.* 5:216–25
10. Dillman, D. A. 1985. The social impacts of information technologies in rural North America. *Rur. Soc.* 50(1):1–26
11. Engel, B. A, Thieme, R. H., Whitaker, A. D. 1989. Representation and reasoning. See Ref. 3, pp. 47–76
12. Foucaut, O., Foucaut, J. F., Landrac, J. Y., Lalin, D. 1988. Expert system to aid identification of natural species, integrating texts and images: application in mycology. In *Appl. Artif. Intell. Agric. Agrochem. Food Process. Ind. Caen, France,* Sept. 29, 30. pp. 125–44
13. Frenzel, L. E. Jr. 1987. *Crash Course in Artificial Intelligence and Expert Systems.* Indianapolis: Sams & Co. 358 pp.
14. Gordon, S. E., Gill, R. T., Dingus, T. A. 1987. Designing for the user: The role of human factors in expert system development. *AI Applic. Environ. Sci.* 1:35–46
15. Harmon, P., Maus, R., Morrissey, W. 1988. *Expert Systems Tools and Applications.* New York: Wiley. 289 pp.
16. Hillman, D. 1988. Bridging acquisition and representation. *AI Expert* 3(11):38–46
17. Hocine, A., Tamine, K. 1988. A knowledge-based system to aid the detection of disease in corn. See Ref. 12, pp. 166–81
18. Hoffman, R. R. 1987. The problem of extracting the knowledge of experts from the perspective of experimental psychology. *AI Mag.* 8(2):53–67
19. Jones, A. L., Lillevik, S. L., Fisher, P. D., Stebbins, T. C. 1980. A microcomputer-based instrument to predict primary apple scab infection periods. *Plant Dis.* 64:69–72
20. Jones, M. J. 1988. Counsellor: a cereal disease advisory program. See Ref. 12, pp. 182–88
21. Jones, P. 1989. Agricultural applications of expert systems concepts. *Agric. Syst.* 31(1):3–18
22. Kemp, R. H., Stewart, T. M., Boorman, A. 1988. An expert system for apple pest and disease diagnosis. See Ref. 12, pp. 189–203
23. Kemp, R., Stewart, T., Boorman, A.

1988. Improving the Expert System Interface. *AI Appls. Nat. Res. Mgnt.* 2(4):48–53

24. Kidd, A. L., Cooper, M. B. 1985. Man-machine interface issues in the construction and use of an expert system. *Int. J. Man-Machine Stud.* 22:91–102
25. Krause, R. A., Massie, L. B., Hyre, R. A. 1975. Blitecast: A computerized forecast of potato late blight. *Plant Dis. Reptr.* 59:95–98
26. Latin, R. X., Miles, G. E., Rettinger, J. C., Mitchell, J. R. 1990. An expert system for diagnosing muskmelon disorders. *Plant Dis.* 74:83–87
27. Michalski, R. S., Davis, J. H., Bisht, V. S., Sinclair, J. B. 1983. A computer-based advisory system for diagnosing soybean diseases in Illinois. *Plant Dis.* 67:459–63
28. O'Keefe, R. M., Balci, O., Smith, E. P. 1987. Validating expert system performance. *IEEE Expert* 2(4):81–90
29. Patton, C. 1985. Knowledge engineering: tapping the experts. *Electron. Des.* 33(10):93–100
30. Payne, E. C., MacArthur, R. C. 1990. *Developing Expert Systems.* New York: Wiley. 401 pp.
31. Plant, R. E., Zalom, F. G., Young, J. A., Rice, R. N. 1989. CALEX/Peaches, an expert system for the diagnosis of peach and nectarine disorders. *HortScience* 24(4):700
32. Prsaye, K. 1988. Acquiring and verifying knowledge automatically. *A. I. Expert*
33. Rajotte, E., Bowser, T. 1990. Expert systems: An aid to the adoption of sustainable agricultural systems. In *Sustainable Agriculture Research and Extension: Proc. Work. Natl. Acad. Press, 1990.* In press
34. Roach, J., Virkar, R., Drake, C., Weaver, M. 1987. An expert system for helping apple growers. *Comp. Elect. Agric.* 2(1):97–108
35. Roach, J. W., Virkar, R. S., Weaver, M. J., Drake, C. R. 1985. POMME: a computer-based consultation system for apple orchard management using PROLOG. *Expert Syst.* 2(2): 56–69
36. Rust, M. 1988. White pine blister rust hazard rating: an expert systems approach. *AI Appl. Nat. Res. Mgnt.* 2(2,3):47–50
37. Saunders, M. C., Haeseler, C. W., Travis, J. W., Miller, B. J., Coulson, R. N., Loh, K. D., Stone, N. D. 1987. GRAPES: an expert system for viticulture in Pennsylvania. *AI Appl. Nat. Res. Mgnt.* 1(2): 13–20
38. Slagele, J. R., Wick, M. R. 1988. A method for evaluating candidate expert system applications. *AI Mag.* 9(3):44–45
39. Stock, M. 1987. AI and expert systems: an overview. *AI Appl. Nat. Res. Mgnt.* 1(1):9–17
40. Suen, C. Y., Grogono, P. D., Shinghal, R., Coallier, F. 1989. Evaluation of expert systems. In *EXPERSYS–89, Proc. ESA World Conf. 1st.*, ed. J. Liebowitz, pp. 103–114. 358 pp.
41. Thiolon, C. 1988. A tool for the diagnosis of disease for plants grown in France. See Ref. 12, pp. 340–54
42. Travis, J. W., Rajotte, E., Bankert, R., Hickey, K. D., Hull, K. D., et al. 1991. Penn State Apple Orchard Consultant Expert System: The design and function of the pest management module. *Plant Dis.* 75: In press
43. Waterman, D. A. 1986. *A Guide to Expert Systems.* Reading, MA: Addison-Wesley
44. Whittaker, A. D., Jones, D. D., Thieme, R. H., Barrett, J. R. 1987. Guidelines for getting started with expert systems. *Agric. Eng.* 68(5):24–27

Annu. Rev. Phytopathol. 1991. 29:361–80

ENVIRONMENTALLY DRIVEN CEREAL CROP GROWTH MODELS*

R. W. Rickman and Betty Klepper

USDA-Agricultural Research Service, Columbia Plateau Conservation Research Center, Pendleton, Oregon 97801

KEY WORDS: wheat pest interactions, mechanisms, pathogen-plant linkage, weed-crop coupling points, process level simulation

---

INTRODUCTION

Crop growth models are becoming common tools for expressing current knowledge of plant physiological processes and the interaction of the whole plant with its environment. Whisler et al (56) list 30 process-level crop simulations for alfalfa, barley, citrus, corn, cotton, peanuts, potato, rice, sugar beet, and wheat, plus one that claims to be adaptable to describe any crop. The IBSNAT (International Benchmark Sites for Agrotechnology Transfer) project is collecting available mechanistic models for 10 major crops worldwide (51). Several statistically based models of epidemic progression and reduction of crop yield are available for assisting in monitoring or anticipating crop damage (6, 7, 26). These models describe effects of diseases or pests (broadly defined to include pathogens, insects, mites, nematodes, or weeds) but they do not describe the mechanics of pest growth and activity or the interrelationship of host crop, pest, and environment. Mechanistic models, which attempt a degree of cause and effect explanation of disease development or pest activity, have recently become available. The linking of mechanistic pest and crop-growth models is a significant step in the development of tools for understanding and improving management of disease or pest effects on crop production (48).

First steps toward an orderly description of interactions between pest- and crop-growth models were provided by Boote et al (5). They listed seven "coupling points" or links by which pests interacted with crops. The interacting mechanisms included: stand reduction, photosynthesis rate reduction, leaf senescence acceleration, light stealing, assimilate sapping, tissue consumption, and turgor reduction. A pest can influence a crop by one or more of these coupling mechanisms. If mechanistic models for the pest and crop are to be combined to describe the influence of the pest on seasonal growth and production by the crop, both models must have algorithms for the coupling mechanisms by which the pest acts. Crop effects on the pest must be considered as well as pest effects on the crop. The interaction of the two models will provide a feedback of information from one model to the other on a computational time-step basis that will modify future computations in both. Generally, the more detailed the physiological or physical processes that are specified in the crop and pest models, and the more knowledge we have of the physiological processes and physical-environmental interactions involved in host-pest association, the better our chances of finding appropriate linkages for joining pest and crop models (25).

The selection of a crop model for linking with a pest model should be made on the basis of availability of the appropriate coupling parameters for the disease(s) or pest(s) of concern. Even the more detailed crop models provide algorithms for only a limited number of coupling mechanisms. The purpose of this paper is to discuss current concepts of the morphological development, phenology, and growth of cereals as characterized by wheat (*Triticum aestivum* L.), to mention some of the more mechanistic wheat growth models available as computer programs, and to describe the coupling mechanisms they can provide.

CEREAL MODELS

General Concepts

This section concentrates on concepts of morphological development, phenology, and growth used to describe cereals. The next section deals with specific models and the algorithms that are used in their computer programs to determine the coupling points provided for linkage to other models.

Cereals, all being in the grass family, are morphologically and phenologically similar. All cereals pass through a series of development phases separated by defined life-cycle events. These phases include emergence, leaf development, root development, tiller appearance, stem (internode) elongation, head development, and kernel growth and filling. The events that mark the beginning and end of these phases include such points as double ridge, jointing, anthesis, and maturity. These phenological events and the development phases that they separate are at the heart of most mechanistic or process-oriented cereal growth models.The timing of phases and the growth of plant tissue during the phases are treated separately in most models. Phases and phenology (development) are timed by temperature and daylength. On the other hand, growth, which involves the capture and partitioning of carbohydrates into new plant tissues, is controlled by light interception, nutrient and water uptake, temperature, and time during the appearance of the plant tissue in question.

As plants progress through the phases of development, they shift partitioning of carbohydrates and other materials from one set of organs to another and produce the yield components that combine to form crop grain yield. The product of plant stand (plants/m^2), heads per plant, kernels per head, and kernel weight determine grain yield. Each component is formed during a different growth phase. Figure 1 illustrates the approximate relationships

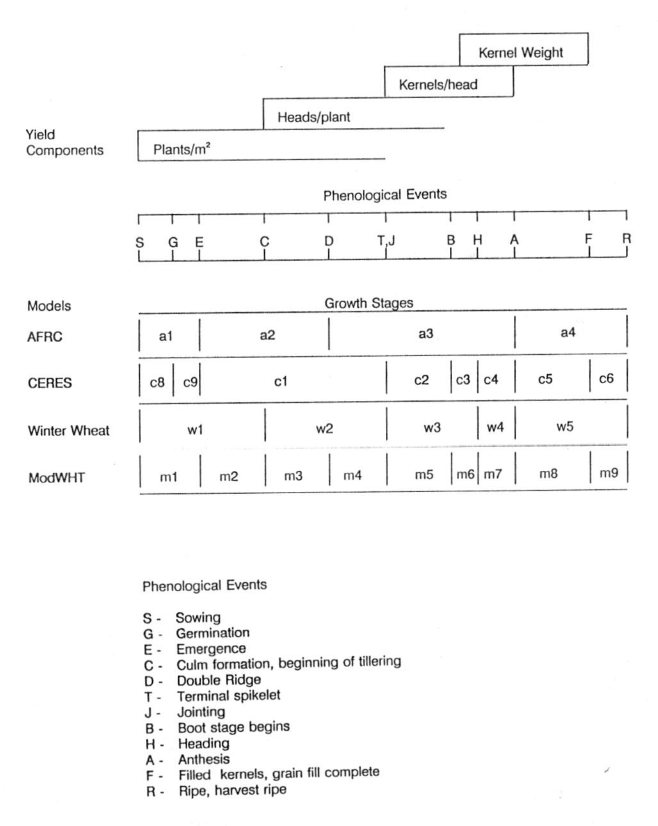

Figure 1 Growth stages as specified in four mechanistic wheat models and the intervals of formation of wheat yield components.

between formation of the yield components and phases of development during the lifetime of a wheat plant. The interaction of any pest with the crop will depend upon the phase of development of the crop when the pest is active and the relative importance of the contribution of the tissue or resource affected to the yield component being formed.

The progression of a plant from phase to phase is observed by noting visible changes in plant morphology. The morphological development of cereals progresses regularly during the passage of growth phases. We illustrate concepts used to observe and model cereal development by referring to the morphological development of winter wheat.

Development of Winter Wheat

The wheat main stem produces leaves alternately on either side of a compressed stem. The stem is comprised of a stack of symmetrical nodes each supporting a leaf, a tiller bud, root buds, and internode tissue. During vegetative development (emergence through tillering phases), internodes do not elongate and the nodes form the crown of the plant where nodal (adventitious, crown) roots originate. The leaves of the main stem and those of the tillers appear in a rigidly regular pattern. Once three full leaves have extended on the main stem, the tiller bud on node 1 (the node supporting leaf 1 (L1)) will extend to expose the first leaf of tiller 1 (L11 of T1). Tillers extend from each successive main stem node, lagging three nodes behind the node producing the currently extending main stem leaf. For example, the first leaf of T2 extends as main stem L5 elongates. Leaves of each tiller extend, in turn, to become visible at the same rate as leaves on the main stem to form a plant as illustrated (Figure 2). Because of the regularity of the leaf extension and tillering pattern, if the number of leaves on the main stem is known, the potential number and location of all tillers on the plant is also known (19, 29). The maximum number of tillers that may be present on a plant at any time follows a Fibonacci series function driven by main-stem leaf number (4). The extension of successive main-stem leaves (and all tillers and their leaves) can be timed accurately using thermal time, also called growing degree days.

Growing degree days are calculated by summing for each day over an interval of days, the average daily celsius temperature corrected for a base temperature. Cool-season grasses and cereals (e.g. winter wheat) have base temperatures near 0°C, and warm-season crops (e.g. maize) have base temperatures near 10°C. One must make it a point to know the base temperature used for any growing degree day or thermal time value, particularly for the classical phenology models that use sums of thermal time for timing of whole the phases of development. Most development models of this type change the base temperature depending on the phase of development and

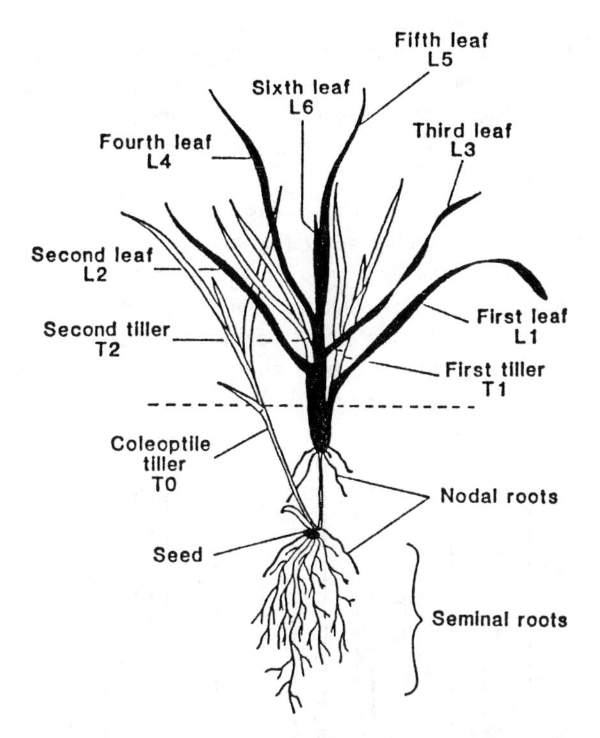

Figure 2 A wheat plant showing the naming system for leaves and tillers.

some change the method of averaging temperature from one phase to the next. Recent models of canopy development based on morphological development, as explained here, use only the interval required to extend one leaf as a basic timing unit (20, 32, 41).

The time interval between the emergence of successive leaves from the whorl of leaves at the top of each culm is called a phyllochron. The phyllochron value is essentially constant throughout much of a growing season if it is measured in units of growing degree days. Winter wheat characteristically has a phyllochron of 100 growing degree days with a base temperature of 0°C (18, 30). Root development also can be anticipated from the passage of phyllochrons (19, 43).

Vegetative development of leaves, tillers, and roots occurs as a series of events all of which take place, in turn, at successively higher nodes on each culm of a the plant (21). For example, tiller 2 (T2) elongates from node 2 (N2), approximately three phyllochrons after the appearance of L2 and coincident with the extension of L5. During the emergence of T2, a pair of nodal roots also extends from N2. About three phyllochrons later, another pair of roots can form from N2 completing the expected set of four nodal roots per node (43). The same sequence of events occurs at N3 just one phyllochron after they occurred at N2. Thus, the plant develops its parts in a rigidly regular procession up the stem during vegetative development. These concepts have been used in various ways to predict the time of appearance of leaves, tillers, and roots (Figures 3 and 4).

The vegetative phase of winter wheat lasts until early spring when the stem apex converts from producing vegetative nodes to producing reproductive nodes, each of which subtends a spikelet. The double ridge stage or event is the first directly observable indication that reproductive development has begun. Once the reproductive phase begins, tiller buds and roots cease extension from active nodes and tiller abortion is induced. Internode tissue elongates to produce jointing, the physical appearance of nodes above the soil

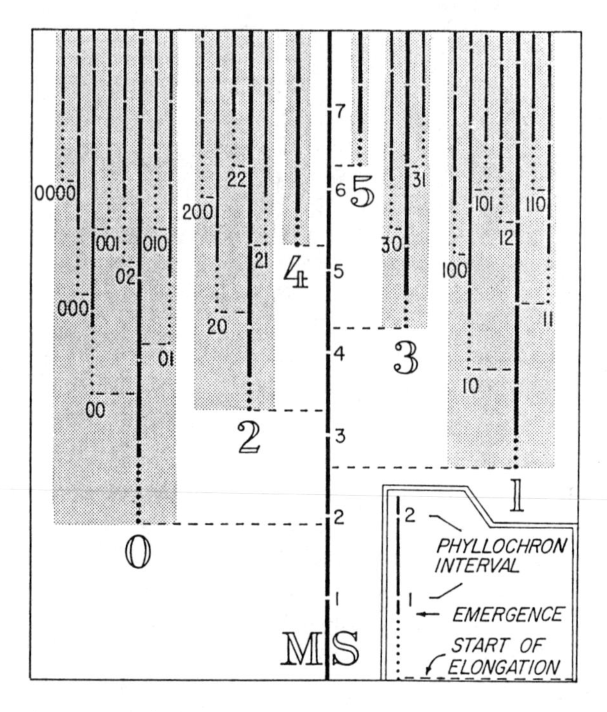

Figure 3 Diagram showing the relationships between leaf and tiller development on the main stem, tillers (single-digit numbers), and subtillers (two- and three-digit numbers) in winter wheat. Each line segment represents the expected interval of extension of each identified leaf relative to all others. Time increases one dimensionally from the bottom to the top of the diagram. Time for formation of each leaf is one phyllochron.

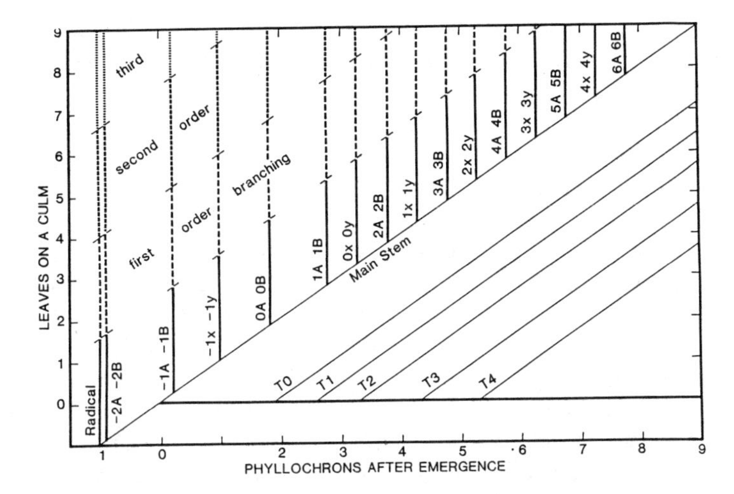

Figure 4 Developmental relationships for each leaf, tiller, and root of wheat. A straight edge laid vertically at a given number of phyllochrons after emergence will predict the number of main-stem tillers that will be present on unstressed plants and the number of leaves on each culm present. When laid horizontally across a main-stem leaf number, the number of root axes and their degree of branching can be determined.

surface. Within the reproductive phase, the yield components spikelet number, kernel number, and kernel weight are determined. Formation of the terminal spikelet occurs about the same time as jointing and sets the number of spikelets in wheat. Kernels are formed as florets on each spikelet between the time of terminal spikelet appearance and anthesis. Following successful anthesis, grain-fill occurs (Figure 1). Floret formation on the head appears to follow a development pattern very similar to that of tillers on the main stem (27, 28).

Each wheat tiller appears at a predefined development stage in normal plants (22), but stresses will delay the appearance of any tiller or, if sufficiently severe, will cause the tiller to be skipped altogether (44). The coleoptilar tiller, by its presence or absence, is a good indicator of seed-bed quality (35). Crown root axes also appear at set times in development (19), and they appear to be skipped in the development process only under situations of severe stress.

Tiller abortion occurs at about the same time that stem internodes begin elongation (jointing). This synchrony is probably not a coincidence because the reallocation of carbohydrates to the developing stem is thought to trigger the abortion. Tiller abortion involves a general rule that the last tiller to be produced is the first tiller to be lost. Tillers with three or fewer leaves visible at jointing are the first to abort and will routinely do so even under favorable growing environments (29). More mature tillers may also abort with higher levels of stress. Senescence of lower leaves is a normal developmental event that proceeds acropetally and can be accelerated by shortages of nitrogen or water.

Model Parameters

Parameters that differ among cereal crops are the base temperature, genetic variability with respect to the timing of appearance and probability of release of tillers, and the length of the phyllochron. For wheat, the degree day is usually calculated with a base temperature of 0°C (44), although individual data sets may generate values ranging from -2 to 4°C. The slope of the line relating leaf number, as observed by the Haun growth stage (13), to cumulative degree days provides the inverse of the phyllochron in units of leaves per degree day. The length of the phyllochron (normally stated with units of degree days per leaf) varies more with planting date than with cultivar (12, 42). Baker et al (2) showed that the rate of change of daylength at emergence was a good predictor of phyllochron length in wheat, but no one knows the mechanism by which plants in the field actually set the length of the phyllochron.

Green leaf area is needed to calculate radiation absorbed by the canopy. Most mechanistic models convert absorbed solar radiation into photosynthate using a constant of from 2 to 8 μg CO_2 per Joule of absorbed photosynthetically active radiation. Photosynthates are distributed to plant tissues using partitioning rules that change with phenological stage to provide carbohydrate, in turn, to leaves, roots, stems, and ears.Grains are generated, pollinated, grown, and finally filled with starch, but few models have specific subroutines to describe the processes involved. Some crop-growth models quantitatively account for water and nutrient budgets. This accounting requires extensive physical and chemical information of soil and a method for predicting root distribution.

Considerations for Model Linkage

When considering the linkage of pest and crop models, the organization of both the models and the program codes implementing the models is of concern. Processes that are easily isolated provide opportunities for simple linking. When a coupling process is influenced by several factors within one or both models, considerably greater planning is required to achieve a successful combined model. For example, the seven coupling points listed by Boote et al (5) can be divided into mechanisms internal and external to the plant. Photosynthesis rate, leaf senescence, and assimilate distribution are internal processes while stand reduction, light stealing, and tissue consumption could occur from external processes. Turgor is an example of a coupling point that is rather complexly related to both internal and external processes and events. A pest that modifies stand could be coupled by the modification of the one variable containing the plant number per unit area for the crop. If turgor is the coupling mechanism, computation of an effect may involve an hourly iterative feedback between the crop model, water budget, and pest model. The following discussion of cereal models attempts to provide information about each model relative to the availability and accessibility of coupling points as listed by Boote et al (5).

Individual Crop Models

In this section we briefly describe the mechanistic features of some models that are available as computer programs. Our goal is to acquaint the reader with the general model framework and to give enough references for details to be obtained from the literature. The models discussed are ARFC WHEAT, developed in Britain for British and Central European conditions, the CERES collection developed for yield prediction globally, WINTER WHEAT developed from a data base from the Central Great Plains of the US, and ModWHT developed in the Pacific Northwest of the US for describing wheat development and growth for global application.

ARFC WHEAT

DESCRIPTION ARFC WHEAT is a no-stress model without limitations of water or nutrient supply. Its principal features are described by Weir et al (54). There are five submodels: (*a*) phenological development, (*b*) tiller and leaf production, (*c*) root production, (*d*) light interception and photosynthesis, and (*e*) dry matter partitioning and grain growth.

Phenological development predicts the beginning and ending dates of four phases: sowing to emergence (a1), emergence to double ridge (a2), double ridge to anthesis (a3), and anthesis to maturity (a4) (Figure 1). Time is measured as a combination of calendar time, temperature, and photoperiod, and the algorithm used differs for each phase (36). Duration of the period between sowing and emergence (a1) is driven by thermal time with a base temperature of 1°C. Thermal time is not computed as a simple mean temperature as usually done for growing degree days, but is a weighted average of eight 3-hour daily time periods whose weighting depends on the temperature of the time period. Prediction of the emergence to double ridge (a2) requires thermal time, vernalization information, and photoperiod. Double ridge to anthesis (a3) is determined simply by photoperiod and anthesis to maturity (a4) is measured in thermal time with a base temperature of 9°C.

The leaf and tiller growth submodel, based on the analysis of Porter (37), uses thermal time computed from mean daily temperature with a 1°C base temperature. Note the thermal time used in this submodel is computed differently than thermal time used in the phenology submodel. Leaves appear on all culms at a rate determined by thermal time using a phyllochron set by the rate of change of daylength at emergence (2). Tillers are generated weekly from an empirically derived relationship using thermal time. Tiller abortion is triggered by the double ridge stage and lasts through anthesis. Probability of tiller survival decreases with increasing tiller population size at the time of tiller appearance and with increasing development stage. Green leaf area index (LAI) for photosynthesis is obtained in this submodel by producing each leaf as a function of a predetermined size and thermal time for appearance, operation, and senescence.

Roots are assumed to require 1.5×10^{-4} g of dry matter per cm of root length for axes and 0.4×10^{-4} g/cm for branches and their rate of elongation is governed by the mean air temperature and the availability of assimilate (38). A total root length in each soil layer is computed.

Gross photosynthesis is calculated hourly using incident radiation, green leaf area index (LAI), and a light intensity- and vapor pressure deficit-dependent stomatal resistance within canopy layers. Radiation is attenuated by successive layers within the canopy. Maintenance and growth respiration are estimated to provide a daily net photosynthesis. No transportation processes are simulated and net photosynthate is assumed to be immediately available for partitioning.

Dry matter is partitioned to roots, leaves, stems and, later, ears differently during each life-cycle phase. For example from emergence to double ridge, roots receive 35%, leaves 55% and stems 10%; but, later in the crop life cycle, from the beginning of ear growth to anthesis, roots receive only 10%, leaves 30%, stems 30%, and ears 30%. After anthesis, all new photosynthate goes to the grain along with an assimilate pool of 30% of the canopy weight at anthesis.

The model has generally done well in simulating wheat growth and development under European conditions (40, 55).

AVAILABLE LINKS Most of the coupling mechanisms are available in the AFRC wheat model. Stand is currently determined by initializing seeds per m^2 and using a fraction of successful emergence as a multiplier. Photosynthesis rate depends upon a variety of environmental factors but the gross computation for photosynthesis is based on a quadratic saturation curve relating photosynthesis rate to incoming photosynthetically active radiation and canopy diffusion resistance with a predefined maximum rate. Leaf senescence is a linear function of thermal time. Light absorption by canopy layers is an explicit part of the model so that sharing part of that light with a competing canopy of a specified shape and density would be possible. Assimilate is determined on an hourly basis and partitioned to the plant daily. Assimilate consumption by a pest could be subtracted but no transport-dependent algorithms are used. Tissue consumption could be dealt with on an individual leaf area basis or by removal of dry matter from appropriate tissue.

There are no tissue-water or osmotic-potential algorithms in the model; therefore turgor is not available. The model was developed for a no stress situation and has no nitrogen budget.

CERES-WHEAT-MAIZE-SORGHUM

DESCRIPTION The CERES models (34) are all similar in that they are based on an accurate water-budgeting procedure. Precipitation in the form of either rain or snow is either stored in the soil or runs off depending on soil-water transmission coefficients that must be provided for the field soil. Redistribution of water among soil layers, extraction by plant roots, and loss from the profile by drainage or evaporation are computed by soil profile layer on a daily basis.

Crop growth is based on the amount of water transpired, daily temperature, and the amount of sunlight intercepted. Daily net photosynthesis is distributed between the roots and shoot depending on which of nine growth stages the crop is in. The leaf area index is computed from a daily estimate of leaf area expansion, leaf senescence, and tiller death. Root growth (both a total length per layer and depth in a soil) is also computed from partitioned photosynthate and a set of rooting-preference factors for each soil layer that must be provided as part of the soil data for the model. After reproductive growth begins, the number of kernels per plant is calculated from plant weight at flowering and kernel weight is calculated from time and temperature available for dry matter distribution, respectively. Final yield is estimated as a product of plant number, kernel number, and kernel size.

While growth and the water budget in all of the CERES models are similar, the phenology for each crop is somewhat different. All CERES crop models use the same nine stages of growth with minor changes in terminology as appropriate for the specific crop: Stage c7, from grain harvest to sowing; c8, sowing to germination; c9, germination to emergence; c1, emergence to terminal spikelet initiation (wheat) or end of juvenile stage (maize and sorghum); c2, terminal spikelet to flag leaf (wheat) or end of juvenile stage to tassel initiation (maize); c3, flag leaf to headed (wheat) or tassel initiation to silking (maize); c4, headed to anthesis (wheat) or silking to beginning grain fill (maize, sorghum); c5, anthesis to grain filled (wheat) or beginning to ending grain fill (maize, sorghum); and c6, grain fill to harvest (Figure 1).

The length of each stage depends upon the accumulation of a required number of thermal time units and/or photoperiod. Thermal time is computed with the mean of maximum and minimum daily air temperature, which is in contrast to the weighted thermal time used in the AFRC wheat model. For winter wheat stage c1, daily thermal time is reduced by a vernalization factor until a vernalization requirement is met. It is also reduced by a photoperiod factor for daylengths shorter than 20 hours (including civil twilight). Adjustments for spring or winter habit and early or late developing wheat cultivars are made by varying the number of the vernalization and/or photoperiod units that must accumulate.

The CERES models have been evaluated most extensively and adapted to more environments than the other models reported here (34).

AVAILABLE LINKS Available linkage points to the CERES models are slightly different from those to the AFRC wheat model. Values for stand are entered directly. Daily dry matter production is calculated as a product of intercepted light and a conversion coefficient, reduced for limiting water or excessive temperature. Total leaf area gained (by growth) and lost (by senescence) are computed daily. Leaf area index is used for a single-layer canopy for intercepting light. Daily assimilate production is partitioned to roots or shoots according to the developing organs of each stage of development. Water availability to the plant is used to reduce the rate of daily leaf-area increase and dry-matter production, but a specific turgor value is not computed. Root length (computed per layer of soil) controls water uptake by the plant and would provide a linkage point for root-damaging pests.

WINTER WHEAT

DESCRIPTION The WINTER WHEAT model (3) contains both a detailed, multiple-layered soil water and nutrient budgeting system (RHIZOS; see 24) and a detailed physiological process model for shoot growth. Within the soil portion of the model, root length is described on a daily basis in a two-dimensional grid. Daily gain and loss of root tissue are both allowed. Soil depth and grid size are selectable according to the detail of knowledge of soil-profile properties and crop-row spacing. Water is budgeted with inflow from the surface, redistribution, surface evaporation and plant extraction accounted for daily.

Growth of the shoot is computed with a combination of light interception by a single-layered canopy and previously established, maximum, potential growth rates for each plant organ. Limitations on growth are computed from shortages of photosynthate, nitrogen, or water. Specific episodes of cold weather can also reduce growth. Green leaf area can be reduced to zero by overwinter freezing, reflecting the environment of the central Great Plains of the U.S.A., the origin of much of the data used to create the model. Canopy leaf area index is regrown from zero in the spring using soluble reserves stored overwinter in crown and root tissue. Nitrogen uptake by the plant is computed daily from both active and passive uptake by the root system. Partitioning of carbohydrate to active plant organs depends upon the organ-specific nitrogen supply and requirement. Shoot growth during the reproductive stage includes the detail of floret development within the cereal spike. Shortages of nitrogen, water, or carbohydrate can limit the number of successful florets as well as the filling rates of successful grains.

Plant phenology controls which organs are active and which tillers survive as the plant develops from vegetative to reproductive growth. Phenological stages are determined from accumulations of mean temperature-based thermal time much as was done in the CERES models, but specific stages of plant development are different. WINTER WHEAT only uses 5 stages: seedling (w1), tillering (w2), jointing (w3), heading (w4), ripening (w5; Figure 1).

Although the model requires a large data set for complete evaluation, it has been tested at several sites within the continental United States (57).

AVAILABLE LINKS Many coupling points are available for WINTER WHEAT. Stand, as with CERES, is entered directly. Photosynthetic rate is computed from light intensity, CO_2 concentration, leaf age, water supply, and nitrogen supply. Leaf senescence is computed more mechanistically than in the above models using leaf age, nitrogen supply, and nitrogen demand by filling grain. The single-layered canopy provides for light stealing by modifying either leaf-area index or incoming light intensity. There is no direct mechanism to deal with relative canopy heights or canopy structure. Assimilate sapping and tissue consumption could both be dealt with on an organ-specific basis. Nitrogen content of consumed materials would also be available. Leaf water potential of the shoot is computed for estimates of water uptake. Turgor of the shoot could be obtained for intervals shorter than 1 day from WINTER WHEAT.

ModWHT

DESCRIPTION The ModWHT winter wheat model (41) is built from six subsystem modules: atmosphere, canopy, shoot, roots, soil, and soil surface (52, 53). Each module contains the algorithms for the processes occurring in that portion of the crop system. Atmosphere contains all weather-related information; canopy contains atmosphere-plant interactions such as light capture and transpiration; shoot contains the development, photosynthate production, and partitioning routines; root contains water and nutrient uptake algorithms and the root-length density profile; soil contains a simple water and nitrogen inventory; and soil surface contains routines relating tillage and management practices to water infiltration and soil evaporation.

An energy balance is used to compute water use. Plant extraction of water from the surface layer downward is weighted by root-length density in each layer. Actual transpiration is determined by potential evapotranspiration, crop LAI, and percent available soil water. Available soil water is the ratio of the current amount of soil water to the amount of water in a soil profile full to field capacity to the depth of rooting.

The shoot module contains the major algorithms of the winter wheat model, which is based on a leaf-by-leaf development of the shoot with timing of tillering and all phenological stages by the seasonal progression of phyllochrons (21). The phyllochron has a constant thermal-time interval for a season. Thermal time is computed as degree days using mean daily air temperature with a 0°C base. Phenological events of emergence, beginning of tillering, single and double ridge, terminal spikelet, jointing, boot, heading, anthesis, and grain filling are estimated (Figure 1). A vernalization requirement is the only factor used to determine the initiation of reproductive growth and number of leaves formed on the main stem. Tiller extension and survival is determined by phenological stage and photosynthate supply during the scheduled time of formation of each tiller.

Photosynthate is produced from light capture by the leaf area index of a single-layer canopy on a daily time step. Temperature, water supply, and nitrogen supply influence daily net assimilate available for growth. The size of each leaf and all other plant parts (roots, stem internodes, spikelets, and kernels) is determined by partitioning photosynthate among an available assimilate pool and all parts currently forming. Water and nitrogen supply both influence rate of photosynthesis. Yield is estimated from the sum of weights of surviving kernels from surviving culms.

The model is not yet published and is presently being assessed based on date and observations available from several sites in the continental United States. It is available as computer code from the senior author of this chapter.

AVAILABLE LINKS Stand is computed as planting rate times a survival factor, which depends on water content of the seedbed. Photosynthesis rate is computed from CO_2 concentration, intercepted light, nitrogen supply, and plant assimilate supply. Leaf senescence is timed by passage of phyllochrons. Light stealing would take place, as in WINTER WHEAT, by reducing LAI or incoming light intensity. Even though all leaf and stem sizes are available, a layered canopy with incremental LAI by layer is not computed. Assimilate sapping from the pool of plant assimilates or from specific plant parts could occur. Tissue consumption of any part of the shoot would be possible. The rate of leaf and stem extension is reduced by a limited water supply, but no plant water potential or turgor values are computed.

Environment and Pests

One of the common deficiencies of all of the above models is a structurally detailed canopy submodel with mechanistic computations of canopy environment pertinent to establishment and survival of pest agents (herbivorous insects or foliar pathogens) within portions of the canopy. For example, humidity is one of the most important environmental characteristics determining the suitability for leaf diseases or pest activity. The duration of leaf wetness controls spore germination and infection. Humidity gradients very near the leaf surface will determine suitability of the environment for reproduction of small insects, both plant pests and their predators. The development of such a detailed canopy model is of comparable complexity to the creation of a crop-growth model itself. At least one comprehensive model that provides the canopy-environment information needed for modeling pest establishment and activity does exist (33). The model, named CUPID, uses within-canopy energy budgeting to obtain light, temperature, and humidity at any of several layers from the soil surface to the top of the canopy on a fraction of a daily basis. Depending on the specific objectives of a pest model, a canopy model by itself may be sufficient or possibly both the crop and a detailed canopy model will be needed to meet all project objectives.

The above models have been compared relative to the seven coupling points proposed in 1983 by Boote et al (5). Since that time, several other possible coupling points have been suggested. In the Workshop on Modeling Pest-Crop Interactions, held Jan. 7–10, 1990 in Manoa, Hawaii (50), the following additional coupling points were listed: metabolic diversion, resource competition, translocation disruption, tissue disruption, leaf appearance reduction, growth rate reduction, respiration acceleration, assimilation rate reduction, toxin production, transpiration enhancement, and disturbance of biochemical processes. It might be argued that some on this list overlap with those of the original seven or with one another. The major implication from this growing list of coupling points is that as models become more complex and fundamentally more mechanistic, additional coupling points will arise.

This observation will be obvious to those who have already tried to couple independently developed models; but, for those who have not, the process is far from trivial. It is usually an exercise in complex programming, but the task can be either straightforward or impossibly complex, depending on the mechanisms of coupling and the way in which each model was originally programmed and documented. Structured program design, independent subroutines or procedures with passed variables (in contrast to common block variables), and construction of modular programs all permit program modification with the knowledge of all points of change. Unstructured programs, large common blocks, GOTO statements (which can divert program flow to anywhere), and large complex routines provide opportunities for unexpected interactions leading to very complex debugging problems (14). Of the above models, the program for ModWHT was designed from its inception in modular form with continued modification and coupling in mind. It is, however, only available in the Microsoft Quick Basic language. The other models are in FORTRAN and have versions to run on either microcomputers or larger machines. The documentation of the AFRC program within the code is very good. That of the CERES models varies somewhat with the specific version of the program one obtains. The WINTER WHEAT program also has reasonably good documentation.

The largest existing effort to simplify the coupling of crop and pest models is the IBSNAT DSSAT (Decision support system for agrotechnology transfer) (51). The pest-to-plant model interface is based on one or more of the seven coupling points of Boote et al (5) and four others: metabolic diversion, resource competition, translocation disruption, and tissue disruption. Although few if any crop models exist to provide the detail of information needed for truly mechanistic pest models, a knowledge of the coupling methods that mimic the mechanisms actually used by pests can guide the creation of next-generation crop models.

Generally, models available today are first-order approximations of our understanding of the phenological and physiological processes that drive crop growth. It remains for the next generation of models to bring in stresses associated with the lack of nutrients, or the presence of weeds, insects, plant parasitic nematodes, and pathogens, and to integrate process modelling to the ecosystem level (50). Relationships such as competition for limiting resources between crop plants and other species, both below and above ground, must be understood in quantitative terms (31).

RESEARCH NEEDS AND PROGRAMMING CHALLENGES

We need to expand our models in two directions: to a shorter time scale, to cover the short time (minutes to hours) during which environment has effects on pest behavior; and to a longer time scale (multiple years to decades) to incorporate the effects of cropping sequences and annual management practices on inoculum potential and weed-seed survival patterns over several generations.

Pests have been combatted with breeding programs and, in recent decades, with relatively selective chemicals. The resulting basic genetic data available for both host and pathogen will eventually become a valuable source of information for guiding mechanistic crop-pathogen models. Many plant morphological and physiological features are controlled by specific gene combinations, and many genes are environmentally regulated. Examples exist for both high- and low-temperature activation of resistance to rust.

Numerous examples can be cited where specific information in a crop model is insufficient to support a pest model. The differential susceptibility of main shoots and tillers of cereals for several diseases has been established (16, 39). The role of soil and plant water potential in predisposing hosts to disease is also clear (1, 10), yet few models generate plant water potentials for the infection periods of concern. Antecedent tillage and residue management practices are implicated in such diseases as bare patch (*Rhizoctonia solani* and take-all (*Gaeumannomyces graminis* (Sacc. Arx and Olivier var. *tritici*)) of wheat (23). However, very few present-day models address more than one year, the crop year, or include tillage in sufficient detail to assess implications for these diseases.

Most crop models do not include information on crop rotations, which can have very important effects on the potential for root diseases. All present-day crop-growth models use local weather scenarios to "drive" the model; none contains global weather models. However, we know that global wind patterns are responsible for moving inoculum of many rust fungi and other wind-borne spores from crop to crop (11).

Simulations of multiple-year or multiple-pathogen effects may require the use of the rule-based decision-analysis tools available from object-oriented programming or from the artificial intelligence programs.

Interactions between plant stress caused by selective herbicides and plant predisposition to disease (47) are hardly well-enough quantified to permit appropriate algorithms to be written, even assuming that sufficient information is available in the crop model to define threshold levels of plant stress and amounts of pathogen inoculum.

Biocontrol organisms can play an important role in the management of plant diseases (9), but modelling these organisms may be unrealistic, considering that this area of science is so new (8). No current crop growth process model contains explicit information on herbicides, yet we know that herbicides can predispose plants to frost damage. We also know that residual herbicides (chlorsulfuron) in soils can favor some root diseases (46). Current models likewise do not address the interactive effects of the pathogen on plant physiology or integrity and on the potential for secondary infections by less aggressive microorganisms. No current crop models specify form of nitrogen (nitrate or ammonium) available to the crop, yet each of these forms of nitrogen influences the severity of specific diseases (15).

Integration of crop and pest models requires careful planning to produce sophisticated ecosystem-level models that couple disease or pest damage and crop growth (45). The coding of a computer program for such a model must also be well planned and well executed because many of the algorithms will be tightly coupled. Independent, loosely coupled algorithms are easy to implement as independent routines, but coupled-interdependent algorithms must be dealt with as a unit, causing the code to become lengthy and complex.

Interdependent algorithms can also become mathematically unstable under certain conditions, so that more extensive testing or restrictive boundary conditions for some variables often must be specified. For the interim, it is reasonable to expect that most integrative efforts will be restricted to regression-type models of the type described by Teng & Gaunt (49), where the loss of potential yield is calculated as a mathematical function of disease intensity at some stage of crop development. For example, King et al (17) described equations to estimate yield losses for England from infection by *Septoria*. They calculated losses from the percentage of infected areas on the flag and immediately preceding leaves, which they obtained in surveys at the milky-ripe stage of crop growth. A model developed to define these percentages of infected leaf area as a function of planting time and weather, could be used to assess yield loss from these regression equations.

Realistic ecosystem-level crop-growth models that incorporate interactive relationships among weather, insect pests, pathogens, weeds, and crops await newer crop models implemented as programs with modular construction and with more detailed information on the spatial and temporal dynamics of the crop-weather system. We have no doubt that the next century will see many computer programs coded from such models.

The programs will require significant computing speed and memory capabilities to operate. In the case of plant diseases, they will require much more explicit information on host-pathogen relations. But most important, both will require interdisciplinary team effort where pathologists, microbiologists, physiologists, soil scientists, and computer specialists all work together to pool their respective expertise in the effort.

Literature Cited

1. Ayres, P. G. 1982. Water stress modifies the influence of powdery mildew on root growth and assimilate import in barley. *Physiol. Plant Pathol.* 21:283–93
2. Baker, C. K., Gallagher, J. N., Monteith, J. L. 1980. Daylength change and leaf appearance in winter wheat. *Plant Cell Environ.* 3:285–87
3. Baker, D. N., Whisler, F. D., Parton, W. J., Klepper, E. L., Cole, C. V., et al. 1985. The development of WINTER WHEAT: A physical physiological process model. See Ref. 57, pp. 176–87
4. Boone, M. Y. L., Rickman, R. W., Whistler, F. D. 1990. Leaf appearance rates of two winter wheat cultivars under high carbon dioxide conditions. *Agron. J.* 82:718–24
5. Boote, K. J., Jones, J. W., Mishoe, J. W., Berger, R. D. 1983. Coupling pests to crop growth simulators to predict yield reductions. *Phytopathology* 73: 1581–87
6. Coakley, S. M., Line, R. F., McDaniel, L. R. 1988. Predicting stripe rust severity on winter wheat using an improved

method for analyzing meteorological and rust data. *Phytopathology* 78:543–49
7. Coakley, S. M., McDaniel, L. R., Shaner, G. 1985. Model for predicting severity of *Septoria tritici* blotch on winter wheat. *Phytopathology* 75:1245–51
8. Cook, R. J. 1986. Plant health and the sustainability of agriculture, with special reference to disease control by beneficial microorganisms. *Biol. Agric. Hort.* 3: 211–32
9. Cook, R. J., Baker, K. F. 1983. *The Nature and Practice of Biological Control of Plant Pathogens*. St. Paul, MN: APS. 439 pp.
10. Cook, R. J., Papendick, R. I. 1972. Influence of water potential of soils and plants on root disease. *Annu. Rev. Phytopathol.* 10:349–74
11. Davis, J. M. 1987. Modeling the long-range transport of plant pathogens in the atmosphere. *Annu. Rev. Phytopathol.* 25:169–88
12. Delecolle, R., Hay, K. M., Guerif, M., Pluchard, P., Varlet-Grancher, C. 1989. A method of describing the progress of apical development in wheat, based on the time-course of organogenesis. *Field Crops Res.* 21:147–60
13. Haun, J. R. 1973. Visual quantification of wheat development. *Agron. J.* 65: 116–19
14. Hodges, T. 1990. Modeling and programming philosophies. In *Predicting Crop Phenology*, ed. T. Hodges, pp. 101–5. Boca Raton, FL: CRC Press
15. Huber, D. M., Watson, R. D. 1974. Nitrogen form and plant disease. *Annu. Rev. Phytopathol.* 12:139–65
16. Jörg, E., Weissenfels, D., Kranz, J. 1987. Diseases and pests on main shoots and tillers of wheat. *J. Plant Dis. Prot.* 94:509–19
17. King, J. E., Jenkins, J. E. E., Morgan, W. A. 1983. The estimation of yield losses in wheat from severity of infection by *Septoria* species. *Plant Pathol.* 32:239–49
18. Kirby, E. J. M., Appleyard, M., Fellows, G. 1985. Leaf emergence and tillering in barley and wheat. *Agronomie* 5:193–200
19. Klepper, B., Belford, R. K., Rickman, R. W. 1984. Root and shoot development in winter wheat. *Agron. J.* 76:117–22
20. Klepper, B., Rickman, R. W. 1989. Phyllome developmental patterns in winter wheat. *Agronomy Abstr.* Am. Soc. Agron., p. 18
21. Klepper, B., Rickman, R. W., Belford, R. K. 1983. Leaf and tiller identification on wheat plants. *Crop Sci.* 23:1002–4
22. Klepper, B., Rickman, R. W., Peterson, C. M. 1982. Quantitative characterization of vegetative development in small cereal grains. *Agron. J.* 74:789–92
23. Kollmorgen, J. F., Ridge, P. E., de Boer, R. F. 1987. Effects of tillage and straw mulches on take-all of wheat in the Northern Wimmera of Victoria. *Aust. J. Exp. Agric.* 27:419–23
24. Lambert, J. R., Baker, D. N., Phene, C. J. 1976. Dynamic simulation of processes in the soil under growing crops: RHIZOS. In *Computers Applied to the Management of Large-Scale Agricultural Enterprises. Proc. US-USSR Sem.* Moscow, Riga, Kishinev. 32 pp.
25. Loomis, R. S., Adams, S. S. 1983. Integrative analyses of host-pathogen relations. *Annu. Rev. Phytopathol.* 21:341–62
26. Madden, L. V. 1983. Measuring and modeling crop losses at the field level. *Phytopathology* 73:1591–96
27. Malvoisin, P., 1984. Organogenesis and growth of the main culm of wheat from sowing to flowering. I. Relationships between leaf growth and the differentiation of young leaves or flowers. *Agronomie* 4:557–64
28. Malvoisin, P., 1984. Organogenesis and growth of the main culm of Triticum aestivum from sowing to flowering. II. An attempted model for the relation between growth and vascularization of culm and leaves. *Agronomie* 4:587–96
29. Masle-Meynard, J., Sebillotte, M. 1981. Etude de l'hétérogenéité d'un peuplement de blé d'hiver. I. Notion de structure du peuplement. *Agronomie* 1:207–16
30. Masle-Meynard, J., Sebillotte, M. 1981. Etude de l'hétérogenéité d'un peuplement de blé. II. Origines des différentes catégories d'individus du peuplement, éléments de description de sa structure. *Agronomie* 1:217–26
31. McGiffen, M. E., Huck, M. G., Spitters, C. J. T. 1990. Physiologically-based simulation of weed crop competition. In *Workshop on Modeling Pest-Crop Interactions*, ed. P. Teng, J. Yuen, Res. Ext. Ser. 120:11–12. Honolulu: Univ. Hawaii
32. McMaster, G. S., Klepper, B., Rickman, R. W., Wilhelm, W. W., Willis, W. O. 1991. Simulation of shoot vegetative development and growth of unstressed winter wheat. *Ecological Modeling*. In press

33. Norman, J. M. 1982. Simulation of microclimates. In *Biometeorology of Integrated Pest Management,* ed. J. L. Hatfield, I. J. Thompson, pp. 65–69. New York: Academic
34. Otter-Nacke, S., Godwin, D. C., Ritchie, J. T. 1986. Testing and validating the CERES-Wheat model in diverse environments. *Publ. YM–15-00407 JSC 20244*. LBJ Space Cent., Houston, TX: Earth Res. Appl. Div.
35. Peterson, C. M., Klepper, B., Rickman, R. W. 1982. Tiller development at the coleoptilar node in winter wheat. *Agron. J.* 74:781–84
36. Porter, J. R. 1983. Modelling stage development in winter wheat. *Aspects Appl. Biol.* 4: 449–55
37. Porter, J. R. 1984. A model of canopy development in wheat. *J. Agric. Sci.* 102:383–92
38. Porter, J. R., Bragg, P. L., Rayner, J. H., Weir, A. H., Landsberg, J. J. 1983. The ARC wheat simulation model-principles and progress. In *Opportunities for Manipulations of Cereal Productivity,* ed. A. Hawkins, B. Jeffcoat. Brit. Plant Growth Regul. Group Monogr. 7:97–108
39. Rees, R. G., Mayer, R. J., Platz, G. J. 1981. Yield losses in wheat from yellow spot: A disease-loss relationship derived from single tillers. *Aust. J. Agric. Res.* 32:851–59
40. Reinink, K., Jorritsma, I., Darwinkel, A. 1986. Adaptation of the AFRC wheat phenology model for Dutch conditions. *Neth. J. Agr. Sci.* 34:1–13
41. Rickman, R. W., Klepper, B. 1989. Dry matter creation and distribution in a development driven winter wheat model. *Agronomy Abstr.,* Am. Soc. Agron., p. 21
42. Rickman, R. W., Klepper, E. L. 1990. Tillering in wheat. See Ref. 14, pp. 73–83
43. Rickman, R. W., Klepper, B., Belford, R. K. 1985. Developmental relationships among roots, leaves, and tillers in winter wheat. In *Wheat Growth and Modeling, NATO ASI Ser. A,* ed. W. Day, R. K. Atkin. 86:406. New York: Plenum
44. Rickman, R. W., Klepper, B. L., Peterson, C. M. 1983. Time distributions for describing appearance of specific culms of winter wheat. *Agron. J.* 75:551–56
45. Rouse, D. I. 1988. Use of crop growth-models to predict the effects of disease. *Annu. Rev. Phytopathol.* 26:183–201
46. Rovira, A. D., McDonald, H. J. 1986. Effects of the herbicide chlorsulfuron on Rhizoctonia bare patch and take-all of barley and wheat. *Phytopathology* 76: 879–82
47. Schoeneweiss, D. F. 1975. Predisposition, stress, and plant disease. *Annu. Rev. Phytopathol.* 13:193–211
48. Teng, P. S. 1988. Pests and pest-loss models. Agrotechnology Transfer No. 8. IBSNAT Project. Univ. Hawaii, Honolulu
49. Teng, P. S., Gaunt, R. E. 1980. Modelling systems of disease and yield loss in cereals. *Agric. Syst.* 6:131–54
50. Teng, P. S., Yuen, J. E., ed. 1990. *Proc. Workshop Modelling Pest-Crop Interactions*. Res. Ext. Ser. 120. Univ. Hawaii, Honolulu
51. Uehara, G. 1985. Decision support system for agrotechnology transfer (DSSAT). *IBSNAT Tech. Rep. 5*. 55 pp.
52. Waldman, S., Rickman, R. W. 1989. Logical organization and implementation of a crop model. *Agronomy Abstr.,* Am. Soc. Agron. p. 25
53. Waldman, S. E., Rickman, R. W. 1990. Programming techniques to expedite communication of scientific models. *Agron. J.* 82:356–59
54. Weir, A. H., Bragg, P. L., Porter, J. R., Rayner, J. H. 1984. A winter wheat crop simulation model without water or nutrient limitations. *J. Agric. Sci.* 102:371–82
55. Weir, A. H., Day, W., Sastry, T. G. 1985. Using a whole crop model. See Ref. 43, pp. 339–55
56. Whisler, F. D., Acock, B., Baker, D. N., Fye, R. E., Hodges, H. P., et al. 1986. Crop simulation models in agronomic systems. *Adv. Agron.* 40: 141–208
57. Willis, W. O., ed. 1985. *ARS Wheat Yield Project*. ARS-38. USDA Agric. Res. Serv.

Annu. Rev. Phytopathol. 1991. 29:381–98

DELIGNIFICATION BY WOOD-DECAY FUNGI

Robert A. Blanchette

Department of Plant Pathology, University of Minnesota, St. Paul, Minnesota 55108

KEY WORDS: wood deterioration, biotechnology, basidiomycetes, lignin degradation, white-rot

---

INTRODUCTION

Wood decay fungi are unique because of their capacity to decompose lignified cell walls. A few species are of special interest because they can selectively remove lignin from wood without extensive cellulose degradation. Lignin is a complex, heterogeneous phenylpropanoid structural polymer that occurs throughout the cell wall (71, 112). Spatially, lignin is intimately interspersed with hemicelluloses forming a matrix that surrounds cellulose microfibrils (67, 75), and provides a formidable physical and chemical barrier to biodegradative systems. Although most saprophytic fungi produce some degradative enzymes, such as cellulases, xylanases, mannanases and others, these enzymes do not permeate and degrade effectively woody substrates unless lignin is unbound, modified, or removed.

Investigations of decomposition processes in forest ecosystems have usually considered lignin to be the most recalcitrant component and the last degraded (10, 88). The decomposition pathway for decay by some fungi in the Basidiomycotina, such as those that cause brown rots, involves the degradation of all wood carbohydrates, including crystalline cellulose. A residual lignin matrix, consisting of chemically modified lignin, is left to be gradually converted to humic substances by long-term processes involving other microbes (38, 41, 106). However, this is not always the sequence of events. Fungi in the Basidiomycotina that cause white rots of wood may simultaneously degrade lignin along with all cell wall carbohydrates, or lignin

0066-4286/91/0901-0381$02:00

may be preferentially removed with little to no loss of cellulose. Recently, many of the lignin compositional trends that have been reported for humic substances in the Amazon river system and other natural environments were shown possibly to originate solely from white-rot degradation (58). When lignin is selectively degraded and removed from wood, large concentrations of cellulose remain (13, 14, 24, 38). The delignified zone may comprise relatively small areas consisting of minute pockets or be extensive, encompassing entire fallen trees (5, 6, 13, 97). The ecological dynamics of selective delignification in forest ecosystems that may result in entire logs of cellulose remaining on the forest floor have not been thoroughly investigated.

The first report of fungi causing wood decay was in 1863 by Hermann Schacht (113), who described the decomposition phenomena of wood. Robert Hartig, a pioneer leader in forest pathology, was the first to promote a general understanding that wood degradation was caused by biological processes (55). He also made many significant discoveries about wood decomposition. He identified different types of degradation and classified them according to the classification developed by his father, Theodore Hartig (57), as either white rot or red rot (the red rots are now referred to as brown rot). He found that some white-rot fungi were able to extract lignin from the woody cell walls but not cellulose (55, 56). In addition, his precise observations showed that two distinct degradative pathways also could be produced by the same white-rot fungus within one substrate (Figure 1). In some regions of the decayed wood, cell walls were delignified and middle lamellae completely removed, whereas in other areas the fungus caused an erosion of all cell wall layers, with the middle lamellae the last to be degraded. These were extraordinary observations for the late 19th century, and unfortunately, they remained relatively unknown until they were rediscovered with the aid of electron microscopy approximately 100 years later.

Many different categories have been used to classify wood decay by white-rot fungi, including the terms white rot, corrosive rot, simultaneous rot, pocket rot, and uniform rot (11, 44, 65, 74, 87, 97). The most appropriate terminology appears to be the grouping of all species in the Basidiomycotina that have the capacity to degrade lignin in the broad category of white-rot fungi (22). Within this group, fungi may (*a*) selectively delignify wood, (*b*) simultaneously degrade lignin and wood polysaccharides throughout the decayed wood, or (*c*) cause a combination of selective delignification and simultaneous decay within the same substrate. The terms simultaneous degradation and selective (or preferential) delignification to describe the two forms of white rot have been widely accepted (3, 14, 22, 38, 58, 69, 97, 106, 126). The proportion of selectively delignified and simultaneous decayed wood varies among the different species of white-rot fungi and within isolates of the same species. Table 1 shows the great diversity that exists among a few representative white-rot fungi.

Table 1 Percent weight, lignin, and sugar residues in wood polysaccharides lost in *Betula papyrifera* after 12 weeks of decay by white-rot fungi[a]

Fungus	% loss in			
	Weight	Lignin[b]	Glucose[c]	Xylose[d]
Dichomitus squalens	44	71	44	43
Ganoderma colossum	35	46	19	45
Heterobasidion annosum	24	55	4	26
Perenniporia medulla-panis	31	73	0	32
Phanerochaete chrysosporium				
BKM-F-1767	39	73	15	55
Fp 102169	55	60	55	66
HHB-11741	46	51	48	58
ME-10	33	41	24	34
Phellinus pini	17	54	5	13
Phlebia tremellosa	34	75	4	39
Trametes versicolor	65	64	65	68

[a]Sound *Betula* contains 20% lignin, 47% glucose, and 24% xylose.
[b]Klason lignin determination using 72% sulfuric acid.
[c]Glucose residue represents hydrolized cellulose from wood.
[d]Xylose residue represents hydrolized xylan from wood.

PATTERNS OF DELIGNIFICATION

All white-rot fungi can cause lignin degradation but some can selectively remove lignin leaving large concentrations of cellulose. These areas of cellulose appear as bright, white zones in the heartwood of living trees or in sapwood and heartwood of downed timber (Figure 2). A well-known fungus, *Phellinus pini,* causes extensive decay in living conifers throughout the world. The decay consists of spindle-shaped pockets of white, delignified wood. It is not surprising that this type of decay, with its striking appearance of white pockets scattered throughout the dark-colored heartwood, was one of the first investigated (56), and continued to receive considerable attention (9, 11, 13, 38, 65, 85–87). Many of the most serious decay-causing fungi result in selective delignification of wood (e. g. *Ganoderma lucidum, Heterobasidion annosum, Inonotus tomentosus, Inonotus dryadeus, Phellinus nigrolimitatus, Phellinus weirii*) (5, 16, 38).

The fossil record also shows that wood with a white-pocket rot, containing spindle-shaped zones of delignified wood, occurred as early as the Triassic in *Araucarioxylon* (117; Figure 2). These samples represent some of the oldest known examples of decay, and demonstrate that little change has occurred in decay patterns over millions of years (118). Since these early gymnosperms most likely had high lignin contents and large concentrations of extractives, it is not surprising to find the decay fungi present were those with the capacity to cause extensive lignin degradation.

In addition to small scattered pockets of delignified wood, large zones of delignification also may occur in wood. Kawase (66) was one of the first investigators to analyze the chemical components of naturally decayed wood. He collected several samples of decayed wood from the forest and found that the cellulose content was exceedingly high and lignin concentration low. In the temperate rain forests of southern Chile, Philippi (103) described a type of decayed wood called *palo podrido,* that was commonly used as a forage feed for cattle and oxen by farmers of the region. The decayed wood apparently was also eaten by the native South Americans (103). Logs, left from clearing forests for pasture, gradually decayed, and during the winter months of subsequent years, farmers split these logs open giving livestock access to the huge concentrations of digestible fiber. This procedure is still practiced today by farmers in southern Chile and it is common to find reamed-out logs scattered over the countryside after cattle have eaten the decayed wood. Recently, *palo podrido* has been characterized and found to consist of extensively delignified wood (6, 37, 52, 128). In some logs, all of the wood except a small outer zone around the circumference was delignified and consisted of 97% cellulose and only 0.9% lignin (6). The term *palo blanco* has been used by farmers to describe this remarkably white wood, whereas *palo podrido* refers to all forms of delignified wood. Collections of decayed wood from this region have shown that there are varying amounts of delignified wood within different logs and that gradations of the two forms of white rot exist, ranging from completely delignified wood to a pocket-rot type of degradation (R. A. Blanchette & E. Agosin, unpublished data).

The temperate rain forests of southern Chile are one of the few areas where exceedingly large amounts of delignified wood are common, and where people have learned to use the resource. Although this phenomenon does not appear to be as prevalent in other regions of the world, there are many examples of large quantities of delignified wood occurring in other forest ecosystems (14, 15, 38, 55, 66). Often, this decay has a mottled appearance, and consists of varying amounts of selectively delignified and simultaneous decayed wood (Figure 3). There are many factors that appear to govern the abundance of each type of decay.

FACTORS THAT INFLUENCE DELIGNIFICATION

Anatomical Considerations

Some species of white-rot fungi, such as *Phellinus pini* and *Ganoderma tsugae,* always cause selective delignification in certain woods. The anatomical and chemical composition of wood, however, appears to influence the distribution and extent of delignified zones. *Phellinus pini* is restricted to the heartwood of living conifers, but may delignify sapwood once the tree dies (13). The pockets of delignified wood are often first associated with latewood regions of annual rings, and elongate within an individual growth ring. In contrast, *Ganoderma tsugae* preferentially degrades the earlywood cells, but not latewood cells (14). In hardwood species, such as *Quercus,* the latewood fibers and rays form borders between pockets of delignification caused by *Inonotus dryophilus* (94). *Phellinus nigrolimitatus* produces large pockets of delignified wood that may form across several annual rings, and decay is not restricted to latewood or earlywood zones. Pockets of delignification also form in tropical species, such as those caused by *Phellinus kawakamii* in *Acacia koa,* where no annual rings are formed in the tree (23). Although the reason for the formation of localized pockets of delignified wood remains unknown, it should be possible to obtain a better understanding of the processes that may be involved with additional investigation. To accomplish this, research should focus on the (*a*) identification of chemical constituents in various wood cells between pockets, (*b*) determination of the spatial distribution and availability of nutrients or elements essential for delignification, (*c*) characterization of accumulated degradation products that may act as enzyme inhibitors, and (*c*) elucidation of the enzymes produced by the fungus and sequence of enzymes within woody substrates that results in selective delignification.

Some fungi that cause white-pocket rots, such as *Trichaptum biforme* (*Hirschioporus pargamenus*) and *Xylobolus frustulatum*, remove the residual cellulose soon after delignification has occurred, leaving the wood with a pitted or honeycomb appearance (95, 96). Often, only incipient stages of decay reveal the presence of delignified cells. Although no definitive studies have been completed, this process also appears to take place in wood decayed by *Phellinus weirii, Trichaptum abietinum, Inonotus tomentosus* and others. For many of the most severe root and trunk rot fungi of trees, little is known about the basic patterns and dynamics of wood decomposition in natural ecosystems.

Greater knowledge of decay mechanisms will provide important insights into fungal colonization strategies and mechanisms used by the fungus to allow it to remain dominant within substrates for long periods of time. This should lead to more effective control procedures. A good example is provided from results of recent basic biochemical studies involving *Heterobasidion annosum* that indicated toxic phenols, produced during wood decomposition, inhibited the fungus (61). This information has led to new strategies that use the toxic features of the phenols to possibly control the fungus in living trees.

Chemical and Ecophysiological Factors

Other factors that influence delignification include chemical composition of the wood and ecophysiological requirements of the fungi. Using *palo podrido* as a model system for studying the factors responsible for delignification, Dill

& Kraepelin (37) concluded that the extremely low nitrogen content of the wood was the primary cause for delignification. Several other ecological factors, including relatively low temperatures (8 to 13°C during the entire year) and high humidity also appeared to favor delignification. Nutrient nitrogen has been found to play a crucial role in wood decay (32, 84, 90). Low levels of nitrogen have been shown to stimulate lignin degradation by several different white-rot fungi, whereas high nitrogen concentrations stimulate polysaccharide degradation (45,78,108). Ligninolytic activity of some white-rot fungi, however, is not affected by nitrogen concentrations (81, 125).

Oxygen concentrations also affect lignin degradation. In general, most investigations have shown that the greatest ligninolytic activity in shake flask cultures of fungi is usually associated with high (100%) oxygen concentrations (78, 110, 127). Oxygen concentrations within tree trunks, however, are extremely low (63), and decomposing logs in a rain forest would not be expected to have an environment of high oxygen. Since fungi apparently do not possess any unique system to tolerate low oxygen concentrations (59), some mechanism must be operative that delivers necessary oxygen for oxidative reactions involved in lignin degradation. What mechanisms fungi use to accomplish this are not known.

Moisture has significant effects on growth of wood decay fungi (25, 26, 106). In a rain forest, decayed wood may be expected to absorb excess moisture and become saturated. However, this does not occur. Instead, pseudosclerotial plates are produced by the fungus around the exterior portions of the decaying wood that regulate the moisture content and apparently limit the entrance of excess water (6). These same structures function to maintain adequate moisture by restricting water loss from decaying wood in regions of low to moderate rainfall (1, 2, 14, 38). The wet environment of southern Chile undoubtedly contributes to the rate and extent of fungal colonization and decay, but its influencing role on selective delignification is questionable. The ability of fungi, once they have colonized a substrate, to prevent desiccation or entrance of excess moisture, suggests that they can partially regulate conditions that are conducive for their continued degradation processes and survival.

The factors that appear most decisive in governing delignification in southern Chile are the quantity and type of lignin present in the native tree species (6). Investigations of wood decayed in the field indicate extensive delignification has been found primarily in *Nothofagus dombeyi* and *Eucryphia cordifolia,* whereas only a simultaneous type of white rot occurred in wood of *Laurelia phillipiana*. The major differences among these trees is that *Nothofagus dombeyi* and *Eucryphia cordifolia* have the lowest lignin contents, largest amount of β-aryl ether bonds in lignin and the highest syringyl/

guaiacyl ratio of all the native woods tested. Indeed, *Nothofagus dombeyi* has one of the highest syringyl lignin contents reported for any wood (6). In vitro decay studies, using a species of *Ganoderma* isolated from delignified wood of *palo podrido* and native woods of southern Chile, confirmed field observations that selective delignification occurred in wood of *Nothofagus dombeyi,* whereas a simultaneous rot was always found in wood of *Laurelia phillipiana* (R. A. Blanchette & E. Agosin, unpublished results).

The lignin of angiosperms contains varying ratios of syringyl (S) to guaiacyl (G) phenylpropane monomer units. *Nothofagus dombeyi* has a S/G ratio of 2.73, whereas the ratio for *Laurelia phillipiana* is 0.35 (6; a correction factor of 2.3 has been used to compare the thioacidolysis determination to the nitrobenzene assay for S/G ratios). The ratio for *Acer saccharum* and *Tilia americana,* for comparison, is approximately 0.75 (23). The lignin composition of native trees from southern Chile helps to explain why delignification is so extensive in some woods of this geographic location and not in others. This is not, however, a universal factor that governs all fungal delignification systems. Many fungi can delignify wood of gymnosperms that have high lignin contents and have lignin composed exclusively of guaiacyl phenylpropane units(93). Apparently, these fungi have evolved ligninolytic systems for selective delignification in conifers. When mycelia from some of these fungi were used to inoculate wood blocks of angiosperms in the laboratory, extensive delignification occurred (5, 21, 98). In nature, these fungi probably would not normally colonize these angiosperm substrates due to the inability of spores to germinate on the substrate (89), or because of fungal inhibition from wood extractives, inappropriate pH, or other unfavorable microenvironmental conditions (106).

Variation Among Fungi

Of the 1600–1700 species of North American wood-rooting fungi, 94% are white-rotters (48). Most of these species occur on angiosperms, but many, as previously mentioned, degrade gymnosperms. The diversity in decay patterns that exist among different species is great. Furthermore, the diversity in decay also can be extraordinarily large among isolates of an individual species (Table 1; 21, 79). There is apparently tremendous genetic diversity in delignification capacity among the fungi found in nature. Methods to screen species and isolates of naturally occurring fungi have included a wide variety of techniques, including chemical and morphological procedures (6, 8, 16, 24, 40, 50, 70, 77, 92, 99, 114). The results from these investigations have provided many superb isolates for experimental purposes with superior capacity for degrading lignin.

Cross-breeding studies of selected homokaryotic strains of *Phanerochaete chrysosporium* indicate that new strains can be obtained with even higher lignin-degrading capacity than the original wild type (64). Several recent studies have examined the genetic factors influencing lignin peroxidase activity (see section on Ligninolytic Systems; 102, 104, 105). Using 53 single basidiospore-derived strains of *Phanerochaete chrysosporium,* lignin peroxidase activity was determined for each monokaryotic isolate and compared to the activity of the parent dikaryon (105). The variation in lignin peroxidase activity varied with many of the haploid progeny exhibiting superior ligninolytic activity to that of the parent. In a study using 20 monokaryotic isolates of *Dichomitus squalens,* a similar experimental procedure was used, but in addition to lignin peroxidase activity, cellulolytic activity also was assayed (102). In that study, all of the haploid isolates had reduced growth rate when compared to the dikaryotic parent, but lignin peroxidase activity and cellulolytic activity varied greatly. Several isolates had low cellulolytic activity while others displayed high ligninolytic activity. It appears possible that significant improvements in ligninolytic activity could be achieved for industrial purposes by selection and cross-breeding of haploid strains.

PHYSICAL ASPECTS OF CELL WALL DELIGNIFICATION

Lignin is distributed throughout the wood cell wall layers, but the greatest concentrations are in the compound middle lamella and cell corner regions (Figure 4). To visualize lignin with electron microscopy, $KMnO_4$ is often used as a fixative and lignin stain to identify the relative distribution of lignin in wood (12, 24). The most lignified layer, the middle lamella, has the greatest electron density. Other methods used to precisely determine the ultrastructural localization of lignin have confirmed the distribution patterns previously determined by $KMnO_4$ staining, and have been used to quantify lignin concentrations in different types of cells as well as within various cell wall layers (38, 99, 111, 123a).

Decay fungi colonize wood via the anatomical paths of least resistance, and quickly move into areas containing easily assimilated nutrients. Ray parenchyma cells are frequently the first to be colonized (31, 124). Movement through pit apertures or direct penetration of cell walls by bore holes allows the fungus to proliferate throughout the wood. With relatively few hyphae within individual xylary cells (i. e. one or two per cell) lignin degradation begins. Lignin degradation by hyphae in cell lumina causes a diffuse reduction in $KMnO_4$ electron density within the secondary wall layers immediately adjacent to the fungus (Figure 4). Gradually, the entire secondary wall becomes less electron dense. The process of lignin removal occurs from the cell lumen toward the middle lamella (Figure 4). Once lignin is removed from the S_1 and S_2 layers of the secondary walls, the middle lamella between cells

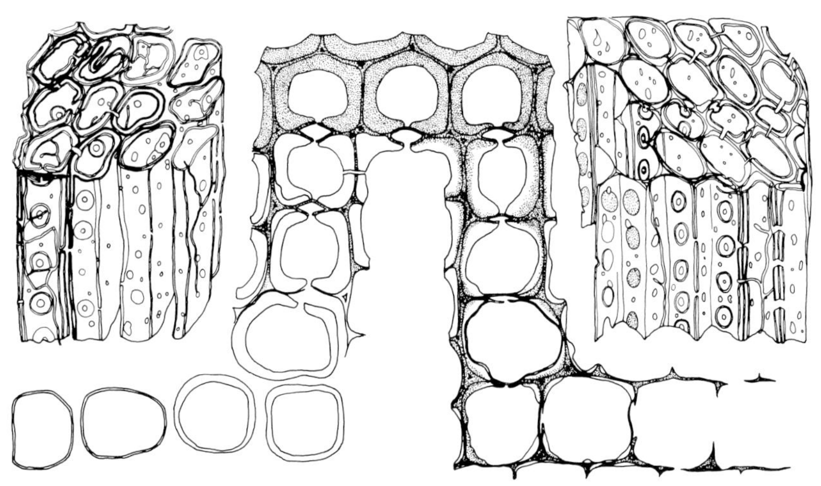

Figure 1 Drawing by Robert Hartig (1878) showing two forms of white rot in wood of *Pinus sylvestris* decayed by *Heterobasidion annosum*. On the left, incipient to advanced stages of selective delignification is shown. Lignin (stippled areas) is progressively removed from tracheid cell walls, resulting in cells void of middle lamellae. Once lignin was removed, cellulolytic activity gradually degraded the delignified secondary wall. On the right a simultaneous decay of all cell wall components is presented. This form of white rot resulted in cell wall erosion of secondary walls followed by portions of the middle lamellae. The last areas to be degraded are the lignified middle lamellae. This drawing represents the first observations showing selective delignification in wood, and demonstrates that white rot fungi can cause different forms of degradation.

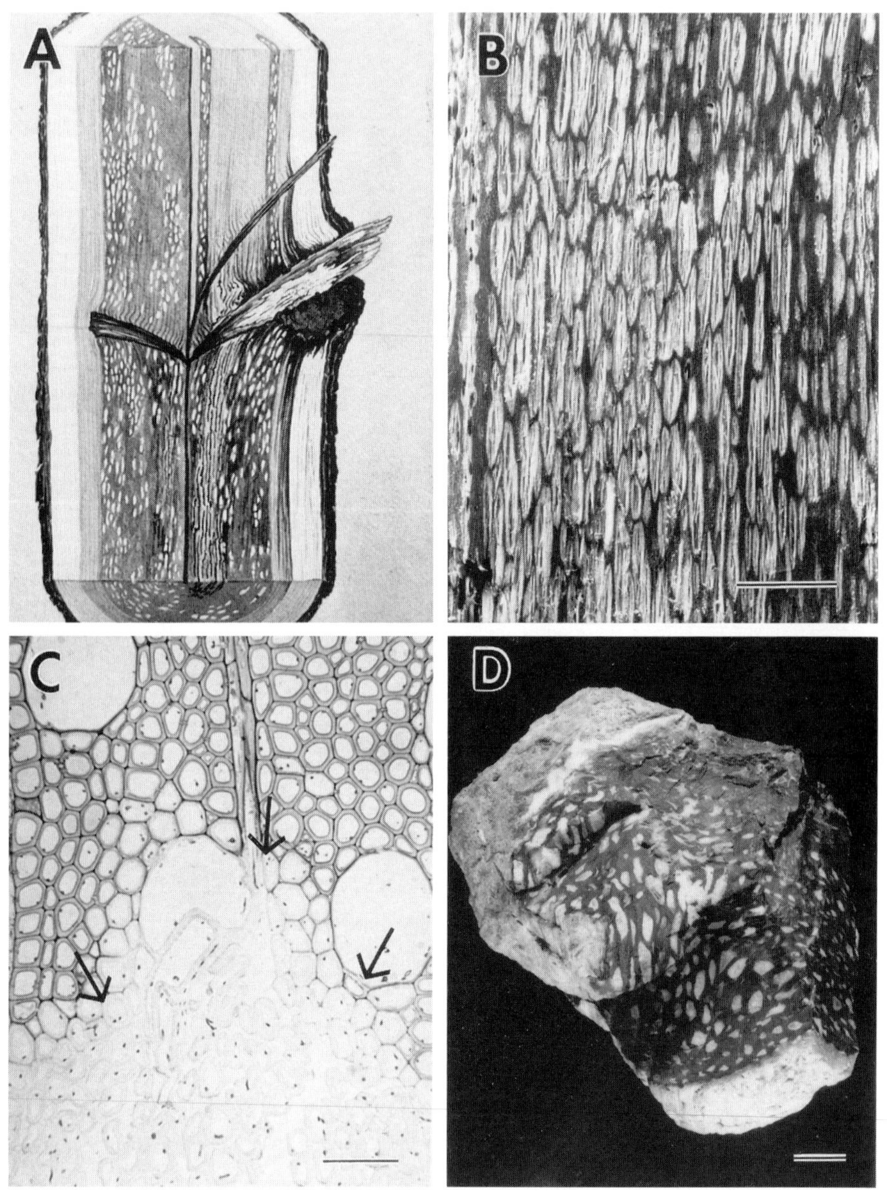

Figure 2 White-pocket rot in living trees, with spindle-shaped areas of delignified wood separated by nondecayed wood. **A** Drawing of a split open section of a conifer showing white-pocket rot within the heartwood. **B** The white-pockets consist of delignified cells where lignin has been completely removed leaving the white, cellulose-rich tracheids remaining. Areas of apparently unaltered wood occur between the pockets. **C** Transverse section and light microscopy micrograph of a white pocket in a deciduous wood (*Betula papyrifera)* inoculated with *Phellinus pini*. Delignified cells (arrows indicate edge of the white pocket) lack middle lamella and are detached from adjacent cells. The hyphae (one or two per cell) can be seen within the delignified cells as well as in nondecayed cells between pockets. **D** Fossilized *Araucarioxylon* wood, a silicified gymnosperm genus from the Triassic age, found in the Fremouw Peak locality, Antarctica. The fossil wood contains a white-pocket rot with spindle-shaped pockets of delignified wood. Bar = 1 cm in **A** and **D**, 50μm in **C.** (**A** reproduced by courtesy of Alex Shigo, US Forest Service, **D** reproduced by courtesy of T.N. Taylor, Ohio State University).

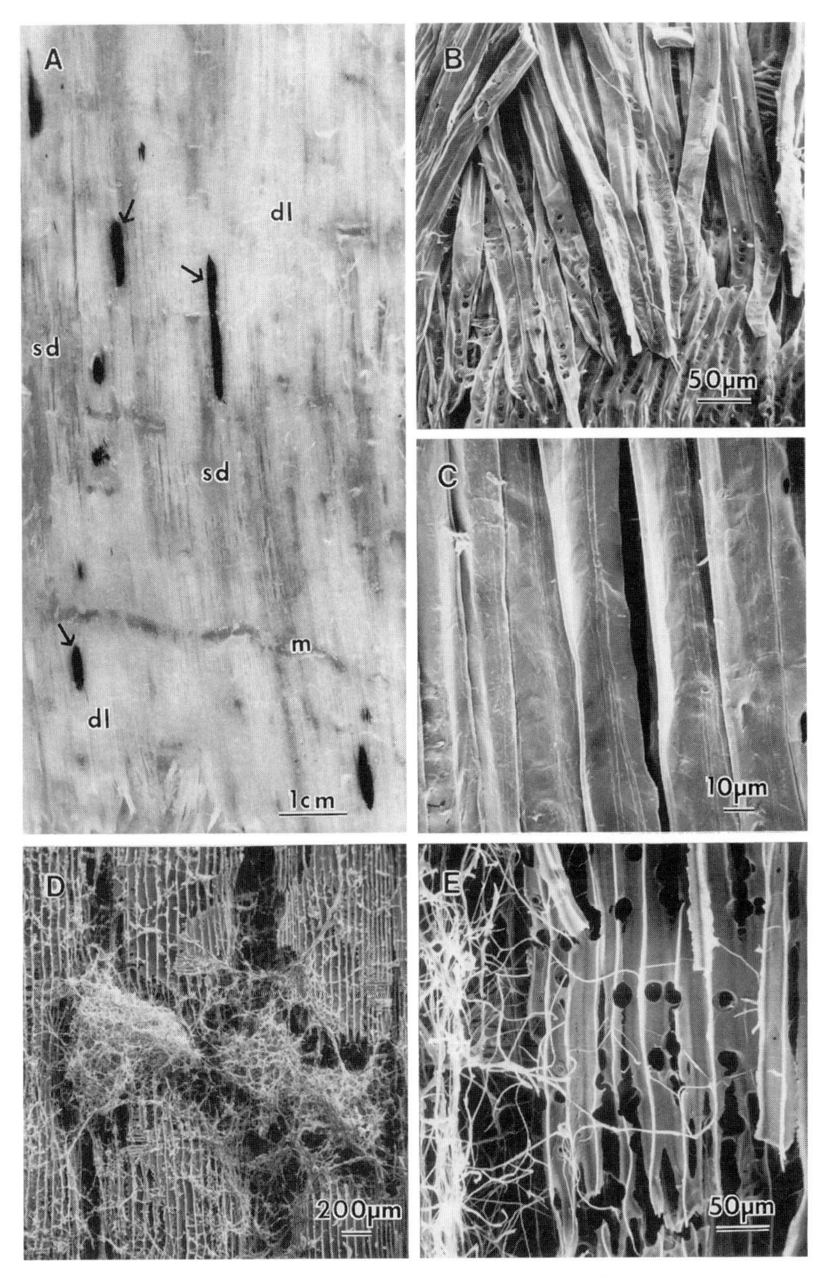

Figure 3 A mottled-rot caused by *Ganoderma tsugae* showing two different forms of white rot, selective delignification and simultaneous decay of all cell wall components, in *Tsuga canadensis*. **A** Mottled appearance of the wood results from a combination of delignified cells (dl) and simultaneous decayed cells (sd). Within the simultaneously decayed zones, large voids were formed that were filled with white fungal mycelia (m). In delignified areas, manganese dioxide deposits were common, appearing as large black zones (arrows). **B** and **C** Wood from the white, delignified zones contained tracheids that lacked middle lamellae and cells easily detached from adjacent cells. **D** and **E** Wood from simultaneously decayed areas had cells with erosion troughs and holes evident within the cell walls. In some areas, large voids formed and were filled with fungal mycelia (14).

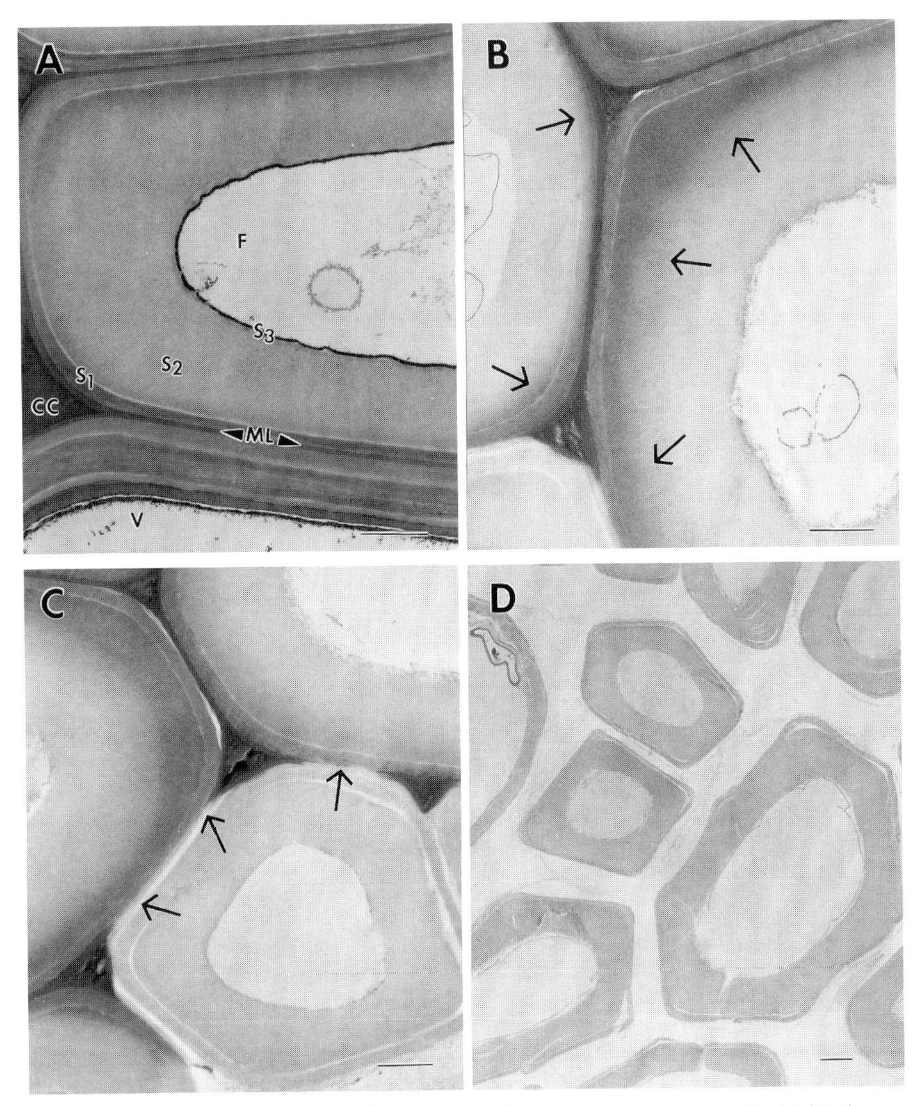

Figure 4 Transmission electron micrographs showing the progressive stages of selective delignification in cell walls of *Nothofagus dombeyi* by *Ganoderma* sp. Wood was fixed with $KMnO_4$ to stain lignin within cell walls. **A** Sound wood showing fiber (F) and vessel (V) with electron-dense cell wall layers. The secondary wall layers (S_1, S_2, and S_3) and middle lamella (ml) are evident. The middle lamella between cells and in cell corners (CC) appear the most electron dense. **B** Incipient stages of delignification showing an electron lucid zone within the S_2 layer of the secondary wall. There is a progressive loss of electron density from the lumen toward the middle lamella (arrows). The area where no stain is apparent represents delignified regions of the cell wall. **C** A more advanced stage shows a cell where secondary wall layers have been delignified and the middle lamella between cells has been degraded (arrows). Other cells show loss of lignin from the S_2 layer of the secondary walls but delignification had not reached the S_1 layer or middle lamella. **D** Advanced stage of delignified wood with lignin removed from the secondary wall layers and middle lamellae completely degraded. Bar = 2 μm (from (6)).

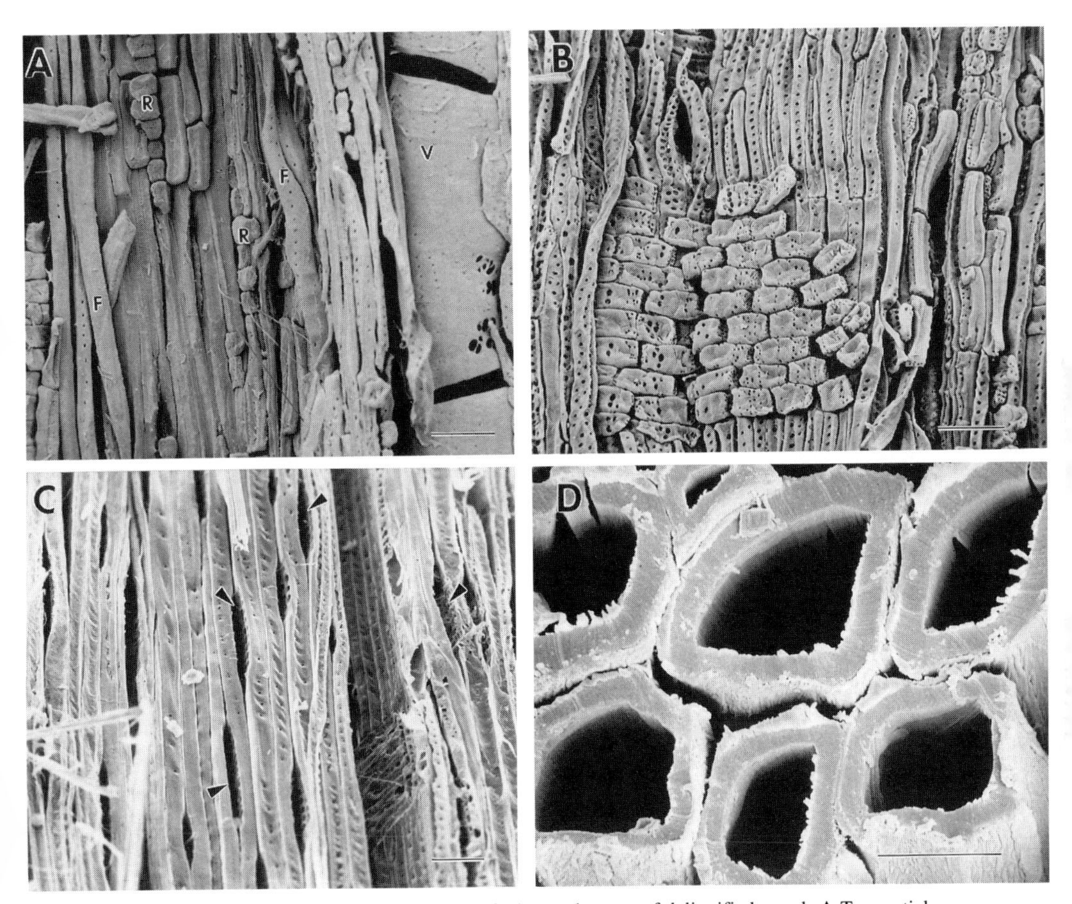

Figure 5 Scanning electron micrographs of advanced stages of delignified wood. **A** Tangential section of wood from *Quercus* sp. delignified by *Ganoderma colossum* showing delignified vessels (V) fibers (F) and ray parenchyma cells (R). All cells are detached and separate from adjacent cells. Delignified vessels separate at perforation plates and ray cells appear as individual cells. **B** Radial section of *Quercus* sp. delignified by *Ganoderma* sp. RLG16162 showing the separation of ray cells and detached fibers. **C** Tangential section of wood from *Pinus* sp. decayed by *Phellinus pini*. Tracheids are delignified and ray parenchyma cells are completely removed (arrowheads). **D** Transverse section of wood from *Betula* sp. delignified by *Phlebia tremellosa* showing the residual secondary wall layers remain intact even after the middle lamellae has been removed. The cellulose-rich secondary walls without middle lamellae readily separate from one another. Bar = 50 μm in **A, B,** and **C;** 10 μm in **D.** (**A** and **B** from (5), **C** from (13)).

is degraded. This layer becomes progressively degraded from the centrally located regions toward the cell corners. The cell corners are the last to be degraded. After most of the middle lamella is degraded, the residual cell walls separate and become detached from adjacent cells. These residual cells are composed of cellulose-rich secondary wall layers (Figure 4). During delignification, hemicelluloses also are degraded. The close spatial relationship of the lignin and hemicellulose matrix apparently necessitates the removal of some hemicelluloses when lignin is removed. Hemicelluloses also appear to provide the energy source for the decay fungi, since lignin alone apparently cannot serve as a growth substrate (75).

Recent studies using colloidal gold cytochemistry with cellulases and xylanases have demonstrated the distribution of cellulose and xylan in sound wood and in wood decayed by fungi (18, 20). In wood cells delignified by fungi, colloidal gold coupled to endo-1, 4-β-glucanase II (EG-II) or 1,4-β-D-glucan cellobiohydrolase I (CBHI) labeled intensely residual cell walls in a pattern similar to that of sound wood, indicating crystalline and amorphous cellulose were still present in large concentrations. Examination of cell walls labeled with endo-1, 4-β-xylanase, however, showed that substantial amounts of xylan had been removed. In another investigation, gold-labeled lignin peroxidase was used to locate lignin within the cell walls of sound and delignified wood (19). In delignified cells, no gold labeling was apparent, except for some in the residual cell corner regions that had maintained a high lignin content. Fluorescence microscopy also has been useful in monitoring the lignin distribution in wood since lignified cell walls fluoresce in UV light of certain wavelengths (9, 47). In a study of *Quercus* wood decayed by *Lentinula edodes,* the Shiitake fungus, delignification was apparent in cells that did not fluoresce (126). In that study, a similar pattern of lignin removal was found as described above. Secondary cell wall layers were progressively delignified from cell lumina toward the middle lamella, followed by degradation of middle lamella between cells, and finally attack of the cell corner regions. Between delignified zones in the wood, a simultaneous type of white rot was evident. *Lentinula edodes,* therefore, appears to be a white-rot fungus exhibiting considerable diversity in delignification capacity among different strains, since previous studies have indicated that this fungus causes only a nonselective attack on the cell wall (R. A. Blanchette, unpublished data; 122).

In addition to differences in delignification observed in various regions of the xylem (latewood vs earlywood), there appears to be a great deal of variability in the types of cells that are delignified by different white-rot fungi. Some fungal species delignify all the different types of cells that occur in angiosperms, including vessels, fibers, and ray parenchyma cells (Figure 5). However, delignification may not occur in vessels of some woods, such as

Acacia koa, where sound vessels occur among completely delignified fibers (23). In decay of *Quercus* species by *Inonotus dryophilus,* the latewood fibers remain free of degradation while earlywood fibers are delignified (94). Fibers delignified by *Ganoderma zonatum* may not be uniformly affected, and irregular zones of lignin may remain in parts of the secondary wall (5). In conifers, it is common to find delignified tracheids among totally degraded ray parenchyma cells (Figure 5). In palm wood, delignification is apparent in the vessel elements but is not observed in other cells (4). Additional studies of delignification processes should provide insight into the factors that allow some cells to resist degradative actions, while others easily succumb to ligninolytic action. The type of lignin occurring within some types of cells and the microfibrillar orientation of the cell wall polymers may explain some of these variations in patterns of resistance and susceptibility to delignification (23).

LIGNINOLYTIC SYSTEMS

The complex structural features of lignin dictate unusual constraints on biological degradation systems (75). Degradation of lignin has been referred to as "enzymatic combustion", consisting of nonspecific enzyme-catalyzed burning of lignin (75). Biochemical investigations using *Phanerochaete chrysosporium* have shown that two heme peroxidases are associated with lignin degradation, lignin peroxidase (LiP) (43, 49, 72, 119, 120) and Mn-dependent peroxidase (MnP) (43, 51, 62, 80, 101). Lignin peroxidases are H_2O_2-dependent enzymes that degrade lignin by a simple initiating reaction: removal of a single electron from aromatic nuclei, producing aryl cation radical species, which undergo nonenzymatic reactions as radicals and as cations. These reactions result in polymer cleavages (54, 68). Mn-dependent peroxidases function to oxidize Mn II to Mn III, which can oxidize phenolic substructures in lignin (46, 51, 100). Although MnP can oxidize phenolic compounds, they have not been demonstrated to mediate degradation of nonphenolic units that dominate in lignin (75). During recent years, both LiP and MnP have been found in extracellular filtrates of many different white-rot fungi (38, 75, 123). Both enzymes have been found in white-rot fungi that simultaneously decay wood cell walls (e. g. *Trametes versicolor*), and in those that selectively remove lignin (e. g. *Phanerochaete chrysosporium* BKM-F-1767). The molecular weight of these enzymes suggests that they do not readily penetrate into sound wood, since the pores of the cell wall would not be large enough for penetration by the enzymes (116). Observations of electron micrographs showing delignification (Figure 4) suggest that whatever the initial agent of delignification is, it must be able to diffuse into the wood cell walls. An enzymatically generated oxidized manganese species (Mn III) would be able to diffuse into wood, and recent studies suggest that Mn III under certain conditions is capable of ligninolytic activity (46). Other low molecular weight compounds, such as the modified porphyrins, also diffuse into wood and degrade lignin. Since LiP is a heme protein containing protoporphyrin IX, various researchers have used synthetic metalloporphyrins to degrade lignin model compounds as well as lignin in wood. Low molecular weight porphyrins appear to mimic the activities of lignin peroxidase and cause delignification in a manner similar to selective delignification by white-rot fungi (34, 101, 115). Some porphryins readily diffuse into the wood and cause an oxidation of lignin. A biomimetic system, such as the one presented by Paszczynski et al (101), provides a model system to study aspects of selective delignification. In such a system, only 1 to 2 days are required to reach the same levels of wood cell wall delignification that usually requires 2 to 3 months of laboratory degradation by white-rot fungi.

It may not be absolutely essential for initial delignification agents to be low molecular weight compounds. Some enzymes, such as the hemicellulases, are produced early in the decay process, and are associated with lignin degradation (17, 38). Hemicellulases may gradually degrade hemicelluloses in the wall immediately adjacent to the lumen and progressively move into the secondary wall. As some of the hemicelluloses are degraded, channels of sufficient size should open, allowing access to LiP and MnP. Immunocytochemical investigations using gold-labeled LiP-antisera and sections of white-rotted wood have shown that LiP does occur in areas of the cell wall undergoing delignification (20, 35, 36). In partially delignified wood decayed by *Phellinus pini* or *Phanerochaete chrysosporium,* LiP was localized within the wall at the edge of electron-dense regions that still contained lignin. The ultrastructural localization of xylanase in decayed wood from *Betula* also was within the cell wall, and apparently in advance of zones showing lignin removal (20). Purified preparations of both LiP and MnP have been infiltrated into white-rotted wood followed by immunogold labeling and were found within cell walls that had a loose, open fibrillar structure (36). No diffusion occurred in nondecayed cell walls.

Many other enzymes (including polyphenol oxidases, laccases, H_2O_2-producing enzymes and quinone-reducing enzymes) are also thought to be important for lignin degradation. (For recent reviews that provide detailed coverage of the enzymes involved in lignin degradation see 30, 38, 60, 72, 73, 75, 119.) Also of interest are the factors that regulate both enzyme production and activity. Manganese, which can be found in exceedingly large concentrations in delignified wood (15; Figure 3), appears to regulate LiP and MnP production (27). In low concentrations of MnII, extracellular LiP predominates, whereas in the presence of higher concentrations of Mn II, MnP is dominant.

Regulation of cellulases is another important aspect, since these enzymes are repressed during selective delignification by some white-rot fungi. Many factors influence various cellulases, for example the presence of cellobiose, glucose and other wood sugars, phenols, and proteases (38). However, very little information is available on the factors involved in regulating cellulases during delignification processes. There are many different avenues that future research can pursue to increase our understanding of the enzymes and interactions associated with lignin breakdown. Although it is clear that our current knowledge of delignification processes is incomplete, the potential for using these fungi, or their enzymes, in biotechnological applications is so great that they are already being suggested for industrial applications (28, 38, 12, 13a, 76, 108).

The surge of research activities in this area will continue, and we can expect to see important new developments that utilize ligninolytic fungi for many commercial purposes. Applications in turn should spur investigations that will provide needed basic information on enzyme mechanisms involved in microbial degradation of living trees and forest products.

CONCLUDING REMARKS

White-rot fungi may cause a simultaneous type of attack that degrades all wood cell wall components or preferential degradation of lignin. Selective delignification may occur throughout the decayed wood (e. g. *palo podrido, palo blanco*), in small localized pockets (e. g. decay by *Phellinus pini)* or in a mottled-rot where delignified cells are surrounded by simultaneously decaying wood (e. g. decay by *Ganoderma tsugae, Heterobasidion annosum*). Fungi that delignify wood include some of our most serious root and butt rot fungi (e. g. *Ganoderma lucidum, Heterobasidion annosum, Inonotus tomentosus, Phellinus weirii, Rigidoporus lignosus, Phellinus noxius*) and trunk rot fungi (e. g. *Ganoderma applanatum, Inonotus dryophilus, Phellinus pini*). Information on basic biological processes, such as degradative mechanisms, survival strategies, and ecological relationships is lacking, but is needed to ascertain potential disease-control measures. Some of this information may become available from investigations that are specifically designed for using selective delignifying fungi for industrial applications.

The biotechnical potential of fungi that delignify wood is enormous. One use currently being tested includes the pretreatment of wood chips with fungi that selectively remove lignin for pulp and paper production. Pretreated chips with selected strains of *Phanerochaete chrysosporium, Ceriporiopsis subvermispora,* and other fungi improved paper strength properties while requiring up to 68% less energy than standard mechanical pulping procedures (82, 83, 91). Another aspect of white-rot fungi that may be of significant use is their ability to "bleach" wood to a white color. The effluents produced during chemical bleaching of kraft pulping are major contributors to water pollution, and new methods of bleaching wood pulp with fungi or their enzymes may provide a viable substitute for some of the chemicals now used (30a, 37a, 39, 121). Delignification processes also may be used to modify wood or agricultural byproducts to increase their nutritional value for ruminant animal feed. The digestibility of straw, bagasse, and wood can be dramatically increased when treated with various white-rot fungi, and have shown no problems of palatability or toxicity (7, 30, 77, 107, 109). Of major significance are the recent investigations showing that fungi with high ligninolytic activity are able to degrade environmental pollutants, such as PCB, DDT, dioxins, industrial dyes, and chlorinated phenols (28–30, 33, 38, 39, 42, 53). The use of these fungi as bioremedial treatments of toxic wastes has immediate application. For example, *Phanerochaete chrysoporium* BKM-F-1767 is currently being used in large scale field tests to detoxify soils (K. Kirk, personal communication).

We can expect to see additional uses for fungi that delignify wood as our knowledge of these organisms increases. One major benefit from these investigations will be the increased comprehension of white-rot degradation mechanisms that will undoubtedly improve our understanding of pathological relationships among decay fungi and living trees.

ACKNOWLEDGMENTS

This article is contribution No. 18,588 of the Minnesota Agricultural Experiment Station. The author thanks André Abad, Kory Cease, Marge Eerdmans and Lewis Otjen for their contributions to research on selective delignification of wood reported here, and to Drs. Kent Kirk, Roberta Farrell, Gary Leatham, and Jim Adaskaveg for their critical reviews of this manuscript. The work of the author has been supported by NSF Grant INT-8900153, Repligen-Sandoz Research Corp., and the Biopulping Consortium. The consortium consists of the University of Wisconsin Biotechnology Center, USDA Forest Products Laboratory, Madison, WI, and 20 member companies involved in pulp and paper production and associated fields.

Literature Cited

1. Abe, Y. 1989. Effect of moisture on the colonization by *Lentinus edodes* and *Hypoxylon truncatum* in wood. *Eur. J. For. Path.* 19:423–34
2. Abe, Y. 1989. Effect of moisture on decay of wood by xylariaceous and diatrypaceous fungi and quantitative changes in the chemical components of decayed wood. *Trans. Mycol. Soc. Jpn.* 30:169–82
3. Adaskaveg, J. E., Gilbertson, R. L. 1986. In vitro decay studies of selective delignification and simultaneous decay by the white-rot fungi *Ganoderma lucidum* and *G. tsugae*. *Can. J. Bot.* 64: 1611–19
4. Adaskaveg, J. E., Blanchette, R. A., Gilbertson, R. L. 1991. Decay of palm wood by white- and brown-rot fungi. *Can. J. Bot.* In press

5. Adaskaveg, J. E., Gilbertson, R. L., Blanchette, R. A. 1990. Comparative studies of delignification caused by *Ganoderma* species. *Appl. Environ. Microbiol.* 56:1932–43
6. Agosin, E., Blanchette, R. A., Silva, H., Lapierre, C., Cease, K. R., et al. 1990. Characterization of Palo Podrido, a natural process of delignification in wood. *Appl. Environ. Microbiol.* 56: 65–74
7. Agosin, E., Tollier, M. T., Brillouet, J. M., Thivend, P., Odier, E. 1986. Fungal pretreatment of wheat straw: effects on the biodegradability of cell walls, structural polysaccharides, lignin and phenolic acids by rumen microorganisms. *J. Sci. Food Agr.* 37:97–106
8. Ander, P., Eriksson, K-E. 1977. Selective degradation of wood components by white-rot fungi. *Physiol. Plant.* 41:239–48
9. Aufess, H. von, Pechmann, H. von, Graessle, H. 1968. Fluoreszenzmikroskpoische Beobachtungen an pilzbefallenem Holz. *Holz Roh Werkst.* 261:50–61
10. Barbour, M. G., Burk, J. H., Pitts, W. D. 1987. *Terrestrial Plant Ecology.* Menlo Park: Benjamin/Cummings. 634 pp.
11. Bjorkman, E., Samuelson, O., Ringstrom, E., Bergelk, T., Malm, E. 1949. Decay injuries in spruce forests and their importance for the production of chemical paper pulp and rayon pulp. *K. Skogshogskolan Skr.* No. 4. Stockholm
12. Bland, D. E., Foster, R. C., Logan, A. F. 1971. The mechanism of permanganate and osimum tetroxide fixation and the distribution of lignin in the cell wall of *Pinus radiata. Holzforschung* 25:137–42
13. Blanchette, R. A. 1980. Wood decomposition by *Phellinus (Fomes) pini*: A scanning electron microscopy study. *Can. J. Bot.* 58:1496–503
14. Blanchette, R. A. 1984. Selective delignification of eastern hemlock by *Ganoderma tsugae. Phytopathology* 64: 153–60
15. Blanchette, R. A. 1984. Manganese accumulation in wood decayed by white rot fungi. *Phytopathology* 74:725–30
16. Blanchette, R. A. 1984. Screening wood decayed by white rot fungi for preferential lignin degradation. *Appl. Environ. Microbiol.* 48:647–53
17. Blanchette, R. A., Abad, A. R. 1988. Ultrastructural localization of hemicellulose in birch wood (*Betula papyrifera*) decayed by brown and white rot fungi. *Holzforschung* 42:393–98
18. Blanchette, R. A., Abad, A. R., Cease, K.R., Farrell, R. L., Lovrien, R. E., Leathers, T. D. 1990. Enzyme immunocytochemistry and ultrastructural localization of cell wall components by enzyme-gold complexes. In *Biotechnology in the Pulp and Paper Manufacture. Applications and Fundamental Investigations,* ed. T. K. Kirk, H-M. Chang, pp. 69–81. Boston: Butterworth-Heinemann
19. Blanchette, R. A., Abad, A. R., Cease, K. R., Farrell, R. L., Skerker, P., Lovrien, R. E. 1991. Cytochemical aspects of wood degradation by white and brown rot basidiomycetes. See Ref. 110a
20. Blanchette, R. A., Abad, A. R., Farrell, R. L., Leathers, T. D. 1989. Detection of lignin peroxidase and xylanase by immunocytochemical labeling in wood decayed by Basidiomycetes. *Appl. Environ. Microbiol.* 55:1457–65
21. Blanchette, R. A., Burnes, T. A., Leatham, G. F., Effland, M. J. 1988. Selection of white rot fungi for biopulping. *Biomass* 15:93–101
22. Blanchette, R. A. Nilsson, T., Daniel, G., Abad, A. 1990. Biological degradation of wood. In *Archaeological Wood: Properties, Chemistry and Preservation,* ed. R. M. Rowell, R. J. Barbour, pp. 141–74. Adv. Chem. Ser. 225, Am. Chem. Soc. Washington DC
23. Blanchette, R. A., Obst, J. R., Hedges, J. I., Weliky, K. 1988. Resistance of hardwood vessels to degradation by white rot basidiomycetes. *Can. J. Bot.* 66:1841–47
24. Blanchette, R. A., Otjen, L., Carlson, M. C. 1987. Lignin distribution in cell walls of birch wood decayed by white rot Basidiomycetes. *Phytopathology* 77: 684–90
25. Boddy, L. 1983. Microclimate and moisture dynamics of wood decomposing in terrestrial ecosystems. *Soil Biol. Biochem.* 15:149–57
26. Boddy, L. 1986. Water and decomposition processes in terrestrial ecosystems. In *Water, Fungi and Plants,* ed. P. G. Ayres, L. Boddy, pp. 375–98. Cambridge: Cambridge Univ. Press
27. Bonnarme, P., Jefferies, T. W. 1990. Mn(II) regulation of lignin peroxidases and manganese-dependent peroxidases from lignin-degrading white rot fungi. *Appl. Environ. Microbiol.* 56:210–17
28. Bumpus, J. A., Aust, S. D. 1986. Biodegradation of environmental pollutants by the white rot fungus *Phanerochaete chrysosporium:* involvement of the lignin degrading system. *BioEssays* 6: 166–70

29. Bumpus, J. A., Tien, M., Wright, D., Aust, S. D. 1985. Oxidation of persistent environmental pollutants by a white rot fungus. *Science* 228:1434–36
30. Buswell, J. A., Odier, E. 1987. Lignin biodegradation. *CRC Crit. Rev. Biotechnol.* 6:1–60
30a. Chang, H-M., Joyce, T. W., Kirk, T. K. 1987. Process of treating effluent from a pulp or papermaking operation. *US Patent No. 4,665,926*
31. Cowling, E. B. 1961. Comparative biochemistry of the decay of sweet gum by white-rot fungi. *US Dept. Agr. Tech. Bull.* 1258, 79 pp.
32. Cowling, E. B., Merrill, W. 1966. Nitrogen in wood and its role in wood deterioration. *Can. J. Bot.* 44:1533–44
33. Cripps, C., Bumpus, J. A., Aust, S. D. 1990. Biodegradation of azo and heterocyclic dyes by *Phanerochaete chrysosporium. Appl. Environ. Microbiol.* 56:1114–18
34. Cui, F., Dolphin, D., Wijesekera, T., Farrell, R., Skerker, P. 1990. In *Biotechnology in the Pulp and Paper Manufacture. Applications and Fundamental Investigations*, ed. T. K. Kirk, H. M. Chang, pp. 481–91. Boston: Butterworth-Heinem. 922 pp.
35. Daniel, G., Nilsson, T., Petterson, B. 1989. Intra- and extracellular localization of lignin peroxidase during the degradation of solid wood and wood fragments by using transmission electron microscopy and immuno-gold labeling. *Appl. Environ. Microbiol.* 55:871–81
36. Daniel, G., Petterson, B., Nilsson, T., Volc, J. 1990. Use of immunogold cytochemistry to detect Mn(II)-dependent and lignin peroxidases in wood degraded by the white rot fungi *Phanerochaete chrysosporium* and *Lentinula edodes. Can. J. Bot.* 68:920–33
37. Dill, I., Kraepelin, G. 1986. Palo podrido: Model for extensive delignification of wood by *Ganoderma applanatum. Appl. Environ. Microbiol.* 52:1305–12
37a. Eaton, D., Kirk, T. K., Chang, H-M. 1983. Process for the decolorization of pulp mill bleach plant effluent. *US Patent No.* 4,420, 369
38. Eriksson, K. E., Blanchette, R. A., Ander, P. 1990. *Microbial and Enzymatic Degradation of Wood and Wood Components*. Berlin/New York: Springer-Verlag. 407 pp.
39. Eriksson, K-E., Kirk, T. K. 1985. Biopulping, biobleaching and treatment of kraft bleaching effluents with white rot fungi. In *Comprehensive Biotechnology* 3(15):271–94. New York: Pergamon
40. Eriksson, I., Lidbrandt, O., Westermark, U. 1988. Lignin distribution in birch (*Betula verrucosa*) as determined by mercurization with SEM- and TEM-EDXA. *Wood Sci. Technol.* 22:251–57
41. Ertel, J. R., Hedges, J. I. 1985. Sources of sedimentary humic substances: Vascular plant debris. *Geochim. Cosmochim. Acta* 49:2097–107
42. Farrell, R. L. 1987. Industrial applications of lignin-transforming enzymes. *Phil. Trans. R. Soc. Lond.* 321:549–53
43. Farrell, R. L., Murtagh, K. E., Tien, M., Mozuch, M. D., Kirk, T. K. 1989. Physical and enzymatic properties of lignin peroxidase isoenzymes from *Phanerochaete chrysosporium. Enzyme Microb. Technol.* 11:322–28
43a. Farrell, R. L., Tien, M., Kirk, T. K. 1987. Novel enzymes which catalyze the degradation and modification of lignin. *US Patent No. 4,687,741*
44. Fengel, D., Wegener, G. 1983. *Wood: Chemistry, Ultrastructure Reactions.* Berlin: Walter de Gruyter. 613 pp.
45. Fenn, P., Kirk, T. K. 1981. Relationship of nitrogen to the onset and suppression of ligninolytic activity and secondary metabolism in *Phanerochaete chrysosporium. Arch. Microbiol.* 130:59–65
46. Forrester, I. T., Grabski, A. C., Burgess, R. R., Leatham, G. F. 1988. Manganese, Mn-dependent peroxidases, and the biodegradation of lignin. *Biochem. Biophys. Res. Commun.* 157: 992–99
47. Fukazawa, K., Imagawa, H., Doi, S. 1976. Histochemical observation of decayed cell walls using ultraviolet and fluoresence microscopy. *Res. Bull. Coll. Exp. For. Hokkaido Univ.* 33:104–14
48. Gilbertson, R. L. 1980. Wood-rotting fungi of North America. *Mycologia* 72:1–49
49. Glenn, J. K., Gold, M. H. 1985. Purification and characterization of an extracellular Mn(II)-dependent peroxidase from the lignin-degrading basidiomycete, *Phanerochaete chrysosporium. Arch. Biochem. Biophys.* 242:329–41
50. Gold, M. H., Glenn, J. K. 1988. Use of polymeric dyes in lignin biodegradation assays. In *Methods in Enzymology-Biomass, Part b, Lignin, Pectin, Chitin,* ed. W. A. Wood, S. T. Kellogg, 161:74–78. San Diego: Academic
51. Gold, M. H., Wariishi, H., Valli, K., Mayfield, M. D., Nipper, V. J., Brown, J. A., Pribnow, D. 1990. Manganese peroxidase from *Phanerochaete chrysosporium:* Biochemical and genetic characterization. In *Biotechnology in*

Pulp and Paper Manufacture, Applications and Fundamental Investigations, ed. T. K. Kirk, H-M. Chong. pp. 429–38. Boston: Butterworth-Heinemann

52. Gonzalez, A., Grinbergs, J., Griva, E. 1986. Biological transformation of wood into feed for cattle - "Palo podrido". *Zbl. Mikrobiol.* 141:181–86
53. Hammel, K. E. 1989. Organopollutant degradation by ligninolytic fungi. *Enzyme Microb. Technol.* 11:776–77
54. Hammel, K. E., Tien, M., Kalyanaraman, B., Kirk, T. K. 1985. Mechanisms of oxidative C - C cleavage of a lignin model dimer by *Phanerochaete chrysosporium* ligninase. *J. Biol. Chem.* 260: 8348–53
55. Hartig, R. 1878. *Die Zersetzungserscheinungen des Holzes der Nadelholzbaume und der Eiche in forstlicher, botanischer und chemischer Richtung.* Berlin/New York: Springer. 151 pp.
56. Hartig, R. 1894. Text-book of the Diseases of Trees. Transl. W. Somerville, H. M.Ward. London: Macmillan. 331 pp.
57. Hartig, T. 1833. Abhandlung uber die Verwandlung der polycotyledonischen Pflanzenzelle in Pilz und Schwammgebilde und der daraus hervorgehenden sogenannten Faulniss des Holzes. Berlin. 46 pp.
58. Hedges, J. I., Blanchette, R. A., Weliky, K., Devol, A. H. 1988. Effects of fungal delignification on the CuO oxidation products of lignin: A controlled laboratory study. *Geochim. Cosmochim. Acta* 52:2717–26
59. Highley, T. L., Bar-Lev, S. S., Kirk, T. K., Larsen, M. J. 1983. Influence of $O_2$and CO_2 on wood decay by heartrot and saprot fungi. *Phytopathology* 73: 630–33
60. Higuchi, T. 1990. Lignin biochemistry: biosynthesis and biodegradation. *Wood Sci. Technol.* 24:23–63
61. Hutterman, A., Cwielong, P. 1991. Biochemistry of lignin degradation as a base for control of *Heterobasidion annosum* inside the living tree. See Ref. 110a
62. Huynh, V-B., Crawford, R. L. 1985. Novel extracellular enzymes (ligninases) of *Phanerochaete chrysosporium. FEMS Microbiol. Lett.* 28:119–23
63. Jensen, K. F. 1969. Oxygen and carbon concentrations in sound and decaying red oak trees. *For. Sci.* 15:246–51
64. Johnsrud, S. C., Eriksson, K-E. 1985. Cross-breeding of selected and mutated homokaryotic strains of *Phanerochaete chrysosporium* K-3: New cellulase deficient strains with increased ability to degrade lignin. *Appl. Microbiol. Biotechnol.* 21:320–27
65. Jurasek, L. 1964. Changes in the microstructure at the destruction of wood by wood destroying fungi. *Drev. Vysk.* 3: 127–44
66. Kawase, K. 1962. Chemical components of wood decayed under natural conditions and their properties. *Hokkaido Diagaku Sapporo J. Fac. Agr.* 52:186–245
67. Kerr, J., Goring, D. A. I. 1975. The ultrastructural arrangement of the wood cell wall. *Cellul. Chem. Technol.* 9: 563–73
68. Kersten, P. J., Tien, M., Kalyanaraman, B., Kirk, T. K. 1985. The ligninase of *Phanerochaete chrysosporium* generates cation radicals from methoxybenzenes. *J. Biol. Chem.* 260:2609–12
69. Kim, Y. S., Park, B. D., Lee, J. K. 1988. Micromorphological features of oak wood cultivated with Shiitake mushroom, *Lentinus edodes* (Berk) Sing. *Korean Wood Sci. Technol.* 16:38–47
70. Kimura, Y., Asada, Y., Kuwahara, M. 1990. Screening of basidiomycetes for lignin peroxidase genes using a DNA probe. *Appl. Microbiol. Biotechnol.* 32: 436–42
71. Kirk, T. K. 1971. Effect of microorganisms on lignin. *Annu. Rev. Phytopathol.* 9:185–210
72. Kirk, T. K. 1987. Lignin-degrading enzymes. *Phil. Trans. R. Soc. Lond.* 321:461–74
73. Kirk, T. K. 1988. Biochemistry of lignin degradation by *Phanerochaete chrysosporium.* In *Biochemistry and Genetics of Cellulose Degradation,* ed. J. P. Aubert, P. Beguin, J. Millet. pp. 315–32. London: Academic
74. Kirk, T. K., Cowling, E. B. 1984. Biological decomposition of wood. In *The Chemistry of Solid Wood*, ed. R. M. Rowell. pp. 455–87. Adv. Chem. Ser. 207. Am. Chem. Soc., Washington DC
75. Kirk, T. K., Farrell, R. L. 1987. Enzymatic "combustion": The microbial degradation of lignin. *Annu. Rev. Microbiol.* 41:465–505
76. Kirk, T. K., Jeffries, T. W., Leatham, G. F. 1983. Biotechnology: Applications and implications for the pulp and paper industry. *Tappi. J.* 66(5):45–51
77. Kirk, T. K., Moore, W. E. 1972. Removing lignin from wood with white-rot fungi and digestibility of resulting wood. *Wood Fiber* 4:72–79
78. Kirk, T. K., Schultz, E., Connors, W. J., Lorenz, L. F., Zeikus, J. G. 1978. Influence of culture parameters on lignin

metabolism by *Phanerochaete chrysosporium*. *Arch. Microbiol.* 117:277–85

79. Kirk, T. K., Tien, M., Johnsrud, S. C., Eriksson, K-E. 1986. Lignin degrading activity of *Phanerochaete chrysosporium* Burds.: Comparison of cellulase-negative and other strains. *Enzyme Microb. Technol.* 8:75–80
80. Kuwahara, M., Glenn, J. K., Morgan, M. A., Gold, M. H. 1984. Separation and characterization of two extracellular H_2O_2-dependent oxidases from ligninolytic cultures of *Phanerochaete chrysosporium*. *FEBS Lett.* 169:247–50
81. Leatham, G. F., Kirk, T. K. 1983. Regulation of ligninolytic activity by nutrient nitrogen in white-rot basidiomycetes. *FEMS Microbiol. Lett.* 16:65–67
82. Leatham, G. F., Myers, G. C., Wegner, T. H., Blanchette, R. A. 1990. Energy savings in biomechanical pulping. In *Biotechnology in Pulp and Paper Manufacture, Applications and Fundamental Investigations,* ed. T. K. Kirk, H. M. Chang. pp. 17–25. Boston: Butterworth-Heinemann
83. Leatham, G. F., Myers, G. C., Wegner, T. H., Blanchette, R. A. 1990. Biomechanical pulping of aspen chips: paper strength and optical properties resulting from different fungal treatments. *Tappi J.* 73(3):249–54
84. Levi, M. P., Cowling, E. B. 1969. Role of nitrogen in wood deterioration. VII. Physiological adaption of wood-destroying and other fungi to substrates deficient in nitrogen. *Phytopathology* 59:460–68
85. Liese, W. 1970. Ultrastructural aspects of woody tissue disintegration. *Annu. Rev. Phytopathol.* 8:231–57
86. Liese, W., Schmid, R. 1966. Untersuchungen uber den Zellwandabbau von Nadelholz durch *Trametes pini*. *Holz Roh-Werkst.* 24:454–60
87. Meier, H. 1955. Uber den Zellwandabbau durch Holzvermorschungspilze und die submikroskopische Strucktur von Fichtentracheiden und Birkenholzfasern. *Holz. Roh-Werkst.* 13:323–38
88. Melillo, J. M., Aber, J. D., Muratore, J. F. 1982. Nitrogen and lignin control of hardwood leaf litter decomposition dynamics. *Ecology* 63:621–26
89. Merrill, W. 1970. Spore germination and host penetration by heartrotting hymenomycetes. *Annu. Rev. Phytopathol.* 8:281–300
90. Merrill, W., Cowling, E. B. 1966. Role of nitrogen in wood deterioration. IV Relationship of natural variation in nitrogen content of wood to its susceptibility to decay. *Phytopathology* 56: 1324–25
91. Myers, G. C., Leatham, G. F., Wegner, T.H., Blanchette, R. A. 1988. Fungal pretreatment of aspen chips improves strength of refiner mechanical pulp. *Tappi. J.* 71(5):105–108
92. Nashida, T. 1989. Lignin biodegradation by wood-rotting fungi. V. A new method for evaluation of the ligninolytic activity of lignin-degrading fungi. *Mokuzai Gakkaishi* 35:675–77
93. Obst, J. R., Landucci, L. 1986. The syringyl content of softwood lignin. *J. Wood Chem. Technol.* 6:311–27
94. Otjen, L., Blanchette, R. A. 1982. Patterns of decay caused by *Inonotus dryophilus* (*Aphyllophorales: Hymenochaetaceae),* a white-pocket rot fungus of oaks. *Can. J. Bot.* 60:2770–79
95. Otjen, L., Blanchette, R. A. 1984. *Xylobolus frustulatus* decay of oak: Patterns of selective delignification and subsequent cellulose removal. *Appl. Environ. Microbiol.* 47:670–76
96. Otjen, L., Blanchette, R. A. 1986. Selective delignification of birch wood (*Betula papyrifera* by *Hischioporus pargamenus* in the field and laboratory. *Holzforschung* 40:183–89
97. Otjen, L., Blanchette, R. A. 1986. A discussion of microstructural changes in wood during decomposion by white rot basidiomycetes. *Can. J. Bot.* 64:905–11
98. Otjen, L., Blanchette, R. A., Effland, M., Leatham, G. 1987. Assessment of 30 white rot basidiomycetes for selective lignin degradation. *Holzforschung* 41: 343–49
99. Otjen, L., Blanchette, R. A., Leatham, G. F. 1988. Lignin distribution in wood delignified by white-rot fungi. X-ray microanalysis of decayed wood treated with bromine. *Holzforschung* 42:281–88
100. Paszczynski, A., Huynh, V-B., Crawford, R. L. 1985. Enzymatic activities of an extracellular, manganese-dependent peroxidase from *Phanerochaete chrysosporium*. *FEMS Microbiol. Lett.* 29: 37–41
101. Paszczynski, A., Crawford, R. L., Blanchette, R. A. 1988. Delignification of wood chips and pulps by using natural and synthetic porphyrins: models of fungal decay. *Appl. Environ. Microbiol.* 54:62–68
102. Pham, T. T. T., Maaroufi, A., Odier, E.1990. Inheritance of cellulose- and lignin-degrading ability as well as endoglucanase isozyme pattern in *Dichomitus squalens*. *Appl. Microbiol. Biotechnol.* 33:99–104
103. Philippi, F. 1893. Die Pilze Chiles,

soweit dieselben als Nahrungsmittel gebraucht werden. *Hedwigia* 32:115–18

104. Raeder, U., Thompson, W., Broda, P. 1989. RFLP-based genetic map of *Phanerochaete chrysosporium* ME446: lignin peroxidase genes occur in clusters. *Mol. Microbiol.* 3:911–18
105. Raeder, U., Thompson, W., Broda, P. 1989. Genetic factors influencing lignin peroxidase activity in *Phanerochaete chrysosporium* ME 446. *Mol. Microbiol.* 3:919–24
106. Rayner, A. D. M., Boddy, L. 1988. *Fungal Decomposition of Wood. Its Biology and Ecology*. New York: Wiley. 587 pp.
107. Reade, A. E., McQueen, R. E. 1983. Investigation of white rot fungi for the conversion of poplar into a potential feedstuff for ruminants. *Can. J. Microbiol.* 29:457–63
108. Reid, I. D. 1983. Effects of nitrogen supplements on degradation of aspen wood lignin and carbohydrate components by *Phanerochaete chrysosporium*. *Appl. Environ. Microbiol.* 45:830–37
109. Reid, I. D. 1989. Solid state fermentations for biological delignification. *Enzyme Microb. Technol.* 11:786–803
110. Reid, I. D., Seifert, K. A. 1982. Effect of an atmosphere of oxygen on growth, respiration and lignin degradation by white rot fungi. *Can. J. Bot.* 60:252–60

110a. Rossmore, H. W., ed. 1991. *Proc. 8th Int. Biodeterior. Biodegrad. Symp.* London: Elsevier. In press

111. Saka, S., Thomas, R. J. 1982. Evaluation of the quantitative assay of lignin distribution by SEM-EDXA technique. *Wood Sci. Technol.* 16:1–8
112. Sarkanen, K. V., Ludwig, C. H. 1971. *Lignins. Occurrence, Formation, Structure, and Reactions*. Wiley-Interscience: New York. 916 pp.
113. Schacht, H. 1863. Uber die Veranderungen durch Pilze in abgestorbenen Pflanzenzellen. *Jahrb. Wiss. Bot.* 3:442–83. (Transl. W. Merrill)
114. Setliff, E. C., Eudy, W. W. 1980. Screening white-rot fungi for their capacity to delignify wood. In *Lignin Biodegradation: Microbiology, Chemistry, and Potential Applications,* ed. T. K. Kirk, H-M. Chang, T. Higuchi. 1:135-49. Boca Raton: CRC
115. Shimada, M., Nakagawa, M., Hattori, T., Higuchi, T. 1989. A new biomimetic lignin degradation system developed with MnCo catalysts and its application to the chlorine-free bleaching of kraft pulp. *Mokuzai Gakkaishi* 35:859–60
116. Strebotnik, E., Messner, K. 1990. Enzymatic attack of wood is limited by the inaccessibility of the substrate. In *Biotechnology in Pulp and Paper Manufacture, Applications and Fundamental Investigations,* ed. T. K .Kirk, H-M. Chang. pp. 111–19. Boston: Butterworth-Heinemann
117. Stubblefield, S. P., Taylor, T. N. 1986. Wood decay in silicified gymnosperms from Antarctica. *Bot. Gaz.* 147:116–25.
118. Stubblefield, S. P., Taylor, T. N. 1988. Recent advances in palaeomycology. *Tansley Rev. No. 12. New Phytol.* 108: 3–25
119. Tien, M. 1987. Properties of ligninase from Phanerochaete chrysosporium and their possible applications. *CRC Crit. Rev. Microbiol.* 15:141–68
120. Tien, M., Kirk, T. K. 1983. Lignin-degrading enzyme from the hymenomycete *Phanerochaete chrysosporium* Burds. *Science* 221:661–63
121. Trotter, P. C. 1990. Biotechnology in the pulp and paper industry: A review. Part 1: Tree improvement, pulping and bleaching, and dissolving pulp applications. *Tappi J.* 73(4):198–204
122. Tsuneda, A., Koshitani, H., Furukawa, I. 1987. Micromorphological patterns of incipient wood decay by Lentinus edodes. *Rep. Tottori Mycol. Inst.* 25: 36–48
123. Waldner, R., Leisola, M. S. A., Fiechter, A. 1988. Comparison of ligninolytic activities of selected fungi. *Appl. Microbial. Biotechnol.* 29:400–7

123a. Westermark, U., Lidbrandt, O., Eriksson, I. 1988. Lignin distribution in spruce (*Picea abies*) determined by mercurization with SEM-EDXA technique. *Wood Sci. Technol.* 22:243–50

124. Wilcox, W. W. 1968. Changes in wood microstructure through progressive stages of decay. *US For. Serv. Res. Pap.* FPL-70. 46 pp.
125. Yoshihara, K., Kamijima, H., Akamatsu, I. 1987. Effect of urea on lignin degradation by *Coriolus hirsutus*. *Mokuzai Gakkaishi* 33:61–70
126. Yoshizawa, N., Watanabe, J., Kobayashi, S., Idei, T. 1989. Comparative histochemistry of wood cell wall degradation by white-rot fungi. *Bull. Utsunomiya. Univ. For.* 25:23–38
127. Yu, H-S., Eriksson, K-E. 1985. Influence of oxygen on the degradation of wood and straw by white-rot fungi. *Svensk. Papperstidn.* 88:57–60
128. Zadrazil, F., Grinbergs, J., Gonzales, A. 1982."Palo Podrido" —decomposed wood used as feed. *J. Appl. Microbiol. Biotechnol.* 15:167–71

Annu. Rev. Phytopathol. 1991. 29:399–420

RESEARCH RELATING TO THE RECENT OUTBREAK OF CITRUS CANKER IN FLORIDA*

Robert E. Stall

Department of Plant Pathology, University of Florida, Gainesville, Florida 32611

Edwin L. Civerolo

USDA, ARS, Agricultural Research Center, Beltsville, Maryland 20707

KEY WORDS: *Xanthomonas campestris* pv. *citri*, bacterial spot of citrus, bacterial diseases of citrus

---

PRESENT STATUS OF CITRUS CANKER IN FLORIDA

Two diseases, both caused by bacteria classified as xanthomonads, occur on citrus in Florida. One is similar to the disease named Asiatic citrus canker (canker A) that was declared eradicated from the state in 1933. Samples of citrus with with this disease were collected and identified in 1986 after an absence from the state of over 50 years. Canker A probably had been in the state a few years before 1986, but just how long is open to speculation. The other disease is currently called bacterial spot of citrus, but has also been called nursery canker and canker E during its short history. It was first identified in 1984, and was then designated as a unique form of citrus canker.

Eradication of the bacterial spot pathogen was attempted, but these efforts were discontinued in 1990. Eradication of canker A is still being pursued actively by the Florida Department of Agriculture and Consumer Services and US Department of Agriculture, Animal and Plant Health Inspection Service (USDA, APHIS). The most recent destruction of infected trees was in 1991.

INTRODUCTION

"Citrus canker" is used in this review as a generic term that includes all diseases of citrus caused by strains of *Xanthomonas*. An extensive analytical bibliography of citrus canker, with abstracts of papers, was published in 1982 by Rossetti et al (58). In addition, Civerolo (10, 11) reviewed literature on citrus bacterial canker (citrus canker) up to 1984. A review of research on citrus canker in Florida, from 1912–1933, was written by K. W. Loucks (50). Schoulties et al (59) reported on the regulatory aspects related to eradication of the most recent outbreak of citrus canker in Florida. Here, we review the research that relates to the recent outbreak of citrus canker in Florida.

Citrus canker is the most notorious of diseases of plants in Florida, because of the widespread attention given to it, because of efforts to eradicate it and because of the importance of citrus to the economy of Florida (51). The eradication effort usually involves the burning of trees to eliminate the pathogen. Photographs of burning trees have been placed on the front pages of most of the major newspapers, and eradication efforts have been sensationalized during new outbreaks by most of the major television stations in Florida. The aggressive actions of burning trees conjures visions of a very contagious disease that would destroy the citrus industry if the disease was not eradicated. This impression, coupled with quarantine measures that would likely be imposed by certain foreign and domestic markets if citrus canker were to become established, causes a great fear of citrus canker in the citrus industry.

The true seriousness of the disease in Florida is unknown, but obviously is somewhere between the concept that the disease would not affect citrus production (41, 71, 72) to the concept that the disease would destroy citrus production in Florida (51). Information on the true effect of citrus canker on citrus production in Florida is not available and extrapolations of disease severity and losses from other areas of the world to Florida are only speculative (64).

Citrus canker has already caused extensive monetary expenditures by growers and state and federal regulators in Florida even though the incidence of the disease was negligible. Eradication measures including the destruction of trees are mandated by state and federal laws to prevent spread of the pathogen. Not only are diseased trees destroyed, but trees with no visible symptoms are also destroyed if they are within certain specified distances of diseased trees (i.e. exposed trees). Significant monetary costs of the disease have resulted from regulations of governmental agencies who imposed additional cultural practices and prevention of sales of fruit in specific regions to prevent the pathogen from spreading to new areas. These regulations are enforced by personnel of the USDA APHIS, and the Department of Agriculture and Consumer Services of the State of Florida who, respectively, regulate interstate and intrastate movement of host plants, fruit, and other regulated articles that may be capable of transmitting the pathogen. The regulations are extensive (1) and the cost is borne by every segment of the citrus industry as well as the general public.

Brief History of Eradication Efforts

That eradication of citrus canker was possible was demonstrated after appearance of the disease in Florida around 1910. Shortly thereafter, a group of growers in Dade County became alarmed about the seriousness and rapid spread of the disease and organized a campaign to eradicate it. This campaign was transferred to a state agency with the formation of the State Plant Board in 1915, and the Federal Government entered the program a short while later. The disease was declared eradicated in Florida in 1933 after an aggressive 20-year program and was successfully eradicated from the USA by 1947 (18). Citrus canker was also eradicated from Australia (14, 46), New Zealand (21), and South Africa (17). Active eradication programs are now in progress in Brazil and Uruguay.

Although many samples of citrus canker have been confiscated at US ports of entry since the declaration of eradication (28), no samples of citrus canker were reported in the state until 1984. Thus, citrus canker was not found in Florida for over 50 years after the eradication campaign ceased. During this 50-year period, state inspectors continued to survey citrus plantings for the disease. Another citrus canker eradication campaign was established in Florida in 1984.

Citrus Canker Diseases

The concept was developed that different forms of citrus canker existed over the years of research and observations on citrus diseases. In addition to Asiatic citrus canker (canker A), cancrosis B (canker B) (8), Mexican lime cancrosis (canker C) (54), and bacteriosis (canker D) (56) were described. These different forms are considered in this review as different diseases, for the sake of discussion, but it is uncertain at this time how the diseases should be classified . The diseases are all caused by xanthomonads, which were considered as unique strains of one taxonomic group of bacteria, now known as *Xanthomonas campestris* pv. *citri*. The major differences in these diseases are the range of susceptible hosts among the Rutaceae and the severity of the diseases.

The symptoms of canker A have been well described (61). This disease was first recognized in Florida and the causal agent was identified and described by Clara Hasse in 1915 as *Pseudomonas citri* (44). Bacterial nomenclature has changed several times since this original description, and the causal agent is now known as *X. campestris* pv. *citri*. The original source of inoculum for canker A in Florida was traced to seedlings of *Poncirus trifoliata* that were imported from Japan for use as rootstocks (3). In subsequent searches, canker A was found throughout Southeastern Asia as well as Japan. Fawcett & Jenkins (25) found samples of the disease in herbarium specimens collected in India in 1823. Canker A now occurs in 30 countries including the USA (9). The causal agent is pathogenic to a wide range of plants among members of the Rutaceae, but grapefruit, limes, and some sweet orange plants are particularly susceptible.

A disease of citrus with symptoms very similar to canker A was reported in Argentina by Carrera in 1933 (8). It differed from citrus canker in the Orient and the USA in that lemon plants were the most susceptible and grapefruit and sweet orange plants were little affected in the field. Fawcett (24) saw the disease during a survey for citrus diseases in South America and agreed that the South American disease was similar to citrus canker. Fawcett called it false canker, and later Bittancourt (4) referred to the disease as cancrosis B. Ducharme (19) reported that a yellow bacterium caused symptoms of cancrosis B on citrus and reported on its host range. Gotuzzo & Rossi (35) charactered the pathogen bacteriologically in 1967 and identified it as a species of *Xanthomonas*. Some strains were said to be atypical of *X. citri,* however. Stall et al (60) reported that isolation of the causal bacterium from canker B lesions was very difficult with routine laboratory media. Occasionally, a colony would appear, but almost all cells from lesions obtained from the field failed to form visible colonies. A special medium was devised for isolation of the causal bacterium (7). The need for a special medium to isolate the canker B bacterium was in contrast to isolation of the canker A bacterium, which could be readily isolated on most laboratory media (51). Goto et al (29) compared strains from cancrosis B and strains from canker A. Strains from cancrosis B differed from strains from canker A in their failure to utilize lactose and maltose, in susceptibility to phage CP3, and in antigenic properties. In addition, the strains from cancrosis B were significantly less virulent than strains from canker A on plants of Unshu, Natsudaidai, Naval orange, and lemon. They concluded that the strains of cancrosis B were sufficiently different from strains of canker A as to be considered a distinct infrasubspecific subdivision of *X. campestris* pv. *citri*.

A bacterium that caused symptoms similar to canker A, but pathogenic only to Mexican lime plants, was reported in Brazil in 1972 (54). The causal bacterium was characterized as a member of *Xanthomonas* by physiological and morphological tests. However, when this bacterium was compared with bacteria from canker A and cancrosis B in pathogenicity tests on 15 citrus species and varieties, the strains from Mexican lime caused disease only on Mexican lime whereas those from canker A and cancrosis B caused lesions on all citrus types. The strains from Mexican lime were also different from those from strains from canker A by being resistant to bacteriophages CP1 and CP2, and serologically distinct from both canker A and cancrosis B strains. It was concluded that the bacteria causing canker on Mexican lime were a *forma specialis* of *X. citri* and were named *X. citri* f. sp. *aurantifolia*. The canker on Mexican lime was later referred to as canker C (57).

Another disease of Mexican lime was reported in the Colima area of Mexico in 1981. This disease resembled canker A in symptoms, except for the absence of lesions on fruit (56). A xanthomonad was isolated from lesions, and strains of it were serologically related to strains of *X. campestris* pv. *citri* from canker A and cancrosis B. The strains from Mexican lime were pathogenic when reinoculated onto plants of Mexican lime. Lesions also developed on trees of Persian lime and grapefruit growing near diseased trees of Mexican lime, but many fewer lesions developed on those trees than on the Mexican lime trees. This disease was named bacteriosis of citrus, and is sometimes referred to as canker D (59).

In addition to differences in the range of citrus hosts affected, the pathogens responsible for these diseases may also differ in aggressiveness on citrus. Undoubtedly, the most aggressive group of strains is associated with canker A. Observations in Argentina allowed comparisons of canker A and cancrosis B in the same area. When strains causing canker A were introduced into lemon groves already infested with strains from cancrosis B, the latter strains disappeared (B. I. Canteros, personal communication). In Brazil, strains of canker C have not been recovered since the original isolations and may not have survived (P. R. Leite Jr., personal communication). The strains of canker D in Mexico are very weakly pathogenic and may have been displaced by an *Alternaria* disease of citrus in the Colima region (66).

Bacterial Spot of Citrus

Bacterial spot was found on citrus plants in a nursery in central Florida and the causal agent was identified as a xanthomonad (59). All diseases of citrus caused by xanthomonads have many symptoms in common, but bacterial spot differs in the type of lesion on leaves and fruit. The difference is most obvious with lesions at 10 to 20 days after infection. The surfaces of lesions of cankers A, B, C, and D become corky and raised. A leaf that contains many lesions may feel like sandpaper. Lesions can often be felt before being seen by the naked eye. In contrast, lesions typical of bacterial spot are usually flat to sunken below the leaf surface. Some strains from bacterial spot cause lesions that are slightly raised above the leaf surface, but not to the extent of cankers A, B, C, or D lesions. Watersoaking of leaf tissues surrounding necrotic areas is prominent with young lesions of bacterial spot, but not with young lesions of canker A. With older lesions, the raised portion of the lesions of canker A develops at the margin of lesions and results in a craterform topography. Lesions of bacterial spot continue to be flat or sunken. Watersoaking occurs around margins of older lesions of both diseases, but the watersoaked margin is more conspicuous with bacterial spot. Lesions of bacterial spot on stems are very similar in appearance to lesions caused by canker A.

Whether or not bacterial spot should be considered as one of the forms of citrus canker (canker E) has been the subject of much controversy, and the answer could be a very important contribution of the current research. Such a question may be primarily academic with other diseases, such as with bacterial canker of stone fruits caused by *Pseudomonas syringae* pv. *syringae* and *P. s.* pv. *morsprunorum,* where two distinct bacteria cause the same disease syndrome (13); or with angular leaf spot and wildfire of tobacco caused by *P. s.* pv. *tabaci,* where two recognizable diseases are caused by variants of the same bacterium (67). However, with a regulated disease such as citrus canker, however, the name of the disease and the causal agent are very important in triggering or suspending regulatory actions by State and Federal agencies.

Locations of Citrus Canker Research in the USA

All research devoted to citrus canker in Florida was suspended in 1933 (51), the year that citrus canker was declared eradicated. No research on citrus canker was allowed in Florida until 1984 when the new introduction occurred. However, a joint project between the Institute of Food and Agricultural Sciences (IFAS) of the University of Florida and the government of Argentina was established in 1978 to study characteristics of the casual bacterium and the epidemiology and control of citrus canker. This project continued until 1982, with additional support from the State Department of Agriculture and Consumer Services and the USDA Tropical Agriculture Grant program. The accomplishments of this project have been reviewed (64). In addition, research was started in 1978 at the facility of the US Department of Agriculture, Agricultural Research Service (USDA-ARS) at Beltsville, MD, under quarantine restrictions. Strains of xanthomonads from citrus around the world were collected and studied. Methods of diagnosis of the diseases by serology were developed (12).

Research on citrus canker in Florida had to be conducted under quarantine conditions after 1984. This prevented many important epidemiological studies that could be done only in the field in the citrus belt in Florida. However, field research was permitted at the USDA-ARS Plant Quarantine facility at Frederick, MD, beginning in 1985. Eventually field research was also permitted at a quarantine facility located at an IFAS Research Center at Hastings, in northern Florida and removed from citrus plantings. Field research with canker A was permitted at Frederick, MD, but only research on bacterial spot was permitted at the Hastings location.

Research Related to Regulation Problems

Early in the research program, answers were sought to questions that were important to the eradication program. Brown & Schubert (6) reported on experiments for chemical treatment of citrus fruit to eliminate contaminating xanthomonads. Because of quarantine restrictions, *X. c.* pv. *vesicatoria* was used as the contaminant in some experiments. Applications of chlorine or sodium orthophenylphenate (SOPP) during washing for 30 seconds eliminated bacteria when 10^6 colony-forming units (cfu) were applied as inoculum. Applications of dual quaternary ammonium compounds, formulations of chlorine dioxide, or peracetic acid as fruit washes for a similar time were equally effective. Chlorine also was effective in reducing the numbers of native bacteria on treated fruit of Mexican lime in Mexico, although a small population remained after treatment (10^2 to 10^3 cfu per cm^2 of fruit surface) (65).

The survival of cells in leaf lesions of canker A and bacterial spot placed on the surface of soil and incorporated into soil was investigated by Graham et al (40) following the 1984 outbreak. Survival under such conditions was also investigated in Florida during the first infestation of canker A (51), but questions arose about the new xanthomonad on citrus. Cells in lesions of canker A survived at least 90 days after tree defoliation and under relatively dry conditions (37). Under similar conditions in Argentina, cells in lesions of canker A survived for 120 days when the leaves were left on the soil surface, but only 85 days when the leaves were buried in moist soil. Cells in lesions of bacterial spot in leaves buried in air-dried sandy soil were detected up to 105 days, but with virtually any moisture in this sandy soil (slightly moist to saturated soil), cells survived less than 24 days. The soil fumigants metam-sodium and dichloropropene-methyl isothiocyanate were more effective than MC-2 (98:2 methylbromide:chloropicrin) in reducing populations of strains from bacterial spot lesions in grapefruit leaves in soils. In tests in Argentina, the latter fumigants did not eradicate the cells of the canker A pathogen in leaves in soil but did shorten the survival time from 85 to 45 days. The fumigants also killed citrus roots, which when not treated resprout and furnish susceptible tissue in the eradication area.

McGuire (53) tested thirteen bactericidal materials for control of canker A in Argentina. The copper-containing products were superior to other compounds in limiting lesion numbers on leaves and fruit, as well as reducing epiphytic populations of *Xanthomonas* on leaves. Copper compounds were previously found to be effective in controlling canker A if sprays were timed to coincide with new flushes of growth (62).

Many other lines of research have been pursued to aid in the eradication effort. Much of the work has been very practically oriented and has been published only in report form. This work has included such investigations as the search for epiphytic survival of the bacterial-spot pathogen on leaves on trees surrounding infested nurseries, treatment of citrus seed for removal of contaminating xanthomonads, heat-treatment of fruit to eradicate cells of xanthomonads in lesions in the fruit, defoliation of trees with an herbicide to accomplish eradication without destruction of trees, investigation of chemicals to decontaminate clothing and equipment, elucidation of the effect of the bacterial-spot pathogen on bud survival in rootstocks, and characterizations of strains of the bacterial-spot pathogen by physiological tests. These reports are available from the Citrus Agricultural Research Center, Lake Alfred, Florida.

COMPARISONS OF BACTERIAL SPOT AND CANKER DISEASES

Etiology

Much of the published work has dealt with the etiology of bacterial spot compared to the etiology of the other diseases of citrus caused by xanthomonads. The differences in lesion types caused by bacteria from canker A and bacterial spot were confirmed by histological studies with the two diseases (16, 47, 49). The blister formation typical of canker A lesions is the result of degradation of host epidermal cell walls, cuticle separation, and protrusion of proliferating host cells through epidermal ruptures. In contrast, there is no host cell proliferation with bacterial spot after infection. The water-soaked flat lesions with bacterial spot are associated with rapid membrane damage that allows leakage of cell fluids into the intercellular spaces. A less aggressive strain of the bacterial-spot pathogen did not cause widespread membrane damage and did not cause extensive water-soaking of tissues. Invasion of leaf tissue by the two groups of pathogens was similar.

Cankers A, B, C, and bacterial spot all occur as lesions on leaves, stems, and fruit (59, 63). Canker D occurs as lesions on leaves and stems but not on fruit (56). Lesions in all susceptible organs continue to expand slowly after infection. The increase in diameter of lesions of canker A in grapefruit leaves in the field in Argentina was determined to be about 0.5 mm per 10 days (61). Under quarantine greenhouse conditions in Florida, lesions caused by a strain of the bacterium from canker A inoculated into leaves of Swingle citrumelo increased in diameter about 0.67 mm per 10 days (22). The rate of lesion expansion was statistically greater with an aggressive strain of the bacterial-spot pathogen compared with a strain of the canker A pathogen on the same host. In another test (38), lesion expansion was more rapid with an aggressive strain of the bacterial-spot pathogen compared with a strain from canker A pathogen in leaves of Swingle citrumelo plants. Expansion of lesions of bacterial spot was slower than lesions of canker A in leaves of Duncan grapefruit. The rate of expansion of canker A lesions is about the same in leaves of several hosts, but tends to be slower in *Poncirus trifoliata*. In contrast, the rate of expansion of lesions of bacterial spot varies considerably with different hosts. Lesions caused by less aggressive strains of the bacterial-spot pathogen expand at a slower rate than canker A.

Population Dynamics

The numbers of cells of the canker A pathogen in actively expanding lesions in grapefruit leaves collected in the field in Argentina were between 10^6 and 10^7 cfu per lesion (61). The populations of cells in lesions dropped to about 10^4 cfu per lesion when the lesions stopped expanding in the winter. No significant differences in viable cell numbers were obtained from cultivars that differed in resistance to the pathogen. Graham et al (38), working under greenhouse conditions, inoculated leaves of several cultivars of citrus by a pinprick method that gave high numbers of cells (10^6 cfu per lesion) in the injured area. A strain of the canker A pathogen maintained the high populations through 40 days of testing on two cultivars in spite of differences in resistance to the bacteria in the cultivars. Populations of three strains of the bacterial-spot pathogen decreased logarithmically with time, except for the most aggressive strain in leaves of Swingle citrumelo and *Poncirus trifoliata*, where the populations remained high, more like the canker A bacterium. Graham et al suggested from this evidence that only the aggressive strains of the bacterial-spot pathogen should be considered as a pathogen of citrus.

The population of cells that exuded from wetted lesions was determined for both canker A and bacterial spot. The population of bacterial cells in rainwater collected from leaves with canker A in the field in Argentina varied between 10^4 and 10^6 cfu per ml of water (61). Timmer et al, in a more exacting study (68), determined bacterial populations in water placed over lesions in leaves detached from trees in the field and found that populations rapidly increased in the water to about the same densities as reported above. They reported that lower populations in water from old lesions, which may have reflected fewer viable cells in the lesions. However, they postulated that bacteria may not be released from older lesions because the lesions become suberized and bacteria may not be liberated freely. Much lower concentrations of bacteria (0 to 10^3 cfu per ml) were obtained from water collected over lesions caused by each of three strains of the bacterial-spot pathogen collected in the field in Florida. As lesions aged, fewer viable bacteria were collected in the water, but no differences could be detected in the numbers of cells from lesions on Swingle citrumelo compared to those on Duncan grapefruit. Fewer cells were collected over lesions caused by the less aggressive strains. The populations of cells in water over the surface of lesions of canker A and bacterial spot formed in a dew chamber were larger than those in water over lesions collected from the field, presumably because the conditions in the dew chamber were more conducive for multiplication of bacteria in the lesions.

Epidemiology

Although spatial and temporal spread of the canker A pathogen was studied in groves in Argentina (15), no information was available on spatial and temporal spread of this pathogen in nurseries. Gottwald et al (31, 33) compared the appearance of canker A and bacterial spot in simulated nurseries and in new grove plantings of two cultivars of citrus at Concordia, Intre Rios, Argentina and Frederick, MD, respectively. A single diseased plant was placed in the center of circularly arranged plants in the simulated new grove plantings and in the center of rectangularly arranged simulated nursery plantings. The primary difference in the two types of plantings was larger plant spacings within rows and between rows in the simulated new grove plantings. In Argentina, canker A was first detected in the simulated new grove planting 49 days after the experiment was started. Final disease incidences at 225 days after the experiment was started in plots of grapefruit and Valencia orange were 97% and 86%, respectively. The rate of increase of disease incidence, based on Gompertz transformations, was 0.009 and 0.005 per day for the grapefruit and Valencia orange plots, respectively. Disease gradients were determined from the focal plant to the edge of the plots. The gradients tended to flatten during the first part of the epidemic, became steeper through the middle of the test, and then flattened again. This fluctuation in gradients was attributed to inoculum fluctuations caused by defoliation of leaves in the plots.

Although the plants grew well at Frederick, MD, the first bacterial spot appeared in the simulated new grove planting 141 days after the focal plant was planted. Two grapefruit trees that were each 4.58 meters from the focal tree became diseased, but no more disease occurred on grapefruit and no disease occurred on any of the Valencia orange trees. In the simulated nursery planting, no bacterial spot occurred in the Valencia orange plantings, but the incidence of disease increased among the grapefruit plants. The first spread was detected on day 94 on three plants, with the farthest plant being 2.37 meters away from the focus. On day 139, three more plants were found with disease with the farthest plant being 5.41 meters away. Disease gradients were not calculated because of the low incidence of disease in the plots.

In spite of the low incidence of bacterial spot in the plots, the pathogen spread throughout the simulated nursery plots in MD, as determined by measurement with immunoflouresence microscopy, DNA-DNA probes, and detection with a semiselective medium (31). Epiphytic xanthomonads were not detected in the simulated nursery planting at the time of planting of the diseased focal plant in plots with either the grapefruit or Valencia orange. However, xanthomonads were detected throughout the grapefruit plots by 50 days after introdution of the focal plant. The presence of the epiphytic xanthomonads was erratic in the Valencia plots throughout the test. The bacterial-spot pathogen was never found on all of the orange plants.

In one naturally infested nursery in central Florida (34), bacterial spot occurred on seedlings of Swingle citrumelo planted near a source of inoculum that consisted of a row of 11-year-old trees of the Flying Dragon cultivar (*Poncirus trifoliata*). The distribution of bacterial spot was aggregated among plants in the row of Flying Dragon trees and in the rows of Swingle citrumelo seedlings. Spread of disease was in a northeastern direction as determined by gradient analyses; this was consistent with patterns of wind-blown rain in the area prior to the spread of the disease.

In four other nurseries (32), three-dimensional response surfaces and isopathic maps were used to visualize disease aggregations and thereby identify probable disease foci. Disease gradients were then used to determine the direction of disease spread.

In a nursery at Frostproof, a few old lesions were seen near the presumed focus of infection. Disease gradients away from the focus were flattest in a northerly direction, which corresponded to the direction of rows. The trees had been hedged and the pathogen presumably was spread down the row during the hedging operation. Lesions were associated with the cut and tattered leaves that had resulted from the hedging operation. In a nursery at Ocoee, the disease probably originated from seedlings purchased as the final of three shipments. The disease was widely scattered thoughout the third planting but not in the other two plantings. Within-row aggregations in the nursery probably resulted from down-the-row spread of the pathogen during routine nursery practices of budding, wrapping, unwrapping, and staking. In a nursery at Lake Wales, the incidence of disease was highly aggregated within rows. A few diseased seedlings were planted and down-the-row spread of the pathogen presumably occurred with subsequent nursery practices. There was very little cross-row spread of the pathogen. Cross-row aggregation occurred in a nursery at Venice. The flattest gradients were both cross-row and down-row from a probable focal point in the western edge of the nursery to the northeast and southeast. Spread was more intense toward the southeast. In this nursery, spread seemed to be caused by weather-related factors rather than by mechanical means.

Observations of symptoms, severity, and spread of bacterial spot in infested nurseries led to the hypothesis that strains from these kinds of lesions varied in aggressiveness. Nearly 500 strains obtained from 25 nurseries were rated for aggressiveness to three citrus types by a detached leaf technique (36, 49). Three groups were distinguished based on the rate of expansion of lesions. Representatives of the three groups caused mild, intermediate, and severe disease in field tests, respectively, and ratings for the three groups in tests on detached leaves consistently correlated with ratings made in field tests. Strains of the three groups differed in causing slow, intermediate, and rapid (in a relative sense) lesion expansion and in low, intermediate, and high populations in inoculated leaves. The most aggressive type occurred in only 4 of 25 nurseries. Only the least aggressive types were isolated from 13 nurseries. From analyses of the spread of the disease in four nurseries and the range in aggressiveness of the strains isolated from them, Graham & Gottwald (36) concluded that only the most aggressive strains were disseminated by wind-blown rain, and the types intermediate and weak in aggressiveness were primarily disseminated by mechanical means. We have found no reports of differences in aggressiveness of strains from canker A.

Isozyme Analyses

The existence of great variations of the different xanthomonads from citrus based on virulence and aggressiveness were also found by Kubicek et al (48) based on isozyme analyses of different strains. Fourteen putative isozyme loci were determined with starch gel electrophoresis of lysates of 116 strains, 80 of which were from bacterial-spot lesions collected in Florida. Four enzymes were monomorphic with all strains. Strains from canker A, canker B, and bacteriosis lesions had little isozymic polymorphism, but those from bacterial spot were extensively polymorphic. Based on calculated coefficients of similarity, the groups of strains were different from each other by isozymic analyses.

Serological Studies

As part of an effort to increase the efficiency of diagnosis of the bacterial-spot pathogen, Brlansky et al (5) developed polyclonal antisera to one strain each from canker A, canker B, and canker C and four strains from bacterial spot. The immunoglobulin G component was isolated from all antisera with the aid of a protein A column. The antisera were then conjugated to tetra-methyl-rhodamine-isothiocyanate. Reactions were determined by fluorescence of treated bacteria. The antisera to the strain from canker A lesions reacted to all strains from canker A lesions (18 strains), but not to any of the strains from bacterial-spot lesions (25 strains), nor to any strain of 14 other pathovars of *Xanthomonas* (31 strains). An antiserum raised to a strain from a bacterial-spot lesion reacted with 14 of 25 strains from bacterial-spot lesions, but not with any of the strains from canker A lesions. It also reacted with two of four strains of *X. c.* pv. *alfalfae,* one of two strains of *X. c.* pv. *dieffenbachiae,* and one of four strains of *X. c.* pv. *manihotis.*

A monoclonal antibody preparation was developed to a strain from a bacterial-spot lesion by Permar & Gottwald (55). The antibody did not react to any of 15 strains of the four xanthomonads pathogenic to citrus, nor to any of five strains of other pathovars of *X. campestris.* It reacted with 14 of 30 strains from bacterial-spot lesions.

A monoclonal antibody was developed by Alvarez et al (2) that reacted with all canker A strains and some bacterial-spot strains. The antibody did not react with strains from lesions of cankers B, C, or D, nor did it react with 130 strains of other pathovars of *X. campestris,* except for a weak reaction with *X. c.* pv. *manihotis,* nor with 89 strains of other genera of bacteria. The bacterial-spot strains did not react with seven other monoclonal antibodies made to various strains from cankers A, B, C, or D. A second antibody reacted only with some of the canker A strains, namely those sensitive to bacteriophage CP1. The strains from canker B lesions seemed to be antigenically heterogeneous since no antibody reacted with all of the strains. However, one antibody was found that provided evidence for a close relationship among strains from cankers B, C, and D.A monoclonal antibody preparation that was specific for canker D strains did not react with any other xanthomonads from citrus.

The bacterial-spot strains were antigenically heterogenous in the studies of Alvarez et al (2). When a panel of eight preparations of monoclonal antibodies were used to characterize 225 bacterial-spot strains, eight distinct groups formed. The most aggressive strains reacted with one antibody that was generated against an aggressive strain of the bacterial spot pathogen. Both the moderately and weakly aggressive strains reacted to another antibody generated to a xanthomonad pathogenic to ti (*Cordyline terminalis*). The moderately and weakly aggressive bacterial-spot strains formed six patterns with a panel of three other antibodies.

The correlation among aggressiveness of the bacterial-spot strains, reactions to monoclonal antibodies, and RFLPs was assessed more fully in a study by Gottwald et al (30). One hundred and ninety-four strains from the four nurseries in central Florida, in which the spread of the bacteria had been studied (discussed above), were rated for aggressiveness by a detached leaf-assay. The same 194 strains were differentiated with a set of eight monoclonal antibodies generated in the study by Alvarez et al (2). A subset of 27 strains were used to determine RFLP profiles. Grouping of strains was accomplished with a cluster analysis of similarity coefficients. There was high goodness-of-fit among comparisons of similarity coefficient matrices for aggressiveness ratings, monoclonal antibody patterns, and RFLP assays. Both monoclonal antibodies and DNA probes distinguished between strains in the more aggressive group and strains from the two less aggressive groups. Interestingly, variations occurred among strains isolated from the same nursery, as well as among strains isolated from the same focus of infestation in a single nursery.

The variation among bacterial-spot strains both in pathogenic aggressiveness to citrus and in genetic characteristics led Graham et al (39) to speculate that some lesions of citrus from which xanthomonads were isolated were actually caused by bacteria from other hosts. They inoculated pathogenic strains from noncitrus hosts onto wounded, detached leaves of Swingle citrumelo and Duncan grapefruit; of 56 strains tested, 13 produced a weakly aggressive reaction in the citrus leaves when inocula were relatively high. Those producing a reaction included strains of *X. c.* pv. *alfalfae, X. c.* pv. *fici, X. c.* pv. *maculifoliigardeniae,* and three strains from an undescribed xanthomonad from plants of *Strelitzia regina.* These noncitrus xanthomonads in leaves of citrus reached populations as high as did the weakly aggressive strains from bacterial-spot lesions. Similarity coefficients for each noncitrus xanthomonad matched with each aggressiveness group of strains from bacterial-spot lesions were calculated for RFLPs or fatty acid profiles. Most of the noncitrus strains could not be separated from the strains from bacterial spot by either analysis. This study raised the possibility that the weakly aggressive strains collected from citrus nurseries may be a mixture of xanthomonads from other plants.

DNA Analyses of Strains

The characteristics of variation in aggressiveness among populations of the bacterial-spot pathogen and the uniformity of aggressiveness among populations of the canker A pathogen matchs up with the genetic variation and uniformity of the two kinds of pathogens, respectively. Hartung & Civerolo (42) determined genomic fingerprints of strains from bacterial-spot and canker A lesions by endonuclease digestion of genomic DNA with *Eco*RI and separation of the fragments by electrophoresis through polyacrylamide gels. The fragments were visualized by stains of ethidium bromide. The shortest fragments formed many discrete bands, and this portion of the genomic DNA could be compared among strains based on the congruousness of the bands. The bands from nine strains obtained from canker A lesions were indistinguishable by this technique. Likewise, the bands from four strains from cancrosis B lesions were indistinguishable, but the bands from strains from canker A and cancrosis B were different. The fingerprint of a bacteriosis strain was indistinguishable from that of the cancrosis B strains.However, bacterial-spot strains exhibited a wide variety of fingerprints. Variation occurred among strains within the same nursery as well as from different nurseries.

Gabriel et al (26) obtained similar findings in RFLP analyses of genomic DNA of strains of xanthomonads from citrus. Four probes ranging from 30 to 37 kb in size were used and two enzyme digests were probed. The data were combined to give similarity coefficients between strains. The analysis revealed three distinct groups made up, respectively, of strains from canker A; strains from cancrosis B, Mexican lime cancrosis, and bacteriosis; and from bacterial-spot lesions. A high degree of similarity occurred among strains within the first two groups, but the strains of each group had low similarity to the strains of the other group. The strains of the third group also had low similarity to the strains of the other two groups, but also had a high degree of variation in similarity coefficients within the group. Two groups with high similarity existed among the bacterial-spot strains, and these were designated as E-1 and E-2. In addition to these two groups, five unique strains existed. Some of the E strains exhibited moderate similarity to some strains of *X. c.* pv. *alfalfae*. In pathogenicity tests, all strains of the E-2 group were "moderately pathogenic" on alfalfa, which distinguished them from the E-1 group of strains. Some strains of *X. c.* pv. *alfalfae* were "somewhat" pathogenic on citrus.

Hartung & Civerolo (43) also reported on RFLPs among the xanthomonads affecting citrus. The number of strains that were common between their studies and those of Gabriel et al (27) could not be determined because of different systems of strain designations, although some strains were used in both studies. Hartung & Civerolo used seven probes and two enzyme digests. They calculated correlation coefficients differently than Gabriel et al (27), but designated similarity groups based on statistical analyses. In spite of differences in the two studies, the conclusions were similar. The group of strains from canker A lesions had similar RFLPs, even though they had been obtained from far-ranging locations. The cancrosis B strains also had similar RFLPs but were different from strains of canker A. Single strains from canker C and canker D lesions were placed in the group with canker B strains. The bacterial-spot strains had only limited relationships to the strains from canker A or cankers B, C, and D lesions. The bacterial-spot strains had considerable polymorphisms, which was in contrast to the limited polymorphism among strains from canker A and canker B, C, and D. Three possible subgroups of bacterial-spot strains were identified. Although there were considerably more polymorphisms among bacterial-spot strains, all of these strains were considered as related because the overall similarity coefficent of RFLPs was relatively high. The bacterial-spot strains were similar to strains of *X. c.* pv. *alfalfae* based on similarity coefficients of their RFLPs, but were not similar to strains of *X. c.* pv. *phaseoli, X. c.* pv. *dieffenbachiae,* or *X. c.* pv. *manihotis*.

Analyses of DNA-DNA homology among strains of the xanthomonads from citrus were performed by Egel (23). The DNA of a canker A strain was

30% homologous to DNA of the type strain of *X. campestris* and was 63% and 62% homologous to DNA from each of two canker B strains. The DNA of an aggressive bacterial-spot strain was 34% homologous to DNA of the type strain of *X. campestris* and 59% and 57% homologous, respectively, to the DNA of the two canker B strains. The DNA of the bacterial-spot strain was 56% homologous to DNA of the canker A strain. Thus, if the strains used in the study are representative of its group, the three xanthomonads from citrus should represent three distinct taxons.

Egel (23) also investigated the variation in DNA homology among strains of the bacterial-spot pathogen. The homologies of DNA from an aggressive bacterial-spot strain ranged from 84% to 91% to DNA extracted from three other strains of the aggressive group, from 80% to 102% with DNA extracted from six moderately aggressive strains; and from 85 to 100% to DNA extracted from five weakly aggressive strains. The homologies of DNA from a moderately aggressive strain ranged from 82% to 96% to DNA from five other strains of the moderately aggressive group: and from 76% to 92% with DNA extracted from five weakly aggressive strains. The homologies of DNA from a weakly aggressive strain ranged from 76% to 92% with DNA from four other weakly aggressive strains. Thus, in 31 determinations of homology of DNA extracted from representatives of the three aggressiveness groups to DNA of strains that represented a cross section of the variability among bacterial-spot strains, the values ranged from 76% to 102%. It would appear that the bacterial-spot strains form a good taxon based on the limits suggested by Wayne et al (70).

Strains of *X. c.* pv. *alfalfae, X. c.* pv. *maculifoliigardeniae,* X. c. pv. *fici,* and an unidentified xanthomonad from *Strelitzia regina* apparently belong to the same DNA homology group as the bacterial-spot strains. The former strains caused lesions on citrus similar to those caused by the weakly aggressive strains (39), and DNA from the strains of the three aggressiveness groups from bacterial-spot lesions had homology to DNA extracted from the noncitrus strains that ranged from 68% to 91%. On the other hand, homologies of DNA from an aggressive strain ranged from 25% to 71% to DNA extracted from *X. c.* pv. *vesicatoria, X. c.* pv. *phaseoli,* and *X. c.* pv. *malvacearum.* The latter strains did not cause lesions in citrus leaves (39) and are probably in other DNA homology groups. The canker A strain had high homology (90%) with *X. c.* pv. *malvacearum.*

Classification of Xanthomonads from Citrus

Obviously, the classification of the xanthomonads from citrus has become very controversial. Gabriel et al (27) proposed to reinstate the canker A strains to species rank and named them *X. citri.* This was done essentially on the basis of clustering of similarity coefficients generated by RFLP analyses. For justification of their proposal, the canker A strains were thought to be distinct in RFLP patterns from the type culture of *X. campestris* and a few other pathovars of *X. campestris.* The strains from cankers B, C, and D also formed a distinct group, and they proposed a new pathovar of *X. campestris,* pv. *aurantifolii.*They also proposed to relegate the bacterial-spot strains to pathovar status within *X. campestris* and provided the name pv. *citrumelo.*

Young et al (73) severely criticized the taxonomic revision proposed by Gabriel et al (27) on technological and philosophical bases. Their major criticism was the use of RFLP data only for the revision of species. Young et al (73) pointed out that historically a range of methods has been used to distinguish one taxon from another. Wayne et al (70) emphasized that species should not be designated without the physiological characterization of it and without determining DNA-DNA homologies.Vauterin et al (69) also criticized the reclassification proposed by Gabriel et al (27) on philosophical grounds. They stated: "The similarities of RFLP patterns refer only to congruence of a restricted fragment of the entire genome. This could lead to the development of an RFLP system, which probably would be as unjustified as a system as one based on pathogenicity. If RFLP data are to be used in taxonomy, they should be complemented with DNA-DNA homology studies."

Young et al (73) also were critical of the designations by Gabriel et al (27) of strains from lesions of cankers B, C, and D and bacterial spot as pathovars of *X. campestris.* They emphasized that the proposals were defective in several criteria in terms of the Standards for Naming Pathovars (20).

Current Perspectives

Before 1984 several forms of citrus canker were described based on symptoms of disease, pathogenicity tests, and a few biochemical properties of the causal bacteria. The bacteria caused similar symptoms on citrus plants, but they had overlapping host ranges. The causal bacteria were all designated as members of *X. campestris,* based on the physiological tests, and they were classified as strains of *X. c.* pv. *citri,* because of pathogenicity to citrus plants. Since all xanthomonad diseases of citrus were called citrus canker and since the causal agent was named *X. c.* pv. *citri,* all of these diseases were subject to state and federal quarantine and eradication protocols in the United States.

The occurrence of a new xanthomonad disease of citrus in Florida triggered quarantine and eradication efforts. During the attempts at eradication, research was focused on characterizations of the new xanthomonad disease of citrus based on comparisons with other xanthomonads causing diseases of citrus and other pathovars of *X. campestris.* New methods generated by the rapid advancement in techniques associated with recombinant DNA research have allowed the characterization of the genetic backgrounds of the strains.

It is now clear that at least three distinct groups of strains of *Xanthomonas* are involved in the diseases of citrus. Strains from lesions typical of canker A form one group, strains from lesions typical of canker B form another group, and strains from bacterial-spot lesions form a third group. Conclusions about canker C and canker D strains are only tentative, because research has included only one canker C strain and usually one or two canker D strains, but those strains seem to be related to canker B strains.

The significance of the variation within each group of strains is still not clear. Based on pathogenic and genetic data, little variation exists among canker A strains, but much variation exists among the bacterial-spot strains. Intermediate levels of variation occur within the strains from lesions of cankers B, C, and D. Speculation has occurred in articles that the great variation among bacterial-spot strains means (*a*) that the strains have wide host ranges (26), (*b*) that they are endemic to Florida (43), or (*c*) that bacterial-spot strains are a mixture of other pathovars that were incidentally disseminated to citrus and caused minor lesions (39). A fourth possibility could be that the bacterial-spot strains are inherently variable and variability is a characteristic of the group.

Possibly, the aggressive strains should be considered as the pathogen responsible for bacterial spot. This group of strains is rather homogeneous by RFLP analyses and antigenic determinants. This group was also the prevalent group in the original outbreak in a nursery in Florida (59), and except for plants taken directly from that nursery, these strains were found only in two other nurseries. The aggressive strains associated with bacterial spot have not be found since 1987 and may have been eliminated by the aggressive citrus canker eradication program that was in effect.

The roles that the moderately and weakly aggressive strains play in the bacterial spot syndrome is not clear and is subject to many interpretations. One possibility is that the moderately aggressive strains are pathogens of some other citrus relative. Two xanthomonads have been isolated from citrus relatives. One was named, *X. c.* pv. *bilvae* (50), and the other was listed as a strain of *X. campestris* isolated from *Feronia elephantatum* (listed in the American Type Culture Collection). During the eradication project, a third xanthomonad was found in Florida on the citrus relative, *Clausena lansium*. These three bacteria share many pathogenic characteristics with the moderately aggressive strains (R. E. Stall, E. L. Civerolo, unpublished data). However, no in-depth comparisons have been made among these groups of bacteria.

Lastly, the weakly aggressive group of strains has been found most often in the nurseries (36). This group of strains, by and large, accounted for the wide degree of variation among the bacterial-spot strains. It is possible that this group of strains is representative of a weakly pathogenic xanthomonad that occurs on many plants. If this bacterium was isolated from many hosts, it could be among culture collections of many pathovars. Such strains in collections could account for the weakly pathogenic reaction on citrus obtained with some cultures of other pathovars. The weakly aggressive strains may have been found with the intense inspection of nurseries and isolations from any lesion that resembled bacterial spot. Future work should include inoculations of a wide range of plants with the weakly aggressive strains from citrus and other hosts.

One argument against the above speculations may be that all of the strains from the bacterial spot lesions formed a distinct DNA homology group. It would be a remarkable coincidence if three distinct groups of strains were to fall into the same DNA homology group, and canker A and canker B strains would fall into two other DNA homology groups. However, this may not be outside the realm of possibility, because a large group of pathovars of *X. campestris* does fall into one DNA homology group as outlined by Hildebrand & Palleroni (45).

The classification of xanthomonads from citrus is still uncertain. Vauterin et al (69) discussed problems in classification of all xanthomonads. Clearly, the strains of *X. campestris* should be reworked and some strains will form new species (69). However, it may not be reasonable to classify individual groups until some recognized pattern is used for the classification of all xanthomonads. Until that is done, The International Society for Plant Pathology Subcommittee on the Taxonomy of Plant Pathogenic Bacteria believes that the strains from citrus should be included under *X. c.* pv. *citri* with subgroups of A, B, C, D, and E as an interim compromise. However, the usage of *X. c.* pv. *citrumelo* to indicate the bacterial-spot strains may continue even though the characterization of the group has not been fully accomplished.

Eradication of the disease known as bacterial spot has ceased in Florida. The strains found in the nurseries since 1987 have been of the moderately to weakly aggressive types, and these are recognized as no threat to citrus production in Florida. The aggressive strains were not found naturally on mature commercial citrus trees in Florida and likewise may not be a significant threat to citrus production. Canker A strains still exist in Florida as of 1991, and eradication efforts for canker A continue.

Literature Cited

1. Anonymous. 1989. Florida citrus canker action plan. Fla. Dept. Agric. Consum. Serv., Gainesville, FL.139 pp.
2. Alvarez, A. M., Benedict, A. A., Mizumoto, C. Y., Pollard, L. W., Civerolo, E. L. 1991. Analysis of *Xanthomonas campestris* pv. *citri* and *X. campestris* pv. *citrumelo* with monoclonal antibodies. *Phytopathology* 81:In press
3. Berger, E. W., Stevens, H. E., Silver, F. 1914. Citrus canker II. *Fla. Agric. Exp. Sta. Bull.* 124. 53 pp.
4. Bittancourt, A. A. 1957. O Cancro citrico. *O Biologico* 23:101–11
5. Brlansky, R. H., Lee, R. F., Civerolo, E. L. 1990. Detection of *Xanthomonas*

campestris pv. *citrumelo* and *X. citri* from citrus using membrane entrapment immunofluor escence. *Plant Dis.* 74: 863–68

6. Brown, G. E., Schubert, T. S. 1987. Use of *Xanthomonas campestris* pv. *vesicatoria* to evaluate surface disinfectants for canker quarantine treatment of citrus fruit. *Plant Dis.* 71:319–23
7. Canteros de Echenique, B. I., Zagory, D., Stall, R. E. 1985. A medium for the cultivation of the B-strain of *Xanthomonas campestris* pv. *citri*, cause of cancrosis B in Argentina and Uruguay. *Plant Dis.* 69:122–23
8. Carrera, C. 1933. Informe preliminar sobre una enfermedad nueva comprobada en los citrus de Bella Vista (Corrientes). *Bol. Mens. Min. Agric. Nac. Buenos Aires* 34:275–80
9. Civerolo, E. L. 1991. CABI-DC-EPPO Data sheet on International organisms: *Xanthomonas campestris* pv. *citri*. CAB-International. Wallingford, England. In Press
10. Civerolo, E. L. 1981. Citrus bacterial canker disease: An overview. *Proc. Intern. Soc. Citric. 1981.* 1:390–94
11. Civerolo, E. L. 1984. Bacterial canker disease of citrus. *J. Rio Grande Valley Hortic. Soc.* 37:127–46
12. Civerolo, E. L., Fan. F. 1982. *Xanthomonas campestris* pv. *citri* detection and identification by enzyme-linked immunosorbent assay. *Plant Dis.* 66:231–36
13. Crosse, J. E. 1966. Epidemiological relations of the pseudomonad pathogens of deciduous fruit trees. *Annu. Rev. Phytopathol.* 4:291–310
14. Curry, D. W. 1989. Eradication of citrus canker from Thursday Island. *Queensland Agric. J.* 115:78–79
15. Danos, E., Berger, R. D., Stall, R. E. 1984. Temporal and spatial spread of citrus canker within groves. *Phytopathology* 74:904–8
16. Dienelt, M. M., Lawson, R. H. 1989. Histopathology of *Xanthomonas campestris* pv. *citri* from Florida and Mexico in wound-inoculated detached leaves of *Citrus aurantifolia:* transmission electron microscopy. *Phytopathology* 79: 336–48
17. Doidge, E. M. 1929. Further studies on citrus canker. *S. Afr. Agric. Dep. B.* 51:1–31
18. Dopson, R. N., Jr. 1964. The eradication of citrus canker. *Plant Dis. Rptr.* 48:30–31
19. DuCharme, E. P. 1950. La causa de la cancrosis del limon. *Inf. Invest. Agric. (IDIA)* 3:27–28
20. Dye, D. W., Bradbury, J. F, Goto, M., Hayward, A. C., Lelliot, R. A., Schroth, M. N. 1980. International standards for naming pathovars of phytopathogenic bacteria and a list of pathovar names and pathotype strains. *Rev. Plant Pathol.* 53:153–68
21. Dye, D. W. 1969. Eradicating citrus canker from New Zealand. *NZ J. Agric.* 118:20–21
22. Egel, D. S., Graham, J. H., Riley, T. D. 1991. Population dynamics of strains of *Xanthomonas campestris* differing in aggressiveness on Swingle citrumelo and grapefruit. *Phytopathology* 81: In press
23. Egel, D. S. 1991. *Pathogenic and genomic characterization of strains of Xanthomonas campestris* causing diseases of citrus. PhD thesis. Univ. Fla., Gainesville. 132 pp.
24. Fawcett, H. S. 1936. *Citrus Diseases and Their Control.* New York/London: McGraw-Hill. 656 pp.
25. Fawcett, H. S., Jenkins, A. E. 1933. Records of citrus canker from herbarium specimens of genus *Citrus* in England and the United States. *Phytopathology* 23:820–24
26. Gabriel, D. W., Hunter, G. E., Kingsley, M. T., Miller, J. W., Lazo, G. R. 1988. Clonal population structure of *Xanthomonas campestris* and genetic diversity among citrus canker strains. *Mol. Plant Micr. Interact.* 1:59–65
27. Gabriel, D. W., Kingsley, M. T., Hunter, J. E., Gottwald, T. 1989. Reinstatement of *Xanthomonas citri* (ex Hasse) and *X. phaseoli* (ex Smith) to species and reclassification of all *X. campestris* pv. *citri* strains. *Int. J. Syst. Bacteriol.* 39:14–22
28. Garnsey, S. M., Ducharme, E. P., Lightfield, J. W., Seymour, C. P., Griffiths, J. T. 1978. Citrus canker: Preventive action to protect the US citrus industry. *Citrus Ind.* 60:5–13
29. Goto, M., Toyoshima, A., Messina, M. A. 1980. A comparative study of the strains of *Xanthomonas campestris* pv. *citri* isolated from citrus canker in Japan and cancrosis B in Argentina. *Ann. Phytopathol. Soc. Jpn.* 46:329–38
30. Gottwald, T. R., Alvarez, A. M., Hartung, J. S., Benedict, A. A. 1991. Diversity of *Xanthomonas campestris* pv. *citrumelo* strains associated with epidemics of citrus bacterial spot in Florida citrus nurseries: Correlation of detached leaf, monoclonal antibody, and

restriction fragment length polymorphism assays. *Phytopathology* 81: In press
31. Gottwald, T. R., Civerolo, E. L., Garnsey, S. M., Brlansky, R. H., Graham, J. H., Gabriel, D. W. 1988. Dynamics and spatial distribution of *Xanthomonas campestris* pv. *citri* group E strains in simulated nursery and new grove situations. *Plant Dis*. 72:781–87
32. Gottwald, T. R., Graham, J. H. 1990. Spatial pattern analysis of epidemics of citrus bacterial spot in Florida citrus nurseries. *Phytopathology* 80:181–90
33. Gottwald, T. R., McGuire, R. G., Garran, S. 1988. Asiatic citrus canker: Spatial and temporal spread in simulated new planting situations in Argentina. *Phytopathology* 78:739–45
34. Gottwald, T. R., Miller, C., Brlansky, R. H., Gabriel, D. W., Civerolo, E. L. 1989. Analysis of the spatial distribution of citrus bacterial spot in a Florida citrus nursery. *Plant Dis*. 73:297–303
35. Gotuzzo, E. A., Rossi, L. A. 1967. Cancrosis de los citricos. *Rev. Invest. Agropecu. Ser*. 5, 4:69–81
36. Graham, J. H., Gottwald, T. R. 1990. Variation in aggressiveness of *Xanthomonas campestris* pv. *citrumelo* associated with citrus bacterial spot in Florida citrus nurseries. *Phytopathology* 80: 190–96
37. Graham, J. H., Gottwald, T. R., Civerolo, E. L., McGuire, R. G. 1989. Population dynamics and survival of *Xanthomonas campestris* in soil in citrus nurseries in Maryland and Argentina. *Plant Dis*. 73:423–27
38. Graham, J. H., Gottwald, T. R., Fardelmann, D. 1990. Cultivar-specific interactions for strains of *Xanthomonas campestris* from Florida that cause citrus canker and citrus bacterial spot. *Plant Dis*. 74:753–56
39. Graham, J. H., Hartung, J. S., Stall, R. E., Chase, A. R. 1990. Pathological, restriction-fragment length polymorphism, and fatty acid profile relationships between *Xanthomonas campestris* from citrus and noncitrus hosts. *Phytopathology* 80:829–36
40. Graham, J. H., McGuire, R. G., Miller, J. W. 1987. Survival of *Xanthomonas campestris* pv. *citri* in citrus plant debris and soil in Florida and Argentina. *Plant Dis*. 71:1094–98
41. Hannon, C. I. 1987. How severe is the citrus canker problem in Florida.? *Plant Dis*. 71:956
42. Hartung, J. S., Civerolo, E. L. 1987. Genomic fingerprints of *Xanthomonas campestris* pv. *citri* strains from Asia, South America, and Florida. *Phytopathology* 77:282–85
43. Hartung, J. S, Civerolo, E. L. 1989. Restriction fragment length polymorphisms distinguish *Xanthomonas campestris* strains isolated from Florida citrus nurseries from *X. c*. pv. *citri*. *Phytopathology* 79:793–99
44. Hasse, C. H. 1915. *Pseudomonas citri*, the cause of citrus canker. *J. Agric. Res*. 4:97–100
45. Hildebrand, D. C., Palleroni, N. J., Schroth, M. N. 1990. Deoxyribonucleic acid relatedness of 24 xanthomonad strains representing 23 *Xanthomonas campestris* pathovars and *Xanthomonas fragariae*. *J. Appl. Bacteriol*. 68:263–69
46. Hill, G. F. 1918. History of citrus canker in the Northern Territory. *N. Territ. Aust. Bull*. 18. 11 pp.
47. Koisumi, M. 1979. Ultrastructure changes in susceptible and resistant plants of citrus following artificial inoculation with *Xanthomonas citri* (Hasse) Dowson. *Ann. Phytopathol. Soc. Jpn*. 45:635–44
48. Kubicek, Q. B., Civerolo, E. L., Bonde, M. R., Hartung, J. S., Peterson, G. L. 1989. Isozyme analysis of *Xanthomonas campestris* pv. *citri*. *Phytopathology* 79:297–300
49. Lawson, R. H., Dienelt, M. M., Civerolo, E. L. 1989. Histopathology of *Xanthomonas campestris* pv. *citri* from Florida and Mexico in wound-inoculated detached leaves of *Citrus aurantifolia:* Light and scanning electron microscopy. *Phytopathology* 79:329–35
50. Leyns. F., DeCleene, M., Swings, J., DeLey, J. 1984. The host range of the genus *Xanthomonas*. *Bot. Rev*. 50:308–56
51. Loucks, K. W. 1934. Citrus canker and its eradication in Florida. Fla. State Dept. Agric. Consum. Serv. Gainesville, Fl. 111 pp.
52. Muraro, R. P. 1986. Observations of Argentina's citrus industry and citrus canker control program with estimations of additional costs to Florida citrus growers under a citrus canker control program. Food Res. Econ. Dep., Univ. Fla., Gainesville. Staff Paper 289. 46 pp.
53. McGuire, R. G. 1988. Evaluation of bactericidal chemicals for control of *Xanthomonas* on citrus. *Plant Dis*. 72: 1016–20
54. Namekata, T., Oliveira, A. R. de. 1972. Comparative serological studies between *Xanthomonas citri* and a bacterium causing canker on Mexican lime. pp.151–52. *Proc. Int. Conf. Plant Pathog. Bact.*,

3rd. Wageningen, The Netherlands. 365 pp.
55. Permar, T. A., Gottwald, T. R. 1989. Specific recognition of a *Xanthomonas campestris* Florida citrus nursery strain by a monoclonal antibody probe in a microfiltration enzyme immunoassay. *Phytopathology* 79:780–83
56. Rodriquez G, S., Garza-Lopez, J. G., Stapleton, J. J., Civerolo, E. L. 1985. Citrus bacteriosis in Mexico. *Plant Dis*. 69:808–10
57. Rossetti, V. 1977. Citrus canker in Latin America: A review. *Proc. Int. Soc. Citric*. 3:918–24
58. Rossetti, V., Feichtenberger, E., Silveria, M. L. 1982. Citrus canker (*Xanthomonas campestris* pv. *citri*): an analytical bibliography. Inst. Biol., Sao Paulo, Brazil. 230 pp.
59. Schoulties, C. L., Civerolo, E. L., Miller, J. W., Stall, R. E., Krass, C. J., et al. 1987. Citrus canker in Florida. *Plant Dis*. 71:388–95
60. Stall, R. E., Miller, J. W., Marco, G. M., Canteros de Echenique, B. I. 1981. Pathogenicity of three strains of the citrus canker organism on grapefruit. *Proc. Int. Conf. Plant Pathog. Bact., 5th*. CIAT, Cali, Colombia, pp. 335–40. 640 pp.
61. Stall, R. E., Miller, J. W., Marco, G. M, Canteros de Echenique, B. I. 1980. Population dynamics of *Xanthomonas citri* causing cancrosis of citrus in Argentina. *Proc. Fla. State Hortic. Soc*. 93:10–14
62. Stall, R. E., Miller, J. W., Marco, G. M., Canteros de Echenique, B. I. 1981. Timing of sprays to control cancrosis of grapefruit in Argentina. *Proc. Int. Soc. Citric*. 1:414–17
63. Stall, R. E. 1988. Canker. In *Compendium of Citrus Diseases,* ed. J. O. Whiteside, S. M. Garnsey, L. W. Timmer, pp. 8–9. St. Paul MN: APS. 80 pp.
64. Stall, R. E., Seymour, C. P. 1983. Canker, a threat to citrus in the Gulf-Coast states. *Plant Dis*. 67:581–85
65. Stapleton, J. J., 1986. Effects of postharvest chlorine and wax treatments on surface microflora of lime fruit in relation to citrus bacteriosis disease. *Plant Dis*. 70:1046–48
66. Stapleton, J. J., Garza-Lopez, J. G. 1988. Epidemiology of a citrus leaf-spot disease in Colima, Mexico. *Phytopathology* 78:440–43
67. Stapp, C. 1961. *Bacterial Plant Pathogens*. Oxford, England: Oxford Univ. Press. 292 pp.
68. Timmer, L. W., Gottwald, T. R., Zitko, S. E. 1991. Bacterial exudation from lesions of Asiatic citrus canker and citrus bacterial spot. *Plant Dis*. 75:192–95
69. Vauterin, L., Swings, J., Kersters, K., Gillis, M., Mew, T. W., et al. 1990. Towards an improved taxonomy of *Xanthomonas*. *Int. J. Syst. Bacteriol*. 40:312–16
70. Wayne, L. G., Brenner, D. L., Colwell, R. R., Grimont, P. A. D., Kandler, O., et al. 1987. Report of the Ad Hoc Committee on Reconciliation of Approaches to Bacterial Systematics. *Int. J. Syst. Bacteriol*. 37:463–64
71. Whiteside, J. O. 1985. How serious a threat is canker to Florida citrus production? *Citrus Ind*. 66:8, 10, 12–14, 16, 17
72. Whiteside, J. O. 1986. Citrus canker: some facts, speculations, and myths about this highly dramatized disease. *Citrus Veg. Mag*. 48:14, 55, 56, 64
73. Young, J. M., Bradbury, J. F., Gardan, L., Gvozdyak, R. I., Stead, D. E., et al. 1991. Comment on the reinstatement of *Xanthomonas citri* (ex Hasse) Gabriel et al 1989 and *X. phaseoli* (ex Smith) Gabriel et al 1989: Indication of the need from minimal standards for the genus *Xanthomonas*. *Int. J. Syst. Bacteriol*. 41:172–77

Annu. Rev. Phytopathol. 1991. 29:421–42

FUNGICIDE RESISTANCE: Practical Experience with Antiresistance Strategies and the Role of Integrated Use

T. Staub

Agricultural Division, CIBA-GEIGY Ltd., 4002 Basel, Switzerland

KEY WORDS: fungicide resistance, antiresistance measures, counter-measures, integrated disease control

---

INTRODUCTION

Early reviews on fungicide resistance and measures to overcome it generally referred to the newness of the resistance problem and to the consequent lack of experience to deal with it. I believe that those days are over and that there is now a considerable body of experience on assessing resistance risk and designing and implementing measures to counter it. Furthermore, for the plant protection community, the occurrence of resistant strains no longer triggers panic reactions or alarmist publications but rather rational appraisals based on monitoring data and, if necessary, changes in the recommendations for use of the particular fungicides. Attitudes toward fungicide resistance in the 1970s were rather naive whereas in the 1980s the plant protection community has tried to face the subject more pragmatically. A concerted effort has been made to meet the challenge of fungicide resistance and numerous workshops have been organized to keep members abreast with experience (12, 16). In industry analysis of the resistance risk and the design of measures to control it have become routine in fungicide development.

Several reviews have described attempts to meet the challenge of fungicide resistance with innovative use strategies (12, 13, 24, 34, 59, 62). Such strategies are largely addressed to the use of resistance-prone fungicides in mixture or in alternation with fungicides from other cross-resistance groups or

0066-4286/91/0901-0421$02:00

with multisite fungicides that are virtually free of resistance risk. These strategies are designed to prevent or retard the development of resistance and at the same time reduce the risk of crop losses should resistant pathogen strains appear.

How successful have these antiresistance strategies been? This review describes some of these strategies, the appearance of field resistance to the major groups of modern fungicides and the evolution of counter-measures (both fortuitously and by design) and how they have worked in practice. The emphasis is less on a comprehensive survey than on highlighting key examples where data are sufficient for conclusions to be drawn for future use strategies. For details on the biological properties and on the mode of action of the fungicide groups discussed here, the reader is referred to the treatise *Modern Selective Fungicides* (40), and for other recent reviews of fungicide resistance to the ACS proceeding of a 1988 symposium (22a).

GENERAL PRINCIPLES FOR ANTIRESISTANCE MEASURES

In any discussion of antiresistance measures two definitions of fungicide resistance must first be considered (15): *field resistance* is characterized by the emergence of resistant strains that can be causally linked to decreased performance of fungicides when used according to label recommendations; *laboratory resistance,* by contrast, can be demonstrated only in special laboratory situations. Typical examples are pathogen strains with decreased sensitivity in vitro. As a rule, such strains would not affect performance of the respective fungicides in the field because they are not sufficiently fit to compete and survive in the real world.

It is important to distinguish in reports on fungicide resistance between these two types of resistance. Antiresistance strategies deal exclusively with the prevention of field resistance, i.e. a breakdown in disease control in practice.

Before specific examples are discussed, it is useful to clarify some of the general principles underlying the design of antiresistance measures and the controversies surrounding them. For fungicides with a certain resistance risk, there are two general areas that affect the emergence of resistance in practice: (*a*) the overall crop management and (*b*) the use of fungicides on the crop. The latter should be integrated with caution into other disease control measures. Concrete measures from both areas are listed in Table 1.

Crop Management and Fungicide Resistance

Good crop production practices are essential for any sound antiresistance strategy. Thus, the introduction of known excessively susceptible cultivars

Table 1 Examples of measures in integrated crop protection and in fungicide use that decrease disease pressure and the selection for fungicide resistance

Crop management and fungicide resistance (ICM)		
use less susceptible cultivars	→	lower, disease incidence
avoid sites with high disease pressure	→	avoids high selection
lower N fertilization	→	lower disease incidence
practise sound sanitation	→	reduces primary inoculum
Fungicide use and fungicide resistance (IPM)		
use only when justified	→	avoids unnecessary selection
use protectively	→	hits small populations
use mixtures with residuals	→	reduces populations exposed to selection
alternate fungicides from different resistance groups	→	reduces selection time
do not use soil against foliar pathogens	→	avoids poorly controlled exposure time

would not be a suitable basis for stable crop production systems. Such susceptible cultivars require the farmers to increase the use of fungicides and consequently increase the risk of resistance for the fungicides used. There are several examples where fungicide resistance first appeared on the most susceptible cultivars (34, 58, 59). Also, in many crops moderating N-fertilization reduces disease susceptibility, which in turn clearly reduces fungicide use and the risk of resistance.

Many other examples can be cited where crop production parameters can influence disease pressure and thus the risk of resistance for fungicides (47). The integration of disease resistance and fungicide use can prolong not only the effectiveness of fungicides but also the useful life of resistant cultivars. Resistance genes, like single-site fungicides, have a history of losing their effectiveness through selection of the respective virulence genes in the pathogen populations.

Fungicide Use and Resistance

Even well-managed crops often need treatments with fungicides. Various treatment options are available with differing effects on the build-up of resistance. At stake are potential losses in crop yields when resistance emerges and the usefulness of fungicides with an inherent risk of resistance. Some of the options shown in Table 1 clearly reduce the risk of resistance by reducing selection intensity or selection time. It is less clear how the size of the treated population, overkill, partial kill, and escape influence the appearance of resistance. The degree of control should be determined primarily by the requirements of the crop and the damage threshold. Higher use rates are not to be recommended purely for resistance reasons. Lower use rates must be compensated for by mixture partners to reach the performance required. Partial kill and escape, which are favored by some mathematical models (29, 54), are risk elements with explosive diseases like late blight of potatoes. It would be a dangerous gamble for a farmer to let such diseases get too much of a head start.

ROLE OF THE FUNGICIDE RESISTANCE ACTION COMMITTEE

The fungicide resistance action committee (FRAC) plays an important role where theoretical considerations and designs have to be implemented to provide farmers with stable disease control. FRAC was founded in 1981 to coordinate efforts by producers of fungicides from the same cross-resistance group to introduce effective antiresistance measures. It is clearly of little use for one fungicide producer to design and implement cautious use-strategies (in the sense of reducing the risk of resistance) if producers of related fungicides with cross-resistance to the first do not exercise similar caution. Thus the general aims of FRAC are *a* to prolong the effective life of resistance-prone fungicides and *b* to avoid crop damage when resistance unpredictably emerges (14, 67; Table 2).

Most work is done in working groups centered around fungicide groups with cross-resistance (25, 67). They are mainly responsible for proposing and validating antiresistance strategies, coordinating sensitivity monitoring programs and ensuring that control strategies are implemented by the various companies and by extension officials. Working groups currently exist for the benzimidazoles, dicarboximides, phenylamides, and DMIs (C14-demethylation inhibitors of sterol biosynthesis). Some of these groups have adopted clear guidelines for resistance management that are being followed, thanks to good communication within industry and with extension people (62).

Table 2 The concept of fungicide resistance management within FRAC (62)

1. Definition of objectives
2. Design of procedures and approaches
   resistance risk assessment
   design and testing of resistance strategies
   determination of monitoring methods
3. Coordination between manufacturers of products with cross resistance
4. Implementation of use stategies and monitoring methods in practice
5. Duration over product lifetime: field monitoring

FRAC, in conjunction with the International Society for Plant Pathology (ISPP), has helped organize workshops on fungicide resistance worldwide, including in developing countries. It has also coordinated a slide set available through APS, with information on mode of action, resistance, and monitoring methods for the different fungicide groups, and produced a video in cooperation with IRAC (insecticide resistance action committee) entitled "The Paradox of Resistance," which presents the concept of resistance and of antiresistance strategies for nonspecialists.

CASES OF FUNGICIDE RESISTANCE AND THE IMPACT OF COUNTER-MEASURES

Let us now turn to some examples where the build-up of fungicide resistance was monitored and where the impact of different use-strategies can be documented. These examples have been chosen from actual case histories and the data are often incomplete. However, they do illustrate certain principles and provide lessons in how to cope with fungicide resistance.

RESISTANCE TO BENZIMIDAZOLES

In the early 1970s it became clear that benzimidazoles are highly prone to the development of resistance. This finding took the plant protection community by surprise because it was the first time that widespread resistance had appeared to a fungicide group. Since antiresistance strategies were not then in place, we can only discuss the circumstances in which resistance appeared more quickly than in others. Delp (13) reported that resistance in peanut leaf spot appeared quickly in the south-eastern US, but not in Texas. In Texas, however, mixtures containing dithiocarbamates were applied to control the added problem of peanut rust. Since rust is not important in the South-East, only benzimidazoles were used there.

Other spectacular examples were the emergence of benzimidazole resistance in *Venturia inaequalis* in Germany (33) and in *Cercospora beticola* in Greece (22). In both cases disease pressure was extremely high and the compounds were used alone. Loss of activity was complete after only two seasons of use. In other areas where disease pressure was lower or where farmers used mixtures with multisite fungicides, resistance emerged much more slowly or not at all (13).

Benzimidazole Resistance in Botrytis *in Swiss Vineyards, 1974–78*

The build-up of resistance to benzimidazole in Botrytis in Swiss vineyards (49) provides a classic illustration of how such resistance appears and persists.

In 1973, after two years of use, Schüepp & Küng observed a complete loss of Botrytis control in many vineyards and benzimidazole products were withdrawn. They compared the resistance in isolates collected in 1974 with a similar collection made in 1978 (Figure 1). One third of the earlier samples from untreated vineyards were found to be resistant. The proportion of resistant strains in intensively treated vineyards was about twice as high. High fitness of the resistant strains is indicated by the stability of overall resistance at about 60% in the absence of selection pressure from 1974 to 1978. Similar circumstances were observed in other northern areas of the European grape-growing region (19, 38). In Southern Europe, where Botrytis pressure is much lower, resistance appeared more slowly and benzimidazoles continued to be useful (23).

The question of antiresistance strategy did not seriously arise in the early 1970s because the plant protection community was taken by surprise and, in any case, no effective alternatives were available. The only possible course of action was to withdraw benzimidazoles from use. Subsequent reintroduction of benzimidazoles has not been possible because the stability of benzimidazole resistance in Botrytis populations as described for Swiss vineyards has also been confirmed in other regions (38, 52).

The following characteristics of benzimidazole resistance in *Botrytis* are common to many other target pathogens in this fungicide group: The high resistance factor and the obviously high fitness of resistant strains represent the most critical challenge in designing antiresistance strategies. Resistance

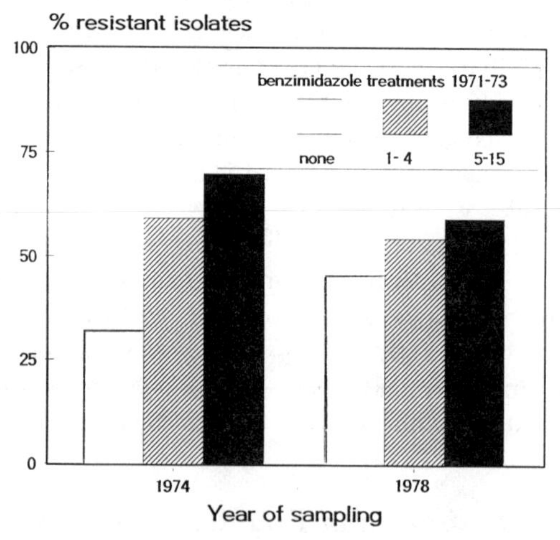

Figure 1 Benzimidazole resistance in *Botrytis cinerea* in Swiss vineyards in relation to previous use of benzimidazoles. In both years 150, 80, and 525 isolates were tested in the treatment categories 0, 1–4, and 5–15, respectively

quickly builds up to high levels in the populations, where it remains stable, and leads to immediate loss of control. It is particularly important in such situations to implement antiresistance strategies right from the start, not only after resistance has already been detected in the field (13).

The combination carbendazim and diethofencarb, which has been available since 1988, is of special interest because it contains two components with negatively correlated cross-resistance. Monitoring data showed an overall increase of resistance to both components between 1988 to 1989 from 4–22%, with an average of 1.1 applications per season. This mixture has to be used with special caution because for antiresistance purposes, it is a single product: Each fungal spore faces only one fungicide of this mixture; by definition it is resistant to the other component. This mixture should only be used alternately with dicarboximides, discussed below (43).

RESISTANCE TO DICARBOXIMIDES

Dicarboximides are particularly effective against Botrytis and they succeeded the benzimidazoles as the major weapon against grey mold of grapes. As had happened with benzimidazoles a few years earlier, resistance to dicarboximides first appeared in the northern part of the European grape-growing region (27). However, loss of activity was slower (1976–1982) and less complete than with benzimidazole resistance and the first reports of resistance detection were not correlated with lower activity in practice (36, 50). Again, a distinct north-south gradient could be observed with resistance appearing much faster in the northern areas where disease pressure is higher (23, 36).

Dicarboximide Resistance in Botrytis *in the Champagne Region, 1980–89*

The Champagne vineyards with their high Botrytis pressure and heavy use of dicarboximides were among the worst hit by dicarboximide resistance. Figure 2 shows how resistance to dicarboximides developed in *Botrytis* on grapes between 1980–89 and how this development was influenced by fungicide use (35, 43, 53, 64). Resistance rapidly appeared and reached high levels in intensively treated vineyards. The appearance of resistance in untreated vineyards lagged considerably behind, but it too had reached 59% by 1983 (41). In the early 1980s, the Champagne region was the first to show a consistent correlation between the presence of resistant strains and decreased performance (38); as a result, the use of dicarboximides was officially discontinued. By 1986, resistance had decreased to below 20% in untreated plots. This development clearly showed that in the absence of selection pressure di-

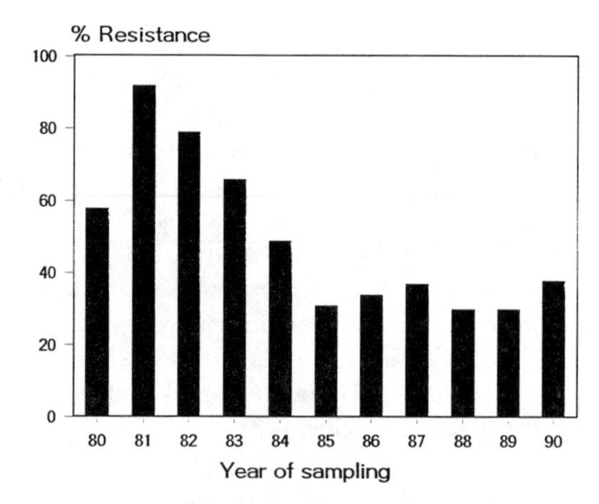

Figure 2 Development of dicarboximide resistance in *Botrytis cinerea* from 1980–1990 in the Champagne region of France.

carboximide-resistant strains of *Botrytis* are less fit than wild-type strains. Such a situation encouraged the design of control strategies that exploit this fitness deficit of resistant strains by limiting the number of applications per season. Subsequently, dicarboximides were used with caution in conjunction with multisite fungicides. In 1989, for instance, dicarboximide treatments averaged 1.3 for the Champagne vineyards (43), of which approximately half were in combination with thiram.

What effect did the use of dicarboximide mixtures have on resistance? Figure 3 shows data, also from the Champagne region, that illustrate the differing results of using dicarboximide alone or in combination with thiram

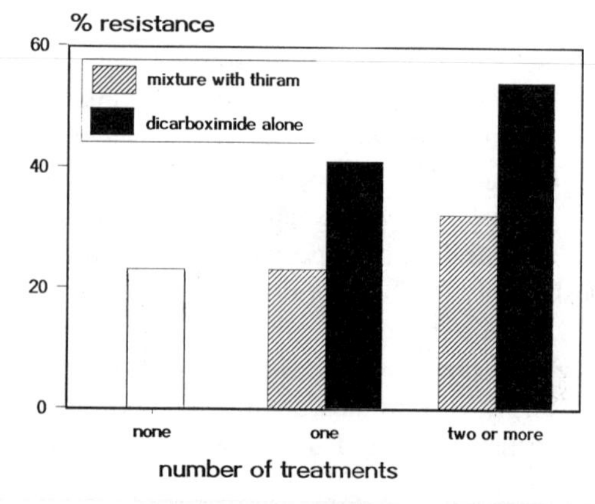

Figure 3 Influence of dicarboximide treatments on dicarboximide resistance in *Botrytis cinerea*.

on the build-up of resistance (41). Selection of resistant strains was much more pronounced with the single product. Even multiple treatments with the combined products resulted in less selection than one application with the single product. The effect of mixtures in preventing the build-up of resistance varied according to local conditions because there were instances where no such effect could be documented (65).

A 1987 survey conducted in Alsace on the sensitivity of *Botrytis* to dicarboximides and benzimidazoles illustrates the difference in the ease with which resistance can develop to these two fungicides (52). Benzimidazole resistance was widespread 10 years after use of this fungicide group had been stopped, whereas resistance to dicarboximides was confined to small pockets even after 10 years of use.

In 1989 and 1990 recommendations for control of*Botrytis* in the Champagne were as follows: the application of a mixture of carbendazin/diethofencarb, followed by a mixture of vinclozolin/thiram (500 + 3200 gr a.i./ha), and a treatment with a dicarboximide at 750 gr a.i./ha (Table 3). In addition, French officials advised such indirect control measures as the reduction of N-fertilization, removal of leaves around grapes, and careful insect and powdery mildew control (42). Downy mildew products with side effects against *Botrytis* were also recommended to lower the selection pressure for both the dicarboximides and the carbendazin/diethofencarb mixture.

RESISTANCE TO PHENYLAMIDES

Phenylamide fungicides are specifically active against the Peronosporales. The first occurrence of resistance was reported in Israel for *Pseudoperonospora cubensis* on plastic-house grown cucumbers (31). The phenylamides first developed resistance problems on a field crop in Europe in 1980, for control of *P. infestans* on potatoes, after having been used as single products, e.g. in Ireland, Holland, and Switzerland. When used in conjunction with

Table 3 French use recommendation for the control of Botrytis in grapes for 1990

Direct Measures	Indirect measures
Fungicide use against Botrytis	less nitrogen to avoid excessive vigor
application of downy mildew products with side effects against Botrytis	removal of leaves around grapes
Botrytis program	careful insect control
1x mixture carbendazim/dietho fencarb followed by	careful powdery mildew control
two treatments mixture dicarboximide/thiran (500–750 + 3200 gr ha)	

residual products, resistance was much slower to emerge and performance problems were rare (58, 59). Resistance in *Plasmopara viticola* appeared first in South Africa and southwestern France (57). In both cases nurseries or experimental vineyards, where disease pressure is usually very high, may have served as the initial foci for the development of resistance.

Phenylamide Resistance in P. infestans

In Ireland, metalaxyl was withdrawn from the market in 1981 and the level of resistance in the population of *P. infestans* receded (17; L. J. Dowley, personal communication). This decrease in resistance and good disease control in field trials led Irish officials in 1985 to ask for the reintroduction of phenylamide-containing products, after other products had proved unsuccessful in controlling heavy attacks of late blight. This time the phenylamides were used in conjunction with residual fungicides. Since then, two recommended treatments have achieved good results, although phenylamide resistance has been detected since 1987 in most fields. Results of monitoring show the prompt response in the population of Phytophthora to decreases and increases in the use of phenylamides; the data further show that resistance can be maintained at a low level with cautious use.

Metalaxyl sensitivity has been closely monitored in the Netherlands since 1980 (10). Use of the fungicide was suspended from 1981–84, after which time it was reintroduced at the request of Dutch officials to fight heavy outbreaks of late blight that could not be controlled with residual products. Since 1985, in critical situations a maximum of two applications of metalaxyl mixed with nearly full strength mancozeb and fentin acetate have been recommended, but use on seed potatoes is prohibited. Monitoring data show that resistance decreased in the absence of selection pressure and no resistant isolates were found in 1986 (Figure 4). Resistance reappeared in 1987 and since 1988 has been found on approximately 40% of all farms (11).

Resistance is consistently higher in the starch potato production area in the north east than in the rest of the Netherlands. Davidse attributes this difference to the use of metalaxyl on seed potatoes, which are grown on the same farms as production crops. Because of the high incidence of resistance on seed potatoes, recommendations were issued in 1989 against the use of metalaxyl mixtures in this region. Elsewhere in the country the mixture performs well when used as recommended.

In several other European potato-growing areas, a similar equilibrium could be reached between some selection of resistant strains during each growing season and the decrease of resistance in the population from one season to the next. FRAC recommendations call for 2–3 early season preventive treatments, using phenylamides mixed with residual partners and use of nonphenylamides thereafter (Table 4). In addition, several countries

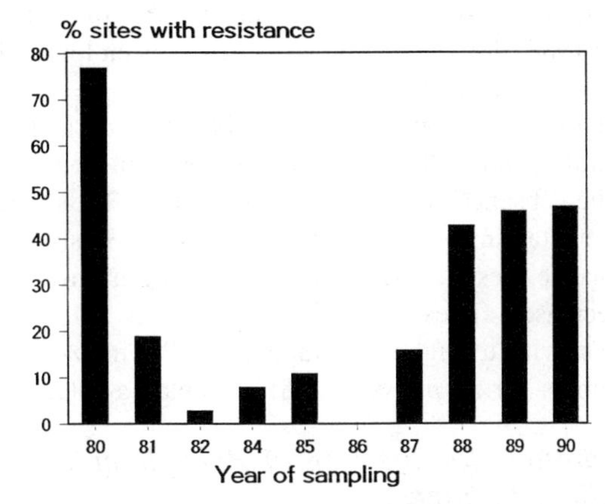

Figure 4 Phenylamide sensitivity situation 1980–90 in *Phytophthora infestans* in Dutch potato fields.

recommend that phenylamides not be used on seed potatoes to avoid seed-transmission of a selected population. In eastern Germany, phenylamides were used intensively on seed potatoes until 1988 when widespread resistance was observed (46). Since the 1989 adoption of the FRAC ban on phenylamide use on seed potatoes, resistance has decreased in some areas.

Phenylamide Resistance in P. viticola *in France*

Similar conclusions can be drawn from the data available on the development of phenylamide resistance in *P. viticola* in the French wine regions from 1983–1987 (44). Problems were first detected in the Charentes region, in western France, where most sites showed resistance in 1983/84. As a result, phenylamide use was reduced and resistance promptly decreased. In other regions, resistance remained stable with the moderate use of phenylamide compounds and good performances were recorded, except in 1987 in the Bourgogne where resistance flared up when disease pressure was high and growers may have relied too heavily on phenylamides. Other French regions,

Table 4 FRAC recommendations for phenylamides against foliar oomycetes

Use of all cultural methods available to reduce disease pressure
Use of only prepacked mixtures with nonphenylamide fungicides
Use of partner at 3/4 to full rate
Use of only 2–4 sprays per season
Restriction to 14-day intervals
(short interval for change to nonphenylamide)
No curative/eradicative use
No soil use against foliar pathogens

including the Champagne and Alsace, remained free of phenylamide-resistant strains. Since 1988, downy mildew pressure has been low and no comparable monitoring data are available.

The French experience with *P. viticola*, like the Irish and Dutch experience with *P. infestans*, shows that phenylamide-resistant populations tend to decrease in the absence of selection pressure. Thus, cautious use of phenylamide-containing products leads to a stable resistance situation and good performance, despite a detectable level of resistance. The basis for this obviously decreased fitness of resistant strains is still poorly understood. Increased sensitivity to high temperatures for *P. viticola* (45) or decreased survival in tubers for *P. infestans* (68) may play a role.

Recommendations for the Use of Phenylamides against Foliar Pathogens

After phenylamide resistance was first encountered, the respective FRAC working group went into action and proposed remedies for the control of foliar Oomycetes. From its inception, the main thrust of the strategy was the use of prepackaged mixtures with residual fungicides. Other proposals have evolved as experience has been gained. Some companies now offer triple and quadruple mixtures that contain a second systemic component. Table 4 summarizes the recommendations for 1990. In addition to the use of prepackaged mixtures, these recommendations include restricting the number of sprays per season and avoiding curative and soil use against foliar pathogens. Preventive use is important because in curative schedules residual mixture partners are ineffective against established infections. Also, the interval for the switch from mixtures to residuals should not exceed the interval for the residual partner in the mixture. Performance problems often occur if this interval is too long (3).

One example of the integration of genetic and chemical control is described by Crute (9) for lettuce downy mildew. Both the pathotypes of *Bremia lactucae* and the sensitivity to phenylamides are monitored to determine the choice of cultivars and chemical treatments.

RESISTANCE TO THE 2-AMINO PYRIMIDINES

Another early but interesting example of fungicide resistance is that to the 2-amino pyrimidines, dimethirimol and ethirimol, in powdery mildews of cucumber and barley, respectively (5, 7). Dimethirimol rapidly lost its effectiveness in Dutch greenhouses after being used exclusively and at high rates to combat continued high disease pressure. After it was withdrawn from use in 1970, sensitivity again increased to the point where, in 1977, it was reintroduced with some success.

Ethirimol, on the other hand, was used only once a season against barley powdery mildew and the small sensitivity shifts observed during its use reverted to normal sensitivity from one season to the next. Interestingly, the most and the least sensitive parts of the pathogen population were lost in this process. The use of ethirimol on winter barley was discontinued to counteract the shifts toward decreased sensitivity and to encourage the population to revert to intermediate sensitivity (5, 7). In the 1980s, ethirimol was used in combination with DMI fungicides as they began to lose their effectiveness against barley powdery mildew. Clearly, the fitness disadvantage of resistant strains, indicated by the reversion to higher sensitivity against the 2-amino pyrimidines from one season to the next, offered much better opportunities to manage resistance situations with this fungicide group than with the benzimidazole examples discussed above. These fitness disadvantages could be used much more effectively on barley where one treatment was sufficient than on greenhouse cucumbers where prolonged protection under high disease pressure was needed and could be achieved with dimethirimol.

RESISTANCE TO DMI FUNGICIDES

For some time it was thought that these compounds were not prone to the development of resistance because the resistant strains found in the laboratory tended to be less fit than wild-type strains (20, 21). However, intensive use in several crops lead to the appearance of less sensitive strains and in a few cases to decreased performance in the field. The first example was cucumber powdery mildew in the Mediterranean region (28) and in Dutch glasshouses (48) in the early 1980s. Around the same time, instances of reduced sensitivity and decreased performance in the field were reported for barley powdery mildew in the UK (8) and for wheat powdery mildew in Germany (51). However, experience in the field and laboratory studies identified significant differences in the genetic basis for and the phenotypic expression of resistance in the DMI fungicides (Table 5). While resistance to benzimidazoles and phenylamides appears to be caused by one genetic with dramatic effects on sensitivity, resistance to DMIs typically appears in many small genetic steps (26, 48, 66). There is only one report of a UV-induced major gene that confers a high degree of resistance to some of the DMIs (30). These steps cumulatively add up to sensitivity differences that can influence the field performance of DMIs. Loss of activity is usually not complete and in some cases decreased performance can be overcome by shortening the intervals between treatments or by higher use rates. It is therefore more difficult to monitor resistance and to relate monitoring data to product performance in the field for the DMIs. Other factors, such as temperature, time of treatment during the epidemic, and the cultivars used can influence performance much more than any observed sensitivity differences.

Table 5 Characterization of resistance for the major fungicide groups

Fungicides	Fungus	R-factor	Fitness of R-strains	Genetic steps
Benzimidazoles	Botrytis *Venturia*	> 1000	high	1
Dicarboximides	Botrytis	10–20**	reduced	1 (few)
Phenylamides	*Plasmopara* *Phytophthora*	10–10000*	reduced in most cases	1
DMIs	*Erysiphe* *Sphaerotheca*	2–50	reduced	many

*depending on strains and compounds in test
**laboratory strains with higher R-factor

Wheat Powdery Mildew in Northern Europe, 1983–90

To illustrate shifts in sensitivity in DMI fungicides, data from the Ciba-Geigy monitoring program for wheat powdery mildew are presented in Figure 5. The data show shifts in sensitivity from 1983 to 1990 in the areas of intensive wheat production in northern Germany and Scandinavia: the average EC50-values increased from 2–4 to 12–14 ppm in 1986. Shifts in DMI sensitivity in the German wheat area were linked mainly to the highly susceptible cv "Kanzler," which required intensive treatments against powdery mildew. Parallel to the sensitivity shifts, the margin of error for deviations from the label recommendations decreased for DMIs in these areas. DMIs still displayed good activity under optimal use conditions. In situations less favorable for the activity of these compounds (cool temperatures or applications after the disease had reached > 5% in the crops) farmers experienced disappointing results andincreasingly switched to morpholine fungicides. Morpholines are also sterol biosynthesis inhibitors, but they interfere at different sites in the synthesis pathway and do not show cross-resistance to DMIs. Moreover they possess a more pronounced eradicative action than the DMIs and their activity is favored by cool temperatures. As a consequence of this substitution of DMI by morpholine treatments, sensitivity to DMIs increased again over the next few years. For four years sensitivity remained stable, slightly lower than at the beginning of the monitoring program. Most current powdery mildew control programs for wheat contain both DMIs and morpholines in mixture or in alternation.

Barley Powdery Mildew Throughout Europe

An important aspect of sensitivity shifts to DMIs is their geographic distribution. A study by Limpert (39) lists 1985 data on sensitivity to triadimenol for barley powdery mildew throughout central Europe. The data show a general gradient with lower sensitivity in the north and increasing sensitivity to the

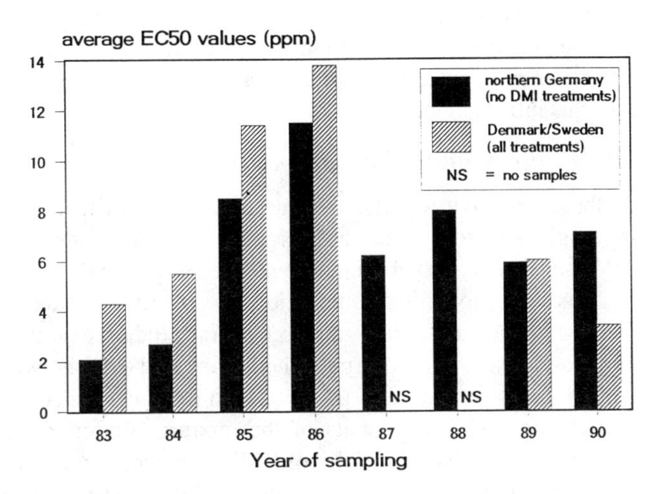

Figure 5 Propiconazole sensitivity of wheat powdery mildew in northern Germany and Scandinavia from 1983–90. EC50-values were determined in large test tubes on wheat seedlings (Ciba-Geigy monitoring data, unpublished).

south. The areas with the lowest sensitivities are also those with the highest disease pressure and with a prevalence of cultivars with high susceptibility to powdery mildew. Similar observations were made in the UK, where sensitivity shift and performance problems first appeared in Scotland on the cv "Golden Promise."After an initial shift toward lower sensitivity to DMI fungicides, barley growers increasingly switched to morpholine treatments. As a response to the decreasing efficacy of the DMIs, combination products of DMIs with morpholines or ethirimol are recommended. These combinations offer excellent disease control while allowing the sensitivity of barley powdery mildew to DMIs to be maintained at an intermediate level.

In most other DMI target pathogens no strains with decreased sensitivity have been found. In a few target pathogens, less sensitive strains have been detected without any clear effect on product performance (34, 56). These strains are either still at very low frequencies or they are not fit enough to cause performance problems. One possible exception is grape powdery mildew; decreased performance and less sensitive strains have been found for some DMIs in vineyards of Portugal, France, and California (60). To avoid further increase of less sensitive strains, French officials have recommended limiting treatments (alone or in mixture with sulfur or dinocap) to a maximum of 3–4 during the 1991 season and using non-DMIs for the rest of the treatments (1).

DMI fungicides have such diverse spectra of activity that the FRAC working group had to develop crop-specific recommendations for preventing or countering sensitivity shifts (25). Each instance of disease control has to be considered separately when formulating antiresistance strategies. The generally slower emergence of strains resistant to DMIs leaves more time to study the response of target populations in practice and to design remedies based on such monitoring information.

Effects of Mixtures and Alternations on Sensitivity

In light of these sensitivity shifts, the question arose whether mixing and alternating fungicides from other groups with the DMI fungicides could stabilize DMI sensitivity. Experimental evidence for preventing or retarding sensitivity shifts with mixtures or alternations is growing. In addition to the circumstantial evidence described above, several studies examining such effects more closely have now been published. Stott et al (61) showed that the use of seed treatments composed of DMIs and ethirimol combined resulted in stable sensitivity to both components of the mixture. Bolton & Smith (4) compared the effects on DMI sensitivity of DMIs used alone and in mixture or in alternation with tridemorph on both powdery mildew sensitivity and barley yield (Figure 6). DMI sensitivity shifts 25 days after the second foliar application were the same for the two DMIs when used alone. Both the

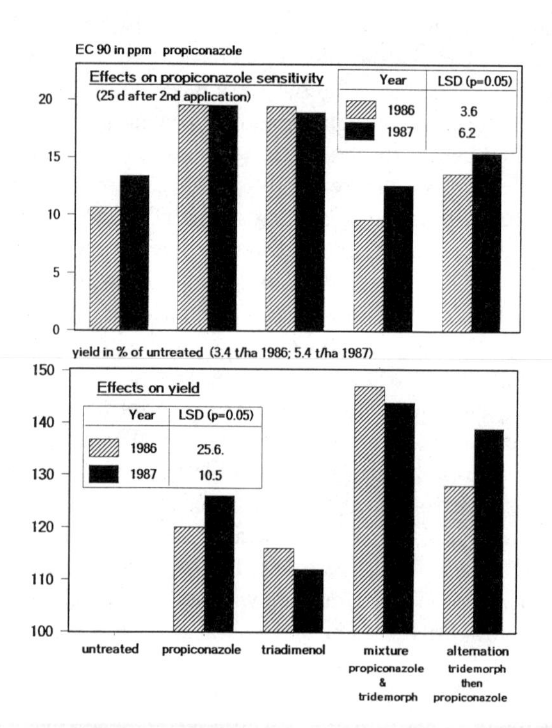

Figure 6 Effects of different DMI treatments on DMI sensitivity of barley powdery mildew and on barley yield.

mixture and the alternation reduced the shifts although the difference was not always statistically significant. In terms of yield increase, the mixture strategy was the most effective treatment, followed by the alternating use. These results show that mixtures and alternations not only prevent or retard sensitivity shifts, but are superior to single products in controlling disease and increasing yields for the farmer. Of the two strategies tested in these two field trials, the mixture strategy was more effective on all counts than the alternation.

Observations on the field performance of the various fungicides groups used for cereal powdery mildew control suggest that with the decrease of DMI sensitivity, sensitivity to morpholines and ethirimol increases (7). When first developed, DMIs were clearly superior to both the morpholines and ethirimol. This ranking was reversed during the shift in DMI sensitivity, which may account for the excellent performance of DMI products combined with these two types of components.

COMPARISON OF MERITS OF MIXTURES AND ALTERNATIONS AS MULTIRESISTANCE MEASURES

Here it may be appropriate to repeat the general arguments favoring the use of mixtures as a key element in antiresistance strategies in many situations (59). If the effect on resistance build-up is assumed to be the same for mixtures and alternations, it can be argued that more effective disease control is usually achieved with the use of mixtures because their components often complement each other in terms of spectrum and mode of action. In addition, mixtures are a safeguard against crop losses should resistance occur. Finally, it is easier to enforce the use of mixtures because they can be packaged together, whereas alternations are only paper recommendations. Therefore, enforcement should also be considered when different control strategies are discussed. Where numerous treatments are required per season, it may be necessary to restrict the use of mixtures to part of the season and to use unrelated chemicals for the rest.

In practice both approaches have been used, depending on the disease, the alternatives available, and the amount of resistance already present in the area. The two approaches are often combined and mixtures are used for only part of the season (phenylamides, dicarboximides). Mixtures often perform surprisingly well even in the presence of substantial levels of resistance (10, 17; L. J. Dowley, personal communication). Where levels of resistance are high it is especially important that sensitivity monitoring be carefully designed to give the true resistance picture in the overall population. Samples need to be taken early in the season to avoid biased samples from crops that have been treated and the sensitivity tests should give a quantitative estimate of the proportion of resistant spores in each sample. Tests used in routine monitoring often give a positive answer for resistance from 1% resistance upwards (55). A population with 1% resistance, however, still responds very well to 1–2 treatments with a particular selective fungicide.

CONCLUSIONS AND OUTLOOK

Practical experience with fungicide resistance and with the antiresistance strategies described above lead to the following conclusions:

Fungicide use is an element of integrated crop production It cannot be stressed strongly enough that the system of crop production has a decisive influence on the risk of resistance. It determines the disease pressure to be controlled by the fungicides and, as many examples cited above illustrate, disease pressure is the one feature common to the first appearance of resistance in any fungicide group.

Fungicide resistance, as a rule, can be managed Rarely must fungicides be withdrawn because of resistance. Withdrawal will probably be even rarer in the future because more cautious use of selective compounds is likely.

Prediction of risk of resistance for new fungicides groups remains difficult With the dicarboximides, resistance developed more slowly than expected because the isolates found in the laboratory did not appear in vineyards (37). Resistance to phenylamides appeared much faster than initial studies had predicted (59). Similarly, DMIs developed resistance in some pathogens, albeit rather slowly, despite predictions to the contrary (21). Nevertheless, laboratory studies and early monitoring can provide useful indications about the risk of resistance and about possible counter-measures.

Monitoring of performance and sensitivity is essential to assess field resistance Only by closely observing the response of the pathogen populations to the widespread use of a new chemical class can an actual assessment of risk of "field resistance" be made. Such monitoring programs automatically take into account all the complex aspects of fitness that are so important for the emergence and the survival of resistant forms in nature. Care is needed in interpreting monitoring data properly (6).

Antiresistance strategies should be designed with areas of high disease pressure in mind Resistance in most fungicides/pathogen combinations regularly appears first in areas or on crops with high disease pressure. If resistance problems can be averted from such danger zones, they are unlikely to appear in areas with lower disease pressure.

Early implementation of counter-measures is important The implementation of counter-measures when a resistance-prone fungicide is introduced is more effective in preventing field resistance and related crop losses than a "wait-and-see" attitude. This precept is especially relevant for new fungicide groups with a proven or an uncertain risk of resistance.

"Old" fungicides remain valuable and new ones are needed for many pathogens With the ban on certain residual partners, the arsenal of alternative antiresistance measures is insufficient for many pathogens (2). Conventional, protective fungicides should not be lightly abandoned and (more challenging) new fungicide groups are badly needed.

Consensus on strategies is necessary for implementation To successfully implement antiresistance strategies, producers, distributors, and extension advisors must all be in agreement. The user would be utterly confused by conflicting advice. The common goal of offering the farmer reliable methods for stable disease control systems should facilitate this task.

The Outlook Is Good

The fungicide groups discussed in this paper have met with some resistance problems because of poor assessment and/or inappropriate implementation of the aspects discussed above. Overall, however, these fungicides remain a successful part of the disease-control arsenal available to farmers. If the lessons learned from experience are applied to future use strategies for existing and as yet undiscovered fungicide groups, the outlook for new fungicide success stories and fewer resistance problems is very promising.

ACKNOWLEDGMENT

Appreciation is extended to several of my colleagues at Ciba-Geigy for their suggestions and critical review of the manuscript.

Literature Cited

1. Anonymous. 1990. Resistance de l'oidium; une note commune INRA, SPV, ITV. *Phytoma* 421:60
2. Baldwin, B. C., Rathmell, W. G. 1988. Evolution of concepts for chemical control of plant disease. *Annu. Rev. Phytopathol.* 26:265–83
3. Bain, R. A., Holmes, S. J. 1990. Phenylamide use on Scottish seed potato crops and resistance in *Phytophthora infestans* isolated from progeny tubers. *Proc. Brighton Crop Prot. Conf. Pests Dis.*, pp. 1121–26
4. Bolton, N. J. E., Smith, J. M. 1988. Strategies to combat fungicide resistance in barley powdery mildew. *Proc. Brighton Crop Prot. Conf. Pests Dis.*, pp. 367–72
5. Brent, K. J. 1978. Chemical control of plant diseases: some relationship to pathogen ecology. In *Plant Disease Epidemiology*, ed. P. R. Scott, A. Bainbridge, pp. 177–86. Oxford: Blackwell
6. Brent, K. J. 1986. Detection and monitoring of resistant forms: An overview. In *Pesticide Resistance-Strategies and Tactics for Management*, NRC

Board Agric., pp. 298–312. Washington, DC: Natl. Acad. Press
7. Brent, K. J. 1988. Resistance experiences in Europe. See Ref. 17, pp. 19–23
8. Butters, J., Clark, J., Hollomon, D. W. 1984. Resistance to inhibitors of sterol biosynthesis in barley powdery mildew. *Meded. Fac. Landbouwwet. Rijksuniv. Gent.* 49/2a:143–51
9. Crute, I. R. 1989. Lettuce downy mildew: a case study in integrated control.In *Plant Disease Epidemiology. Genetics, Resistance, and Management,* ed. K. J. Leonard, W. E. Fry, 2: 30–53. New York: McGraw-Hill. 377 pp.
10. Davidse, L. C., Henken, J., Van Dalen, A., Jaspers, A. B. K., Mantel, B. C. 1989. Nine years of practical experience with phenylamide resistance in *Phytophtora infestans* in the Netherlands. *Neth. J. Plant Pathol.* (Suppl.1)95:197–213
11. Davidse, L. C., van den Berg-Velthuis, G. C. M., Mantel, B. C., Jaspers, A. B. K. 1990. Phenylamides and *Phytophthora. Proc. Int. Symp. Phytophthora Dublin Sept. 89*
12. Dekker, J., Georgopoulos, S. G., eds. 1982. *Fungicide Resistance in Crop Protection.* Wageningen, Neth: PUDOC. 265 pp.
13. Delp, C. J. 1980. Coping with resistance to plant disease control agents. *Plant Dis.* 64:652–57
14. Delp, C. J. 1984. Industry's response to fungicide resistance. *Crop Prot.* 3:3–8
15. Delp, C. J., Dekker, J. 1985. Fungicide resistance: Definitions and use of terms. *OEPP/EPPO Bull.* 15:333–35
16. Delp, J. C., ed. 1988. *Fungicide Resistance in North America.* St. Paul, MN:APS Press. 133 pp.
17. Dowley, L. J., O'Sullivan, E. 1987. Survey of metalaxyl resistant strains of *Phytophthora infestans* in Ireland. *Rep. Oak Res. Cen., Carlow, Ireland*
18. Deleted in proof
19. Ehrenhardt, H., Eichhom, K. W., Thate, R. 1973. Zur Frage der Resistenzbildung von *Botrytis cinerea* gegenüber systemischen Fungiziden. *Nachrichtenbl. Dtsch. Pflanzenschutzdienstes* 25:49–50
20. Fuchs, A., de Waard, M. A. 1982. Resistance to ergosterol-biosynthesis inhibitors. I. Chemistry and phenomenological aspects. See Ref. 12, pp. 71–86
21. Fuchs, A., de Ruig, S. P., van Tuyl, J. P., de Vries, F. W. 1977. Resistance to Triforine: a nonexistent problem. *Neth. J. Plant Pathol.* 83(Suppl. 1):189–205
22. Georgopoulos, S. G., Dovas, C. 1973. A serious outbreak of strains of *Cercospora beticola* resistant to benzimidazole fungicides in northern Greece. *Plant Dis. Rep.* 62:205–8
22a. Green, M. B., LeBaron, H. M., Moberg, W. K., eds. 1990. *Managing Resistance to Agrochemicals.* ACS Ser. 421. Washington, DC: Am. Chem. Soc.
23. Gullino, M. L., Garibaldi, A. 1985. Present situation of resistance to fungicides in Italian vineyards. In *Fungicides for Crop Protection, BCPC Monog.* 31:319–22
24. Heany, S. P., Martin, T. J., Smith, J. M. 1988. Practical approaches to managing anti-resistance strategies with DMI fungicides. *Proc. Brighton Crop Prot. Conf. Pests Dis.*, pp. 1097–1106
25. Highwood, D. P. 1990. Fungicide resistance action committee. *Pestic. Outlook* 1:30–32
26. Hollomon, D. W., Butters, J., Clark, J. 1984. Genetic control of triadimenol resistance in barley powdery mildew. *Proc. Brighton Crop Prot. Conf.*, pp. 477–82
27. Holz, B. 1979. Ueber eine Resistenzerscheinung von *Botrytis cinerea* an Reben gegen die neuen Kontaktbotrytizide im Gebiet der Mittelmosel. *Weinberg Keller* 26:18–25
28. Huggenberger, F., Collins, M. A., Skylakakis, G. 1984. Decreased sensitivity of *Sphaerotheca fuliginea* to fenarimol and other ergosterol-biosynthesis inhibitors. *Crop Prot.* 32:137–49
29. Kable, P. F., Jeffrey, H. 1980. Selection for tolerance in organisms exposed to sprays of biocide mixtures: A theoretical model. *Phytopathology* 70:8–12
30. Kalamarakis, A. E., Demopoulos, V. P., Ziogas, B. N., Georgopoulos, S. G. 1989. A highly mutable major gene for triadimenol resistance in *Nectria haematococca* var. *cucurbitae. Neth. J. Plant Pathol.* 95(Suppl. 1):109–20
31. Katan, T., Bashi, E. 1981. Resistance to metalaxyl in isolate of *Pseudoperonospora cubensis,* the downy mildew of cucurbit. *Plant Dis.* 65:798–800
32. Kato, T. 1988. Resistance experiences in Japan. See Ref. 16, p. 40
33. Kiebacher, H., Hoffmann, G. M. 1976. Benzimidazol-Resistenz bei *Venturia inaequalis. Z. Pflanzenkr. Pflanzenschutz* 83:352–58
34. Köller, W., Scheinpflug, H. 1987. Fungal resistance to sterol biosynthesis inhibitors: A new challenge. *Plant Dis.* 71:1066–74
35. Leroux, P., Fritz, R., Gredt, M. 1977. Etudes en laboratoire de souches de *Botrytis cinerea* Pers. resistantes á la Dichlozoline, au Dicloran, au Quintozene, á la Vinclozoline et au 26 019 RP

(Glycophene). *Phytopathol. Z.* 89:347–58
36. Leroux, P., Besselat, B. 1984. Pourriture grise: La résistance aux fongicides de *Botrytis cinerea. Phytoma Défense des cultures - Juin*, pp. 25–31
37. Leroux, P., Gredt, M. 1984. Resistance of *Botrytis cinerea* Pers. to fungicides. *Tagungsber. Akad. Landwirtschaftswiss. DDR* 222:323–33
38. Leroux, P., Clerjeau, M. 1985. Resistance of *Botrytis cinerea* Pers. and *Plasmovara viticola* (Berk. and Curt) Berl. and de Toni to fungicides in French vineyards. *Crop Prot.* 4:137–60
39. Limpert, E. 1987. Frequencies of virulence and fungicide resistance in the European barley mildew population in 1985. *J. Phytopathol.* 119:298–311
40. Lyr, H., ed. 1987. *Modern Selective Fungicides*. Jena, Germany: VEB Gustav Fischer Verlag. 383 pp.
41. Moncomble, D. 1988. Rapport général sur les activités techniques en Champagne en 1987. *Vign. Champen.* 109 (2):74–100
42. Moncomble, D. 1988. La lutte contre la pourriture grise de la vigne en 1988. *Vign. Champen. 109* (5):315–17
43. Moncomble, D. 1990. Rapport général sur les activités techniques techniques en 1989. *Vign. Champen. Fév.* (2):9–30
44. Moreau, C., Clerjeau, M., Morzières, J. P. 1987. Bilan des essais détection de souches de *Plasmopara viticola* résistantes aux anilides anti-oomycetes (Metalaxyl, Ofurace, Oxadixyl et Benalaxyl) dans le vignoble français en 1987. Rapport G. R. I. S. P. de Bordeaux, Pont de la Maye
45. Piganeau, B., Clerjeau, M. 1985. Influence différentielle de la température sur la sporulation et la germination des sporocystes de souches de *Plasmopara viticola* sensibles et resistantes aux phenylamides. In *Fungicides for Crop Protection, BCPC Monogr.* 31:327–30
46. Pluschkell, H. J., Oeser, J. 1990. Weitere Untersuchungsergebnisse zur Metalaxylresistenz bei *Phytophthora infestans* an Kartoffeln. *Nachrichtenbl. Pflanzenschutz DDR* 44(7):158–59
47. Royle, D. J. 1990. Advances towards integrated control of cereal diseases. *Pestic. Outlook* 1(3):20–25
48. Schepers, H. T. A. M. 1985. Changes during a three-year period in the sensitivity to ergosterol biosynthesis inhibitors of *Sphaerotheca fuliginea* in the Netherlands. *Neth. J. Plant Pathol.* 91:105–18
49. Schüepp, H., Küng, M. 1981. Stability of tolerance to MBC in populations of *Botrytis cinerea* in vineyards of northern and eastern Switzerland. *Can. J. Plant Pathol.* 3:180–81
50. Schüepp, H., Siegfried, W. 1983. Die Traubenfäule 1982 und die teilweise ungenügenden Bekämpfungserfolge mit den Dicarboximid-Fungiziden. *Schweiz. Z. Obst- Weinbau* 119:61–70
51. Schulz, U., Scheinpflug, H. 1986. Investigations on sensitivity- and virulence-dynamics of *Erysiphe graminis* f. sp. *tritici* with and without triadimenol treatment. *Proc. Brighton Crop Prot. Conf.*, pp. 531–38
52. Service Protection Végétaux Alsace 1988. Report on *Botrytis* situation on grapes for 1987
53. Service Protection Végétaux Champagne. 1990. Résultats viticoles 1990, pp. 38–41
54. Skylakakis, G. 1984. Quantitative evaluation of strategies to delay fungicide resistance. *Proc. Brighton Crop Prot. Conf.*, pp. 565–72
55. Sozzi, D., Staub, T. 1987. Accuracy of methods to monitor sensitivity of *Phytophthora infestans* to phenylamide fungicides. *Plant Dis.* 71:422–25
56. Stanis, V. F., Jones, A. L. 1985. Reduced sensitivity to sterol-inhibiting fungicides in field isolates of *Venturia inaequalis. Phytopathology* 75:1098–101
57. Staub, T., Diriwiichter, G. 1986. Status and handling of fungicide resistance in pathogens of grapevine. *Proc. Brighton Crop Prot. Conf.*, pp. 771–80
58. Staub, T., Sozzi, D. 1981. Résistance au métalaxyl en pratique et les conséquence pour son utilisation. *Phytiatr. Phytopharm.* 30:283–91
59. Staub, T., Sozzi, D. 1984. Fungicide resistance: A continuing challenge. *Plant Dis.* 68:1026–31
60. Steva, H., Clerjeau, M. 1990. Cross resistance to sterol biosynthesis inhibitor fungicides in strains of *Uncinula necator* isolated in France and Portugal. *Meded. Fac. Landbouwwet. Rijksuniv. Gent.* 55:983–88
61. Stott, I. P. H., Noon, R. A. 1990. Flutriafol, ethirimol and thiabendazole seed treatment—an update on field performance and resistance monitoring. *Proc. Brighton Crop Prot. Conf. Pests Dis.*, pp. 1169–74
62. Urech, P. A. 1990. Industry's achievements in meeting the needs of society and farmers for modern plant protection. *Proc. Brighton Crop Prot. Conf. Pests Dis.*, pp. 283–92
63. Urech, P. A., Staub, T. 1985. The resistance strategy for acylalanine fungicides. *EPPO Bull.* 15:539–43

64. Valentine G., Moncomble, D. 1988. Expérimentation et recherche sur le *Botrytis*. *Vign. Champen. 109* 6:368–85
65. Valentin, G., Moncomble, D. 1990. Pourriture 1989. *Vign. Champen.* 6:31–49
66. van Tuyl, J. M. 1977. Genetics of fungal resistance to systemic fungicides. *Meded. Landbouwhogesch. Wageningen* 77–2:1–137
67. Wade, M., Delp, C. J. 1985. Aims and activities of industry's fungicide resistance action committee (FRAC). *EPPO Bull.* 15:577–83
68. Walker, A. S. L., Cooke, L. R. 1990. The survival of *Phytophthora infestans* in potato tubers—The influence of phenylamide resistance. *Proc. Brighton Crop Prot. Conf. Pests Dis.*, pp. 1109–14

Annu. Rev. Phytopathol. 1991. 29:443–67

MOLECULAR GENETIC ANALYSIS OF THE RICE BLAST FUNGUS, *MAGNAPORTHE GRISEA*

Barbara Valent and Forrest G. Chumley

Central Research and Development, The Du Pont Company, P.O. Box 80402, Experimental Station, Wilmington, Delaware 19880-0402

KEY WORDS: pathogenicity, host specificity, avirulence genes, *Pyricularia oryzae*, plant-fungus interactions

---

INTRODUCTION

Ever more powerful molecular genetic analysis of the rice blast fungus, *Magnaporthe grisea* (Hebert) Barr (anamorph, *Pyricularia grisea* Sacc., formerly *P. oryzae* Cav.) (58, 62) promises to yield a detailed understanding of the molecular basis of host-pathogen interactions in rice blast. Blast is thus increasingly attractive as a model system for the study of interactions between an economically important crop plant and a damaging fungal pathogen (76). The potent combination of classical genetics, molecular biology, cytology, and cell biology is now permitting the following critical elements of host-pathogen interactions to be analyzed. (*a*) *Mechanisms of pathogenesis: M. grisea* undergoes a complex sequence of developmental and metabolic events during pathogenesis, including elaboration of an appressorium, which supports direct penetration through the cuticle of the host plant (6, 22, 25, 26). (*b*) *Host species specificity:* As a species, *M. grisea* includes pathogens not only of rice *(Oryza sativa)*, but of many other grasses as well. However, individual isolates have a limited host range. (*c*) *Host cultivar specificity:* Races (pathotypes) of the fungus are distinguished among isolates that parasitize rice, depending on the rice cultivars they successfully infect. (*d*) *Extent and mechanism of spontaneous genetic variation:* The epidemiological

0066-4286/91/0901-0443$2:00

and evolutionary dynamics of the rice blast fungus have long been a matter of interest to plant pathologists. Pathogens of rice show a high degree of variability in the field; new races frequently appear with the ability to attack previously resistant rice cultivars. Molecular techniques offer new approaches for the critical evaluation and analysis of genetic variation in the blast fungus.

In this review, we discuss the recent application of tools for molecular genetic analysis of *M. grisea*. We also review past and current research in the problem areas outlined above.

FROM THE FIELD INTO THE LABORATORY

For a plant pathogen model system to be useful in understanding and controlling plant disease, the pathogen must be manipulated in the laboratory in a state as close to the wild state as possible. Isolates of *M. grisea* tend to be unstable with regard to colony appearance, fertility, and pathogenicity during subculture in the laboratory (50, 69, 78), as is typical of many fungal pathogens. Therefore, it is critical that subculturing of the fungus be minimized, preferably by maintaining permanent stocks in a frozen, nonmetabolizing condition (79). A watchful eye must be kept on all strains to avoid "degraded" forms described below. Future efforts to understand and eliminate undesirable sources of instability will facilitate molecular genetic studies of the rice blast fungus.

The instability of isolates fresh from the field is apparent in the "mosaic" pattern of colony growth on certain standard agar media (37, 78). Mosaic appearance of cultures is associated with several desirable characteristics, including female fertility in genetic crosses. Mosaic cultures derived from a single conidium growing on oatmeal agar are at first smooth and conidiated. Approximately ten days after conidiation, papillae appear in the mycelium. The papillae take the form of very heavily conidiated clumps or nonconidiating mycelial outgrowths that exhibit gray pigmentation in a range of intensities. The mosaic trait appears to be easily lost, often within a few generations of subculturing on agar plates. A predictable progression of morphological changes occurs until cultures reach a final "degraded" state characterized by female infertility, lack of pigment, and poor conidiation. These degraded cultures can be propagated indefinitely and they do not revert to mosaic appearance after single spore isolation or after passage through the host (B. Valent and F. G. Chumley, unpublished observations).

Two types of visible mutations appear spontaneously at a high frequency when the rice blast fungus is subcultured on agar media. About 0.5% of the conidia produced by some field isolates of the rice pathogen carry mutations in the *BUF1* melanin biosynthetic gene, which is essential for infecting unwounded plants (7). Certain normally pigmented isolates of the pathogen, including isolates fresh from the field, cross with other normally pigmented strains to produce 5–25% Buf^- progeny (7, 40, 79). Most isolates of the fungus, including pathogens of rice as well as pathogens of other grasses, show frequent spontaneous appearance of mutants with altered spore morphology due to mutations in the *SMO1* gene (20). High mutability is not a general feature of genes of *M. grisea* because auxotrophs, drug resistant mutants, and other visible mutants (such as those with defects in the *ALB1* melanin biosynthetic gene) occur at normal frequencies.

Female fertility, together with mosaic appearance of colonies, can decrease after being subcultured even a few times (50, 69, 78). The highest levels of fertility are observed with cultures derived directly from ascospores. Some degraded strains still serve as males in genetic crosses. Pathogenicity is more stable than colony appearance and female fertility. However, loss of pathogenicity occasionally occurs during subculture, and loss of ability to conidiate abundantly can seriously restrict pathogenicity testing.

Classical Genetic Analysis

M. grisea is a filamentous, heterothallic Ascomycete, order Pyrenomycetes (23, 31, 73, 84). Compatibility for mating is governed by alternate alleles of the mating type locus, *Mat1*. Between compatible strains, mating is characterized by an additional degree of complexity in which male and female roles are assumed by each parent. Fertile strains should thus be regarded as self-sterile hermaphrodites.

Early genetic studies were conducted using strains of *M. grisea* that parasitize finger millet *(Eleusine coracana)* or weeping lovegrass *(Eragrostis curvula),* because field isolates infecting these grasses are typically hermaphrodites that cross to produce viable ascospores (42, 49, 68–70, 77, 81). Although rice pathogens occur as both mating types, isolates in single geographic locations typically are of only one mating type; where both mating types occur, the strains are not interfertile (30, 52, 86). Field isolates pathogenic to rice typically are female sterile (28, 78). Kato & Yamaguchi (30) reported rare crosses between field isolates pathogenic to rice, but the fertility was poor and not reproducible. A unique hermaphroditic field isolate of the rice pathogen, named GUY11, was isolated in French Guiana (51, 52). GUY11 produces low numbers of viable ascospores when mated either with other field isolates of the rice pathogen (40, 51, 52) or with field isolates of *M. grisea* from other grasses (B. Valent, F. G. Chumley, unpublished results).

DEVELOPMENT OF FERTILE LABORATORY STRAINS Fertile laboratory strains that mate to produce abundant ascospores with high viability are critical for rigorous genetic analyses. Highly fertile laboratory strains that infect weeping lovegrass and goosegrass *(Eleusine indica)* were easily produced (77). The interfertility of rice pathogens and hermaphroditic strains pathogenic to other grasses (71, 78, 82) suggested that it would be possible to generate recombinant progeny with both female fertility and pathogenicity to rice. Early attempts to produce such recombinants were hampered by the scarcity of rice pathogens among the progeny (34, 78, 79), The failure to recover female fertile rice pathogens in the early studies suggests that genes required for pathogenicity to rice might be linked, in repulsion, to genes required for fertility. However, hermaphroditic rice pathogens were obtained among the fifth and sixth generation progeny in a backcrossing scheme whereby fertility from a pathogen of weeping lovegrass (lab strain 4091-5-8) (77, 78) was introgressed into a pathogen of rice isolated from the field in China (strain 0-135) (79). Another series of crosses, involving two rice pathogens from China (0-111 and 0-137), produced a few hermaphroditic rice pathogens among first and second generation progeny, and led to strain 4224-7-8, the source of several avirulence genes described below (B. Valent, L. Farrall & F. G. Chumley, manuscript in preparation). The isolation of GUY11 has facilitated genetic analysis (11, 40) and should result in further improvements in the fertility of *M. grisea* laboratory strains pathogenic to rice.

ISOLATION OF MUTANTS An extensive collection of mutants, carefully stored in a nonmetabolizing state, enhances the utility of a genetic system. Numerous nutritional and drug-resistant mutants of *M. grisea* have been reported (16, 49, 68–70); unfortunately, these mutants have not been preserved. Crawford et al (8) isolated and retained a large number of auxotrophs following UV mutagenesis. The yield of auxotrophic colonies following UV treatment was approximately 0.35% of the survivors of mutagenesis in which about 95% of the starting population was killed. Melanin-deficient mutants were recovered after UV mutagenesis at frequencies ranging from 0.01% to 0.1% of the survivors (7). Spontaneous mutants resistant to 6-methylpurine, the sulfonylurea chlorimuron ethyl (Du Pont Agricultural Products; "DPX-F6025") or benomyl were isolated in independent experiments at frequencies of 10^{-6} to 10^{-7} (F. G. Chumley, B. Valent, unpublished results). These "normal" frequencies for recovery of mutants in a haploid fungus stood in sharp contrast to the very high frequency of spontaneous Buf^{-} and Smo^{-} mutants.

CHARACTERIZATION OF THE VEGETATIVE DIPLOID PHASE As with *Verticillium* (61), heterokaryons of *M. grisea* produce conidia that are haploid and homokaryotic, exhibiting parental arrangements of genetic markers (8). Heterokaryons occasionally give rise to fast-growing sectors with the properties of unstable vegetative diploids that produce conidia with all expected combinations of auxotrophic and morphological markers. Therefore, *M. grisea* does not produce either heterokaryotic or diploid conidia as necessary for complementation and dominance tests for determinants of pathogenicity and host specificity (8). Heterokaryons of *M. grisea* have been useful in tests for complementation of auxotrophic mutations (8). The parasexual cycle in *M. grisea* is of interest both as a tool for genetic analysis and as a potential source of variation for the pathogen in the field (8, 16).

Molecular Biology

CLONED GENES Much effort has been devoted to applying the methods of molecular biology to *M. grisea* (9, 27, 41, 59, 60, 67), and several genes have been cloned. The *ILV1* gene, which encodes the enzyme acetolactate synthase (the first enzyme in the biosynthetic pathway leading to branched chain amino acids), was cloned using the *ILV2* gene of *Saccharomyces cerevisiae* as a heterologous probe (F. G. Chumley & B. Valent, unpublished results). A benomyl-resistance allele of the beta-tubulin gene of *M. grisea, TUB1,* was cloned using the *tub-2* gene of *Neurospora crassa* as a heterologous probe (J. A. Sweigard, B. Valent & F. G. Chumley, unpublished results). The *LYS1* gene of *M. grisea* was cloned from a cosmid library by complementation of the *lys-1* mutation, which affects an unknown step in lysine biosynthesis (59). The *RSY1* gene, which encodes scytalone dehydratase, a melanin biosynthetic enzyme (7), was cloned by screening a cDNA expression library with an antibody to the purified enzyme (J. Pierce, unpublished results). The cutinase gene of *M. grisea, CUT1,* was cloned using a cDNA clone for the cutinase gene of *Colletotrichum capsici* as a heterologous probe (J. A. Sweigard, F. G. Chumley, B. Valent, submitted for publication). Sequencing of the *RSY1* and *CUT1* genes has shown that they have characteristics typical of genes from other filamentous ascomycetes (2).

TRANSFORMATION In the first successful transformations of *M. grisea,* the recipients were arginine auxotrophs with a mutation *(arg3–12)* in the gene encoding ornithine carbamoyl-transferase, and donor plasmids carried the *ArgB* gene of *Aspergillus nidulans* (59, 60). The original vector carried no DNA from *M. grisea;* donor sequences were integrated into chromosomes by nonhomologous recombination. As with other filamentous fungi (13), integration of linear tandem arrays, sometimes with sequence rearrangements, was common, although some transformants contained single integrated copies of the donor plasmid. Prototrophic transformants arose at a frequency of about 35 per microgram of plasmid DNA when protoplasts were spread on the surface of minimal regeneration agar. The efficiency of transformation was elevated to approximately 100 transformants per microgram by embedding the transformed protoplasts in a top agar overlay. Similar efficiencies were obtained when isoleucine-valine auxotrophs with the *ilv1–4* mutation were transformed with plasmids carrying the *ILV1* gene of *M. grisea* (F. G. Chumley & B. Valent, unpublished results). Homologous recombination at the *ILV1* locus occurred in about 50% of the transformants, with approximately half of the events involving insertion of the plasmid into the genome by homologous crossing over and half of them involving gene conversion or double crossovers in which vector sequences were not integrated. Daboussi et al (9) described transformation of *M. grisea* using the nitrate-reductase gene of *Aspergillus* to complement chlorate-resistant, nitrate-reductase-deficient mutants.

Selection of transformants using a dominant drug-resistance marker has wide application, because any strain can be used as a recipient. *M. grisea* has been transformed using dominant genes for resistance to chlorimuron ethyl, benomyl, and, most importantly, hygromycin B. Sulfonylureas such as chlorimuron ethyl block the growth of plants and microorganisms by inhibiting the biosynthesis of branched chain amino acids. Rapid cloning of dominant alleles of the *ILV1* gene of *M. grisea* that confer resistance to chlorimuron ethyl (F. G. Chumley & B. Valent, unpublished results) was accomplished by standard eviction techniques developed for yeast (80). Spontaneous mutants resistant to sulfonylurea were isolated from Ilv^+ transformants that contained a duplication of the *ILV1* gene. The vector plus mutant gene were then reisolated by digesting genomic DNA with an appropriate restriction enzyme, ligating, and transforming bacteria to create new chimeric plasmids. Such plasmids transformed prototrophic strains of *M. grisea* to sulfonylurea resistance, or $ilv1^-$ auxotrophs to prototrophy. At low efficiency, strains of *M. grisea* have been transformed to benomyl resistance using the *Bml* gene of *N. crassa* (53) or a cloned *BEN* gene of *M. grisea* (M. J. Orbach, J. A. Sweigard, F. G. Chumley & B. Valent, unpublished results). Selection of transformants for resistance to hygromycin B, first reported for *M. grisea* by Leung et al (41), is now in common use. The hygromycin resistance genes used contained the coding sequence of a gene from the drug resistance transposon Tn903 fused to control elements of fungal genes such as the *gpd* and *trpC* genes of *A. nidulans* (65) or the *cpc-1* gene of *N. crassa* (M. Plamann, personal communication). In all cases of selecting for drug resistance, transformation frequencies have ranged from 1 to 15 transformants per microgram of donor DNA. Efforts are underway in various laboratories to increase the efficiency of transformation by modifying current transformation protocols and by using new techniques for electroporation (15) and particle bombardment (32).

GENE DISRUPTION EXPERIMENTS Genetic engineering of mutants involves the use of transformation technology to inactivate the genomic copy of a cloned gene, thereby permitting a test of its importance for a particular phenotype. The cutinase gene of *M. grisea, CUT1,* present in one copy per haploid genome, was used to evaluate the potential for gene disruption in *M. grisea,* because *cut*$^-$ mutants were needed to test the importance of cutinase in pathogenesis (see below). *M. grisea CUT1*-defective mutants were constructed using a one-step gene disruption strategy (63). Two independent systems were described (J. A. Sweigard, F. G. Chumley, B. Valent, submitted for publication). In the first, strain 4170-1-3 *(Mat1–1 arg3–12)* (60), a pathogen of weeping lovegrass and barley *(Hordeum vulgare)* was the recipient for transformation. The disrupted cutinase gene was constructed in vitro by replacing an internal restriction fragment in the *CUT1* gene with a fragment of DNA of *A. nidulans* containing the *ArgB* gene. The disruption vector contained 4.4 kb of homologous DNA of *M. grisea,* with approximately equal-length DNA segments flanking the inserted *Aspergillus* DNA. In the second system, the rice pathogen CP983, a prototroph, was tranformed with a disruption vector that carried the Tn903 hygromycin phosphotransferase gene fused to control elements of the *trpC* gene of *A. nidulans.* This plasmid had the chimeric hygromycin resistance gene inserted into a single restriction site in the middle of the *CUT1* gene. Although this vector contained 11 kb of DNA of *M. grisea,* the vector *M. grisea* sequences were derived from a pathogen of weeping lovegrass and may not have been totally homologous to the corresponding DNA in the rice pathogen recipient. Results for both systems were similar. From 2 to 4% of the transformants generated with circular plasmid showed a neat replacement of the resident cutinase gene with the engineered copy. When the donor plasmid was linearized by digesting with a restriction enzyme that cleaved sequences from *M. grisea* to create one "recombinogenic end" (55), about 10% of the transformants contained simple *cut*$^-$ disruptions. The efficiency of generating gene disruptions might be further improved by transformation using linear molecules with two recombinogenic ends, as is true for yeast (63).

All potential *cut*$^-$ insertion mutants were shown by Northern analysis and by esterase activity stains on nondenaturing polyacrylamide gels to lack a functional *CUT1* gene (J. A. Sweigard, F. G. Chumley, B. Valent, submitted for publication). Clearly, gene disruption strategies are feasible in *M. grisea.*

CHARACTERIZATION OF MIDDLE REPETITIVE DNA SEQUENCES An analysis of middle repetitive DNA sequences in the genome of *M. grisea* led to the description of interesting and useful elements specific for the rice pathogen and named "MGR", for *M. grisea* repeat (17). Using ^{32}P-labeled total genomic DNA as a plaque hybridization probe, recombinant clones containing repeated DNA sequences were identified in a λEMBL3 library containing genomic DNA from a rice pathogen. Clones containing repeated DNA sequences gave an intense signal (in response to probe) relative to single copy clones because of the enrichment of repeated sequences in the probe. Approximately 10% of the recombinant phage (with an average insert size of 17 kb) contained MGR sequences. Subsequent analysis showed that rice pathogens from all regions of the world have 40–50 copies of MGR per haploid genome. MGR sequences are either absent or present in low copy number in field isolates that infect other grasses. The results suggest that rice pathogens from around the world are descendents from a small ancestral population and that rice pathogens are now reproductively isolated from nonpathogens of rice in the same geographic area. Strains pathogenic to grasses other than rice contain middle repeat sequences with little homology to MGR probes ((17); K. Dobinson & J. E. Hamer; F. G. Chumley, J. A. Sweigard, & B. Valent, unpublished results).

MGR sequences in a single strain are polymorphic, both with respect to restriction sites and to sequence arrangement (17). Two MGR-sequence elements have been studied in detail: MGR583 defined by the subclone pCB583, and MGR586 defined by the subclone pCB586. These non-homologous repeated sequences are often, but not always, adjacent to one another in the genome of rice pathogens. The sequence MGR586 is commonly used for "MGR fingerprinting", because it identifies a highly polymorphic series of *Eco*R I restriction fragments (Figure 1) that is unique for every rice pathogen tested to date (17, 45). MGR fingerprints of rice pathogens have 50 or more resolvable *Eco*R I fragments ranging from 0.7 kb to 20 kb in length. As discussed below, MGR fingerprints are extremely useful for identifying strains, confirming parent-mutant relationships, mapping the genome, and studying the origin and evolution of populations of *M. grisea* in the field.

In contrast to the highly polymorphic hybridization patterns detected by MGR586, sequence MGR583 detects a few highly conserved restriction fragments in addition to a polymorphic array of bands. Recent results provide insight into the nature of MGR583 sequences (M. J. Orbach, L. Farrall, F. G. Chumley & B. Valent, unpublished results). *Bam*H I digests of genomic DNA from pathogens of rice and pathogens of other grasses contain a conserved 2-kb fragment with MGR583 homology. In addition, all rice pathogens contain conserved 4.0-kb and 5.6-kb *Hin*d III fragments that include the 2-kb *Bam*H I fragment. Sequencing of a 4.2-kb fragment of the conserved region has identified a 928 amino acid open reading frame with significant homology to the reverse transcriptase genes of LINE1-like ("polyA-type") retrotransposons (12). The conserved region hybridizes to four mRNAs, 7.5, 2.6, 2.4, and 0.5 kb in length (M. J. Orbach, L. Farrall, F. G. Chumley & B. Valent, unpublished results; 17). Thus, MGR583 may represent a family of poly A-type retrotransposons present in the genome of *M. grisea*. The origin and identity of MGR586-repeated sequences remains an intriguing question because MGR586 has no detectable homology to the MGR583 transcripts. This

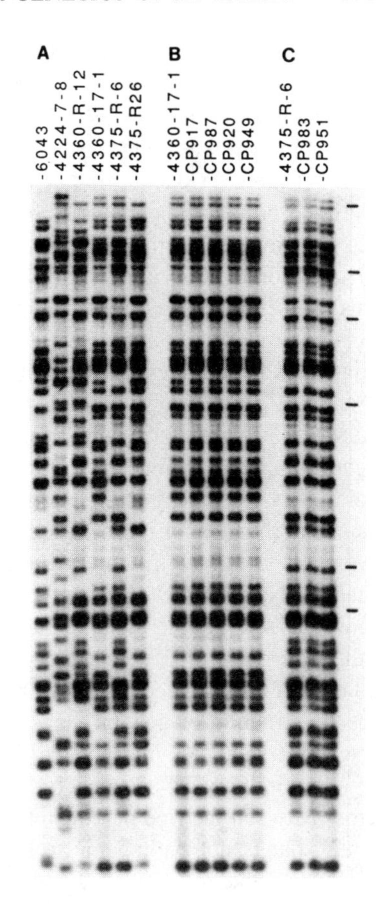

Figure 1 DNA Fingerprints based on MGR sequences. Genomic DNA from the strains indicated was digested with *Eco*R I, subjected to high resolution agarose gel electrophoresis, blotted to a hybridization membrane and probed with radiolabeled MGR586 (pCB586) (17). (A) Mendelian segregation of MGR sequences. Strains 6043 and 4224-7-8 were crossed to produce strains 4360-R-12 and 4360-17-1. These latter two strains were then crossed to produce strains 4375-R-6 and 4375-R-26. Strains 6043 and 4224-7-8 are the parents of a large RFLP-mapping population. (B) Parent-mutant family marked by MGR fingerprints. Strain 4360-17-1 is a rice pathogen that is avirulent on cultivar Yashiro-mochi and a nonpathogen of weeping lovegrass. Strains CP917 and CP920 are independent spontaneous mutants of 4360-17-1, selected for virulence on Yashiro-mochi. Strains CP987 and CP949 are spontaneous mutants selected for pathogenicity on weeping lovegrass; CP987 was derived from CP917, and CP949 was derived from 4360-17-1. (C) Parent-mutant family marked by MGR fingerprints. Strain 4375-R-26 is a rice pathogen that is avirulent on cultivar Yashiro-mochi and a nonpathogen of weeping lovegrass. Strains CP983 and CP951 are independent spontaneous mutants of 4375-R-26 that were selected for virulence on Yashiro-mochi and pathogenicity on weeping lovegrass, respectively.

suggests that, although MGR586 and MGR583 sequences are often contiguous, they are unlikely to be part of the same transposable element.

ELECTROPHORETIC KARYOTYPING Electrophoretic separation of chromosome-size DNAs of *M. grisea* by contour-clamped homogeneous electric field (CHEF) gel electrophoresis (54) has demonstrated the existence of two classes of chromosomes (17; M. J. Orbach, F. G. Chumley, B. Valent, manuscript in preparation). All strains analyzed by CHEF electrophoresis contain five or six chromosome-size DNAs ranging from 3 to greater than 10 megabases in length. Significant chromosome-length polymorphisms are apparent between different strains of *M. grisea*. In addition to five or six large chromosomes, many strains also contain one to four small, apparently nonessential, chromosomes ranging in length from 0.5 to 3 megabases. These small chromosomes, which resemble the "B" chromosomes of some higher eukaryotes (29), do not segregate normally in genetic crosses. All rice pathogens tested, except for the hermaphroditic rice pathogen GUY11, contain B-like chromosomes. All cloned genes of *M. grisea* have been assigned to chromosomes by blotting CHEF gels and probing with ^{32}P-labeled DNA. This approach will yield a clearer picture of differences in genome organization between various strains of the pathogen.

Based on cytological studies of developing asci (44, 71, 83), a consensus has developed that the haploid number of chromosomes is six. Results of electrophoretic karyotype studies of diverse strains of *M. grisea* agree with this consensus. These results are not consistent with reported extensive "heteroploidy" and extreme variation in chromosome number (56, 57, 72), as suggested by early cytogenetic studies.

GENETIC MAPPING Cloning genes of unknown biochemical function by "chromosome walking" involves isolation of overlapping adjacent chromosomal DNA segments that span the region from a previously cloned marker linked to the gene of interest. A reliable and densely marked genetic map is a prerequisite to having closely linked genetic markers that flank any gene of interest. Nagakubo et al (49) reported four cases of genetic linkage among 11 loci analyzed, but the mutant strains used in these crosses have been lost. Few other examples of genetic linkage have been detected (8; B. Valent, F. G. Chumley, unpublished results). Therefore, major efforts have been devoted to the construction of genetic maps using physical markers, namely *r*estriction *f*ragment *l*ength *p*olymorphisms (RFLPs) (48).

A preliminary RFLP map for *M. grisea* was published (64), based on a cross between the rice pathogen GUY11 and strain 2539, a fertile laboratory strain that is not pathogenic to rice (40). Fifty-two physical markers were placed on the map either by detection of linkage or by hybridization to electrophoretically separated chromosomes. Using the same criteria, two lactate dehydrogenase loci (*ldh1* and *ldh3*), the mating-type locus *(Mat1),* and the major ribosomal DNA repeat were placed on the map. No host-specificity genes were placed on the map, although two such genes reportedly segregated in the cross (40).

The observations that MGR sequences are found on all chromosome-size DNAs (17) and that MGR polymorphisms segregate as stable Mendelian markers in genetic crosses (for example see Figure 1; 18, 79) suggested that MGR sequences would be valuable physical markers for genetic mapping. Two linked MGR sequences flank the *SMO1* locus (18), and one particular MGR sequence is linked to an avirulence gene (79). A linkage map based on segregation of MGR sequence polymorphisms has been constructed in a cross between a pathogen of rice and one of weeping lovegrass (J. Romao & J. E. Hamer, personal communication).

A major RFLP mapping effort is now underway for a genetic cross between two rice pathogens (Figure 1) in which at least five host specificity genes have segregated (see below). Analysis of a RFLP mapping population consisting of 67 individuals has defined six linkage groups with a total of 189 physical and genetic markers (J. A. Sweigard, A. Walter, M. J. Orbach, F. G. Chumley, B. Valent, unpublished results). Using eight restriction enzymes, 30% of the cosmid clones used as probes identified scoreable RFLPs. MGR sequence polymorphisms were also mapped: 20 of approximately 50 MGR polymorphisms detected by MGR586 were easily scored, and they all segregated simply. Only two showed 100% linkage to one another, as expected if more than one band in Southern hybridization corresponded to a single MGR element. MGR polymorphisms mapped to widely dispersed locations on all of the six linkage groups that were defined. Additional polymorphisms were mapped using MGR583 and a grass pathogen-specific repeat as probes. Chromosome-size DNAs resolved on CHEF gels have been blotted and hybridized with the mapping probes to establish correspondence between linkage groups and separated chromosomes. The parents chosen for this effort appear to differ by one chromosomal translocation. All five host specificity genes mapped as normal Mendelian markers.

Molecular Mechanisms of Pathogenesis

THE DISEASE CYCLE To succeed as a pathogen, the fungus must execute a series of developmental and metabolic pathways as it attacks its host. Such critical pathways constitute the "disease cycle," which proceeds as follows. A conidium attaches to the host surface using spore-tip mucilage (19) and germinates. The germ tube subsequently differentiates a specialized infection organ, the appressorium, which attaches very tightly to the cuticle. The appressorium elaborates a penetration peg that pierces the cuticle and epidermal cell wall. In a successful interaction, the fungus grows intracellularly, invading adjacent epidermal cells as well as underlying mesophyll cells. Five to seven days later, conidiophores differentiate, and thousands of new conidia are released from the lesion to reinitiate the disease cycle. Potential genes for pathogenesis include those important for the developmental pathway that the fungus undergoes during infection (6, 22, 25) and those that encode enzymes that might be directly involved in the process of infection (e.g. enzymes that degrade plant cell walls; 6a, 24, 66).

PENETRATION The blast fungus undergoes differentiation of functional appressoria on various artificial surfaces (6, 26), thus permitting analysis of the developmental sequence in isolation from the host plant. Figure 2 illustrates the morphological changes that occur as the fungus differentiates on cellophane membranes. Between 24 and 31 hours, the fungus penetrates into the cellophane membrane where it forms structures that resemble infection

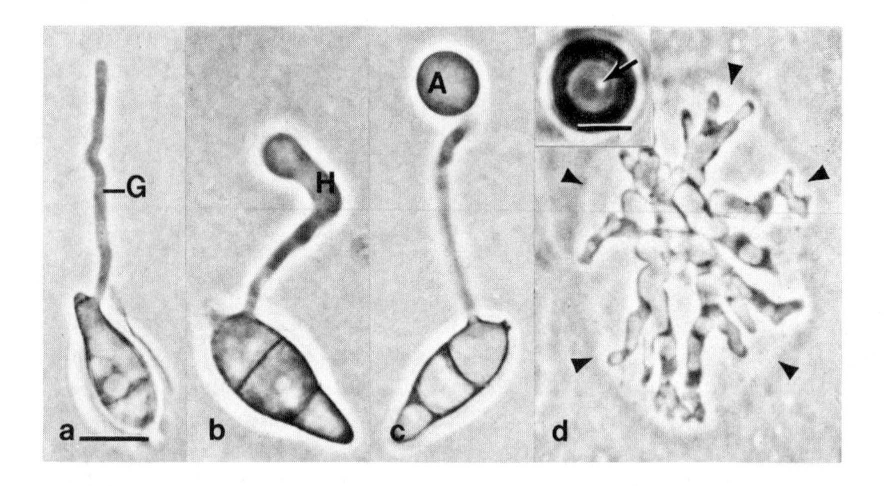

Figure 2 A series of phase contrast light micrographs depicting morphological changes of *M. grisea* during in vitro development of a penetration structure. (a) Conidial germination from an apical cell. G, germ tube. (b) Early appressorium development included the formation of a hook and a terminal swelling. (c) A, appressorium. Apical inflation caused the hook to be raised off the substratum. (d-Inset) After the appressorium became melanized, a single penetration peg (inset, arrow) placed eccentrically within the appressorium pore extended into the substratum. (d) Infection hypha-like structures were produced laterally within cellophane from the distal end of the penetration peg and were surrounded by a refractile zone (area outlined by arrowheads) where the cellophane appeared degraded. Scale bar in (a)–(d) = 10μm. Bar in inset = 5μm. (After Bourett & Howard (6), with permission.)

hyphae (6). The formation and function of appressoria on surfaces such as Mylar® and polyvinylchloride suggest the cues that elicit infection-structure development are primarily physical and mechanical, not chemical.

It is now clear that appressoria of *M. grisea* penetrate by a mechanism that involves a critical mechanical component. The genes *ALB1, RSY1,* and *BUF1,* which encode enzymes essential for biosynthesis of the gray pigment DHN-melanin, are required for penetration (7). Howard & Ferrari (26) have presented convincing evidence that DHN-melanin mediates the build-up of hydrostatic pressure in the appressorium and that this high pressure provides the essential driving force for a mechanical component of penetration. The physical force that penetration pegs can exert is illustrated by the observation that appressoria can penetrate artificial surfaces such as Mylar® or polyvinylchloride (25, 26). It seems unlikely that any substratum-modifying enzyme could play a role in penetration of these surfaces, and therefore penetration must be primarily by mechanical means.

Whether or not enzymes play a role in penetration of host surfaces remains unknown for the rice-blast pathogen. In *Colletotrichum* and *Fusarium,* cutinase appears to be essential for penetration into unwounded host tissue (33a). The obvious prediction that cutinase might play a role in penetration by *M. grisea* was tested with the *CUT1*-deficient insertion mutants described in the previous section. *CUT1* is the only gene in *M. grisea* that is homologous to the cutinase genes of *Colletotrichum* and *Fusarium* (J. A. Sweigard, F. G. Chumley & B. Valent, submitted for publication). In sharp contrast to the critical role in penetration demonstrated for the *ALB1*, *RSY1*, and *BUF1* genes (7), no apparent role for the *CUT1* gene was detected under the laboratory conditions tested (J. A. Sweigard, F. G. Chumley, B. Valent, submitted for publication). Independently derived *CUT1*-defective strains were indistinguishable from wild-type strains in pathogenicity tests on three different hosts (rice, barley, and weeping lovegrass), even when older plants, which presumably have thicker and possibly more complex cuticles, were inoculated. These results do not rule out a possible role for cutinase during infection of "hardened" host plants growing in the field. The *CUT1* gene clearly is not expressed in the *cut1*$^-$ mutants, yet the mutants retain some activity in the standard cutinase enzyme assay. This remaining activity may indicate that *M. grisea* produces an additional cutinase unlike any previously described.

The isolation and analysis of mutants that fail to form functional appressoria could lead to better understanding of the penetration process. One interesting class of mutants, named Smo$^-$ for abnormal *s*pore *mo*rphology, was first identified while screening mutagenized cultures for apressorium-defective mutants (20). The fifteen independent *smo*$^-$ alleles tested define a single genetic locus, *SMO1*. Smo$^-$ mutants exhibit dramatic pleiomorphic alterations in the shapes of conidia and appressoria, and germinating Smo$^-$ conidia form appressoria on surfaces not normally conducive to formation of infection structures. These mutants retain pathogenicity to their original hosts, weeping lovegrass or rice, although pathogenicity toward rice appears to be reduced with regard to the number and size of lesions (18, 20). Cloning *SMO1* (18) will be critical for understanding its role in differentiation and for understanding the frequent spontaneous occurrence of *smo*$^-$ mutants during subculture on agar medium.

PROLIFERATION Heath et al (22) have reported a previously undescribed dimorphism in the infection hyphae of the blast fungus growing on weeping lovegrass, goosegrass, and rice (Figure 3). Typically, the very narrow penetration peg enlarges to form a primary hypha. This primary hypha subsequently differentiates into an enlarged bulbous secondary hypha that proliferates throughout the host tissue. Secondary hyphae often grow at right angles to primary hyphae. The bulbous form of the fungus inside plant tissue differs dramatically from the slender filamentous growth on agar media.

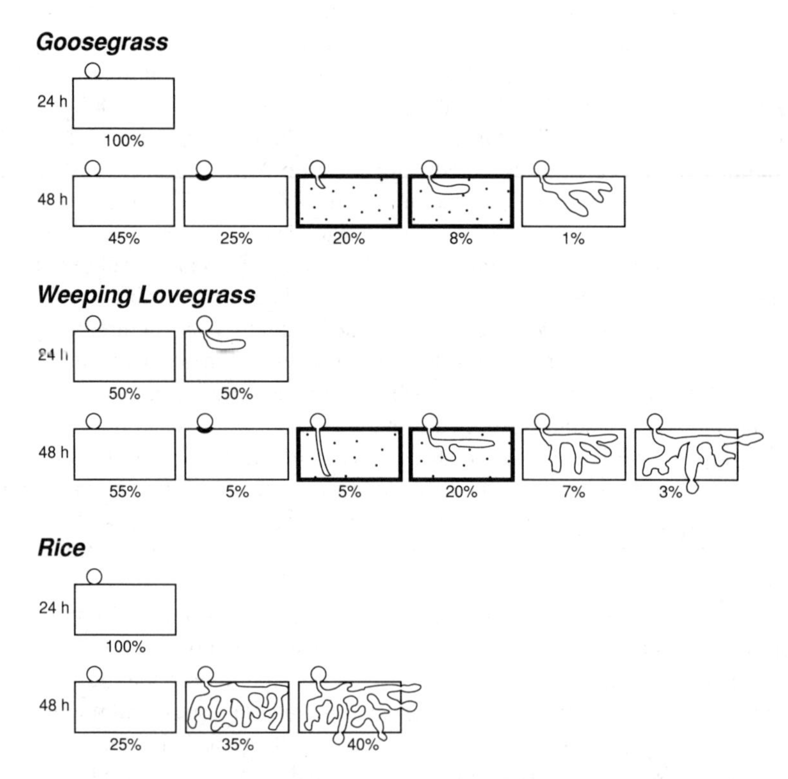

Figure 3 Diagrammatic representation comparing the frequency of the most typical combinations of fungal growth and plant responses seen in the three host species 24 and 48 hrs after inoculation with strain O-42. Each rectangle represents the epidermal cell underlying the fungal appressorium (small circle). A thick plant cell wall represents browning or autofluorescence, and granular cytoplasm is represented by stippled areas. Thin vertical hyphae represent primary hyphae, and thicker horizontal hyphae represent bulbous secondary hyphae. Values show the approximate percentage of appressorium sites exhibiting each type of interaction. (After Heath et al (22), with permission.)

Formation of secondary hyphae thus represents another poorly understood aspect of fungal development that is intimately associated with pathogenesis. Perhaps these intracellular structures are specialized for nutrient acquisition, or they may provide some protection from host-defense responses. Because structures resembling secondary hyphae have been observed after penetration into cellulose membranes (Figure 2; 6), the formation of these structures may be independent of cues specific to host tissue.

A variety of phytotoxic metabolites have been detected in cultures of the rice blast fungus, and several have been suggested to play a role in rice blast pathogenesis (39). These include piricularine (a substance of unknown structure), pyriculol and pyriculariol (polyketides), coumarin, picolinic acid, tyrosol, and tenuazonic acid. The significance of many of these reports has been difficult to determine because of a lack of reproducibility in the isolation of the reported substance, in the assay of its phytotoxic effects, or both. In most cases it has been difficult to detect the putative toxin at phytotoxic concentrations in diseased tissue. Tenuazonic acid (a cyclized acetoacetyl derivative of isoleucine) stands out among the possible phytotoxins because it has reproducibly been associated with the fungus in culture (38), and, although it has rather low specific activity, it has been detected at toxic concentrations in a variety of diseased tissues (74, 75). Isolation and genetic analysis of mutants deficient in production of putative toxins will be required to establish whether or not tenuazonic acid or other metabolites play a role in pathogenesis.

Genetics and Cytology of Host Specificity

Host specificity has been the basis for classifying pathogens of rice *(P. oryzae)* separately from pathogens of other grasses *(P. grisea)*. Even though MGR fingerprinting has now clearly distinguished the world population of rice pathogens from pathogens of other grasses (17), some field isolates of the rice pathogen cause disease on other grass species (47, 58). Because host ranges overlap, and because strains with diverse host specificities are interfertile and morphologically indistinguishable, *P. oryzae* and *P. grisea* have now been synonymized, with the earlier name, *P. grisea,* having priority (62). Studies by Leung & Williams (43), which revealed very little isozyme polymorphism among isolates of diverse host specificities and from diverse geographic locations, support this conclusion. Races of the rice pathogen are identified using differential cultivars that were selected to give, to the extent possible, all-or-nothing responses to field isolates of the pathogen (1, 36, 46, 87). However, some rice pathogens produce intermediate symptoms on some cultivars including small to intermediate size lesions (5, 46) or reproducibly clear differences in lesion density (inoculum efficiency) with no differences in lesion size (3).

Genetic analysis of complex phenotypes is fraught with danger. Pathogenicity is an extremely complex property involving such distinct components as infection efficiency (determines lesion number), rate of lesion development, extent of colonization (determines lesion size), and efficiency of sporulation. In addition, many aspects of this host-pathogen interaction are notoriously sensitive to environmental variation (4, 36). Therefore, it is critical that pathogenicity tests be clearly described, documented with photographs or preserved specimens (35, 79), and proven to be reproducible. The pathogenicity test represents the greatest challenge associated with molecular genetic analysis of rice blast.

HISTOLOGICAL COMPLEXITY A detailed cytological study of O-42, a field isolate of the rice pathogen (6, 7, 26, 78), and 4091-5-8, a laboratory strain that infects weeping lovegrass and goosegrass, was undertaken on rice, weeping lovegrass, and goosegrass (22). Figure 3 illustrates the variability in fungal development observed at different appressorial sites in a single plant-fungus interaction. The variability is especially pronounced for low to moderately compatible interactions. Macroscopic lesions can represent successful growth of the fungus at very few infection sites. The visible symptoms produced by O-42, a moderately successful pathogen of weeping lovegrass but a poor pathogen of goosegrass, resulted from successful growth at 10% and 1% of the infection sites of the two kinds of plants, respectively. At unsuccessful sites, failure was often due to a strikingly high number of appressoria that failed to penetrate. At other unsuccessful sites, failure of the pathogen was accompanied by cytoplasmic granulation of the invaded epidermal cell and by cell wall fluorescence. For O-42 on the youngest leaf of rice cultivar M201, a highly compatible interaction, 25% of all appressoria failed to penetrate but growing colonies developed at all appressorial sites where penetration occurred. In older rice leaves, which are known to be less susceptible than younger leaves (36), about 70% of all appressoria failed to penetrate and the fungus grew noticeably more slowly at the 20% of appressorial sites where growing colonies formed. Results similar to those shown in Figure 3 were observed for strain 4091-5-8 (22).

GENES FOR PATHOGENICITY AND VIRULENCE Genetic analysis of differences in host species specificity was first reported by Yaegashi (81). A pathogen of weeping lovegrass appeared to differ genetically from three pathogens of finger millet at two linked loci, one that controlled pathogenicity toward weeping lovegrass and the other that controlled pathogenicity toward finger millet. Using two of the same strains, Valent et al (77, 78) defined a single gene, named *Pwl1*, that determined pathogenicity toward weeping lovegrass, and they confirmed segregation of *Pwl1* in two subsequent crosses. Heath et al (21) reported the first efforts toward identifying cytological feature(s) affected by *Pwl1* by analyzing two full tetrads and the parents. Pathogenicity, defined as total withering and death of inoculated tissue, versus nonpathogenicity, defined as rare limited lesions, correlated perfectly with the absence or presence, respectively, of brown cells around developing colony margins. Other cytological features clearly did not cosegregate with *Pwl1*. Further analysis of a statistically significant number of progeny will determine whether or not *Pwl1* controls lesion expansion through an effect on browning of cells at the lesion margin.

In contrast to the simple genetic differences detected in the analysis of pathogens of weeping lovegrass and finger millet (78, 81), field isolates of the rice pathogen appear to differ from pathogens of weeping lovegrass and finger millet by several genes that control pathogenicity toward rice. Most of the progeny from crosses between such strains produce either no visible symptoms or small restricted lesions on rice (78, 79, 82, 85). An extensive study was undertaken to determine the genetic differences between O-135, a pathogen of both rice and of weeping lovegrass, and strain 4091-5-8, a pathogen of weeping lovegrass that produced no visible symptoms on any of 15 rice cultivars tested (79). Light-microscope studies showed that half of the appressoria produced by 4091-5-8 on rice cultivar M201 failed to penetrate (22). When penetration was successful, the fungus failed to grow out of the first-invaded epidermal cell, which exhibited granular cytoplasm, browning, and autofluorescence. In crosses between O-135 and 4091-5-8, all progeny infected weeping lovegrass but very few were pathogens of rice and those produced only small lesions ringed with brown.

A backcrossing scheme with the rice pathogen O-135 as the recurrent parent was pursued for six generations (79). The dual goals of these studies were to introgress the high fertility of 4091-5-8 into rice pathogens and to elucidate the genetic differences that govern the ability to infect rice. An analysis of the MGR sequences in a sample of the backcross progeny showed progressive restoration of the rice pathogen genetic background with each generation. Progeny from the backcross generations were assayed for pathogenicity toward rice cultivars CO39, M201, and Yashiro-mochi. Pathogenicity was assessed using a scale with six "lesion types" (type 0–type 5) defined by lesion size (79), with the division between pathogens and nonpathogens corresponding to whether or not lesions capable of supporting sporulation were produced. Significantly, normal distributions of lesion types often occurred among progeny that were pathogens of rice, with the number of progeny that produced the large type 5 lesions increasing as the backcrosses progressed. Multiple infection assays for lesion types demonstrated that individual strains were ranked consistently relative to other strains in the progeny sets. This reproducibility suggested that environmental variation was not the major factor determining lesion size in these experiments, but rather that the distributions of lesion types were due to the segregation of minor genes for lesion size on rice. These minor genes did not appear to play a role in pathogenicity toward weeping lovegrass because all strains were pathogens of weeping lovegrass.

In addition to the minor genes for lesion size, backcrossing identified a second type of gene that differed between the parents, which were weeping lovegrass and rice pathogens, respectively (79). In the later backcross generations, it became apparent that major genes with an all-or-nothing effect on the ability to infect rice were segregating. Three unlinked major genes, namely *Avr1-CO39*, *Avr1-M201*, and *Avr1-YAMO*, were identified. They were called "avirulence genes" because they had cultivar-specific effects (79), i.e. they determined virulence or avirulence on rice cultivars CO39, M201, and Yashiro-mochi, respectively. The avirulence alleles of these major genes appear to be derived from 4091-5-8, the pathogen of weeping lovegrass. Yaegashi & Asaga (82) reported strong circumstantial evidence that a pathogen of finger millet carries an avirulence gene corresponding to the rice resistance gene *Pi-a*. These results suggest that avirulence genes specific for rice cultivars are common in nonpathogens of rice.

Avirulence genes were identified more easily in crosses involving only rice pathogens, where minor genes did not segregate. In the RFLP mapping cross 4360, parent strain 4224-7-8 was avirulent on rice cultivars Yashiro-mochi, Maratelli, Tsuyuake, and Minehikari, and virulent on rice cultivars CO39, M201, and Sariceltik (B. Valent, L. Farrall, F. G. Chumley, manuscript in preparation). The second parent, strain 6043 (40), was virulent on all seven of these rice cultivars. All-or-nothing differences in the ability to infect specific rice cultivars segregated in this cross. Strain 4224-7-8, but not 6043, was a weeping lovegrass pathogen. Four avirulence genes, namely, *Avr2-YAMO*, *Avr1-MARA*, *Avr1-TSUY*, and *Avr1-MINE*, corresponding to rice cultivars Yashiro-mochi, Maratelli, Tsuyuake, and Minehikari, respectively, and a single gene that controls pathogenicity to weeping lovegrass, *Pwl2*, were shown to segregate in several subsequent crosses. Genetic analysis demonstrated that the *Avr2-YAMO* gene is not linked to the previously identified avirulence gene, *Avr1-YAMO* (79; B. Valent, L. Farrall, F. G. Chumley, manuscript in preparation). Likewise, *Pwl2*, the second major gene that controls pathogenicity to weeping lovegrass, is not linked to the previously identified *Pwl1* (77, 78).

Leung et al (40) identified four major genes controlling cultivar specificity in *M. grisea*. Two linked genes, named *Pos1* and *Pos2*, were identified in a RFLP mapping cross (64) between the rice pathogen GUY11 and strain 2539, which did not cause lesions on any of the rice cultivars tested. Pathogenicity was rated by use of six "interaction phenotypes" (IP) that were defined on a discontinuous scale involving lesion size and lesion density, two distinct aspects of pathogenicity. IP 0, 1, and 3 were defined by lesion size and IP 5, 7, and 9 were defined by lesion density. The distinction between avirulent and virulent progeny was drawn between IP 3, with small lesions that supported spare sporulation and IP 5, with lesions that supported profuse sporulation. Because one parent in this cross was not a rice pathogen, the segregation of minor genes for pathogenicity toward rice may have complicated scoring of segregation (79). Such genetic complexity makes it imperative that any hypothesis regarding single-gene segregation must be supported by several subsequent crosses to confirm segregation in different genetic backgrounds. The model of single gene segregation of host specificity was successfully tested for *Pos1* in a subsequent cross; but this same cross produced an aberrant segregation pattern for *Pos2* and called into question the genetic hypothesis by which it was named (40). No subsequent crosses were attempted for the two remaining genes reported by Leung et al (40), for the seven genes reported by Ellingboe et al (11), or for several genes reported by Tharreau (71a). Thus, these can only be considered as preliminary reports, awaiting further genetic analysis.

New Insights on Variation in Host Specificity

Reports on the degree of variability in cultivar specificity for races of *M. grisea* have been the subject of ongoing debate. Ou and his colleagues (56, 57), working mainly with Philippine strains of the fungus, reported that vegetative cultures derived from a single conidium produced conidia of so many different races that it was impossible to define a "race" of the fungus. Latterell (35–37) presented evidence that races were stable, although mutants with altered specificity were detected in some cases. Working in the Philippines, Bonman et al (4) did not observe the continuous variation reported by Ou. Those advocating that variation for cultivar specificity is continuous made no attempt to prove that the frequent "mutants" were not merely contaminants. Problems with differences in interpretation of borderline symptoms produced in environmentally sensitive infection assays and the potential effect of contaminated seed likewise were not addressed. In the future, an important standard must be imposed on discussions of mutability in host specificity: using MGR fingerprints (Figure 1) and other markers, contamination must be ruled out as a source of "variation," the raw results of pathogenicity assays must be reported, and the reproducibility of the assays and the purity of the host seed must be documented. Molecular genetic approaches are now resolving some of the major issues concerning variability.

Levy et al (45) have used DNA fingerprints based on MGR sequence polymorphisms to resolve issues of pathotype stability and diversity in US populations of the rice blast pathogen. They predicted that if races are stable, as suggested by Latterell, strains with similar patterns of cultivar specificity should represent distinct clonal lineages that could be marked by lineage-associated MGR fingerprints. They recognized that lineage-associated MGR fingerprints should be detectable only if MGR sequence polymorphisms are sufficiently variable to distinguish between lineages, but not so variable as to obscure lineages. MGR fingerprints were analyzed for 42 field isolates collected throughout the southern US rice-growing region over a period of 30 years. Strains were grouped into putative clonally derived lineages based on similarities in MGR fingerprints and without prior knowledge of pathotype identity. Seven MGR lineages proved to be diagnostic for pathotypes (International races IB-1, IC-17, IG-1, IB-45, ID-13, IH-1, and IB-49A). Two MGR lineages exhibited multiple pathotypes (IB-49B/IB54 and IB-45/ID13). Those cases where MGR lineages were not diagnostic for pathotype could be explained by mutation in specificity for a small number of cultivars. Two pathotypes (IB-49 and IG-1) were represented by a pair of lineages. Thus, it appears that the US blast-fungus population contains a limited number of clonal lineages. These results provide strong support for the view that US pathotypes are relatively stable and reliably distinguishable by carefully standarized pathogenicity assays (4, 37).

Rice pathogens isolated from the field in Central China frequently change specificity toward certain rice cultivars in laboratory studies (B. Valent, L. Farrall, F. G. Chumley, manuscript in preparation). That is, spontaneous virulent mutants appear frequently in standard infection assays. Typically, the virulent mutants have undergone a change in specificity toward only one cultivar. For example, mutants of strain O-137 became virulent on rice cultivar Yashiro-mochi with no change in specificity toward several other cultivars tested. Similarly, multiple independent mutants changed specificity toward cultivar Tsuyuake. O-137 appears to be the source of avirulence genes *Avr2-YAMO* (corresponding to Yashiro-mochi) and *Avr1-TSUY* (corresponding to Tsuyuake) contributed by strain 4224-7-8 to the RFLP cross 4360. The avirulence alleles of these two genes are unstable and this instability is retained through genetic crosses. The allele of the gene *Pwl2* that determines nonpathogenicity toward weeping lovegrass is also unstable. MGR fingerprints (Figure 1), as well as fertility, specificity toward other hosts and colony characteristics confirmed that presumptive mutants were indeed derived from expected parental strains. These three unstable genes for host specificity were clearly derived from rice pathogens. Several genes for host specificity derived from pathogens of other grasses do not appear to mutate at the same high frequency. The degree of instability observed in these rice pathogens is greater than that described by Latterell (35, 36, 37) in her extensive studies of other strains. It appears that stability for host specificity may vary depending on the specific strains and genes involved.

Blast caused by *M. grisea* has become a serious problem for wheat farmers in Brazil, starting in the 1980s. Because the disease first occurred in the state of Parana, where rice blast was prevalent, it seemed possible that the rice-blast fungus had mutated to become a pathogen of wheat. However, MGR fingerprinting of Brazilian rice and wheat pathogens has clearly demonstrated that the Brazilian wheat pathogens are not derived from rice pathogens indigenous to the area (F. G. Chumley, L. Farrall, B. Valent, unpublished results). Brazilian rice pathogens resembled rice pathogens collected from various parts of the world in having many copies of MGR sequences. In contrast, the wheat pathogens exhibited MGR fingerprints typical of non-pathogens of rice, having very few bands with MGR sequence homology.

A Look to Future Efforts

Classical genetics, cytology, and cell biology are providing new insights into the molecular basis of pathogenicity and host specificity in *M. grisea*. The next challenge is to clone critical genes and analyze their modes of action. Technology is now in place for both approaches to cloning fungal genes for which the gene product is unknown: (*a*) cloning genes by expression of function ("shotgun" cloning), and (*b*) cloning genes based on map position, through chromosome walking. Cloning genes that control host cultivar and host species specificity should provide clues to how these genes function. These clues could come from cytological localization of the gene product in infected tissue, or from comparisons of the sequences of the cloned genes to sequence databases. Understanding how pathogen genes function should provide insight on how plant resistance genes function. Because techniques for the molecular genetic manipulation of rice are becoming more sophisticated, direct analysis of host resistance mechanisms is an ever more attractive possibility.

As in bacterial plant pathogens (33), avirulence genes of *M. grisea* corresponding to specific rice cultivars appear to be common both in pathogens and in nonpathogens of rice. It remains to be determined whether or not the rice-blast system represents a "gene-for-gene" system as first defined by Flor (14). Classical avirulence genes are dominant, suggesting that the expressed form of the gene confers the avirulence phenotype. Since dominance cannot be determined for avirulence genes of *M. grisea* by classical methods (8), dominance will be confirmed only after the genes have been cloned. It must also be determined if avirulence genes identified in the fungus correspond to dominant resistance genes in the host as predicted by the gene-for-gene model. It will be interesting to learn if the function of major genes controlling pathogenicity toward weeping lovegrass is analogous to the function of avirulence genes in the rice pathogen. The mode of action of minor genes controlling lesion size on rice is equally interesting. It should be determined whether or not these minor genes in the pathogen correspond functionally to "minor resistance genes" in rice and what if any cultivar specific effects they may have.

Continuous variation in cultivar specificity (56, 57) and extreme variation in chromosome number (56, 57, 72) reported for *M. grisea* have not been supported by subsequent studies. Nevertheless, this fungus exhibits interesting variability in colony appearance, fertility, pathogenicity and host specificity, in addition to high mutability at specific genetic loci such as *BUF1* and *SMO1*. Analysis of stable and unstable avirulence genes will provide insight into the mechanisms that give rise to new races of the rice pathogen in the field. Understanding the potential for variation in the fungus should have high priority. The high frequency of repeated sequences in the fungal genome may account for some variability via homologous recombination leading to deletion of genomic segments. Transposons and B- like chromosomes are potential sources of variation. Polymorphisms in the lengths of double-strand RNAs (27), while apparently not correlated with cultivar specificity, may account for other types of variability. The parasexual cycle, which appears to be quite active (8, 16), should be further explored as a potentially important source of variation in the field.

Differences in genome organization between strains of the rice-blast fungus will limit genetic analysis. Therefore, it is critical to focus on a small number of highly fertile laboratory strains with limited diversity in genetic background, and to understand in detail the molecular genetics, cytology, cell biology and biochemistry of the interaction of these strains with key host plants. The isolation and analysis of nonpathogenic mutants derived from fertile, highly pathogenic strains should receive a high priority, because it will permit the identification of genes that play key roles in pathogenesis. The products of such genes could serve as targets for the design of novel, environmentally safe fungicides. Understanding the mode of action of essential pathogenicity determinants should also provide insights for engineering disease-resistant crop cultivars. The complexity emerging in the host-pathogen system of *M. grisea* increases the challenge and the anticipated satisfaction when molecular mechanisms of pathogenicity and host specificity are unravelled.

Literature Cited

1. Atkins, J. G., Robert, A. L., Adair, C. R., Goto, K., Kozaka, T., et al. 1967. An international set of rice varieties for differentiating races of *Pyricularia oryzae*. *Phytopathology* 57:297–301
2. Ballance, D. J. 1986. Sequences important for gene expression in filamentous fungi. *Yeast* 2:229–36
3. Bonman, J. M., Bandong, J. M., Lee, Y. H., Lee, E. J., Valent, B. 1989. Race-specific partial resistance to blast in temperate japonica rice cultivars. *Plant Dis*. 73:496–99
4. Bonman, J. M., Vergel De Dios, T. I., Bandong, J. M., Lee, E. J. 1987. Pathogenic variability of monoconidial isolates of *Pyricularia oryzae* in Korea and in the Philippines. *Plant Dis*. 71:127–30
5. Bonman, J. M., Vergel De Dios, T. I., Khin, M. M. 1986. Physiologic specialization of *Pyricularia oryzae* in the Philippines. *Plant Dis*. 70:767–69
6. Bourett, T. M., Howard, R. J. 1990. In vitro development of penetration structures in the rice blast fungus *Magnaporthe grisea*. *Can. J. Bot*. 68:329–42.

6a. Bucheli, P., Doares, S. H., Albersheim, P., Darvill, A. 1990. Host-pathogen interactions 36. Partial purification and characterization of heat-labile molecules secreted by the rice blast pathogen that solubilize plant cell wall fragments that kill plant cells. *Physiol. Mol. Plant Pathol*. 36:159–73

7. Chumley, F. G., Valent, B. 1990. Genetic analysis of melanin-deficient, nonpathogenic mutants of *Magnaporthe grisea*. *Mol. Plant-Micr. Interact*. 3: 135–43
8. Crawford, M. S., Chumley, F. G., Weaver, C. G., Valent, B. 1986. Characterization of the heterokaryotic and vegetative diploid phases of *Magnaporthe grisea*. *Genetics* 114:1111–29
9. Daboussi, M. J., Djeballi, A., Gerlinger, C., Blaiseau, P. L., Bouvier, I., et al. 1989. Transformation of seven spe-

cies of filamentous fungi using the nitrate reductase gene of *Aspergillus nidulans*. *Curr. Genet.* 15:453–56

10. Deleted in proof
11. Ellingboe, A. H., Wu, B. C., Robertson, W. 1990. Inheritance of avirulence/virulence in a cross of two isolates of *Magnaporthe grisea* pathogenic to rice. *Phytopathology* 80:108–11
12. Fanning, T. G., Singer, M. F. 1987. LINE-1: a mammalian transposable element. *Biochim. Biophys. Acta* 910:203–12
13. Fincham, J. R. S. 1989. Transformation in fungi. *Microbiol. Rev.* 53:148–70
14. Flor, H. H. 1971. Current status of the gene-for-gene concept. *Annu. Rev. Phytopathol.* 9:275–96
15. Foster, W., Neumann, E. 1989. Gene transfer by electroporation. A practical guide. In *Electroporation and Electrofusion in Cell Biology,* ed. E. Neumann, E. A. Sowers, C. A. Jordan, pp. 299–318. New York: Plenum
16. Genovesi, A. D., Magill, C. W. 1976. Heterokaryosis and parasexuality in *Pyricularia oryzae* Cavara. *Can. J. Microbiol.* 22:531–36
17. Hamer, J. E., Farrall, L., Orbach, M. J., Valent, B., Chumley, F. G. 1989. Host species-specific conservation of a family of repeated DNA sequences in the genome of a fungal plant pathogen. *Proc. Natl. Acad. Sci. USA* 86:9981–85
18. Hamer, J. E., Givan, S. 1990. Genetic mapping with dispersed repeated sequences in the rice blast fungus: Mapping the *SMO* locus. *Mol. Gen. Genet.* 223:487–95
19. Hamer, J. E., Howard, R. J., Chumley, F. G., Valent, B. 1988. A mechanism for surface attachment in spores of a plant pathogenic fungus. *Science* 239:288–90
20. Hamer, J. E., Valent, B., Chumley, F. G. 1989. Mutations at the *SMO* genetic locus affect the shape of diverse cell types in the rice blast fungus. *Genetics* 122:351–61
21. Heath, M. C., Valent, B., Howard, R. J., Chumley, F. G. 1990. Correlations between cytologically detected plant-fungal interactions and pathogenicity of *Magnaporthe grisea* toward weeping lovegrass. *Phytopathology* 80:1382–86
22. Heath, M. C., Valent, B., Howard, R. J., Chumley, F. G. 1990. Interactions of two strains of *Magnaporthe grisea* with rice, goosegrass and weeping lovegrass. *Can. J. Bot.* 68:1627–37
23. Hebert, T. T. 1971. The perfect stage of *Pyricularia grisea*. *Phytopathology* 61:83–87
24. Hirayama, T., Sudo, T., Nagayama, H., Matsuda, K., Tamari, K. 1976. Number and interrelation of components of carboxymethyl cellulase enzyme from *Pyricularia oryzae*. *Agr. Biol. Chem.* 40:2137–42
25. Howard, R. J., Bourett, T. M., Ferrari, M. A. 1991. Infection by *Magnaporthe grisea:* An in vitro analysis. In *Electron Microscopy of Plant Pathogens,* ed. K. Mendgen, D-E. Lesemann, pp. 251–64. Berlin:Springer-Verlag
26. Howard, R. J., Ferrari, M. A. 1989. Role of melanin in appressorium function. *Exp. Mycol.* 13:403–18
27. Hunst, P. L., Latterell, F. M., Rossi, A. E. 1986. Variation in double-stranded RNA from isolates of *Pyricularia oryzae*. *Phytopathology* 76:674–78
28. Itoi, S., Mishima, T., Arase, S., Nozu, M. 1983. Mating behavior of Japanese isolates of *Pyricularia oryzae*. *Phytopathology* 73:155–58
29. Jones, R. N., Rees, H. 1982. *B Chromosomes*. London: Academic. 266 pp.
30. Kato, H., Yamaguchi, T. 1982. The perfect state of *Pyricularia oryzae* Cav. from rice plants in culture. *Ann. Phytopathol. Soc. Jpn* 48:607–12
31. Kato, H., Yamaguchi, T., Nishihara, N. 1976. The perfect state of *Pyricularia oryzae* Cav. in culture. *Ann. Phytopathol. Soc. Jpn.* 42:507–10
32. Klein, T. M., Wolf, E. D., Wu, R., Sanford, J. C. 1987. High velocity microprojectiles for delivering nucleic acids into living cells. *Nature* 327:70–73
33. Kobayashi, D. Y., Tamaki, S. J., Keen, N. T. 1989. Cloned avirulence genes from the tomato pathogen *Pseudomonas syringae* pv. tomato confer cultivar specificity. *Proc. Natl. Acad. Sci. USA* 86:157–61

33a. Kolattukudy, P. E. 1985. Enzymatic penetration of the plant cuticle by fungal pathogens. *Annu. Rev. Phytopathol.* 23:223–50

34. Kolmer, J. A., Ellingboe, A. H. 1988. Genetic relationships between fertility and pathogenicity and virulence to rice in *Magnaporthe grisea*. *Can. J. Bot.* 66:891–97
35. Latterell, F. M. 1975. Phenotypic stability of pathogenic races of *Pyricularia oryzae,* and its implications for breeding of blast resistant rice varieties. In *Proc. Sem. Horizontal Resistance to the Blast Disease of Rice,* pp. 199–234. Cali, Colombia: Cent. Int. Agric. Trop.
36. Latterell, F. M., Marchetti, M. A.,

Grove, B. R. 1965. Co-ordination of effort to establish an international system for race identification in *Pyricularia oryzae*. In *The Rice Blast Disease. Proc. Symp. Intl Rice Res. Inst., July, 1963,* pp. 257–74. Baltimore: Johns Hopkins Press

37. Latterell, F. M., Rossi, A. E. 1986. Longevity and pathogenic stability of *Pyricularia oryzae*. *Phytopathology* 76:231–35
38. Lebrun, M-H., Dutfoy, F., Gaudemer, F., Kunesch, G., Gaudemer, A. 1990. Detection and quantification of the fungal phytotoxin tenuazonic acid produced by *Pyricularia oryzae*. *Phytochemistry* 29:3777–83
39. Lebrun, M-H., Orcival, J., Duchartre, C. 1984. Resistance of rice to tenuazonic acid, a toxin from *Pyricularia oryzae*. *Rev. Cytol. Biol. Végét. Bot.* 7:249–59
40. Leung, H., Borromeo, E. S., Bernardo, M. A., Notteghem, J. L. 1988. Genetic analysis of virulence in the rice blast fungus *Magnaporthe grisea*. *Phytopathology* 78:1227–33
41. Leung, H., Lehtinen, U., Karjalainen, R., Skinner, D., Tooley, P., et al. 1990. Transformation of the rice blast fungus *Magnaporthe grisea* to hygromycin B resistance. *Curr. Genet.* 17:409–11
42. Leung, H., Williams, P. H. 1985. Genetic analyses of electrophoretic enzyme variants, mating type, and hermaphroditism in *Pyricularia oryzae* Cavara. *Can. J. Genet. Cytol.* 27:697–704
43. Leung, H., Williams, P. H. 1986. Enzyme polymorphism and genetic differentiation among geographic isolates of the rice blast fungus. *Phytopathology* 76:778–83
44. Leung, H., Williams, P. H. 1987. Nuclear division and chromosome behavior during meiosis and ascosporogenesis in *Pyricularia oryzae*. *Can. J. Bot.* 65:112–23
45. Levy, M., Romao, J., Marchetti, M. A., Hamer, J. E. 1991. DNA fingerprinting with a dispersed repeated sequence resolves pathotype diversity in the rice blast fungus. *Plant Cell* 3:95–102
46. Ling, K. C., Ou, S. H. 1969. Standarization of the international race numbers of *Pyricularia oryzae*. *Phytopathology* 59:339–42
47. Mackill, A. O., Bonman, J. M. 1986. New hosts of *Pyricularia oryzae*. *Plant Dis.* 70:125–27
48. Michelmore, R. W., Hulbert, S. H. 1987. Molecular markers for genetic analysis of phytopathogenic fungi. *Annu. Rev. Phytopathol.* 25:383–404
49. Nagakubo, T., Taga, M., Tsuda, M., Ueyama, A. 1983. Genetic linkage relationships in *Pyricularia oryzae*. *Mem. Coll. Agric., Kyoto Univ.* 122:75–83
50. Nagakubo, T., Taga, M., Tsuda, M., Ueyama, A. 1983. Sexuality in *Pyricularia oryzae*. *Mem. Coll. Agric., Kyoto Univ.* 122:53–59
51. Notteghem, J. L. 1990. Results and orientation of EC projects on rice resistance to blast. *Summa Phytopathol.* 16:57–72
52. Notteghem, J. L., Silue, D., Milazza, J. 1991. Mating type analysis of *Magnaporthe grisea* populations pathogenic on rice. *Phytopathology*. In press
53. Orbach, M. J., Porro, E. B., Yanofsky, C. 1986. Cloning and characterization of the gene for beta-tubulin from a benomyl-resistant mutant of *Neurospora crassa* and its use as a dominant selectable marker. *Mol. Cell. Biol.* 6:2452–61
54. Orbach, M. J., Vollrath, D., Davis, R. W., Yanofsky, C. 1988. An electrophoretic karyotype of *Neurospora crassa*. *Mol. Cell. Biol.* 8:1469–73
55. Orr-Weaver, T. L., Szostak, J. W. 1983. Yeast recombination: the association between double-strand gap repair and crossing-over. *Proc. Natl. Acad. Sci. USA* 80:4417–21
56. Ou, S. H. 1980. A look at worldwide rice blast disease control. *Plant Dis.* 64:439–45
57. Ou, S. H. 1980. Pathogen variability and host resistance in rice blast disease. *Annu. Rev. Phytopathol.* 18:167–87
58. Ou, S. H. 1985. *Rice Diseases*. Kew, Surrey: Commonw. Mycol. Inst. 380 pp. 2nd ed.
59. Parsons, K. A. 1988. *The development of genetic transformation in the ascomycete* Magnaporthe grisea *and the cloning of the* LYS1$^+$ *gene*. PhD thesis. Univ. Colo., Boulder
60. Parson, K. A., Chumley, F. G., Valent, B. 1987. Genetic transformation of the fungal pathogen responsible for rice blast disease. *Proc. Natl. Acad. Sci. USA* 84:4161–65
61. Puhalla, J. E., Mayfield, J. E. 1974. The mechanism of heterokaryotic growth in *Verticillium dahliae*. *Genetics* 76:411–22
62. Rossman, A. Y., Howard, R. J., Valent, B. 1990. *Pyricularia grisea,* the correct name for the rice blast disease fungus. *Mycologia* 82:509–12
63. Rothstein, R. J. 1983. One-step gene disruption in yeast. In *Recombinant DNA,* ed. R. Wu, G. Lawrence, K. Moldave, pp. 202–11. New York: Academic
64. Skinner, D. Z., Leung, H., Leong, S. A. 1990. Genetic map of the blast fun-

gus, *Magnaporthe grisea*. In *Genetic Maps; Locus Maps of Complex Genomes*, ed. S. J. O'Brien, pp. 3.82–3.83. Cold Spring Harbor, NY: Cold Spring Harbor Press. 5th ed.

65. Staben, C., Jensen, B., Singer, M., Pollock, J., Schechtman, M., et al. 1989. Use of a bacterial Hygromycin B resistance gene as a dominant selectable marker in *Neurospora crassa* transformation. *Fungal Genet. Newsl.* 36:79–81
66. Sumizu, K., Yoshikawa, M., Tanaka, S. 1961. Studies of xylanase of *Pyricularia oryzae*. *J. Biochem.* 50:538–43
67. Sweigard, J. A., Orbach, M. J., Valent, B., Chumley, F. G. 1990. A miniprep procedure for isolating genomic DNA from *Magnaporthe grisea*. *Fungal Genet. Newsl.* 37:4
68. Taga, M., Nakagawa, H., Tsuda, M., Ueyama, A. 1978. Ascospore analysis of kasugamycin resistance in the perfect stage of *Pyricularia oryzae*. *Phytopathology* 68:815–17
69. Taga, M., Nakagawa, H., Tsuda, M., Ueyama, A. 1979. Identification of three different loci controlling kasugamycin resistance in *Pyricularia oryzae*. *Phytopathology* 69:463–66
70. Taga, M., Waki, T., Tsuda, M., Ueyama, A. 1982. Fungicide sensitivity and genetics of IBP-resistant mutants of *Pyricularia oryzae*. *Phytopathology* 72:905–8
71. Tanaka, Y., Murata, N., Kato, H. 1979. Behavior of nuclei and chromosomes during ascus development in the mating between either rice-strain or weeping lovegrass strain and ragi-strain of *Pyricularia*. *Ann. Phytopathol. Soc. Jpn.* 45:182–91

71a. Tharreau, D. 1990. *Contribution a l'étude de l'hérédité de l'avirulence chez* Pyricularia oryzae *Cavara, et de l'hérédité de la résistance de son hôte* (Oryza sativa *L.*). DEA thesis. Univ. Paris XI, VI.

72. Tolmsoff, W. J. 1983. Heteroploidy as a mechanism of variability among fungi. *Annu. Rev. Phytopathol.* 21:317–40
73. Ueyama, A., Tsuda, M. 1975. Formation of the perfect state in culture of *Pyricularia* sp. from some graminaceous plants (preliminary report). *Trans. Mycol. Soc. Jpn.* 16:420–22
74. Umetsu, N., Kaji, J., Aoyama, K., Tamari, K. 1974. Toxins in blast-diseased rice plants. *Agric. Biol. Chem.* 38:1867–74
75. Umetsu, N., Kaji, J., Tamari, K. 1973. Isolation of tenuazonic acid from blast diseased rice plants. *Agric. Biol. Chem.* 37:451–52
76. Valent, B. 1990. Rice blast as a model system for plant pathology. *Phytopathology* 80:33–36
77. Valent, B., Chumley, F. G. 1987. Genetic analysis of host species specificity in *Magnaporthe grisea*. In *Molecular Strategies for Crop Protection*, ed. C. J. Arntzen, C. Ryan, pp. 83–93. New York: Liss
78. Valent, B., Crawford, M. S., Weaver, C. G., Chumley, F. G. 1986. Genetic studies of fertility and pathogenicity in *Magnaporthe grisea (Pyricularia oryzae)*. *Iowa State J. Res.* 60:569–94
79. Valent, B., Farrall, L., Chumley, F. G. 1991. *Magnaporthe grisea* genes for pathogenicity and virulence identified through a series of backcrosses. *Genetics* 127:87–101
80. Winston, F., Chumley, F. G., Fink, G. R. 1983. Eviction and transplacement of mutant genes in yeast. See Ref. 63, pp. 211–28
81. Yaegashi, H. 1978. Inheritance of pathogenicity in crosses of *Pyricularia* isolates from weeping lovegrass and finger millet. *Ann. Phytopathol. Soc. Jpn.* 44:626–32
82. Yaegashi, H., Asaga, K. 1981. Further studies on the inheritance of pathogenicity in crosses of *Pyricularia oryzae* with *Pyricularia* sp. from finger millet. *Ann. Phytopathol. Soc. Jpn.* 47:677–79
83. Yaegashi, H., Hebert, T. T. 1976. Perithecial development and nuclear behavior in *Pyricularia*. *Phytopathology* 66:122–26
84. Yaegashi, H., Nishihara, N. 1976. Production of the perfect stage in *Pyricularia* from cereals and grasses. *Ann. Phytopathol. Soc. Jpn.* 42:511–15
85. Yaegashi, H., Yamada, M. 1984. Possible genetic factor conditioning the size of blast lesions induced on rice leaves with *Pyricularia oryzae*. *Ann. Phytopathol. Soc. Jpn.* 50:664–67
86. Yaegashi, H., Yamada, M. 1986. Pathogenic race and mating type of *Pyricularia oryzae* from Soviet Union, China, Nepal, Thailand, Indonesia and Colombia. *Ann. Phytopathol. Soc. Jpn.* 52:225–34
87. Yamada, M., Kiyosawa, S., Yamaguchi, T., Hirano, T., Kobayashi, T., et al. 1976. Proposal of a new method for differentiating races of *Pyricularia oryzae* Cavara in Japan. *Ann. Phytopathol. Soc. Jpn.* 42:216–19

Annu. Rev. Phytopathol. 1991. 29:469–90

CURRENT OPTIONS FOR NEMATODE MANAGEMENT

Larry W. Duncan

Citrus Research and Education Center, University of Florida, 700 Experiment Station Road, Lake Alfred, Florida 33850

KEY WORDS: crop loss assessment, modeling, IPM sampling, cultural control, biological control, chemical control

---

INTRODUCTION

Methods currently used to control losses due to plant-parasitic nematodes have changed very little in a generation. While research during the past decade in such areas as nematode physiology, biological control, system simulation, and molecular biology has progressed rapidly, results are only beginning to influence programs for nematode management. This slow pace of change belies the rate at which new management options can be expected in the future. As in all areas of biology, advances in biochemistry and molecular biology will directly affect nematode management through improvements in host plant resistance (56, 104, 148), the use of nematode antagonists (71), and new methods to interrupt the normal life-cycle of nematodes (38). Technical methods derived partly from such research are being applied to other areas of nematode management such as species identification (16) and the quantification of nematode populations in the soil (63).

A public desire to change many current methods of managing plant pests in ways that do not pollute or otherwise degrade the environment has increased concomitantly with progress in research (70, 139). The concepts of integrated pest management (IPM) and sustainable agriculture (146) evolved in response to environmental concerns. Because IPM provides a working methodology for pest management in sustainable agricultural systems, methods to manage nematodes have evolved and will continue to evolve in response to IPM

0066-4286/91/0901-0469$02:00

principles or, paraphrasing Geier (52), *a* determining how the life system of a pest must be modified to reduce it to tolerable levels *b* affecting the modification through biological knowledge combined with current technology and *c* developing new methods of control with current technology that are compatible with economic and environmental requirements. The appeal of IPM is its emphasis on reducing the role of pest management in environmental degradation by using the safest tactics available in the context of economic and environmental needs, and by invoking management only when it is determined to be necessary through biomonitoring and use of economic thresholds. In this review some current options for managing nematodes are discussed within the guidelines of IPM. My emphasis is on methods by which decisions are made to manage nematode problems with currently available tactics.

ELEMENTS OF NEMATODE MANAGEMENT PROGRAMS

The principles of IPM outlined above emphasize the need for accurate biomonitoring and predictions of crop losses to provide a rational basis for selecting nematode management tactics (9, 46). Accordingly, the fundamental elements of nematode management programs considered below include sampling, prediction, and management tactics.

Sampling

Many methods used to manage nematodes do not require regular sampling to achieve economic benefits. Nevertheless, for most management tactics, systems can be developed to optimize economic inputs and returns and minimize environmental risk, if accurate low-cost methods of sampling are available to determine the species and in some cases the population density of nematodes in a field (42, 96). Sampling is critical to evaluate the need to use nematicides. Where varieties of crops are available that are resistant to major nematode pests, population monitoring is desirable if a cultivar is not the best one available in the absence of the pest (110), or if other nematode species may be present for which resistance is not available (31). Other cultural practices, such as the use of crop rotations or fallow, are employed to best advantage in response to information about the species and numbers of nematodes in a field. Notable exceptions to the need for sampling are systems in which nematodes are controlled due to practices used primarily for other reasons. Commercial tomato growers in Florida have no need to consider options to manage *Meloidogyne spp.* as long as soil fumigation with methyl bromide is used for broad spectrum control of weeds and soilborne pathogens other than nematodes (99a).

Direct and indirect methods are commonly used to estimate nematode levels in fields. With most annual crops, sampling should occur before sowing and may occur before or after an overwinter period that precedes sowing. Preplant sampling is necessary when management options cannot be used (rotation, resistance, soil fumigation) or are less effective (some nematicides) after planting. In some systems it is useful to estimate nematode levels in fields after harvest and before overwintering (8, 74). Population density and thus sample accuracy is higher at this period than following overwintering. Preoverwinter soil sampling practiced by soybean growers in Florida provides adequate time to prepare to grow resistant lines of soybean if *M. incognita* is detected or to grow nonhost crops if the more virulent *M. arenaria* is found (75). In crops attacked by *Meloidogyne spp.*, indirect estimation of nematode densities based on root-gall severity can be the most cost-effective sampling method if performed in the field at or prior to harvest.

A Canadian IPM program to manage *M. hapla* on carrot is based on assessment of nematode damage potential via root damage indices obtained from the previous carrot crop (8). Root damage assessment requires less time and labor than processing soil and root samples for nematode counts and permits sequential sampling to optimize sample collection, because samples are analyzed individually, unlike samples for nematode counts that are usually bulked and often subsampled. When each sample is evaluated, damage foci within a field are apparent and may be treated selectively. This detection of damage foci has resulted in widespread adoption of spot treatments with nematicides and subdivision of fields for cropping to nonhost crops by carrot growers in southwestern Quebec (G. Belair, personal communication).

Despite advantages, some aspects of data obtained from preoverwinter sampling require further characterization. A logarithmic relationship between gall indices on carrots during two cropping seasons (8) suggests a measurable relationship between population density and damage symptoms at harvest. However, early season damage to the plant due to heavy nematode pressure often limits population development and may result in similar end-of-season nematode levels in areas of light and heavy root damage (31). Interpreting the significance of samples taken prior to overwintering can also be confounded due to unpredictable, annual variation in survivorship (34, 45, 66).

A second indirect sample method uses seedlings planted in soil from samples as a bioassay based on symptoms of nematode infection (13, 90, 91). The method requires effective planning by growers to obtain samples early enough to permit bioassay development before planting. It is useful only with nematodes that quickly cause distinctive root symptoms. Bioassays can involve significantly less labor than methods requiring extraction and counting of nematodes and are thus particularly useful in areas where the demand for the assay is seasonally high. The method can also reduce the detection limits of species that may be present in low numbers. Wheat growers in Australia estimate the need for fumigation to manage *Heterodera avenae* based on bioassay results from samples taken several months previously (13).

In most monitoring systems nematodes are extracted from samples of soil or plant material and identified and enumerated directly (5, 85). Generally, 10–25 samples from a defined area are composited and mixed to reduce processing expense. Nematodes may be recovered from the entire sample or from subsamples, depending on the extraction technique.

Estimates of the time necessary to obtain and process typical soil samples range from 45 min to several hours (51, 122). A number of standard techniques are used to optimize sample expense in terms of sample quality. Systematic sampling patterns are generally used within areas selected for soil uniformity or symptom expression to control obvious sources of variability in count data (51, 60). In perennial crops, timing and sample location are selected to maximize population density and reduce sample variability(27, 49). Eventually, a generalized optimum sampling pattern and sample size for the selected conditions must be estimated from count data.

Studies of nematode spatial patterns have been used to optimize sampling either by computer simulation using actual counts or mathematical formulae descriptive of the spatial patterns (51, 57, 98, 99, 127). The last approach was used recently in The Netherlands to estimate optimum sampling patterns and sample size to reliably detect single, low density foci of *Globodera rostochiensis* in potato fields (123). Results of preliminary surveys suggest that adoption of the newly developed sampling protocols could result in reducing by 80% the current level of government-required soil fumigation treatments (C. H. Schomaker, personal communication). Other approaches used to estimate an optimum sample size given a particular sampling pattern include analyses of sample precision and accuracy in specific fields (24, 84, 122) or utilization of Taylor's Power Law (137) to estimate the general relationships between the sample mean and variance (11, 32, 50, 87). The mean density for which an optimum sample size is estimated depends on whether the objective is detection (123) or quantification with reference to an economic threshold (44).

It is useful to investigate the effect of mean population density on general sample requirements since economic thresholds vary greatly, based on host crops, edaphic and climatic conditions, and crop prices and management costs. Although there is evidence that mean and variance relationships from nematode sample counts are relatively stable (32, 50), they are likely to be more variable than those for many nonsoil-dwelling organisms. This is due in part to the low mobility of nematodes and the extreme heterogeneity of soil environments compared to those above ground. The relationship of the variance to the mean can be strongly influenced by factors (historical incidence of infestation, cultural practices, edaphic conditions, and cropping patterns) that concomitantly effect the nematode spatial patterns and the carrying capacity of the soil. Where variance to mean relationships of *T. semipenetrans* in citrus were derived from Taylor's Power Law, differences were identified based on the size of nematode patches (32). This suggests that a sampling program in orchards might profitably begin by sampling relatively small (1–2 ha) areas to determine whether future sampling of larger (5 ha) areas is appropriate. Where patch size is small, their number will help determine the degree to which stratified sampling is worthwhile. Heterogeneous soil textures within orchards was also found to strongly influence the variance to mean relationship (29a), indicating that easily recognizable factors may be useful for refining estimates of optimum sample size.

Samples sufficiently large to reliably measure population density within 25–50% of the true mean have been suggested as acceptable for diagnostic use and are often the norm in field research (24, 84). Low sample accuracy is imposed by the high cost of sampling with respect to the current predictive value of samples using current technology. Nevertheless, ranges of nematode population density between fields are often large and, unlike many pest organisms, fairly stable over short time intervals. Therefore, population estimates from samples are often sufficiently high or low to permit sound management decisions. Uncertainty over whether to invoke management tactics when population measurements are near the economic threshold is currently as much due to unmeasured variables that confound crop-loss predictions as to low sample accuracy. Simple detection of highly pathogenic nematodes can be a basis for invoking management, and regular, frequent sampling is one method used by commercial tomato growers in California to enhance the reliability of detecting *Meloidogyne spp*. (P. Roberts, personal communication).

Nematology sample sizes recommended by advisory agencies are often too small to estimate mean density for management purposes according to some published estimates (24, 32, 50, 87). However, the ability to accurately estimate the variability of composite samples from data sets of counts from individual cores of soil was questioned recently (122).A commonly used formula (39, 131) in entomology and nematology to derive optimum sample size tends to overestimate by an expected value of three units (35). The observation is trivial when the optimum number of samples is large, for example when damage indices are measured in the field (8) or when estimating an optimum number of individual soil cores in a sample (57). However, when means and variances associated with replicate composite samples from the same field are used to derive an optimum sample size, the derived value tends to be small and may overestimate sampling requirements severalfold.

Predicting Nematode Population Change and Crop Loss

Most applied prediction in nematology is confined to qualitative descriptions of expected crop loss based on broad density classes of nematodes (5, 142). Relationships between nematode population densities at two points in time have also been measured in a large number of systems, and to some extent have influenced general recommendations concerning cropping patterns (68, 75, 111, 141a). Although specific recommendations may be based on data that illustrate very definite relationships, it surprises few people when expected results are well off the mark. For this reason advisory programs currently tend to be conservative in predicting management thresholds (46).

Forecasting crop loss due to nematodes has relied primarily on critical point (34) damage functions derived by fitting experimental yield data to preplant population densities of nematodes under defined conditions (e.g. 4, 31, 47, 68, 73, 76, 88, 101, 125, 128). Data for these models are obtained from experiments in pots and microplots where nematode densities are controlled, and also from field plots where densities are estimated. The limited number of studies to develop simulation models of nematode host relationships (41, 89) is due partly to recognition that few options are available for nematode control during the course of a growing season. Although simulation models provide a method to investigate the nature of nematode effects on plants, research requirements are much more extensive than those for single point models. An advantage in a program to develop both types of models is that factors found through simulation to significantly affect the damage function could become part of the environmental definition for future critical point models.

Theoretical foundations for using damage functions to optimize management inputs based on probable crop response have been developed (42, 44, 102) and will become increasingly useful as sample precision improves and as data bases are developed to support their use. Critical point damage functions are useful because most management options must be initiated prior to planting a crop and because nematode migration is relatively slow. However, these models are loosely based on the seldom-met assumptions that, for given edaphic conditions, key environmental variables such as soil temperature and moisture are reasonably stable. Predictably, variability of damage functions is highest in regions with unstable weather patterns and variable soil conditions, such as nonirrigated agriculture in Florida (78, 88), and less so in irrigated agriculture in more stable climates as found in California's San Joaquin Valley (46). Estimates of variability in damage functions due to uncontrolled factors such as weather and to the heterogeneity of edaphic factors and cultural practices that occur in all agricultural systems (43, 76, 77, 88, 141a) are needed to begin to evaluate economic risks associated with management tactics and to help identify the most important regional determinants of the shape of the damage function.

Since the relationship between yield and population density is generally nonlinear (125), the problem of predicting crop loss using samples that average the counts from fields in which nematode populations are aggregated is recognized (106, 126) but unresolved for most systems. The use of computer simulation to integrate crop losses across a field by estimating the nematode microdistribution based on a composite sample has been proposed (43). For example, a measurement of the population mean from a field sample can be used to estimate sample variance from Taylor's Power Law. Mean and variance estimates can be further used to derive an estimate of the parameter k from the negative binomial distribution (39), which is often descriptive of the frequency distribution of nematode counts from single cores of soil. A similar approach was used to derive formulae relating the degree of tobacco loss overestimation to composite sample means of the density of *M. incognita* (97). There is currently insufficient information on whether frequency distribution parameters for most nematodes can be accurately estimated from global samples to know whether these methods provide a robust solution to the problem of crop-loss overestimation.

Critical point models of the relationship of final population density (Pf) to initial density (Pi) are commonly used in nematology for many of the same reasons as for the damage function. Logistic models have been derived with parameters descriptive of nematode biology, and simple linear regression models often approximate density-dependent population change when data are log-transformed (34, 46, 68, 88). For management purposes, population models are used primarily to predict the influence of a given management tactic on the density of nematodes that will remain to attack subsequent crops (46, 75). Damage functions and population models have been used to investigate problems related to management including differences in long-term efficacy of fumigant and nonfumigant nematicides (107), interspecific competition (68), overwinter survival (66), factors affecting the rate of development of resistance-breaking races (69), and effects of biocontrol agents on nematode populations (105).

The value of long-term optimization of nematode management tactics was recognized early in work to use models as management tools (68). Practices that are valuable during the current growing season may needlessly contribute to problems in subsequent years. Models have been used in temporal sequence to simulate the effects of management options on nematode populations and crop growth over more than one growing season and in perennial crops (26, 31, 46, 68, 75, 96, 100, 101). Cumulative profits based on management costs and crop yield and value are one criterion to evaluate the relative value of large numbers of management options such as sequences of rotation crops, resistant varieties, fallow, and nematicide treatments. For example, a multiple-season algorithm was recently used to derive different strategies for commercial growers and family farmers to control *Heterodera goetingiana* on crops in Italy (47). Models of pesticide movement and fate in the soil provide the option for using other optimization criteria in mathematical simulations of management systems (36, 136).

The use of deterministic models to simulate management sequences permits efficient organization of large amounts of information to help develop management strategies. However, a recent study of the effect of compounding sample errors, model errors, and seasonal variability on prediction of nematode population change, well-illustrated the qualitative nature of current long-term forecasts in nematology (86). Probabilities associated with simulated events were too low after more than 1–2 crops to have predictive value. These results emphasize the problems currently faced both in measuring nematode population densities and identifying key predictive components of the host-parasite interaction. Nevertheless, the relative values of several management options may be more accurately estimable than the economic value of a specific management option. One reason is that uncontrolled edaphic and climatic factors affecting a given management option are the same for all other options and will have similar effects in many cases. Further, when management options are compared, sampling error is not a direct component of the error as it is when a prediction is made for an individual field based on a sample estimate of population density. Sample errors do seriously affect predictions insofar as they contribute to the experimental error in model derivation, and progress in forecasting depends heavily on improved methods to quantify nematode populations. Improved methods to quantify nematode populations would also provide incentive to develop the data bases which are necessary to extend models based on edaphic and climatic variables (45).

Management Tactics

An objective of IPM is to reduce the use of pesticides by substituting other management tactics. In entomology, the fact that levels of natural biological control are often related to patterns or intensity of pesticide use has been extensively exploited. Biological control is not currently an option to manage most nematode pests, but other tactics such as crop rotations and resistant varieties are commonly used and permit the cultivation of susceptible crops without the use of nematicides. Since nematodes do not have the migratory and reproductive capacity of many arthropod pests, careful scheduling of management tactics may reduce nematode populations to nondamaging levels for a period of time. Most tactics considered below are traditional methods to manage nematodes. Their relative values were formerly determined based on convenience and short-term profitability. As IPM systems develop, there is renewed interest in the possibilities offered by many of the methods to manage nematodes without environmental side-effects. Concurrently, a steady reduction in the number of available nematicides has resulted in a pronounced shift in the relative profitability of some methods.

EXCLUSION For some nematodes, regulatory management practices occur outside of the field at a small fraction of the cost of controlling the pest in the soil. A number of nematode species are routinely excluded from nonendemic areas that extend beyond governmental boundaries where it is possible for quarantine personnel to examine plant material at a few entry points (80). Effective, long-term exclusion of nematodes from fields within political boundaries usually requires that a species be nonendemic to regions within a state or country and that it have a narrow host range so that efforts are focused on just a few plant species. These valuable opportunities arise when crops are newly introduced into an area. Citrus production is shifting southward in Florida in response to recent climatic changes. Although *Tylenchulus semipenetrans* is present in most citrus growing areas of the state, it is not anticipated to be a problem in the large area of new plantings because of a state-operated program to certify planting material as nematode-free and because the other hosts of this nematode are not grown in newly planted areas (30). Government-sponsored measures to foster movement of citrus from the Nile river valley in Egypt into newly reclaimed desert regions has fostered similar interest in restricting the movement of *T. semipenetrans* into these areas (M. Eissa, personal communication). Nursery certification in Florida requires that they be established on noninfested sites that are isolated from obvious sources of contamination. Periodic inspection and sampling is used to insure integrity of the plant material.

Sanitation measures to prevent movement of nematodes between fields on implements or in irrigation water are practiced to a limited degree (21, 40). Preventing migration of nematodes between adjacent fields requires physical barriers and is rarely attempted. The best-known examples of the use of physical exclusion were chemical (ethylene dibromide) barriers between citrus orchards in Florida, which effectively prevented migration of *Radopholus citrophilus* the causal agent of the disease spreading decline (108). Detection of ethylene dibromide in groundwater in these areas resulted in elimination of chemical barriers in 1984 (70). Mechanical pruning of the root systems may be an option for barrier maintenance (33) but is currently not widely practiced.

USE OF SUBOPTIMAL HOSTS OR ENVIRONMENTS Crop rotation is the most important method currently practiced to control plant-parasitic nematodes, based on the extent to which it is employed, its value in maintaining soil quality, and its lack of side effects. However, constraints to the use of rotation for nematode control can be formidable. A number of agricultural systems exist whose climate or infrastructure can support only monocultures or limited numbers of crops. Examples include the grain belts worldwide (13), and the Sahelian zone of subsaharan Africa, which supports only the cultivation of dryland crops such as millet, peanut and cowpea (28). The occurrence of polyphagous nematodes (e.g. Meloidogyne, Pratylenchus) or nematode communities can restrict the selection of suitable host plants. Crop value profoundly affects the use of rotation for nematode management. Even where crop rotation can be profitably employed, use of nematicides to permit monoculturing may bring greater returns in high value crops such as tobacco (67) and even those of low value such as wheat (13). Nevertheless, in areas of greater crop diversity, rotation schedules are commonly designed and employed for nematode management (2, 53, 114, 115, 145). When the availability of nematicides are restricted in specific crops or localities, rotation is often an important option. Suspension of the use of 1–3 dichloropropene nematicides in California in 1990 is expected to increase the practice of crop rotation to manage nematodes in crops such as cotton and tomato (P. Roberts, personal communication).

Resistant varieties offer many of the same advantages for nematode management as rotation crops, with the additional feature of permitting production of crops best suited to the needs of the grower. Resistance to nematodes refers to the relative ability of a nematode to reproduce on a crop variety, whereas tolerance refers to the relative growth and yield of a variety when grown in infested soil. With notable exceptions, resistant varieties of most major agronomic and horticultural crops are available for use against some of the important nematode pathogens (22). The weakest aspects of resistance against nematodes may be heavy reliance on single-gene resistance and the lack of multispecies resistance in many crops. Tomato well illustrates the problem because a single genetic source of resistance to several species of Meloidogyne has been introduced into a large number of varieties with good marketing characteristics and adapted to a range of climates. Growers in California increasingly choose to use these varieties rather than other means to manage *Meloidogyne spp*. Resistant nematode races that may develop on any variety will reproduce on all resistant lines. The use of resistance in northern parts of the state may also be restricted because resistance is lacking for *M. hapla,* a species with a range that is generally confined to temperate climates (141a).

Single-gene resistance may be more durable against some nematodes than it is for other pests and pathogens because nematodes disperse slowly and often reproduce parthenogenetically and at relatively low numbers. Nevertheless, the value of methods to reduce selection pressures such as alternation of resistant varieties with nonhost crops or susceptible varieties (67, 69, 110, 141a) is recognized. Some resistant varieties are nontolerant of invasion by nematodes and yield poorly when preplant nematode densities are high. These varieties are useful in management programs to maintain nematodes at nondamaging levels if preplant nematode densities are reduced through tactics such as crop rotation or use of nematicides (67). Resistance to nematodes in a number of varieties declines or disappears as temperature increases. It was proposed that this may be a plant adaptation to reduce the selection of virulent nematode races by permitting plants to become vigorously established early in a season before high soil temperatures reduce resistance mechanisms and allow reproduction by the remaining nematodes (22). This observation may have some value with regard to the use of planting dates to manage nematodes and conserve resistant varieties.

Single major genes usually impart resistance to nematodes that is qualitative in nature; nematode reproduction is prevented, or nearly so. Quantitative differences in cultivar resistance and tolerance to nematodes that result from multigenic factors require more effort to identify and characterize because of the influence of environment and intraspecific competition on nematode reproduction (31, 77, 140) and because field trials are required to measure tolerance in the form of crop yield (22). Simple tolerance to nematode damage is less desirable than resistance if populations can develop to such high densities that they affect subsequent crops. Nevertheless, crops such as small grains that may tolerate and maintain high densities of some species of Meloidogyne (141a) permit extension of crop rotation cycles before tactics are necessary to reduce the populations. The value of multigenic resistant lines in reducing the development of nematode virulence, particularly in perennial tree crops, makes this an important area of future research that could be facilitated by faster methods to evaluate nematode reproduction in large screening trials. Other research areas that can help identify useful sources of resistance involve aspects of the preinfection process such as host stimulation of egg hatch, attraction, recognition, and penetration (22).

Fallow (often bare fallow) is commonly practiced in some rotation systems to reduce nematode populations through starvation (18, 58, 108, 120). Use of bare fallow is limited by the expense of maintaining weed control while forgoing income from crop-production and it is well-recognized that soil erosion due to fallow should preclude its use under many conditions. Possible exceptions are areas with little or no topsoil such as semidesert regions where cropping seasons are very brief (28). Similarly, in the central ridge region of Florida's citrus industry, where sandy soils contain almost no organic topsoil, bare fallow is used to reduce levels of *Radopholus citrophilus* prior to fumigation and replanting (108).

Starvation of nematodes via fallow or use of resistant or nonhost crops prior to overwintering may be particularly useful where nematodes must expend unusual amounts of energy to survive the overwinter period. In the Sahelian zone of Senegal, maintenance of bare fallow during the brief growing season reduced population densities of five genera of plant-parasitic nematodes to barely detectable levels (28). Bare fallow served to reduce the energy reserves necessary for nematodes to survive as anhydrobiotes (148a) during the long dry season. However, the treatment has the side effect of severely reducing productivity of crops planted in the subsequent season (6), possibly due to elimination of beneficial biota (119) such as mycorrhizae and *Rhizobium spp.* The narrow range of crop plants that can be grown in the region, all of which are good hosts for many of the nematode genera, makes nematicides the only means of nematode control used by the small plot farmers. Discovery of nonhost forage plants might eliminate detrimental effects to the soil and would provide an additional economic incentive for use of starvation to manage nematodes in this dryland agricultural system.

Flooding routinely controls several species of Meloidogyne capable of attacking paddy rice. Even species such as *M. graminicola,* which can survive in paddies, can be managed by growing seedlings in flooded conditions that prevent nematode penetration (12). Vegetables (138) and tobacco (25) can often be grown successfully following paddy rice without damage from *Meloidogyne spp.* Flooding employed specifically as a management tactic is also recommended for nematode control (82, 120) in some systems.

A few susceptible crops can be protected from nematode invasion by sowing when soil temperatures are lower than required for nematode activity. Wheat yields are higher in the presence of *Heterodera avenae* in Australia when planted early, but the practice is complicated by the chance of frost damage and requires sufficient early-season rainfall (13). In California, sugar beet losses to *H. schachtii* are lower in early planted crops (134) and reproduction of *Meloidogyne spp.* on winter small grains is prevented by planting crops after soil temperatures fall below 18° C (141a).

DIRECT MANAGEMENT TACTICS Natural biological control of nematodes by parasites or predators has been documented in three systems involving perennial crops or monocultured cereals (10, 72, 135). No system for production and release of a biological control agent in the field is widely used, although some products are commercially available (19). Nevertheless, a number of fungal and bacterial antagonists of nematodes have good potential for biological control (71, 83), and interest in the area is high. Twenty percent of the research papers presented at the Second International Nematology Congress in 1990 dealt with classical biological control. Progress toward using some of these organisms is impeded by insufficient knowledge of their ecology to permit reliable establishment of populations in agricultural soils

(71); for others by the inability to mass rear them in the absence of hosts (109).

Recent reviews of this topic emphasize the need for methods to quantify populations of biocontrol agents in the field (93) to permit reliable ecological studies and the necessity to consider biotic factors in addition to predation and parasitism that interfere with the nematode life cycle (38). Examples include use of lectins that bind to chemoreceptors (23, 81) and have caused reduced root-galling by root-knot nematodes, attractants to interfere with host-finding (38), and hormone inhibitors to block nematode development (20) as well as chemicals that block host responses necessary for nematode development (55).

Soil amendments are commonly used to control nematodes in parts of the tropics (95, 113). Manures, compost, oil cakes, and other industrial byproducts can affect population levels of plant-parasitic nematodes. The use of chitin to control nematodes is under active investigation and a commercial product (Clandosan) containing chitin and urea was recently registered for nematode management in the United States. However, the high rate required for nematode control by most organic amendments (tons/ha) limit use of these compounds primarily to small plot agricultural systems (71). Various organic amendments can stimulate growth of nematode predators and parasites (121), and as such are indirectly involved in biological control. Toxic concentrations of chemicals such as ammonia and phenolic compounds are also released by many soil amendments (113) but the specific mechanisms of nematode control are poorly understood (83). For example, chitin amendments release ammonia upon degradation at rates that can be nematicidal, but the buildup of chitinolytic organisms may also contribute to mortality of nematode eggs that contain chitin (132). However, chitin is not found in juvenile nematodes that in one study were killed in chitin-amended soil at higher rates than would occur due to concentrations of ammonia released by the amendment (83).

An important consideration in the use of biological control agents may be a requirement to integrate them into programs with other management strategies. Because of the time necessary to reduce population densities of nematodes to nondamaging levels it may be advantageous to effect control when nematode numbers are relatively low (83, 135). For some species of cyst and root-knot nematodes, these levels may be lower than current detection thresholds. Reduction of heavy nematode infestations with other management practices, prior to implementing biocontrol, has also been suggested (71).

Some physical methods to manage nematodes include burning crop stubble, which has long been used to kill some seed (*Anguina agrostis,*) stem (*Ditylenchus angustus,*) and leaf (*Aphelenchoides besseyi*) nematodes (61, 64); and heat treatment of plant material that is widely employed to disinfest a variety of endoparasitic nematodes (108, 130). Soil solarization is not widely practiced for nematode management, although research activity in this area has been high recently (54, 62, 103, 133). Use of plastic mulches in many agricultural systems makes the concept of widespread use of solarization for nematode control attractive and development of improved methods to use solar energy to heat soil at sufficient depth (117) may eventually foster use of the method. Solarization is currently feasible as an alternative to steam and chemicals to disinfest small amounts of soil, for example in nursery and greenhouse operations (54).

Chemical management of nematodes is a diverse subject dealt with in several recent reviews (1, 60, 79, 94, 139). The means by which chemicals are used to control nematodes has changed considerably in a decade, and there is a sense of urgency in the recognized need for additional changes. Several of the most widely used fumigant nematicides such as dibromochloropropane (DBCP), dichloropropene/dichloropropane mixture (D-D) and ethylene dibromide (EDB) are no longer available for use in the USA and a number of other countries. Deregistration of these compounds was largely due to groundwater contamination. The same problem is associated with several nonfumigant organophosphate and carbamate chemicals such as aldicarb and carbofuran, whose use is restricted regionally. Recognition that decomposition of these compounds is primarily a biological process (17) highlighted problems associated with rapid leaching of the compounds below zones of high biological activity. Thus, many recent changes in strategies for using nematicides involve reducing the probability of groundwater contamination.

Movement of nematicides through the soil profile is regulated in part by soil properties and the timing and amount of precipitation. Integration of irrigation schedules with nematicide use is increasingly recognized as a key factor in managing nematicidal efficacy and reducing groundwater contamination (139). Optimal irrigation patterns following nematicide application can significantly reduce movement of water below the rooting zone before the chemicals are removed by root absorption or normal degradation processes. Use of aldicarb in California and Florida is restricted to low-rainfall months when the management of chemical movement via irrigation has the highest chance for success (65). Increasingly, use of many nematicides is legally restricted based on soil texture, organic matter content, depth to water table, and proximity to drinking water wells.

Some chemicals that can be applied to crops in the field without significant phytotoxicity or residue accumulation in plant tissue are suited to multiple application during a season. The practice has the advantage of sometimes providing the same or better degree of control as single applications of higher dosages while reducing chemical movement below the rooting zone (1). Use of drip irrigation is recognized as an economical delivery means for single or multiple applications of many chemicals. Drippers release chemicals exclusively in the irrigated zone of highest root and nematode activity. Similarly, application patterns of chemicals by other delivery methods are being modified to target soil regions of highest root and nematode abundance. Granular nematicides applied in bands under the citrus tree canopy resulted in better control of *T. semipenetrans* and higher yields than conventional treatments applied at the canopy dripline (29). Systemic nematicides move less through the soil profile when applied beneath the canopy due to greater root density and restricted rainfall in this soil zone.

Treatment of seeds, planting stock and above-ground plant parts with systemic nematicides (30, 59, 92, 116, 141) is another strategy to reduce the risk of groundwater contamination by avoiding direct application of chemicals to the soil. However, nematicide effects in the roots of many plants are lower when they are transported from other organs than when they are applied to the soil (29).

The use of economic thresholds and the timing of nematicide applications to coincide with periods of maximum nematode activity reduce the potential for groundwater contamination by restricting nematicide use. It was recently estimated that use of soil sampling can reduce soil treatment to control potato cyst nematodes by 80% in some parts of The Netherlands (C. H. Schomaker, personal communication). Nematicide use in perennial crops is timed to coincide with seasonal periods of new root infection and nematode population development (36, 118, 142).

IMPROVING NEMATODE IDENTIFICATION AND MEASUREMENT AS A BASIS FOR PROGRESS IN NEMATODE MANAGEMENT

Quantitative methods will be used increasingly in nematode management programs because public pressure to reduce the use of agricultural pesticides is creating the demand that growers be able to justify pesticide use on a per case basis. Any benefit that this trend may confer to the science of nematology will be enhanced if the new data reflect improved methods to sample, identify, and quantify nematodes. The need for efficient mechanized sampling devices that permit the economic acquisition of greater numbers of soil samples has been noted (5, 50). However, quantitative nematology will progress very little from its present state without fundamental change in the way nematode populations are identified and measured. The methods by which nematodes are extracted from samples, identified, and enumerated are expensive. Identification can be difficult and subjective, and extraction is often inefficient (14, 144). Alternatively, the ability to more accurately assess nematode populations in the field at lower cost could result in rapid improvement in management-related predictions. Improving forecast accuracy must occur partly as an iteration of three processes: evaluation of the accuracy of predictions that are based on biological models; identification of sources of error; and model rederivation. Sampling is currently the most intractable requirement of this process because of its contribution to the high cost of conducting the requisite number of field studies. Few research programs can commit such resources to collecting data in more than a few locations or for more than several years.

Improved methods to identify and quantify nematodes will also result in faster development of the tactical options available for control. Cultural, biological, and chemical control tactics are developed and refined based on measurement of their efficacy against nematodes in the field. Screening for resistance to nematodes requires large-scale estimation of population development on germplasm lines. Similarly, screening new chemicals to kill or disrupt nematode life cycles is often restricted to a few species such as root-knot nematodes that cause visible symptoms which are easily evaluated. Rapid, reliable species and race identification is critical for selection of appropriate rotation crops or resistant cultivars and for regulatory management of plant shipments (37).

Major improvements in sample processing will initially occur as kits are developed to detect specific nematode proteins or nucleic acids. Monoclonal antibodies and DNA probes have been developed that permit identification of a number of species including *Globodera pallida* and *G. rostochiensis,* several species of Heterodera and four species of Meloidogyne (3, 7, 15, 16, 63, 112, 124). Both methods can provide resolution at a level of one egg/sample in test preparations (63, 124). Research to efficiently homogenize and detect nematode life stages in soil matrices is active and recent work using mitochondrial nucleic acid hybridization reliably detected Meloidogyne eggs at a level of 5 eggs/100 g soil in 4 types of soil (63).

Use of chemical methods to detect and quantify nematodes in soil and plant samples should greatly decrease the time currently spent in processing and microscopy. Detection thresholds will be lowered significantly and quantitation improved because homogenization of the nematodes provides a more uniform distribution of the detectable propagules in a soil suspension. It will be possible to evaluate the status of other soilborne pests and pathogens using a single sample preparation. Improvements such as these will increase the value of sample information and have the potential to reduce processing costs. The feasibility of obtaining such information will provide the incentive to develop equipment to efficiently obtain high quality samples in the field. The ability to accurately assess nematode populations in a field will introduce a new phase in nematode management in which the potential of current tactics is more fully attainable, and new tactics become possible.

Literature Cited

1. Apt, W. J., Caswell, E. P. 1988. Application of nematicides via drip irrigation. *Ann. Appl. Nematol.* 2:1–10
2. Aung, T., Prot, J-C. 1990. Effects of crop rotations on *Pratylenchus zeae* and on yield of rice cultivar UPL Ri-5. *Rev. Nematol.* 13(4):445–47
3. Backett, K. D., Atkinson, H. J., Forrest, J. M. 1990. Development of a rapid immunological screen for resistance in the potato to *Globodera pallida*. *Int. Nematol. Congr., 2nd,* Aug. 11–17, Veldhoven, The Netherlands, p. 52 (Abstr.)
4. Barker, K. R. 1989. Yield relationships and population dynamics of *Meloidogyne* spp. on flue-cured tobacco. *Ann. Appl. Nematol.* 21:597–603
5. Barker, K. R., Imbriani, J. L. 1984. Nematode advisory programs—Status and prospects. *Plant Dis.* 68(8):735–41
6. Baujard, P., Bodian, Y., Duncan, L. W., Martiny, B., Pariselle, A., Sarr, E. 1989. Nouvelles études au champ sur les effets de deux nematicides fumigants bormés (DBCP et EDB) sur les rendements des cultures dans le bassin arachidier du Sénégal. *Rev. Nematol.* 12(1): 85–90
7. Beckenbach, K., Xue, B., Kachinka, B., Baillie, D. 1989. Use of Tx-1 related sequences to identify races in the Phyla nematoda *J. Nematol.* 21(4):551 (Abstr.)
8. Belair, G., Boivin, G. 1988. Spatial pattern and sequential sampling plan for *Meloidogyne hapla* in muck-grown carrots. *Phytopathology* 78(5):604–7
9. Bird, G. W. 1987. Role of nematology in integrated pest management programs. See Ref. 142a, pp. 114–21
10. Bird, A. F., Brisbane, P. G. 1988. The influence of *Pasteuria penetrans* in field soils on the reproduction of root-knot nematodes. *Rev. Nematol.* 11:75–81
11. Boag, B., Brown, D. J. F., Topham, P. B. 1987. Vertical and horizontal distribution of virus-vector nematodes and implications for sampling proceudres. *Nematologica* 33:83–96
12. Bridge, J., Page, S. L. J. 1982. The rice root-knot nematode, *Meloidogyne graminicola,* on deep water rice (*Oryza sativa* subsp. *indica*). *Rev. Nematol.* 5:225–32
13. Brown, R. H. 1987. Control strategies of low value crops. See Ref. 13a, pp. 351–87

13a. Brown, R. H., Kerry, B. R., eds. 1987. *Principles and Practice of Nematode Control in Crops.* Sydney, Australia: Academic

14. Brown, S. M., Miller, M. F., Viglierchio, D. R. 1987. Laboratory consistency in extraction of nematodes from soil and roots. *Nematropica* 17(2):179–92
15. Burrows, P. R. 1990. DNA hybridisation probes to identify pathotypes of *Globodera rostochiensis* and *G. pallida.* See Ref. 3, p. 60
16. Burrows, P. R. 1990. The rapid and sensitive detection of the plant parasitic nematode *Globodera pallida* using a non-radioactive biotinylated DNA probe. *Rev. Nematol.* 13:185–90
17. Castro, C. E., Belser, M. O. 1968. Reductive dehalogenation of the biocides ethylene dibromide, 1,2-dibromo-3-chloropropane, and 2,3-dibromobutane in soil. *Environ. Sci. Technol.* 2:779–83
18. Caswell, E. P., Apt, W. J. 1989. Pineapple nematode research in Hawaii: Past, present and future. *J. Nematol.* 21:147–57
19. Cayrol, J-C. 1983. Lutte biologique contre les *Meloidogyne* au moyen d'*Arthrobotrys irregularis. Rev. Nematol.* 6(2):265–73
20. Chitwood, D. J. 1987. Inhibition of steroid or hormone metabolism or action in nematodes. See Ref. 142a, pp. 122–30
21. Cohn, E. 1976. Report of investigations on nematodes of citrus and subtropical fruit crops in South Africa. *Citrus and Subtropical Fruit Res. Inst.,* Nelspruit. 41 pp.
22. Cook, R., Evans, K. 1987. Resistance and tolerance. See Ref. 13a, pp. 179–231
23. Davis, E. L., Kaplan, D. T., Dickson, D. W., Mitchell, D. J. 1989. Reproduction of lectin-treated *Meloidogyne* spp. in two related soybean cultivars. *Rev. Nematol.* 12(3):257–63
24. Davis, R. M. 1984. Distribution of *Tylenchulus semipenetrans* in a Texas grapefruit orchard. *J. Nematol.* 16:313–17
25. de Guiran, G. 1970. Le problème *Meloidogyne* sur tabac à Madagascar. *Cah. ORSTOM, Ser. Biol.* 11:187–208
26. Duncan, L. W. 1983. *Predicting effects of plant-parasitic nematode communities on crop growth.* PhD. thesis. Univ. Calif., Riverside
27. Duncan, L. W. 1986. The spatial distribution of citrus feeder roots and of the citrus nematode, *Tylenchulus semipenetrans. Rev. Nematol.* 9(3):233–40
28. Duncan, L. W. 1986. Effects of bare fallow on plant-parasitic nematodes in

the Sahelian zone of Senegal. *Rev. Nematol.* 9(1):75–81
29. Duncan, L. W. 1989. Effect of fenamiphos placement on *Tylenchulus semipenetrans* and yield in a Florida citrus orchard. *J. Nematol.* 21:703–6
29a. Duncan, L. W. 1988. Citrus fibrous root and *Tylenchulus semipenetrans* sample optimization in Florida flatwoods citrus. *J. Nematol.* 20:633
30. Duncan, L. W., Cohn, E. 1990. Nematode parasites of citrus. See Ref. 79a, pp. 321–46
31. Duncan, L. W., Ferris, H. 1983. Effects of *Meloidogyne incognita* on cotton and cowpeas in rotation. *Proc. Beltwide Cotton Prod. Res. Conf.*, pp. 22–26
32. Duncan, L. W., Ferguson, J. F., Dunn, R. A., Noling, J. W. 1989. Application of Taylor's Power Law to sample statistics of *Tylenchulus semipenetrans* in Florida citrus. *J. Nematol.* 21:707–11
33. Duncan, L. W., Kaplan, D. T., Noling, J. W. 1990. Maintaining barriers to the spread of *Radopholus citrophilus* in Florida citrus orchards. *Nematropica* 20:71–88
34. Duncan, L. W., McSorley, R. 1988. Modeling nematode populations. See Ref. 142a, pp. 377–89
35. Duncan, L. W., McSorley, R. T. 1990. A conservative aspect of sample size optimization. See Ref. 3, p. 72
36. Duncan, L. W., Noling, J. W. 1988. Computer simulated management of *Tylenchulus semipenetrans* on citrus. *Int. Citrus Congr. Middle East,* Tel-Aviv, Israel, pp. 262 (Abstr.)
37. Dunn, R. A., Eisenback, J. D., Radewald, J. D., Westerdahl, B. B. 1989. Extension nematology: Taxonomy and the dilemma of diagnosis. *J. Nematol.* 21(4):558–9 (Abstr.)
38. Dusenberry, D. B. 1987. Prospects for exploiting sensory stimuli in nematode cointrol. See Ref. 142a, pp. 131–35
39. Elliott, J. M. 1977. Some methods for the statistical analysis of samples of benthic invertebrates. *Freshwater Biol. Assoc. Sci. Publ. No. 25.* Ambleside, Cumbria, UK. 160 pp. 2nd ed.
40. Esser, R. P. 1984. How nematodes enter and disperse in Florida nurseries via vehicles. *Fla. Dep. Agric. Consum. Serv., Div. Plant Ind., Nematol. Circ.* 109:1–2
41. Ferris, H. 1976. Development of a computer-simulation model for a plant-nematode system. *J. Nematol.* 8:255–63
42. Ferris, H. 1978. Nematode economic thresholds: derivation, requirements and theoretical considerations. *J. Nematol.* 10:341–50
43. Ferris, H. 1984. Nematode damage functions: The problems of experimental and sampling error. *J. Nematol.* 16:1–9
44. Ferris, H. 1984. Probability range in damage predictions as related to sampling decisions. *J. Nematol.* 16:246–51
45. Ferris, H. 1985. Density-dependent nematode seasonal multiplication rates and overwinter survivorship: A critical point model. *J. Nematol.* 17:93–100
46. Ferris, H., Ball, D. A., Beem, L. W., Gudmunson, L. A. 1986. Using nematode count data in crop management decisions. *Calif. Agric.* 40:12–14
47. Ferris, H., Greco, N. 1990. Management strategies for *Heterodera goettingiana* in a vegetable cropping system in Itlay. *Rev. Nematol.* In press
48. Ferris, H., Jaffee, B. A., McKenry, M. V., Juurma, A. 1989. Analysis of nematode stress on peach trees due to *Criconemella xenoplax. J. Nematol.* 21 (4):560 (Abstr.)
49. Ferris, H., McKenry, M. V. 1974. Seasonal fluctuations in the spatial distribution of nematode populations in a California vineyard. *J. Nematol.* 6:203–10
50. Ferris, H., Mullens, T. A., Foord, K. E. 1990. Stability and characteristics of spatial description parameters for nematode populations. *J. Nematol.* 22:427–39
51. Francl, L. J. 1986. Improving the accuracy of sampling field plots for plant-parasitic nematodes. *J. Nematol.* 18(2):190–95
52. Geier, P. W. 1966. Management of insect pests. *Annu. Rev. Entomol.* 11:471–90
53. Germani, G., Reversat, G., Luc, M. 1983. Effect of *Sesbania rostrata* on *Hirschmanniella oryzae* in flooded rice. *J. Nematol.* 15:269–71
54. Giblin-Davis, R. M., Verkade, S. D. 1988. Solarization for nematode disinfestation of small volumes of soil. *Ann. Appl. Nematol.* 2:41–45
55. Glazer, I., Apelbaum, A., Orion, D. 1985. Effect of inhibitors and stimulators of ethylene production on gall development in *Meloidogyne javanica*-infected tomato roots. *J. Nematol.* 17(2):145–49
56. Gleddie, S., Keller, W. A., Setterfield, G. 1986. Production and characterization of somatic hybrids between *Solanum melongena* L. and *S. sisymbriifolium* Lam. *Theor. Appl. Genet.* 71:613–21
57. Goodell, P. B., Ferris, H. 1981. Sample optimization for five plant-parasitic nematodes in an alfalfa field. *J. Nematol.* 13:304–13

58. Gowen, S., Queneherve, P. 1990. Nematode parasites of bananas, plantains, and abaca. See Ref. 79a, pp. 431–60
59. Guerout, R. 1975. Banana corms coating with nematicidal mud: a preplant treatment. *Nematropica* 5:22
60. Hague, N. G. M., Gowen, S. R. 1987. Chemical control of nematodes. See Ref. 13a, pp. 131–78
61. Hardison, J. R. 1980. Role of fire for disease cointrol in grass seed production. *Plant Dis*. 64:641–45
62. Heald, C. H., Robinson, A. F. 1987. Effects of soil solarization on *Rotylenchulus reniformis* in the lower Rio Grande Valley of Texas. *J. Nematol*. 19(1):93–103
63. Hyman, B. C., Peloquin, J. J., Platzer, E. G. 1990. Optimization of mitochondrial DNA-based hybridization assays to diagnostics in soil. *J. Nematol*. 22 (3):273–78
64. Ichinohe, M. 1972. Nematode diseases of rice. See Ref. 145a, pp. 127–43
65. Inst. Food Agric. Sci. (IFAS). 1989. *Florida Nematode Control Guide*, Entomol. Nematol. Dep. Coop. Ext. Serv., Univ. Fla., Gainesville. 138 pp.
66. Jeger, M. J., Starr, J. L. 1985. A theoretical model of the winter survival dynamics of *Meloidogyne* spp. eggs and juveniles. *J. Nematol*. 17:257–60
67. Johnson, C. S., Komm, D. A., Jones, J. L. 1989. Control of *Globodera tabacum solanacearum* by alternating host resistance and nematicide. *J. Nematol*. 21: 16–23
68. Jones, F. G. W., Kempton, R. A. 1978. Population dynamics, population models and integrated control. See Ref. 130, pp. 333–61
69. Jones, F. G. W., Parrott, D. M., Ross, G. J. S. 1967. The population genetics of the potato cyst-nematode, *Heterodera rostochiensis:* Mathematical models to simulate the effects of growing eelworm-resistant potatoes bred from *Solanum tuberosum* spp. *Andigena*. *Ann. Appl. Biol*. 60:151–71
70. Kaplan, D. T. 1988. Future considerations for nematode management in citrus. In *Proc. Int. Citrus Congr., 6th, Tel-Aviv, Israel*, pp. 969–75
71. Kerry, B. A. 1990. An assessment of progress toward microbial control of plant-parasitic nematodes. *J. Nematol*. 22(4S):621–31
72. Kerry, B. R., Crump, D. H. 1977. Observations of fungal parasites of females and eggs of the cereal cyst-nematode, *Heterodera avenae* Woll., and other cyst-nematodes. *Nematologica* 23:193–201
73. Kimpinski, J., McRae, K. B. 1988. Relationship of yield and *Pratylenchus* spp. population densities in Superior and Russet Burbank potato. *Ann. Appl. Nematol*. 2:29–33
74. Kinloch, R. A. 1982. The relationship between soil populations of *Meloidogyne incognita* and yield reduction of soybean in the Coastal Plain. *J. Nematol*. 14:162–68
75. Kinloch, R. A. 1986. Soybean and maize cropping models for the management of *Meloidogyne incognita* in the Coastal Plain. *J. Nematol*. 18:451–58
76. Koenning, S. R., Anand, S. C., Wrather, J. A. 1988. Effect of within-field variation in soil texture on *Heterodera glycines* and soybean yield. *J. Nematol*. 20(3):373–80
77. LaMondia, J. A., Brodie, B. B. 1986. Effects of initial nematode density on population dynamics of *Globodera rostochiensis* on resistant and susceptible potatoes. *J. Nematol*. 18(2):159–64
78. LaMondia J. A., Brodie, B. B., Brucato, M. L. 1985. Management of *Globodera rostochiensis* as influenced by nematode population densities and soil type. *J. Nematol*. 18:74–78
79. Lembright, H. W. 1990. Soil fumigation: Principles and application technology. *Ann. Appl. Nematol*. 22:632–44
79a. Luc, M., Sikora, R. A., Bridge, J., eds. 1990. *Plant Parasitic Nematodes in Subtropical and Tropical Agriculture*. Wallingford: CAB Inst.
80. Maas, P. W. Th. 1987. Physical methods and quarantine. See Ref. 13a, pp. 265–91
81. Marban-Mendoza, N., Jeyaprakash, A., Jansson, H-B., Damon, R. A. Jr., Zuckerman, B. M. 1987. Control of root-knot nematodes on tomato by lectins. *J. Nematol*. 19(3):331–35
82. Mateille, T., Foncelle, B., Ferrer, H. 1988. Lutte contre les nematodes du bananier par submersion du sol. *Rev. Nematol*. 11:235–38
83. McClure, M. A. 1990. Biological control of nematodes. In *New Directions in Biological Control: Alternatives for Suppressing Agricultural Pests and Diseases*, pp. 255–69. New York: Liss
84. McSorley, R. 1982. Simulated sampling strategies for nematodes distributed according to a negative binomial model. *J. Nematol*. 14:517–22
85. McSorley, R. 1987. Extraction of nematodes and sampling methods. See Ref. 13a, pp. 13–47
86. McSorley, R. 1990. Applied population

modeling: Fact or fiction? See Ref. 3, p. 107
87. McSorley, R., Dankers, W. H., Parrado, J. L., Reynolds, J. S. 1985. Spatial distribution of the nematode community on perrine marl soils. *Nematropica* 15:77–92
88. McSorley, R., Dickson, D. W. 1989. Effects and dynamics of a nematode community on soybean. *J. Nematol.* 21(4):490–99
89. McSorley, R., Ferris, J. M. 1979. PHEX: A simulation of lesion nematodes in corn roots. *Purdue Univ. Agric. Exp. Stn. Res. Bull. 959*
90. McSorley, R., Parrado, J. L. 1983. A bioassay sampling plan for *Meloidogyne incognita*. *Plant Dis.* 67:182–84
91. McSorley, R., Pohronezny, K. L. 1981. A simple bioassay as a supplement to soil extraction for detection or root knot nematodes. *Proc. Soil Crop Sci. Soc. Fla.* 40:121–23
92. Milne, D. L., DeVilliers, E. A., Smith, B. L. 1977. Response of nematode populations and pineapple yields to foliar application of systemic nematicides. *Citrus Subtrop. Fruit J.* 525:12–14
93. Mitchell, D. J., Kannwischer-Mitchell, M. E., Dickson, D. W. 1987. A semiselective medium for the isolation of *Paecilomyces lilacinus* from soil. *J. Nematol.* 19:255–56
94. Morton, H. V. 1986. Modification of proprietary chemicals for increasing efficacy. *J. Nematol.* 18:123–28
95. Muller, R., Gooch, P. S. 1982. Organic amendments in nematode control, an examination of the literature. *Nematropica* 12(2):319–26
95a. Nickle, W. R., ed. 1984. *Plant and Insect Nematodes*. New York: Marcel Dekker
96. Noe, J. P. 1988. Theory and practice of the cropping systems approach to reducing nematode problems in the tropics. *J. Nematol.* 20(2):204–13
97. Noe, J. P., Barker, K. R. 1985. Overestimation of yield loss of tobacco caused by the aggregated spatial pattern of *Meloidogyne incognita*. *J. Nematol.* 17:245–51
98. Noe, J. P., Barker, K. R. 1985. Relation of within-field spatial variation of plant parasitic nematode population densities and edaphic factors. *Phytopathology* 75:1247–52
99. Noe, J. P., Campbell, C. L. 1985. Spatial pattern analysis of plant-parasitic nematodes. *J. Nematol.* 17:86–93
99a. Noling, J. W. 1987. Multiple pest problems and control on tomato. *Fla. Dep. Agric. Consum. Serv. Div. Plant Ind., Nematol. Circ. No. 139*. 4 pp.
100. Noling, J. W., Duncan, L. W. 1988. Estimation of citrus nematode stress and yield losses in a mature citrus grove. *J. Nematol.* 20(4):653 (Abstr.)
101. Noling, J. W., Ferris, H. 1987. Nematode-degree days, a density-time model for relating epidemiology and crop losses in perennials. *J. Nematol.* 19 (1):108–18
102. Osteen, C., Johnson, A. W., Dowley, C. C. 1981. Applying the economic threshold concept to control of lesion nematodes on corn. *USDA Tech. Bull. 1670*. 32 pp.
103. Overman, A. J., Jones, J. P. 1986. Soil solarization, reaction, and fumigation effects on double-cropped tomato under full-bed mulch. *Proc. Fla. State Hortic. Soc.* 99:315–18
104. Paul, H., Van Deelen, J. E. M., Henken, B., De Bock, Th. S. M., Lange, W., Kerns, F. A. 1990. Expression of resistance to *Heterodera schachtii* in hairy roots of a nematode-resistant monotelosomic addition plant of beet *(Beta vulvaris)*. See Ref. 3, p. 122
105. Perry, J. N. 1978. A population model for the effect of parasitic fungi on numbers of the cereal cyst-nematode, *Heterodera avenae*. *J. Appl. Ecol.* 15:781–87
106. Perry, J. N. 1983. Effects of spatial heterogeneity on Jones' model for cyst-nematode population dynamics and crop root damage. *J. Appl. Ecol.* 20:849–56
107. Perry, J. N., Clark, S. J. 1983. Use of population models to estimate the proportion of cyst-nematode eggs killed by fumigation. *J. Appl. Ecol.* 20:857–64
108. Poucher, C., Ford, H. W., Suit, R. F., DuCharme, E. P. 1967. Burrowing nematode in citrus. *Fla. Dept. Agric. Div. Plant Ind. Bull. No. 7,* Gainesville
109. Reise, R. W., Hackett, R. J., Sayre, R. M., Huettel, R. N. 1988. Factors affecting cultivation of three isolates of *Pasteuria* spp. *J. Nematol.* 20:657 (Abstr.)
110. Roberts, P. A. 1982. Plant resistance in nematode pest management. *J. Nematol.* 14:24–33
111. Roberts, P. A., Thomason, I. J., McKinney, H. E. 1981. Influence of non-hosts, crucifers, and fungal parasites on field populations of *Heterodera schachtii*. *J. Nematol.* 13:165–71
112. Robinson, M. P. 1989. Quantification of soil and plant populations of meloidogyne using immunoassay techniques. *J. Nematol.* 21(4):583–84 (Abstr.)
113. Rodriguez-Kabana, R. 1986. Organic and inorganic nitrogen amendments to

soil as nematode suppressants. *J. Nematol.* 18(2):129–35

114. Rodriguez-Kabana, R., Ivey, H., Backman, P. A. 1987. Peanut-cotton rotations for the management of *Meloidogyne arenaria*. *J. Nematol.* 19(4):484–86
115. Rodriguez-Kabana, R., Robertson, D. G., Backman, P. A., Ivey, H. 1988. Soybean-peanut rotations for the management of *Meloidogyne arenaria*. *Ann. Appl. Nematol.* 2:81–85
116. Roman, J., Oramas, D., Green, J. 1984. Use of oxamyl for the control of the nematodes in yam (*Dioscorea rotundata* Poir). *J. Agric. (Univ. Puerto Rico)* 68:383–86
117. Saleh, H., Abu-Gharbieh, W. I., Al-Banna, L. 1988. Effect of solarization combined with solar-heated water on *Meloidogyne javanica*. *Nematologica* 34:290–91
118. Sarah, J. L. 1987. Banana nematodes and their control in Africa. *Nematropica* 17(2):199–216
119. Sarah, J. L. 1987. Utilisation d'une jachère travaillée pour lutter contre les nematodes parasites de l'ananas. *Fruits* 42:357–60
120. Sarah, J. L., Lassoudière, A., Guerout, R. 1983. La jachère nue et l'immersion du sol: deux méthodes intéressantes de lutte integrée contre *Radopholus similis* (Cobb) dans les bananeraies de sols tourbeux de Côte d'Ivoire. *Fruits* 38:35–42
121. Schlang, J., Stendel, W., Muller, J. 1988. Influence of resistant green manure crops on the population dynamics of *Heterodera schachtii* and its fungal egg parasites. *Proc. Eur. Soc. Nematol., 19th Int. Nematol. Symp., Uppsala, Sweden,* p. 69 (Abstr.)
122. Schmitt, D. P., Barker, K. R., Noe, J. P., Koenning, S. R. 1990. Repeated sampling to determine the precision of estimating nematode population densities. *J. Nematol.* 22:552–59
123. Schomaker, C. H., Been, T. H. 1990. Sampling strategies for potato cyst nematodes. See Ref. 3, p. 136
124. Schots, A., Gommers, F. J., Bakker, J., Egberts, E. 1990. Serological differentiation of plant-parasitic nematode species with polyclonal and monoclonal antibodies. *J. Nematol.* 22(1):16–23
125. Seinhorst, J. W. 1965. The relationship between nematode density and damage to plants. *Nematologica* 11:137–54
126. Seinhorst, J. W. 1973. The relation between nematode distribution in a field and loss in yield at different average nematode densities. *Nematologica* 19: 421–27
127. Steinhorst, J. W. 1982. The distribution of cysts of *Globodera rostochiensis* in small plots and the resulting sampling errors. *Nematologica* 28:285–97
128. Smolik, J. D., Evenson, P. D. 1987. Relationship of yields and *Pratylenchus* spp. population densities in dry land and irrigated corn. *Ann. Appl. Nematol.* 1:71–73
129. Southards, C. J., Nusbaum, C. J. 1967. Genetic variability of tobacco response to *Pratylenchus brachyurus*. *Phytopathology* 57:18–21
130. Southey, J. F., ed. 1978. *Plant Nematology, Bull. 201,* pp. 302–12. London: Min. Agric., Fish. Food
131. Southwood, T. R. E. 1978. *Ecological Methods with Particular Reference to the Study of Insect Populations*. London: Chapman & Hall. 2nd. ed.
132. Spiegel, Y., Cohn, E., Chet, I. 1989. Use of chitin for controlling *Heterodera avenae* and *Tylenchulus semipenetrans*. *J. Nematol.* 21(3):419–22
133. Stapleton, J. J., De Vay, J. E. 1984. Thermal components of soil solarization as related to changes in soil, and root microflora and increased plant growth responses. *Phytopathology* 74:255–59
134. Steele, A. E. 1984. Nematode parasites of sugarbeet. See Ref. 95a, pp. 507–69
135. Stirling, G. R., McKenry, M. V., Mankau, R. 1979. Biological control of root-knot nematodes (*Meloidogyne* spp.) on peach. *Phytopathology* 69:806–8
136. Swartz, M. S., Bird, G. W., Ritchie, J. T. 1989. A modeling-monitoring approach to the risk/benefit assessment of aldicarb in potato production. *J. Nematol.* 21(4):590 (Abstr.)
137. Taylor, L. R. 1961. Aggregation, variance, and the mean. *Nature* 189:732–35
138. Thames, W. H., Stoner, W. N. 1953. A preliminary trial of lowland culture rice in rotation with vegetable crops as a means of reducing root-knot nematode infestations in the Everglades. *Plant Dis. Reptr.* 37:187–92
139. Thomason, I. J. 1987. Challenges facing nematology: Environmental risks with nematicides and the need for new approaches. See Ref. 142a, pp. 469–76
140. Todd, T. C., Pearson, C. A. S. 1988. Establishment of *Heterodera glycines* in three soil types. *Ann. Appl. Nematol.* 2:57–60
141. Townshend, J. L. 1990. Growth of carrot and tomato from oxamyl-coated seed and control of *Meloidogyne hapla*. *J. Nematol.* 22(2):170–75

141a. Univ. Calif. 1990. Integrated pest management for tomatoes. *Statewide Integrated Pest Manage. Proj., Div. Agric. Natl. Res. Publ.* No. 3274, pp. 60–65
142. Van Gundy, S. D. 1984. Nematodes. In *Integrated Pest Management for Citrus,* pp. 129–31. Riverside: Univ. Calif.
142a. Veech, J. A., Dickson, D. W., eds. 1987. *Vistas on Nematology*. Hyattsville, MD: Soc. Nematol.
143. Viglierchio, D. R. 1990. The impact of nematode adaptability on the prospect for their control. *Rev. Nematol.* 13:3–9
144. Viglierchio, D. R., Schmitt, R. V. 1983. On the methodology of nematode extraction from field samples: Comparison of methods for soil extraction. *J. Nematol.* 15(3):450–54
145. Weaver, D. B., Rodriguez-Kabana, R., Robertson, D. G., Akridge, R. K., Carden, E. L. 1988. Effect of crop rotation on soybean in a field infested with *Meloidogyne arenaria* and *Heterodera glycines*. *Ann. Appl. Nematol.* 2:106–9
145a. Webster, J. M., ed. 1972. *Economic Nematology*. London: Academic
146. Weil, R. R. 1990. Defining and using the concept of sustainable agriculture. *J. Agron. Educ.* 19(2):126–30
147. Wheeler, T. A., Kenerley, C. M., Jeger, J. J., Starr, J. L. 1987. Effect of quadrat and core sizes on determining the spatial pattern of *Criconemella sphaerocephalus*. *J. Nematol.* 19(4):413–19
148. Williamson, V. M., Colwell, G., Mei, H., Ho, J-Y. 1990. Molecular analysis of the region of the tomato genome containing Mi, a gene conferring resistance to root-knot nematodes. See Ref. 3, p. 153
149a. Wonersley, C. 1987. A reevaluation of strategies employed by nematode anhydrobiotes in relation to their natural environment. See Ref. 142a.

SUBJECT INDEX

G

H

CUMULATIVE INDEXES

CONTRIBUTING AUTHORS, VOLUMES 20–29

CHAPTER TITLES, VOLUMES 20–29